T0181881

Universitext

Universitext is a series of textbooks that presents material from a wide variety of mathematical disciplines at master's level and beyond. The books, often well class-tested by their author, may have an informal, personal even experimental approach to their subject matter. Some of the most successful and established books in the series have evolved through several editions, always following the evolution of teaching curricula, into very polished texts.

Thus as research topics trickle down into graduate-level teaching, first textbooks written for new, cutting-edge courses may make their way into *Universitext*.

Universitext

Universitext is a series of textbooks that presents material from a wide variety of mathematical disciplines at master's level and beyond. The books, often well class-tested by their author, may have an informal, personal even experimental approach to their subject matter. Some of the most successful and established books in the series have evolved through several editions, always following the evolution of teaching curricula, to very polished texts.

Thus as research topics trickle down into graduate-level teaching, first textbooks written for new, cutting-edge courses may make their way into *Universitext.*

More information about this series at http://www.springer.com/series/223

Mark J.D. Hamilton

Mathematical Gauge Theory

With Applications to the Standard Model of Particle Physics

Mark J.D. Hamilton
Department of Mathematics
Ludwig-Maximilian University of Munich
Munich, Germany

ISSN 0172-5939 ISSN 2191-6675 (electronic)
Universitext
ISBN 978-3-319-68438-3 ISBN 978-3-319-68439-0 (eBook)
https://doi.org/10.1007/978-3-319-68439-0

Library of Congress Control Number: 2017957556

Mathematics Subject Classification (2010): 55R10, 53C05, 22E70, 15A66, 53C27, 57S15, 22E60, 81T13, 81R40, 81V19, 81V05, 81V10, 81V15

Printed on acid-free paper

This Springer imprint is published by Springer Nature
The registered company is Springer International Publishing AG
The registered company address is: Gewerbestrasse 11, 6330 Cham, Switzerland

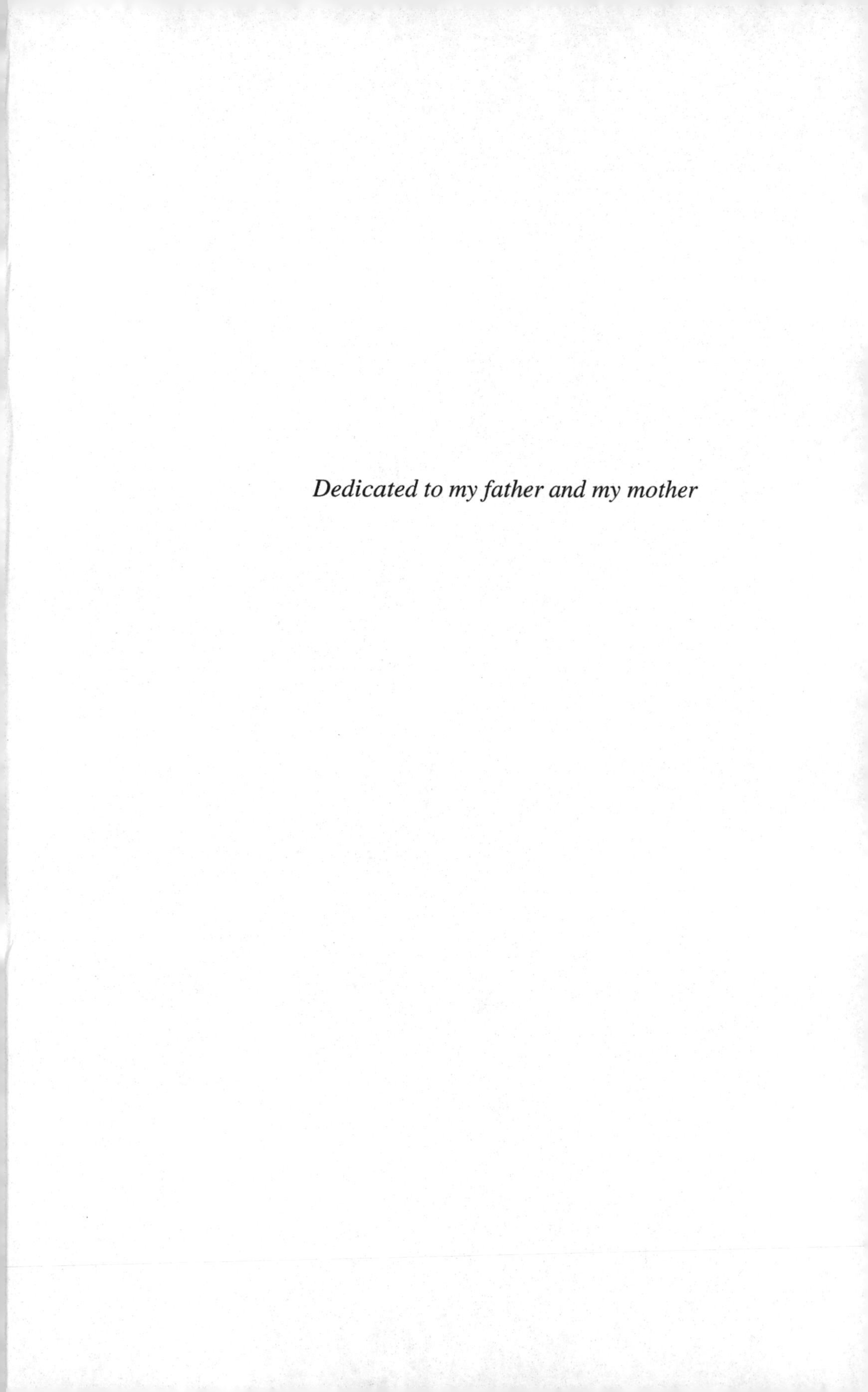

Dedicated to my father and my mother

Preface

With the discovery of a new particle, announced on 4 July 2012 at CERN, whose properties are "consistent with the long-sought Higgs boson" [31], the final elementary particle predicted by the classical Standard Model of particle physics has been found. The aim of this book is to explain the mathematical background as well as some of the details of the Standard Model. It is directed both at students of mathematics, who are interested in applications of gauge theory in physics, and at students of physics, who would like to understand more of the mathematics behind the Standard Model.

The book is based on my lecture notes for graduate courses held at the University of Stuttgart and the LMU Munich in Germany. A selection of the material can be covered in one semester. Prerequisites are an introductory course on manifolds and differential geometry as well as some basic knowledge of special relativity, summarized in the appendix. The first six chapters of the book treat the mathematical framework of gauge theories, in particular Lie groups, Lie algebras, representations, group actions, fibre bundles, connections and curvature, and spinors. The following three chapters discuss applications in physics: the Lagrangians and interactions in the Standard Model, spontaneous symmetry breaking, the Higgs mechanism of mass generation, and some more advanced and modern topics like neutrino masses, CP violation and Grand Unification.

The background in mathematics covered in the first six chapters of the book is much more extensive than strictly needed to understand the Standard Model. For example, the Standard Model is formulated on 4-dimensional Minkowski spacetime, over which all fibre bundles can be trivialized and spinors have a simple explicit description. However, this book is also intended as an introduction to modern theoretical physics as a whole, and some of the topics (for instance, on spinors or non-trivial fibre bundles) may be useful to students who plan to study topics such as supersymmetry or superstring theory. Depending on the time, the interests and the prior knowledge of the reader, he or she can take a shortcut and immediately start at the chapters on connections, spinors or Lagrangians, and then go back if more detailed mathematical knowledge is required at some point.

Since we focus on the Standard Model, several topics related to gauge theory and fibre bundles could not be covered, such as characteristic classes, holonomy theory, index theorems, monopoles and instantons as well as applications of gauge theory in pure mathematics, like Donaldson and Seiberg–Witten theory. For those topics a number of textbooks exist, some of which can be found in the bibliography.

An interesting and perhaps underappreciated fact is that a substantial number of phenomena in particle physics can be understood by analysing representations of Lie groups and by rewriting or rearranging Lagrangians. Examples of such phenomena, which we are going to study, are:

- symmetries of Lagrangians
- interactions between fields corresponding to elementary particles (quarks, leptons, gauge bosons, Higgs boson), determined by the Lagrangian
- the Higgs mechanism of mass generation for gauge bosons as well as the mass generation for fermions via Yukawa couplings
- quark and neutrino mixing
- neutrino masses and the seesaw mechanism
- CP violation
- Grand Unification

On the other hand, if precise predictions about scattering or decay of particles should be made or if explicit formulas for quantum effects, such as anomalies and running couplings, should be derived, then quantum field calculations involving Green's functions, perturbation theory and renormalization are necessary. These calculations are beyond the scope of this book, but a number of textbooks covering these topics can be found in the physics literature.

The references I used during the preparation of the book are listed in each chapter and may be useful to the reader for further studies (this is only a selection of references that I came across over the past several years, sometimes by chance, and there are many other valuable books and articles in this field).

It is not easy to make a recommendation on how to fit the chapters of the book into a lecture course, because it depends on the prior knowledge of the audience. A rough guideline could be as follows:

- *One-semester course:* Often lecture courses on differentiable manifolds contain sections on Lie groups, Lie algebras and group actions. If these topics can be assumed as prior knowledge, then one could cover in gauge theory the unstarred sections of Chaps. 4 to 7 and as much as possible of Chap. 8, perhaps going back to Chaps. 1 to 3 if specific results are needed.
- *Two-semester course:* Depending on the prior knowledge of the audience, one could cover in the first semester Chaps. 4 to 6 in more detail and in the second semester Chaps. 7 to 9. Or one could cover in the first semester Chaps. 1 to 5 and in the second semester Chaps. 6 to 8 (and as much as possible of Chap. 9).

Munich, Germany
July 2017

Mark J.D. Hamilton

Acknowledgements

There are several people and institutions I would like to thank. First, I am grateful to Dieter Kotschick and Uwe Semmelmann for their academic and scientific support since my time as a student. I want to thank Tian-Jun Li for our mathematical discussions and the invitation to conferences in Minneapolis, and the Simons Center for Geometry and Physics for the invitation to a workshop in Stony Brook. I would also like to take the opportunity to thank (belatedly) the German Academic Scholarship Foundation (Studienstiftung) for their generous financial support during my years of study.

I am grateful to the LMU Munich and the University of Stuttgart for the opportunity to give lecture courses on mathematical gauge theory, which formed the basis for this book. I want to thank the students who attended the lectures, in particular, Ismail Achmed-Zade, Anthony Britto, Simon-Raphael Fischer, Simon Hirscher, Martin Peev, Alexander Tabler, Danu Thung, Juraj Vrábel and David Wierichs, as well as my course assistants Nicola Pia and Giovanni Placini for reading the lecture notes and commenting on the manuscript. Furthermore, I would like to thank Bobby Acharya for his excellent lectures on the Standard Model and Robert Helling and Ronen Plesser for our interesting discussions about physics.

Special thanks to Catriona Byrne, my first contact at Springer, to Rémi Lodh for his excellent editorial support and suggestions while I was writing the manuscript, to the anonymous referees, the editors and the copyeditor for a number of comments and corrections, and to Anne-Kathrin Birchley-Brun for assistance in the production and publication of the book.

Finally, I am grateful to John, Barbara and Patrick Hamilton, Gisela Saalfeld and Inge Schmidbauer for their encouragement and support over the years, and to Guoshu Wang for her friendship.

Conventions

We collect some conventions that are used throughout the book.

General

- Sections and subsections marked with a * in front of the title contain additional or advanced material and can be skipped on a first reading. Occasionally these sections are used in later chapters.
- A word in *italics* is sometimes used for emphasis, but more often to denote terms that have not been defined so far in the text, like *gauge boson*, or to denote standard terms, like *skew field*, whose definition can be found in many textbooks. A word in **boldface** is usually used for definitions.
- Diffeomorphisms of manifolds and isomorphisms of vector spaces, groups, Lie groups, algebras and bundles are denoted by $\cong$.
- We often use the Einstein summation convention by summing over the same indices in an expression, without writing the symbol $\sum$ (we also sum over two lower or two upper indices).
- If A is a set, then $\mathrm{Id}_A : A \rightarrow A$ denotes the identity map.
- A disjoint union of sets is denoted by $\dot{\cup}$ or $\dot{\bigcup}$.
- The symbols Re and Im denote the real and imaginary part of a complex number (and sometimes of a quaternion).

Linear Algebra

- We denote by $\mathrm{Mat}(n \times m, R)$ the set of $n \times m$-matrices with entries in a ring R.

- $n \times n$-unit matrices

$$\begin{pmatrix} 1 & & & \\ & 1 & & \\ & & \ddots & \\ & & & 1 \end{pmatrix}$$

are denoted by I_n or I.
- The conjugate of a complex number or quaternion q is denoted by $\bar{q}$ and occasionally by q^*. The conjugate $\bar{A}$ of a complex or quaternionic matrix A is defined by conjugating each entry: $(\bar{A})_{ij} = \overline{A_{ij}}$.
- The transpose of a matrix A is denoted by A^T. A matrix is called symmetric if $A^T = A$ and skew-symmetric or antisymmetric if $A^T = -A$. For a complex or quaternionic matrix A we set $A^\dagger = (\bar{A})^T$. A complex or quaternionic matrix A is called Hermitian if $A^\dagger = A$ and skew-Hermitian if $A^\dagger = -A$.

  Note that the Dirac conjugate $\overline{\psi}$ of a spinor ψ, defined in Definition 6.7.15, has a special meaning and is in general not equal to $\psi^\dagger$ or ψ^*.
- For a matrix A we denote by $\det(A)$ the determinant and by $\mathrm{tr}(A)$ the trace.
- In $\mathbb{K}^n$, where $\mathbb{K} = \mathbb{R}, \mathbb{C}, \mathbb{H}$, we denote by $e_1, e_2 \ldots, e_n$ the standard basis vectors with n entries

$$e_1 = \begin{pmatrix} 1 \\ 0 \\ 0 \\ \vdots \\ 0 \end{pmatrix}, \quad e_2 = \begin{pmatrix} 0 \\ 1 \\ 0 \\ \vdots \\ 0 \end{pmatrix}, \quad \ldots, \quad e_n = \begin{pmatrix} 0 \\ 0 \\ 0 \\ \vdots \\ 1 \end{pmatrix}.$$

- If $f: V \to W$ is a linear map between vector spaces, we sometimes write fv instead of $f(v)$ for a vector $v \in V$ to reduce the number of brackets in formulas.

Groups

- The neutral element of a group G is usually denoted by $e \in G$ (if it is not a matrix group, where the neutral element is usually denoted by I).
- By a group without preceding words like "topological" or "Lie" we mean a group in the algebraic sense, without the additional structure of a topological space or smooth manifold.
- We usually write the group operation as $\cdot$ (multiplication). Occasionally, we write the operation for abelian groups as $+$ (addition). The neutral element is then sometimes denoted by 1 or 0, respectively.

Manifolds

- All manifolds in this book are smooth ($\mathscr{C}^\infty$), unless stated otherwise.
- On a manifold M we denote by $\mathscr{C}^\infty(M)$ the set of smooth functions on M with values in $\mathbb{R}$ and by $\mathscr{C}^\infty(M, W)$ the set of smooth functions with values in a vector space W.
- The differential of a smooth map $f: M \to N$ between manifolds M and N at a point $p \in M$ is denoted by D_pf or f_* (push forward). If $g: N \to Q$ is another smooth map, then the chain rule in $p \in M$ is

$$D_p(g \circ f) = D_{f(p)}g \circ D_pf$$

or simply

$$(g \circ f)_* = g_* \circ f_*.$$

- If X is a vector field on a manifold M and $p \in M$ a point, then we denote by X_p or $X(p)$ the value of X in p.
- By a curve γ through a point p in a manifold M tangent to a vector $X \in T_pM$ we mean a smooth curve $\gamma: I \to M$, defined on an open interval I around 0, such that $\gamma(0) = p$ and $\dot{\gamma}(0) = \frac{d}{dt}\big|_{t=0} \gamma(t) = X$.
- Suppose f is a real or complex-valued function on a manifold M, $p \in M$ a point and $X \in T_pM$ a tangent vector. Then we denote the directional derivative of f along X by $(L_Xf)(p)$, $df(X)$ or $D_pf(X)$. The same notation is used if f takes values in a real or complex vector space.

Diagrams

Feynman diagrams can be read with time increasing from left to right. Arrows on fermion lines indicate particle-flow and point in the direction of momentum-flow for particles, but opposite to the direction of momentum-flow for antiparticles [125, Sect. 9.2]. The Feynman diagrams in this book have been prepared with feynMF/feynMP. Commutative diagrams have been prepared with TikZ and function plots with MATLAB.

Contents

Part I
Mathematical Foundations

Chapter 1
Lie Groups and Lie Algebras: Basic Concepts

Gauge theories are field theories of physics involving *symmetry groups*. Symmetry groups are groups of transformations that act on something and leave something (possibly something else) invariant. For example, symmetry groups can act on geometric objects (by rotation, translation, etc.) and leave those objects invariant. For the symmetries relevant in field theories, the groups act on fields and leave the *Lagrangian* or the *action* (the spacetime integral over the Lagrangian) invariant.

Concerning symmetry groups or groups in general we can make a basic distinction, without being mathematically precise for the moment: groups can be *discrete* or *continuous*. Both types of symmetry groups already occur for elementary geometric objects. Equilateral triangles and squares, for example, appear to have discrete symmetry groups, while other objects, such as the circle S^1, the 2-sphere S^2 or the plane $\mathbb{R}^2$ with a Euclidean metric, have continuous symmetry groups. A deep and less obvious fact, that we want to understand over the course of this book, is that similar symmetry groups play a prominent role in the classical and quantum field theories describing nature.

From a mathematical point of view, continuous symmetry groups can be conceptualised as *Lie groups*. By definition, Lie groups are groups in an algebraic sense which are at the same time smooth manifolds, so that both structures – algebraic and differentiable – are compatible. As a mathematical object, Lie groups capture the idea of a continuous group that can be parametrized locally by coordinates, so that the group operations (multiplication and inversion) are smooth maps in those coordinates. Lie groups also cover the case of discrete, i.e. 0-dimensional groups, consisting of a set of isolated points.

In theoretical physics, Lie groups like the Lorentz and Poincaré groups, which are related to spacetime symmetries, and gauge groups, defining *internal* symmetries, are important cornerstones. The currently accepted *Standard Model of elementary particles*, for instance, is a gauge theory with Lie group

$$\mathrm{SU}(3) \times \mathrm{SU}(2) \times \mathrm{U}(1).$$

M.J.D. Hamilton, *Mathematical Gauge Theory*, Universitext,
https://doi.org/10.1007/978-3-319-68439-0_1

There are also *Grand Unified Theories* (GUTs) based on Lie groups like SU(5). We shall see in later chapters that the specific kind of Lie group in a gauge theory (its dimension, whether it is abelian or not, whether it is *simple* or splits as a product of several factors, and so on) is reflected in interesting ways in the physics. For example, in the case of the Standard Model, it turns out that:

- The fact that there are 8 *gluons*, 3 *weak gauge bosons* and 1 *photon* is related to the dimensions of the Lie groups SU(3) and SU(2) × U(1) (the SU(5) Grand Unified Theory has 12 additional gauge bosons).
- The fact that the strong, weak and electromagnetic interactions have different strengths (*coupling constants*) is related to the product structure of the gauge group SU(3) × SU(2) × U(1) (GUTs built on simple Lie groups like SU(5) have only a single coupling constant).
- The fact that gluons interact directly with each other while photons do not is related to the fact that SU(3) is non-abelian while U(1) is abelian.

Our main mathematical tool to construct non-trivial Lie groups will be *Cartan's Theorem*, which shows that any subgroup (in the algebraic sense) of a Lie group, which is a closed set in the topology, is already an embedded Lie subgroup.

Besides Lie groups, *Lie algebras* play an important role in the theory of symmetries. Lie algebras are vector spaces with a bilinear, antisymmetric product, denoted by a bracket $[\cdot,\cdot]$, satisfying the *Jacobi identity*. As an algebraic object, Lie algebras can be defined independently of Lie groups, even though Lie groups and Lie algebras are closely related: the tangent space to the neutral element $e \in G$ of a Lie group G has a canonical structure of a Lie algebra. This means that Lie algebras are in some sense an infinitesimal, algebraic description of Lie groups. Depending on the situation, it is often easier to work with linear objects, such as Lie algebras, than with non-linear objects like Lie groups. Lie algebras are also important in gauge theories: *connections on principal bundles*, also known as *gauge boson fields*, are (locally) 1-forms on spacetime with values in the Lie algebra of the gauge group.

In this chapter we define Lie groups and Lie algebras and describe the relations between them. In the following chapter we will study some associated concepts, like representations (which are used to define the actions of Lie groups on fields) and invariant metrics (which are important in the construction of the gauge invariant *Yang–Mills Lagrangian*). We will also briefly discuss the structure of simple and semisimple Lie algebras.

Concerning symmetries, we will study in this chapter Lie groups as symmetry groups of vector spaces and certain structures (scalar products and volume forms) defined on vector spaces (in Chap. 3 on group actions we will study Lie groups as symmetry groups of manifolds). Symmetry groups of vector spaces are more generic than it may seem: it can be shown as a consequence of the *Peter–Weyl Theorem* that any compact Lie group can be realized as a group of rotations of some finite-dimensional Euclidean vector space $\mathbb{R}^m$ (i.e. as an embedded Lie subgroup of $SO(m)$).

We can only cover a selection of topics on Lie groups. The main references for this and the following chapter are [24, 83, 142] and [153], where more extensive discussions of Lie groups and Lie algebras can be found. Additional references are [14, 34, 70, 77] and [129].

1.1 Topological Groups and Lie Groups

We begin with a first elementary mathematical concept that makes the idea of a continuous group precise.

Definition 1.1.1 A **topological group** G is a group which is at the same time a topological space so that the map

$$\begin{aligned} G \times G &\longrightarrow G \\ (g, h) &\longmapsto g \cdot h^{-1} \end{aligned}$$

is continuous, where $G \times G$ has the canonical product topology determined by the topology of G.

Remark 1.1.2 Here and in the following we shall mean by a **group** just a group in the algebraic sense, without the additional structure of a topological space or smooth manifold.
We usually set e for the neutral element in G. An equivalent description of topological groups is the following.

Lemma 1.1.3 *A group G is a topological group if and only if it is at the same time a topological space so that both of the maps*

$$\begin{aligned} G \times G &\longrightarrow G \\ (g, h) &\longmapsto g \cdot h \\ G &\longrightarrow G \\ g &\longmapsto g^{-1}, \end{aligned}$$

called **multiplication** *and* **inversion**, *are continuous.*

Proof Suppose that multiplication and inversion are continuous maps. Then the map

$$\begin{aligned} G \times G &\longrightarrow G \times G \\ (g, h) &\longmapsto \left(g, h^{-1}\right) \end{aligned}$$

is continuous and hence also the composition of this map followed by multiplication. This shows that

$$\begin{aligned} G \times G &\longrightarrow G \\ (g, h) &\longmapsto g \cdot h^{-1} \end{aligned}$$

is continuous, hence the group G is a topological group.

Conversely, assume that G is a topological group. Then the map

$$\begin{aligned} G &\longrightarrow G \times G \longrightarrow G \\ g &\longmapsto (e, g) \longmapsto e \cdot g^{-1} = g^{-1} \end{aligned}$$

is continuous and hence the map

$$\begin{aligned} G \times G &\longrightarrow G \times G \longrightarrow G \\ (g, h) &\longmapsto (g, h^{-1}) \longmapsto g \cdot \left(h^{-1}\right)^{-1} = g \cdot h \end{aligned}$$

is also continuous. This proves the claim. □

The concept of topological groups is a bit too general to be useful for our purposes. In particular, general topological spaces can be very complicated and do not have to be, for example, locally Euclidean, like topological manifolds. We now turn to the definition of Lie groups, which is the type of continuous groups we are most interested in.

Definition 1.1.4 A **Lie group** G is a group which is at the same time a manifold so that the map

$$\begin{aligned} G \times G &\longrightarrow G \\ (g, h) &\longmapsto g \cdot h^{-1} \end{aligned}$$

is smooth, where $G \times G$ has the canonical structure of a product manifold determined by the smooth structure of G.

Remark 1.1.5 Note that we only consider Lie groups of finite dimension.

Remark 1.1.6 Here and in the following we mean by a **manifold** a smooth manifold, unless stated otherwise.

Of course, every Lie group is also a topological group. We could define Lie groups equivalently as follows.

Lemma 1.1.7 *A group G is a Lie group if and only if it is at the same time a manifold so that both of the maps*

$$\begin{aligned} G \times G &\longrightarrow G \\ (g,h) &\longmapsto g \cdot h \\ G &\longrightarrow G \\ g &\longmapsto g^{-1} \end{aligned}$$

are smooth.

Proof The proof is very similar to the proof of Lemma 1.1.3. □

A Lie group is thus a second countable, Hausdorff, topological group that can be parametrized locally by finitely many coordinates in a smoothly compatible way, so that multiplication and inversion depend smoothly on the coordinates.

Remark 1.1.8 (Redundancy in the Definition of Lie Groups) A curious fact about Lie groups is that it suffices to check that multiplication is smooth, because then inversion is automatically smooth (see Exercise 1.9.5).

Remark 1.1.9 (Hilbert's Fifth Problem) It was shown by Gleason [65], Montgomery and Zippin [97] in 1952 that a topological group, which is also a topological manifold, has the structure of a Lie group. This is the solution to one interpretation of *Hilbert's fifth problem* (see [135] for more details). We will not prove this theorem (the existence of a smooth structure), but we will show in Corollary 1.8.17 that on a topological group, which is a topological manifold, there is at most one smooth structure that turns it into a Lie group.

We will see in Sect. 1.5 that there is a deeper reason why we want symmetry groups to be smooth manifolds: only in this situation can we canonically associate to a group a *Lie algebra*, which consists of certain *left-invariant* smooth vector fields on the group. Vector fields are only defined on smooth manifolds (they need tangent spaces and a tangent bundle to be defined). This explains why we are particularly interested in groups having a smooth structure.

We consider some simple examples of Lie groups (more examples will follow later, in particular, in Sect. 1.2.2).

Example 1.1.10 Euclidean space $\mathbb{R}^n$ with vector addition is an n-dimensional Lie group, since addition

$$(x,y) \longmapsto x+y$$

and inversion

$$x \longmapsto -x$$

for $x, y \in \mathbb{R}^n$ are linear and hence smooth. The Lie group $\mathbb{R}^n$ is connected, non-compact and abelian.

Remark 1.1.11 Euclidean spaces $\mathbb{R}^n$ can also carry other (non-abelian) Lie group structures besides the abelian structure coming from vector addition. For example,

$$\mathrm{Nil}^3 = \left\{ \begin{pmatrix} 1 & x & y \\ 0 & 1 & z \\ 0 & 0 & 1 \end{pmatrix} \in \mathrm{Mat}(3 \times 3, \mathbb{R}) \,\middle|\, x, y, z \in \mathbb{R} \right\}$$

with matrix multiplication is an example of a (so-called *nilpotent*), non-abelian Lie group structure on $\mathbb{R}^3$, also known as the **Heisenberg group**.

Example 1.1.12 Every countable group G with the manifold structure as a discrete space, i.e. a countable union of isolated points, is a 0-dimensional Lie group, because every map $G \times G \rightarrow G$ is smooth (locally constant). A discrete group is compact if and only if it is finite. In particular, the integers $\mathbb{Z}$ and the finite cyclic groups $\mathbb{Z}_k = \mathbb{Z}/k\mathbb{Z}$ for $k \in \mathbb{N}$ are discrete, abelian Lie groups. The Lie group $\mathbb{Z}_2$ can be identified with the 0-sphere

$$S^0 = \{x \in \mathbb{R} \mid |x| = 1\} = \{\pm 1\}$$

with multiplication induced from $\mathbb{R}$.

Example 1.1.13 The circle

$$S^1 = \{z \in \mathbb{C} \mid |z| = 1\}$$

is a 1-dimensional Lie group with multiplication induced from $\mathbb{C}$: multiplication on $\mathbb{C}$ is quadratic and inversion on $\mathbb{C}^* = \mathbb{C} \setminus \{0\}$ is a rational function in real and imaginary parts, thus smooth. Both maps restrict to smooth maps on the embedded submanifold $S^1 \subset \mathbb{C}$. The Lie group S^1 is connected, compact and abelian.

Example 1.1.14 The following set of matrices together with matrix multiplication is a Lie group:

$$\mathrm{SO}(2) = \left\{ \begin{pmatrix} \cos\alpha & -\sin\alpha \\ \sin\alpha & \cos\alpha \end{pmatrix} \in \mathrm{Mat}(2 \times 2, \mathbb{R}) \,\middle|\, \alpha \in \mathbb{R} \right\}.$$

As a manifold $\mathrm{SO}(2) \cong \mathbb{R}/2\pi\mathbb{Z} \cong S^1$. We have

$$\begin{pmatrix} \cos\alpha & -\sin\alpha \\ \sin\alpha & \cos\alpha \end{pmatrix} \cdot \begin{pmatrix} \cos\beta & -\sin\beta \\ \sin\beta & \cos\beta \end{pmatrix} = \begin{pmatrix} \cos(\alpha+\beta) & -\sin(\alpha+\beta) \\ \sin(\alpha+\beta) & \cos(\alpha+\beta) \end{pmatrix}$$

and

$$\begin{pmatrix} \cos\alpha & -\sin\alpha \\ \sin\alpha & \cos\alpha \end{pmatrix}^{-1} = \begin{pmatrix} \cos(-\alpha) & -\sin(-\alpha) \\ \sin(-\alpha) & \cos(-\alpha) \end{pmatrix},$$

showing that multiplication and inversion are smooth maps and that SO(2) is closed under these operations. The Lie group SO(2) is one of the simplest examples of a whole class of Lie groups, known as **matrix** or **linear Lie groups** (the Heisenberg group is another example). The Lie group SO(2) is *isomorphic* to S^1.
We want to generalize the examples of the Lie groups S^0 and S^1 and show that the 3-sphere S^3 also has the structure of a Lie group. This is a good opportunity to introduce (or recall) in a short detour the skew field of **quaternions** $\mathbb{H}$.

1.1.1 Normed Division Algebras and the Quaternions

The real and complex numbers are known from high school and the first mathematics courses at university. There are, however, other types of "higher-dimensional" numbers, which are less familiar, but still occur in mathematics and physics. It is useful to consider the following, general, algebraic notions (a nice reference is [8]):

Definition 1.1.15

1. A real **algebra** is a finite-dimensional real vector space A with a bilinear map
$$A \times A \longrightarrow A$$
$$(a, b) \longmapsto a \cdot b$$
and a unit element $1 \in A$ such that $1 \cdot a = a = a \cdot 1$ for all $a \in A$. In particular, the multiplication on A is distributive, but in general not associative.
2. We call the algebra A **normed** if there is a norm $||\cdot||$ on the vector space A such that $||ab|| = ||a|| \cdot ||b||$.
3. We call the algebra A a **division algebra** if $ab = 0$ implies that either $a = 0$ or $b = 0$.
4. The algebra has **multiplicative inverses** if for any non-zero $a \in A$ there is an $a^{-1} \in A$ such that $aa^{-1} = a^{-1}a = 1$.

It follows that every normed algebra is a division algebra and every associative division algebra has multiplicative inverses (this is not true in general for non-associative division algebras).

The following is a classical theorem due to Hurwitz.

Theorem 1.1.16 (Hurwitz's Theorem on Normed Division Algebras) *There are only four normed real division algebras:*

1. *The real numbers* $\mathbb{R}$ *of dimension* 1.
2. *The complex numbers* $\mathbb{C}$ *of dimension* 2.
3. *The quaternions* $\mathbb{H}$ *of dimension* 4.
4. *The octonions* $\mathbb{O}$ *of dimension* 8.

We want to describe the quaternions in this section and leave the (non-associative) octonions to Exercise 3.12.15.

Recall that there is an algebra structure on the vector space $\mathbb{R}^2$, so that this vector space becomes a *field*, called the complex numbers $\mathbb{C}$: the multiplication is associative and commutative. Furthermore, every non-zero element of $\mathbb{C}$ has a multiplicative inverse.

The complex plane is spanned as a real vector space by the basis vectors 1 and i, with $i^2 = -1$. By distributivity this determines the product of any two complex numbers. We define the **conjugate** of a complex number $z = a + ib$ as $\bar{z} = a - ib$ and the **norm** squared as $||z||^2 = z\bar{z} = a^2 + b^2$. The multiplicative inverse of a non-zero complex number is then

$$z^{-1} = \frac{\bar{z}}{||z||^2}.$$

We also have

$$||uv|| = ||u|| \cdot ||v||$$

for all complex numbers $u, v \in \mathbb{C}$, so that $\mathbb{C}$ is a normed division algebra.

There is a similar construction of an algebra structure on the vector space $\mathbb{R}^4$, so that this vector space becomes a *skew field*, called the **quaternions** $\mathbb{H}$: the multiplication is associative and every non-zero element of $\mathbb{H}$ has a multiplicative inverse. However, the multiplication is not commutative. As a real vector space $\mathbb{H}$ is spanned by the basis vectors 1, i, j and k. The product satisfies

$$i^2 = j^2 = k^2 = ijk = -1.$$

Using associativity of multiplication this determines all possible products among the basis elements i, j, k and thus by distributivity the product of any two quaternions. We have

$$\begin{aligned} ij &= -ji = k, \\ jk &= -kj = i, \\ ki &= -ik = j, \end{aligned}$$

showing in particular that the product is not commutative. The products among i, j and k can be memorized with the following diagram:

We define the **real quaternions** by

$$\operatorname{Re}\mathbb{H} = \{a = a1 \in \mathbb{H} \mid a \in \mathbb{R}\}$$

and the **imaginary quaternions** by

$$\operatorname{Im}\mathbb{H} = \{bi + cj + dk \in \mathbb{H} \mid b, c, d \in \mathbb{R}\}.$$

We also define the **conjugate** of a quaternion $w = a + bi + cj + dk$ as

$$\bar{w} = a - bi - cj - dk$$

and the **norm** squared as

$$||w||^2 = a^2 + b^2 + c^2 + d^2 = w\bar{w} = \bar{w}w.$$

The multiplicative inverse of a non-zero quaternion is then

$$w^{-1} = \frac{\bar{w}}{||w||^2}.$$

We also have

$$||wz|| = ||w|| \cdot ||z||$$

for all $w, z \in \mathbb{H}$, so that $\mathbb{H}$ is a normed division algebra.

Remark 1.1.17 One has to be careful with division of quaternions: the expression $\frac{z}{w}$ for $z, w \in \mathbb{H}$ is not well-defined, even for $w \neq 0$, since multiplication is not commutative. One rather has to write zw^{-1} or $w^{-1}z$.
Multiplication of quaternions defines a group structure on the 3-sphere:

Example 1.1.18 The 3-sphere

$$S^3 = \{w \in \mathbb{H} \mid ||w|| = 1\}$$

of unit quaternions is a 3-dimensional embedded submanifold of $\mathbb{H} \cong \mathbb{R}^4$ and a Lie group with multiplication induced from $\mathbb{H}$. As a Lie group S^3 is connected, compact and non-abelian (it contains, in particular, the elements $1, i, j, k \in \mathbb{H}$). The 3-sphere and thus the quaternions have an interesting relation to the **rotation group** $\mathrm{SO}(3)$ of $\mathbb{R}^3$, to be discussed in Example 1.3.8.

Remark 1.1.19 We shall see in Exercise 3.12.15 that there is an algebra structure on the vector space $\mathbb{R}^8$, so that this vector space becomes a normed division algebra, called the **octonions** $\mathbb{O}$. This multiplication induces a multiplication on S^7. However, the multiplication does not define a group structure on S^7, because it is not associative.

*1.1.2 *Quaternionic Matrices*

Certain properties of matrices with quaternionic entries cannot be proved in the same way as for matrices with real or complex entries, because the quaternions are only a skew field. In particular, it is not immediately clear how to make sense of the inverse and a determinant for quaternionic square matrices. Even if this is possible, it is not clear that such a determinant would have the nice properties we expect of it, like multiplicativity and the characterization of invertible matrices as those of non-zero determinant. Since we are going to consider groups of quaternionic matrices as examples of Lie groups, like the so-called **compact symplectic group** $\mathrm{Sp}(n)$, we would like to fill in some of the details in this section (we follow the exposition in [83, Sect. I.8] and [152]). Like any section with a star * this section can be skipped on a first reading.

Definition 1.1.20 We denote by $\mathrm{Mat}(m \times n, \mathbb{H})$ the set of all $m \times n$-matrices with entries in $\mathbb{H}$.
The set $\mathrm{Mat}(m \times n, \mathbb{H})$ is an abelian group with the standard addition of matrices. We can also define **right** and **left multiplication** with elements $q \in \mathbb{H}$:

$$\begin{aligned} \mathrm{Mat}(m \times n, \mathbb{H}) \times \mathbb{H} &\longrightarrow \mathrm{Mat}(m \times n, \mathbb{H}) \\ (A, q) &\longmapsto Aq \end{aligned}$$

and

$$\begin{aligned} \mathbb{H} \times \mathrm{Mat}(m \times n, \mathbb{H}) &\longrightarrow \mathrm{Mat}(m \times n, \mathbb{H}) \\ (q, A) &\longmapsto qA. \end{aligned}$$

This gives $\mathrm{Mat}(m \times n, \mathbb{H})$ the structure of a right or left module over the quaternions $\mathbb{H}$; we call this a **right** or **left quaternionic vector space** (since $\mathbb{H}$ is not commutative, left and right multiplication differ). In particular, the spaces of row and column vectors, denoted by $\mathbb{H}^n$, each have the structure of a right and left quaternionic vector space.

We can define matrix multiplication

$$\begin{aligned} \mathrm{Mat}(m \times n, \mathbb{H}) \times \mathrm{Mat}(n \times p, \mathbb{H}) &\longrightarrow \mathrm{Mat}(m \times p, \mathbb{H}) \\ (A, B) &\longmapsto A \cdot B \end{aligned}$$

in the standard way, where we have to be careful to preserve the ordering in the products of the entries. All of the constructions so far work in exactly the same way for matrices over any ring.

We now restrict to quaternionic square matrices $\mathrm{Mat}(n \times n, \mathbb{H})$. It is sometimes useful to have a different description of such matrices. The following is easy to see:

Lemma 1.1.21 *Let $q \in \mathbb{H}$ be a quaternion. Then there exist unique complex numbers $q_1, q_2 \in \mathbb{C}$ such that*

$$q = q_1 + jq_2.$$

Let $A \in \mathrm{Mat}(n \times n, \mathbb{H})$ be a quaternionic square matrix. Then there exist unique complex square matrices $A_1, A_2 \in \mathrm{Mat}(n \times n, \mathbb{C})$ such that

$$A = A_1 + jA_2.$$

Definition 1.1.22 For a matrix $A = A_1 + jA_2 \in \mathrm{Mat}(n \times n, \mathbb{H})$, with A_1, A_2 complex, we define the **adjoint** to be the complex square matrix

$$\chi_A = \begin{pmatrix} A_1 & -\overline{A_2} \\ A_2 & \overline{A_1} \end{pmatrix} \in \mathrm{Mat}(2n \times 2n, \mathbb{C}).$$

Example 1.1.23 For the special case of a quaternion $q = q_1 + jq_2 \in \mathbb{H}$ with $q_1, q_2 \in \mathbb{C}$, considered as a 1×1-matrix, we get

$$\chi_q = \begin{pmatrix} q_1 & -\overline{q_2} \\ q_2 & \overline{q_1} \end{pmatrix}.$$

The quaternions $\mathbb{H}$ are sometimes *defined* as the subset of $\mathrm{Mat}(2 \times 2, \mathbb{C})$ consisting of matrices of this form.

Remark 1.1.24 Another definition of the adjoint, used in [152], is to write $A = A_1 + A_2 j$ with A_1, A_2 complex and define

$$\chi_A = \begin{pmatrix} A_1 & A_2 \\ -\overline{A_2} & \overline{A_1} \end{pmatrix}.$$

We continue to use our Definition 1.1.22, which seems to be more standard.

Using the adjoint we get the following identification of $\mathrm{Mat}(n \times n, \mathbb{H})$ with a subset of $\mathrm{Mat}(2n \times 2n, \mathbb{C})$.

Proposition 1.1.25 (Quaternionic Matrices as a Subspace of Complex Matrices) *We define an element $J \in \mathrm{Mat}(2n \times 2n, \mathbb{C})$ by*

$$J = \begin{pmatrix} 0 & I_n \\ -I_n & 0 \end{pmatrix}.$$

Then $J^{-1} = -J$ and the image of the injective map

$$\begin{aligned} \chi\colon \mathrm{Mat}(n \times n, \mathbb{H}) &\longrightarrow \mathrm{Mat}(2n \times 2n, \mathbb{C}) \\ A &\longmapsto \chi_A \end{aligned}$$

consists of the set

$$\mathrm{im}\,\chi = \left\{X \in \mathrm{Mat}(2n \times 2n, \mathbb{C}) \mid JXJ^{-1} = \overline{X}\right\}.$$

The proof is a simple calculation. The following proposition can also be verified by a direct calculation, where for the second property we use that $Cj = j\overline{C}$ for a complex matrix C:

Proposition 1.1.26 (Properties of the Adjoint) *The adjoints of quaternionic $n\times n$-matrices A, B satisfy*

$$\begin{aligned} \chi_{A+B} &= \chi_A + \chi_B \\ \chi_{AB} &= \chi_A \chi_B \\ \mathrm{tr}(\chi_A) &= 2\mathrm{Re}(\mathrm{tr}(A)). \end{aligned}$$

Corollary 1.1.27 *If quaternionic $n \times n$-matrices A, B satisfy $AB = I$, then $BA = I$.*

Proof If $AB = I_n$, then

$$\chi_A \chi_B = \chi_{AB} = I_{2n}.$$

Hence by a property of the inverse of complex matrices

$$\chi_{BA} = \chi_B \chi_A = I_{2n}$$

and thus $BA = I_n$. □

We can therefore define:

Definition 1.1.28 A matrix $A \in \mathrm{Mat}(n \times n, \mathbb{H})$ is called **invertible** if there exists a matrix $B \in \mathrm{Mat}(n \times n, \mathbb{H})$ with $AB = I = BA$. The matrix B is called the **inverse** of A.

Corollary 1.1.29 *A matrix $A \in \mathrm{Mat}(n \times n, \mathbb{H})$ is invertible if and only if its adjoint χ_A is invertible.*

Proof If A is invertible, then χ_A is invertible by Proposition 1.1.26. Conversely, assume that χ_A is invertible. We can write χ_A^{-1} in the form

$$\chi_A^{-1} = \begin{pmatrix} B_1 & B_2 \\ B_3 & B_4 \end{pmatrix}$$

for some complex $n \times n$-matrices $B_1, \ldots, B_4$. Then

$$A_1 B_1 - \overline{A_2} B_3 = I_n,$$
$$A_2 B_1 + \overline{A_1} B_3 = 0.$$

Setting

$$B = B_1 + jB_3 \in \mathrm{Mat}(n \times n, \mathbb{H})$$

we get

$$\begin{aligned} AB &= A_1 B_1 + jA_2 B_1 + A_1 jB_3 + jA_2 jB_3 \\ &= (A_1 B_1 - \overline{A_2} B_3) + j(A_2 B_1 + \overline{A_1} B_3) \\ &= I_n. \end{aligned}$$

We conclude that A is invertible with inverse B. □

Definition 1.1.30. The **determinant** of a quaternionic matrix $A \in \mathrm{Mat}(n \times n, \mathbb{H})$ is defined by

$$\det(A) = \det(\chi_A).$$

Remark 1.1.31 Note that the determinant is not defined for matrices over a general non-commutative ring.

The determinant on $\mathrm{Mat}(n \times n, \mathbb{H})$ has the following property.

Proposition 1.1.32 (Quaternionic Determinant Is Real and Non-negative) *The determinant of a matrix $A \in \mathrm{Mat}(n \times n, \mathbb{H})$ is real and non-negative,* $\det(A) \geq 0$.

Proof According to Proposition 1.1.25

$$J\chi_A J^{-1} = \overline{\chi_A}$$

for all $A \in \mathrm{Mat}(n \times n, \mathbb{H})$. Since $\det(J) = 1$ we get

$$\det(\chi_A) = \det\left(J\chi_A J^{-1}\right) = \overline{\det(\chi_A)},$$

hence $\det(\chi_A) \in \mathbb{R}$.

The proof that $\det(A)$ is non-negative is more involved. Any quaternionic matrix $A \in \mathrm{Mat}(n \times n, \mathbb{H})$ can be brought to the form

$$\begin{pmatrix} I_k & 0 \\ 0 & 0 \end{pmatrix}$$

for some $k \leq n$ by applying elementary row and column operations. The corresponding elementary quaternionic matrices E act by left or right multiplication on A, thus the adjoint χ_E acts by left or right multiplication on χ_A. The elementary matrix E for switching two rows (columns) or adding a row (column) to another one are real, thus

$$\chi_E = \begin{pmatrix} E & 0 \\ 0 & E \end{pmatrix}$$

and $\det(\chi_E) = \det(E)^2 = 1$. It remains to consider the elementary matrix E that multiplies a row (column) by a non-zero quaternion a. Writing $a = a_1 + ja_2$ with a_1, a_2 complex, the elementary matrix E is of the form $E = E_1 + jE_2$ with

$$E_1 = \begin{pmatrix} 1 & & & & & & \\ & \ddots & & & & & \\ & & 1 & & & & \\ & & & a_1 & & & \\ & & & & 1 & & \\ & & & & & \ddots & \\ & & & & & & 1 \end{pmatrix}, \quad E_2 = \begin{pmatrix} 0 & & & & & & \\ & \ddots & & & & & \\ & & 0 & & & & \\ & & & a_2 & & & \\ & & & & 0 & & \\ & & & & & \ddots & \\ & & & & & & 0 \end{pmatrix}$$

and

$$\chi_E = \begin{pmatrix} E_1 & -\overline{E_2} \\ E_2 & \overline{E_1} \end{pmatrix}.$$

Interchanging twice a pair of rows and twice a pair of columns we can bring χ_E to the following form, without changing $\det(\chi_E)$:

$$\chi_E' = \begin{pmatrix} a_1 & -\bar{a}_2 & & & \\ a_2 & \bar{a}_1 & & & \\ & & 1 & & \\ & & & \ddots & \\ & & & & 1 \end{pmatrix}.$$

It follows that

$$\det(\chi_E) = \det(\chi'_E) = |a_1|^2 + |a_2|^2 > 0.$$

This proves the claim, since χ_A is the product of matrices of non-negative determinant. □

Remark 1.1.33 A second, independent proof of the second assertion in Proposition 1.1.32 uses that the subset $\mathrm{GL}(n, \mathbb{H})$ of invertible matrices in $\mathrm{Mat}(n \times n, \mathbb{H})$ is connected, cf. Theorem 1.2.22. Therefore its image under the determinant is contained in $\mathbb{R}_+$. The claim then follows, because $\mathrm{GL}(n, \mathbb{H})$ is dense in $\mathrm{Mat}(n \times n, \mathbb{H})$.

Example 1.1.34 According to Exercise 1.9.8 the determinant of a matrix

$$(a) \in \mathrm{Mat}(1 \times 1, \mathbb{H}), \quad \text{with } a \in \mathbb{H},$$

is equal to

$$\det(a) = ||a||^2.$$

The next corollary follows from what we have shown above.

Corollary 1.1.35 (Properties of the Quaternionic Determinant) *The determinant is a smooth map*

$$\det\colon \mathrm{Mat}(n \times n, \mathbb{H}) \longrightarrow \mathbb{R},$$

where $\mathrm{Mat}(n \times n, \mathbb{H})$ *has the manifold structure of* $\mathbb{R}^{4n^2}$. *For all matrices* $A, B \in \mathrm{Mat}(n \times n, \mathbb{H})$ *the following identity holds:*

$$\det(AB) = \det(A)\det(B).$$

A matrix $A \in \mathrm{Mat}(n \times n, \mathbb{H})$ *is invertible if and only if* $\det(A) \neq 0$.
The determinant of quaternionic matrices, defined above via the adjoint, therefore has at least some of the expected properties.

Remark 1.1.36 Note that the determinant of quaternionic matrices does *not* define an $\mathbb{H}$-multilinear, alternating map on $\mathbb{H}^n \times \ldots \times \mathbb{H}^n$ (n factors).

1.1.3 Products and Lie Subgroups

There are certain well-known constructions that yield new groups (and manifolds) from given groups (manifolds). Some of these constructions are compatible for both groups and manifolds and can be employed to generate new Lie groups from known ones. We discuss two important examples of such constructions: **products** and **subgroups**.

Proposition 1.1.37 (Products of Lie Groups) *Let G and H be Lie groups. Then the product manifold $G \times H$ with the direct product structure as a group is a Lie group, called the* **product Lie group**.

Proof This follows, because the maps

$$\begin{aligned} G \times H &\longrightarrow G \times H \\ (g,h) &\longmapsto \left(g^{-1}, h^{-1}\right) \end{aligned}$$

and

$$\begin{aligned} (G \times H) \times (G \times H) &\longrightarrow G \times H \\ (g_1, h_1, g_2, h_2) &\longmapsto (g_1 g_2, h_1 h_2) \end{aligned}$$

are smooth. □

Example 1.1.38 The **n-torus**

$$T^n = \underbrace{S^1 \times \cdots \times S^1}_{n}$$

is a compact, abelian Lie group. Similarly multiple (finite) products of copies of S^3 and S^1 are compact Lie groups. A particularly interesting case is $S^3 \times S^3$, because it can be identified with the Lie group Spin(4), to be defined in Chap. 6 (see Example 6.5.17).

Definition 1.1.39 Let G be a Lie group.

1. An **immersed Lie subgroup** of G is the image of an injective immersion $\phi\colon H \to G$ from a Lie group H to G such that ϕ is a group homomorphism.
2. An **embedded Lie subgroup** of G is the image of an injective immersion $\phi\colon H \to G$ from a Lie group H to G such that ϕ is a group homomorphism and a homeomorphism onto its image.

We call the map ϕ a **Lie group immersion** or **Lie group embedding**, respectively. In both cases, the set $\phi(H)$ is endowed with the topology, manifold structure and group structure such that $\phi\colon H \to \phi(H)$ is a diffeomorphism and a group isomorphism. Then $\phi(H)$ is a Lie group itself. The difference between embedded and immersed Lie subgroups $\phi(H) \subset G$ is whether the topology on $\phi(H)$ coincides with the subspace topology on $\phi(H)$ inherited from G or not. The group structure on $\phi(H)$ is in both cases the subgroup structure inherited from G.
An embedded Lie subgroup can be described equivalently as an embedded submanifold which is at the same time a subgroup. Most of the time we will consider

embedded Lie subgroups (immersed Lie subgroups appear naturally in Sect. 1.6). Note the following:

Proposition 1.1.40 *If $\phi: H \to G$ is a Lie group immersion where H is compact, then ϕ is a Lie group embedding.*

Proof Since H is compact and G is Hausdorff, it follows that the injective immersion $\phi: H \to G$ is a closed map, hence a homeomorphism onto its image. □
Hence immersed Lie subgroups which are not embedded can only be non-compact.

Example 1.1.41 Consider the Lie group $G = S^1$ and an element $x = e^{2\pi i\alpha} \in S^1$ with $\alpha \in \mathbb{R}$. The number α is rational if and only if there exists an integer N such that $N\alpha$ is an integer. This happens if and only if $x^N = 1$. Hence if α is rational, then x generates an embedded Lie subgroup in S^1, isomorphic to the finite cyclic group $\mathbb{Z}_K$, where K is the smallest positive integer such that $K\alpha$ is an integer. If α is irrational, then x generates an immersed Lie subgroup in S^1, isomorphic to $\mathbb{Z}$.

Example 1.1.42 Similarly, consider the Lie group $G = T^2$, which we think of as being obtained by identifying opposite sides of a square $[-1, 1] \times [-1, 1]$. It can be shown that the straight lines on the square of rational slope through the neutral element $e = (0, 0)$ define embedded Lie subgroups, diffeomorphic to S^1, while the straight lines of irrational slope through $(0, 0)$ define immersed Lie subgroups, diffeomorphic to $\mathbb{R}$.

Example 1.1.43 The sets $\mathbb{K}^* = \mathbb{K} \setminus \{0\}$ of invertible elements, for $\mathbb{K} = \mathbb{R}, \mathbb{C}, \mathbb{H}$, together with multiplication are Lie groups of dimension $1, 2, 4$, respectively. The spheres S^0, S^1, S^3 are embedded Lie subgroups of codimension 1.
The proof of the following (non-trivial) theorem is deferred to Sect. 1.8.

Theorem 1.1.44 (Cartan's Theorem, Closed Subgroup Theorem) *Let G be a Lie group and suppose that $H \subset G$ is a subgroup in the algebraic sense. Then H is an embedded Lie subgroup if and only if H is a closed set in the topology of G.*

By a **closed subgroup** of a Lie group G we always mean a subgroup which is a closed set in the topology of G. The difficult part in the proof of this theorem is to show that a closed subgroup of a Lie group is an embedded submanifold. Cartan's Theorem allows us to construct many new interesting Lie groups by realizing them as closed subgroups of given Lie groups.

Remark 1.1.45 Groups can also be constructed by taking the **quotient** of a group G by a normal subgroup H. However, for a Lie group G and a subgroup H, the quotient space G/H will not be a smooth manifold in general (at least not canonically). We will show in Sect. 3.8.3 that if H is a *closed* subgroup of a Lie group G, then G/H is indeed a smooth manifold. In this case, if H is normal, then G/H will again be a Lie group.

1.2 Linear Groups and Symmetry Groups of Vector Spaces

The most famous class of examples of Lie groups are the **general linear groups** $\mathrm{GL}(n, \mathbb{K})$ over the real, complex and quaternionic numbers $\mathbb{K}$ as well as the following groups:

Definition 1.2.1 A closed subgroup of a general linear group is called a **linear group** or **matrix group**.

By Cartan's Theorem 1.1.44 linear groups are embedded Lie subgroups and, in particular, Lie groups themselves. We are especially going to study the following families of linear groups, which are called **classical Lie groups**:

- the **special linear groups** in the real, complex and quaternionic case
- the **(special) orthogonal group** in the real case
- the **(special) unitary group** in the complex case
- the **compact symplectic group** (also called the **quaternionic unitary group**) in the quaternionic case
- the real **pseudo-orthogonal groups for indefinite scalar products**, like the Lorentz group.

The general linear groups are the (maximal) symmetry groups of vector spaces. The families of linear groups above arise as automorphism groups of certain structures on vector spaces. They can also be understood as *isotropy groups* in certain *representations* of the general linear groups.

There are two classes of Lie groups we are interested in which are not (at least *a priori*) linear:

- the **exceptional compact Lie groups** $\mathrm{G}_2, \mathrm{F}_4, \mathrm{E}_6, \mathrm{E}_7, \mathrm{E}_8$ (we will discuss G_2 in detail in Sect. 3.10)
- the **spin groups**, which are certain double coverings of (pseudo-)orthogonal groups.

All Lie groups that we will consider belong to one of these classes or are products of such Lie groups. Most linear groups are non-abelian and certain classes of linear groups – the (special) orthogonal, (special) unitary and symplectic groups – are compact. Lie groups like the Lorentz group and its spin group are not compact.

There are several reasons why linear groups are important, in particular with regard to *compact* Lie groups:

1. First, it is possible to prove that any compact Lie group is *isomorphic* as a Lie group to a linear group, see Theorem 1.2.7. In particular, the compact exceptional Lie groups and the compact spin groups are isomorphic to linear groups.
2. Secondly, there is a classification theorem which shows that (up to finite coverings) any compact Lie group G is isomorphic to a product

$$G = G_1 \times \ldots \times G_r$$

of compact Lie groups, all of which belong to the classes mentioned above (classical linear, spin and exceptional Lie groups); see Theorem 2.4.23 and Theorem 2.4.29 for the classification of compact Lie groups.

Even though the classical linear groups look quite special, they are thus of general significance, in particular for gauge theories with compact gauge groups.

Spin groups, such as the universal covering of the Lorentz group and its higher-dimensional analogues, are also important in physics, because they are involved in the mathematical description of *fermions*.

Finally, the exceptional Lie groups appear in several places in physics: E_6, for example, is the gauge group of certain *Grand Unified Theories*, E_8 plays a role in *heterotic string theory* and G_2 is related to *M-theory compactifications*.

Remark 1.2.2 It is an interesting fact that there are non-compact Lie groups which are *not* isomorphic to linear groups. One example is the universal covering of the Lie group $\mathrm{SL}(2,\mathbb{R})$ (see [70, Sect. 5.8] for a proof).

1.2.1 Isomorphism Groups of Vector Spaces

The simplest and fundamental linear groups, perhaps already known from a course on linear algebra, are the general linear groups themselves. Let $\mathbb{K} = \mathbb{R}, \mathbb{C}, \mathbb{H}$.

Definition 1.2.3 For $n \geq 1$ the **general linear group** is defined as the group of linear isomorphisms of $\mathbb{K}^n$:

$$\mathrm{GL}(n,\mathbb{K}) = \{A \in \mathrm{Mat}(n \times n, \mathbb{K}) \mid A \text{ is invertible}\},$$

where $\mathrm{Mat}(n \times n, \mathbb{K})$ denotes the ring of $n \times n$-matrices with coefficients in $\mathbb{K}$. Group multiplication in $\mathrm{GL}(n,\mathbb{K})$ is matrix multiplication.
For $\mathbb{K} = \mathbb{H}$ we act with matrices on the left of the right vector space $\mathbb{H}^n$, so that the maps are indeed $\mathbb{H}$-linear:

$$A(vq) = (Av)q \quad \forall A \in \mathrm{Mat}(n \times n, \mathbb{H}), v \in \mathbb{H}^n, q \in \mathbb{H}.$$

The following alternative description of the general linear group follows immediately:

Proposition 1.2.4 *The general linear group is given by*

$$\mathrm{GL}(n,\mathbb{K}) = \{A \in \mathrm{Mat}(n \times n, \mathbb{K}) \mid \det(A) \neq 0\}.$$

Clearly, for $n = 1$, we have

$$\mathrm{GL}(1,\mathbb{K}) = \mathbb{K}^*.$$

We first consider the real general linear group.

Proposition 1.2.5 (The Real General Linear Group Is a Lie Group) $\mathrm{GL}(n, \mathbb{R})$ *is a non-compact n^2-dimensional Lie group. It is not abelian for $n \geq 2$.*

Proof It is clear by properties of the determinant that $\mathrm{GL}(n, \mathbb{R})$ is a group in the algebraic sense.

Note that by continuity of the determinant, $\mathrm{GL}(n, \mathbb{R})$ is an open subset of

$$\mathrm{Mat}(n \times n, \mathbb{R}) \cong \mathbb{R}^{n^2}.$$

In particular, $\mathrm{GL}(n, \mathbb{R})$ is a smooth manifold of dimension n^2. Multiplication of two matrices A, B is quadratic in their coordinates, hence a smooth map. According to Remark 1.1.8 this shows that $\mathrm{GL}(n, \mathbb{R})$ is indeed a Lie group (we can also see directly that inversion of a matrix is, by Cramer's rule, a rational map in the coordinates, hence smooth).

The manifold $\mathrm{GL}(n, \mathbb{R})$, as a subset of $\mathbb{R}^{n^2}$ with the Euclidean norm, is not bounded, because it contains the unbounded set of diagonal matrices of the form rI_n with $r \in \mathbb{R}^* = \mathbb{R} \setminus \{0\}$ (these elements actually define a subgroup of the general linear group, isomorphic to $\mathbb{R}^*$). By Heine–Borel, $\mathrm{GL}(n, \mathbb{R})$ is not compact.

To show that $\mathrm{GL}(n, \mathbb{R})$ is not abelian for $n \geq 2$, note that $\mathrm{GL}(n, \mathbb{R})$ contains the subgroup H isomorphic to $\mathrm{GL}(2, \mathbb{R})$, consisting of matrices of the form

$$\begin{pmatrix} A & 0 \\ 0 & I_{n-2} \end{pmatrix}, \quad A \in \mathrm{GL}(2, \mathbb{R}).$$

It therefore suffices to show that $\mathrm{GL}(2, \mathbb{R})$ is not abelian: it is easy to find two matrices in $\mathrm{GL}(2, \mathbb{R})$ which do not commute. □

Using similar arguments it can be shown:

Proposition 1.2.6 (Complex and Quaternionic General Linear Groups Are Lie Groups) *Over the complex numbers and quaternions we have:*

1. $\mathrm{GL}(n, \mathbb{C})$ *is a non-compact $2n^2$-dimensional Lie group. It is not abelian for $n \geq 2$.*
2. $\mathrm{GL}(n, \mathbb{H})$ *is a non-compact $4n^2$-dimensional Lie group. It is not abelian for $n \geq 1$.*

As an application of the *Peter–Weyl Theorem*, the following can be shown (for a proof, which is beyond the scope of this book, see [24, 83, 129]):

Theorem 1.2.7 (Compact Lie Groups Are Linear) *Let G be a compact Lie group. Then there exists a smooth, injective group homomorphism ϕ of G into a general linear group* $\mathrm{GL}(n, \mathbb{C})$ *for some n.*

According to Corollary 1.8.18 the map ϕ is a *Lie group isomorphism* onto a linear group (see Sect. 1.3 for the formal definition of Lie group homomorphisms and isomorphisms).

The Peter–Weyl Theorem shows that every compact Lie group can be considered as a linear group. If we assume this result, we shall see later as a consequence of Theorem 2.1.39 that a compact Lie group G can even be embedded as a closed subgroup in a unitary group $\mathrm{U}(n)$ for some n and thus in the rotation group $\mathrm{SO}(2n)$ by Exercise 1.9.10. As a consequence any compact Lie group G can be literally thought of as a group whose elements are rotations on some $\mathbb{R}^{2n}$.

Remark 1.2.8 By comparison, recall that *Cayley's Theorem* says that any finite group of order n can be embedded as a subgroup of the symmetric group S_n.

1.2.2 Automorphism Groups of Structures on Vector Spaces

We want to consider specific classes of linear groups that arise as automorphism groups of certain structures on vector spaces.

Definition 1.2.9 We define the following scalar products:

1. On $\mathbb{R}^n$ the **standard Euclidean scalar product**

$$\langle v, w\rangle = v^T w = \sum_{k=1}^{n} v_k w_k.$$

2. On $\mathbb{C}^n$ the **standard Hermitian scalar product**

$$\langle v, w\rangle = v^\dagger w = \sum_{k=1}^{n} \overline{v_k} w_k.$$

3. On $\mathbb{H}^n$ the **standard symplectic scalar product**

$$\langle v, w\rangle = v^\dagger w = \sum_{k=1}^{n} \overline{v_k} w_k.$$

Here

$$v = \begin{pmatrix} v_1 \\ v_2 \\ \vdots \\ v_n \end{pmatrix}, \quad w = \begin{pmatrix} w_1 \\ w_2 \\ \vdots \\ w_n \end{pmatrix}$$

are column vectors in $\mathbb{R}^n$, $\mathbb{C}^n$ and $\mathbb{H}^n$, respectively.

Definition 1.2.10 Let $\mathbb{K} = \mathbb{R}, \mathbb{C}, \mathbb{H}$. We define the **standard volume form** vol on $V = \mathbb{K}^n$ by

$$\begin{aligned} \mathrm{vol}\colon V \times \ldots \times V &\longrightarrow \mathbb{K} \\ (v_1, \ldots, v_n) &\longmapsto \det(v_1, \ldots, v_n), \end{aligned}$$

where $(v_1, \ldots, v_n)$ is the $n \times n$-matrix with columns $v_1, \ldots, v_n$.

Remark 1.2.11 For $\mathbb{K} = \mathbb{R}, \mathbb{C}$ this form is $\mathbb{K}$-multilinear and alternating, but not for $\mathbb{K} = \mathbb{H}$.

Definition 1.2.12 Let $n \geq 1$.

1. The **special linear groups** for $\mathbb{K} = \mathbb{R}, \mathbb{C}, \mathbb{H}$ are defined as the automorphism groups of the standard volume forms:

$$\mathrm{SL}(n, \mathbb{K}) = \{A \in \mathrm{GL}(n, \mathbb{K}) \mid \mathrm{vol}(Av_1, \ldots, Av_n) = \mathrm{vol}(v_1, \ldots, v_n) \quad \forall v_1, \ldots, v_n \in \mathbb{K}^n\}.$$

2. The **orthogonal, unitary** and **(compact) symplectic** (also called **quaternionic unitary**) **groups** are defined as the automorphism groups of the standard Euclidean, Hermitian and symplectic scalar products:

$$\begin{aligned} \mathrm{O}(n) &= \{A \in \mathrm{GL}(n, \mathbb{R}) \mid \langle Av, Aw \rangle = \langle v, w \rangle \quad \forall v, w \in \mathbb{R}^n\}, \\ \mathrm{U}(n) &= \{A \in \mathrm{GL}(n, \mathbb{C}) \mid \langle Av, Aw \rangle = \langle v, w \rangle \quad \forall v, w \in \mathbb{C}^n\}, \\ \mathrm{Sp}(n) &= \{A \in \mathrm{GL}(n, \mathbb{H}) \mid \langle Av, Aw \rangle = \langle v, w \rangle \quad \forall v, w \in \mathbb{H}^n\}. \end{aligned}$$

3. The **special orthogonal** and **special unitary groups** are defined as

$$\begin{aligned} \mathrm{SO}(n) &= \mathrm{O}(n) \cap \mathrm{SL}(n, \mathbb{R}), \\ \mathrm{SU}(n) &= \mathrm{U}(n) \cap \mathrm{SL}(n, \mathbb{C}). \end{aligned}$$

These Lie groups are called **classical groups**.

Remark 1.2.13 There are additional classes of classical groups, called *(non-compact) symplectic groups*, which are defined as automorphism groups of skew-symmetric forms. We will not consider these Lie groups in the subsequent discussions.

Remark 1.2.14 In the physics literature the Lie group Sp(n) is sometimes denoted by USp(n) (and occasionally by Sp($2n$) or USp($2n$)). For example, [149] uses the notation Sp(n), whereas [90] uses the notation Sp($2n$) and [138] uses the notation USp($2n$). We continue to use the notation Sp(n).
It is often useful to have the following alternative description of these groups.

Proposition 1.2.15 (Matrix Description of Classical Groups) *Let $n \geq 1$.*

1. *The special linear groups are given by*

$$\mathrm{SL}(n, \mathbb{K}) = \{A \in \mathrm{Mat}(n \times n, \mathbb{K}) \mid \det(A) = 1\}.$$

2. *The orthogonal, unitary and symplectic groups are given by*

$$\mathrm{O}(n) = \left\{A \in \mathrm{Mat}(n \times n, \mathbb{R}) \mid A \cdot A^T = I\right\},$$
$$\mathrm{U}(n) = \left\{A \in \mathrm{Mat}(n \times n, \mathbb{C}) \mid A \cdot A^\dagger = I\right\},$$
$$\mathrm{Sp}(n) = \left\{A \in \mathrm{Mat}(n \times n, \mathbb{H}) \mid A \cdot A^\dagger = I\right\},$$

where A^T denotes the transpose of A and $A^\dagger = (\bar{A})^T$.

3. *The special orthogonal and special unitary groups are given by*

$$\mathrm{SO}(n) = \left\{A \in \mathrm{Mat}(n \times n, \mathbb{R}) \mid A \cdot A^T = I, \det(A) = 1\right\},$$
$$\mathrm{SU}(n) = \left\{A \in \mathrm{Mat}(n \times n, \mathbb{C}) \mid A \cdot A^\dagger = I, \det(A) = 1\right\}.$$

Proof

1. For $A \in \mathrm{Mat}(n \times n, \mathbb{K})$ and column vectors $v_1, \ldots, v_n \in \mathbb{K}^n$ the following identity holds:

$$\mathrm{vol}(Av_1, \ldots, Av_n) = \det(A) \cdot \mathrm{vol}(v_1, \ldots, v_n).$$

This follows because as matrices

$$(Av_1, \ldots, Av_n) = A \cdot (v_1, \ldots, v_n)$$

and the determinant is multiplicative: $\det(AB) = \det(A)\det(B)$. This implies the formula for $\mathrm{SL}(n, \mathbb{K})$.

2. A matrix $A \in \mathrm{GL}(n, \mathbb{R})$ satisfies

$$\langle Av, Aw \rangle = \langle v, w \rangle \quad \forall v, w \in \mathbb{R}^n$$

if and only if

$$v^T \left(A^T A\right) w = v^T w \quad \forall v, w \in \mathbb{R}^n.$$

Choosing $v = e_i, w = e_j$ this happens if and only if $A^T A = I$ and thus $AA^T = I$. The complex and quaternionic case follow similarly.

3. This is clear by the results above.

□

We did not define a quaternionic *special* unitary group, because it turns out that there is no difference between such a group and the quaternionic unitary group. This follows from the next proposition.

Proposition 1.2.16 *Let* $A \in \mathrm{Sp}(n)$. *Then* $\det(A) = 1$.

Proof Let $A = A_1 + jA_2 \in \mathrm{Mat}(n \times n, \mathbb{H})$ with A_1, A_2 complex. According to Proposition 1.1.32 the determinant of A is real and non-negative. We have

$$A^\dagger = A_1^\dagger - jA_2^T$$

and

$$\chi_{A^\dagger} = \chi_A^\dagger.$$

Thus for $A \in \mathrm{Sp}(n)$

$$1 = \det(I) = \det(\chi_A)\det\left(\chi_A^\dagger\right) = |\det(\chi_A)|^2 = (\det(A))^2.$$

Therefore $\det(A) = 1$. □

We now want to prove the main result in this subsection.

Theorem 1.2.17 (Classical Groups Are Linear) *The special linear, (special) orthogonal, (special) unitary and symplectic groups are closed subgroups of general linear groups, i.e. linear groups. They have the following properties:*

1. *The special linear groups have dimension*

$$\dim \mathrm{SL}(n, \mathbb{R}) = n^2 - 1,$$
$$\dim \mathrm{SL}(n, \mathbb{C}) = 2n^2 - 2,$$
$$\dim \mathrm{SL}(n, \mathbb{H}) = 4n^2 - 1.$$

These Lie groups are not compact for $n \geq 2$.

2. *The orthogonal, unitary and symplectic groups have dimension*

$$\dim \mathrm{O}(n) = \frac{1}{2}n(n-1),$$
$$\dim \mathrm{U}(n) = n^2,$$
$$\dim \mathrm{Sp}(n) = 2n^2 + n.$$

(continued)

Theorem 1.2.17 (continued)

These Lie groups are compact for all $n \geq 1$.

3. *The special orthogonal and special unitary groups have dimension*

$$\dim \mathrm{SO}(n) = \frac{1}{2}n(n-1),$$

$$\dim \mathrm{SU}(n) = n^2 - 1.$$

These Lie groups are compact for all $n \geq 1$.

Proof

- **Closed subgroups:** It is clear that all of these subsets are subgroups of general linear groups. The maps

$$\det\colon \mathrm{Mat}(n \times n, \mathbb{K}) \longrightarrow \mathbb{K},$$

for $\mathbb{K} = \mathbb{R}, \mathbb{C}$, and

$$\det\colon \mathrm{Mat}(n \times n, \mathbb{H}) \longrightarrow \mathbb{R}$$

are continuous (polynomial in the coordinates), hence the preimages $\det^{-1}(1)$ are closed subsets. This shows that $\mathrm{SL}(n, \mathbb{K})$ is a closed subgroup of the general linear group $\mathrm{GL}(n, \mathbb{K})$. Similarly, the map

$$\mathrm{Mat}(n \times n, \mathbb{R}) \longrightarrow \mathrm{Mat}(n \times n, \mathbb{R})$$

$$A \longmapsto A \cdot A^T$$

is continuous (quadratic in the coordinates), hence the preimage of I is a closed subset. This shows that $\mathrm{O}(n)$ and the intersection $\mathrm{SO}(n) = \mathrm{O}(n) \cap \mathrm{SL}(n, \mathbb{R})$ are closed subgroups of the general linear group $\mathrm{GL}(n, \mathbb{R})$. Similarly for $\mathrm{U}(n)$, $\mathrm{SU}(n)$ and $\mathrm{Sp}(n)$.

- **Compactness:** To show that $\mathrm{SL}(n, \mathbb{K})$ is not compact for $n \geq 2$ it suffices to show that the subgroup $\mathrm{SL}(2, \mathbb{K})$ is not compact. This follows by considering the unbounded subset of matrices of the form

$$\begin{pmatrix} 1 & a \\ 0 & 1 \end{pmatrix} \in \mathrm{SL}(2, \mathbb{K}), \quad a \in \mathbb{R}.$$

To show that $\mathrm{O}(n)$ and hence $\mathrm{SO}(n)$ are compact it suffices to show by Heine–Borel that $\mathrm{O}(n)$ is a bounded subset of the Euclidean space

$$\mathrm{Mat}(n \times n, \mathbb{R}) \cong \mathbb{R}^{n^2}.$$

Let $A \in \mathrm{O}(n)$. For fixed $i = 1, \ldots, n$ we have

$$1 = \left(AA^T\right)_{ii} = A_{i1}^2 + A_{i2}^2 + \ldots + A_{in}^2,$$

hence $|A_{ij}| \leq 1$ for all indices i, j. This implies the claim. Compactness of $\mathrm{U}(n)$, $\mathrm{SU}(n)$ and $\mathrm{Sp}(n)$ follows similarly.

- **Dimensions of special linear groups:** Finally we calculate the dimensions. We claim that the smooth map

$$\det\colon \mathrm{Mat}(n \times n, \mathbb{K}) \longrightarrow \mathbb{K}$$

for $\mathbb{K} = \mathbb{R}, \mathbb{C}$ has 1 as a regular value. This implies again that $\mathrm{SL}(n, \mathbb{R})$ and $\mathrm{SL}(n, \mathbb{C})$ are smooth manifolds and also the formulas $n^2 - 1$ and $2n^2 - 2$ for the dimensions (the argument for $\mathrm{SL}(n, \mathbb{H})$ is slightly different and is given below).

To prove the claim, let $A \in \mathrm{SL}(n, \mathbb{K})$ for $\mathbb{K} = \mathbb{R}, \mathbb{C}$ and write A as

$$A = (v_1, v_2, \ldots, v_n),$$

where the v_i are column vectors in $\mathbb{K}^n$. For $a \in \mathbb{K}$ fixed consider the curve

$$C(t) = ((1 + at)v_1, v_2, \ldots, v_n)$$

in $\mathrm{Mat}(n \times n, \mathbb{K})$. Then $C(0) = A$ and since the determinant is multilinear

$$\det C(t) = (1 + at) \det A = (1 + at),$$

thus

$$\left.\frac{d}{dt}\right|_{t=0} \det C(t) = a.$$

This shows that the differential of the determinant is surjective in every $A \in \mathrm{SL}(n, \mathbb{K})$ and 1 is a regular value.

We now prove the formula for the dimension of $\mathrm{SL}(n, \mathbb{H})$. We want to show that

$$\det\colon \mathrm{Mat}(n \times n, \mathbb{H}) \longrightarrow \mathbb{R}$$

has 1 as a regular value. This implies that $\mathrm{SL}(n, \mathbb{H})$ is a smooth manifold of dimension $4n^2 - 1$. Let $A = A_1 + jA_2 \in \mathrm{SL}(n, \mathbb{H})$ with A_1, A_2 complex, written in terms of complex column vectors as

$$A_1 = (v_1, v_2, \ldots, v_n),$$
$$A_2 = (w_1, w_2, \ldots, w_n).$$

For $b \in \mathbb{R}$ consider the curve

$$D(t) = ((1+bt)v_1, v_2, \ldots, v_n) + j((1+bt)w_1, w_2, \ldots, w_n)$$

in $\mathrm{Mat}(n \times n, \mathbb{H})$. Then $D(0) = A$ and the adjoint of $D(t)$ is

$$\chi_{D(t)} = \begin{pmatrix} (1+bt)v_1 & v_2 & \ldots & v_n & -(1+bt)\bar{w}_1 & -\bar{w}_2 & \ldots & -\bar{w}_n \\ (1+bt)w_1 & w_2 & \ldots & w_n & (1+bt)\bar{v}_1 & \bar{v}_2 & \ldots & \bar{v}_n \end{pmatrix}$$

with determinant

$$\det D(t) = \det \chi_{D(t)} = (1+bt)^2 \det \chi_A = (1+bt)^2.$$

It follows that

$$\left.\frac{d}{dt}\right|_{t=0} \det D(t) = 2b.$$

This shows that the differential of the determinant is surjective in every $A \in \mathrm{SL}(n, \mathbb{H})$ and 1 is a regular value.

- **Dimensions of $\mathrm{O}(n)$, $\mathrm{U}(n)$ and $\mathrm{Sp}(n)$:** To calculate the dimension of the orthogonal group $\mathrm{O}(n)$ consider the map

$$f\colon \mathrm{Mat}(n \times n, \mathbb{R}) \longrightarrow \mathrm{Sym}(n, \mathbb{R})$$
$$A \longmapsto A \cdot A^T,$$

where $\mathrm{Sym}(n, \mathbb{R})$ denotes the space of symmetric, real $n \times n$-matrices. Then $\mathrm{O}(n) = f^{-1}(I)$. The differential of this map at a point $A \in \mathrm{O}(n)$ in the direction $X \in \mathrm{Mat}(n \times n, \mathbb{R})$ is

$$(D_A f)(X) = XA^T + AX^T.$$

Let $B \in \mathrm{Sym}(n, \mathbb{R})$ and set

$$X = \frac{1}{2}BA.$$

Then $(D_A f)(X) = B$ and thus I is a regular value of f. This shows that $\mathrm{O}(n)$ is a smooth manifold of dimension

$$\begin{aligned} \dim \mathrm{O}(n) &= \dim \mathrm{Mat}(n \times n, \mathbb{R}) - \dim \mathrm{Sym}(n, \mathbb{R}) \\ &= n^2 - \frac{1}{2}n(n+1) \\ &= \frac{1}{2}n(n-1). \end{aligned}$$

We can calculate the dimensions of $\mathrm{U}(n)$ and $\mathrm{Sp}(n)$ in a similar way, utilizing for $\mathbb{K} = \mathbb{C}, \mathbb{H}$ the map

$$f\colon \mathrm{Mat}(n \times n, \mathbb{K}) \longrightarrow \mathrm{Herm}(n, \mathbb{K})$$
$$A \longmapsto A \cdot A^{\dagger},$$

where $\mathrm{Herm}(n, \mathbb{K})$ denotes the space of Hermitian $n \times n$-matrices (the set of all matrices B with $B^{\dagger} = B$). Again I is a regular value and thus $\mathrm{U}(n)$ and $\mathrm{Sp}(n)$ are smooth manifolds of dimension

$$\begin{aligned}\dim_{\mathbb{R}} \mathrm{Mat}(n \times n, \mathbb{K}) - \dim_{\mathbb{R}} \mathrm{Herm}(n, \mathbb{K}) &= kn^2 - \left(\frac{1}{2}k(n-1)n + n\right) \\ &= \frac{1}{2}kn(n+1) - n,\end{aligned}$$

where $k = 2, 4$ for $\mathbb{K} = \mathbb{C}, \mathbb{H}$. This implies

$$\dim \mathrm{U}(n) = n^2$$

and

$$\dim \mathrm{Sp}(n) = 2n^2 + n.$$

- **Dimensions of** $\mathrm{SO}(n)$ **and** $\mathrm{SU}(n)$**:** We claim that $\mathrm{SO}(n)$ is a submanifold of codimension zero in $\mathrm{O}(n)$. The determinant on $\mathrm{O}(n)$ has values in $\{+1, -1\}$,

$$\det\colon \mathrm{O}(n) \longrightarrow \{+1, -1\}.$$

This map obviously has 1 as a regular value.

Similarly, we claim that $\mathrm{SU}(n)$ is a submanifold of codimension one in $\mathrm{U}(n)$. The determinant on $\mathrm{U}(n)$ has values in S^1,

$$\det\colon \mathrm{U}(n) \longrightarrow S^1.$$

We claim that this map has 1 as a regular value. Let $A \in \mathrm{SU}(n)$, written in terms of complex column vectors as

$$A = (v_1, v_2, \ldots, v_n).$$

For $\alpha \in \mathbb{R}$ consider the curve

$$C(t) = \left(e^{i\alpha t} v_1, v_2, \ldots, v_n\right).$$

It is easy to check that $C(t)$ is a curve in $\mathrm{U}(n)$ and

$$\det C(t) = e^{i\alpha t}\det A = e^{i\alpha t}$$

with

$$\left.\frac{d}{dt}\right|_{t=0} \det C(t) = i\alpha.$$

This proves the claim. □

Example 1.2.18 It follows directly from the matrix description that

$$\begin{aligned}\mathrm{SL}(1,\mathbb{R}) &= \mathrm{SL}(1,\mathbb{C}) = \{1\},\\ \mathrm{SL}(1,\mathbb{H}) &= S^3\end{aligned}$$

and

$$\begin{aligned}\mathrm{O}(1) &= S^0,\\ \mathrm{U}(1) &= S^1,\\ \mathrm{Sp}(1) &= S^3.\end{aligned}$$

We also saw that

$$\mathrm{SO}(2) = \left\{ \begin{pmatrix} \cos\alpha & -\sin\alpha \\ \sin\alpha & \cos\alpha \end{pmatrix} \in \mathrm{Mat}(2\times 2,\mathbb{R}) \;\middle|\; \alpha \in \mathbb{R} \right\}$$

and it is not difficult to check that

$$\mathrm{SU}(2) = \left\{ \begin{pmatrix} a & -\bar{b} \\ b & \bar{a} \end{pmatrix} \in \mathrm{Mat}(2\times 2,\mathbb{C}) \;\middle|\; a,b \in \mathbb{C},\ |a|^2 + |b|^2 = 1 \right\}.$$

We will discuss less trivial identifications between Lie groups, like $\mathrm{SU}(2) \cong S^3$, after we have defined the notion of a Lie group isomorphism.

Remark 1.2.19 Some of the linear groups appear directly in gauge theories: For instance, as mentioned before, the gauge group of the current Standard Model of particle physics is the product

$$\mathrm{SU}(3) \times \mathrm{SU}(2) \times \mathrm{U}(1).$$

There are Grand Unified Theories based on Lie groups like $\mathrm{SU}(5)$ and $\mathrm{SO}(10)$ (or rather its *universal covering group* $\mathrm{Spin}(10)$).

Remark 1.2.20 (Classical Linear Groups as Isotropy Groups) Note that $\mathrm{GL}(n, \mathbb{K})$ and its subgroups act canonically on the (column) vector space $\mathbb{K}^n$ by matrix multiplication from the left. This is an example of a representation, called the **fundamental representation**. We will consider representations in more detail in Sect. 2.1.

Representations are special classes of *group actions*, see Chap. 3. It is not difficult to see that the linear groups can be realized as *isotropy groups* of certain elements in suitable representation spaces of the general linear groups. Representations and isotropy groups will also be used in Sect. 3.10 to define the exceptional Lie group G_2 as an embedded Lie subgroup of $\mathrm{GL}(7, \mathbb{R})$.

1.2.3 Connectivity Properties of Linear Groups

Proposition 1.2.21 (Connected Components of Lie Groups) *Let G be a Lie group. Then all connected components of G are diffeomorphic to the connected component G_e of the neutral element $e \in G$. In particular, all connected components have the same dimension.*

Proof Let $g \in G$ and denote the connected component of G containing g by G_g. Consider the *left translation*, given by

$$\begin{aligned} L_g \colon G_e &\longrightarrow G \\ x &\longmapsto gx \end{aligned}$$

restricted to the connected component G_e containing e. The image of this smooth map is connected and contains g, therefore the image is contained in G_g. By the same argument

$$L_{g^{-1}} \colon G_g \longrightarrow G$$

has image contained in G_e. It follows that

$$L_g \colon G_e \longrightarrow G_g$$

is a diffeomorphism and thus all connected components of G are diffeomorphic. □

We want to understand how many connected components the classical Lie groups have.

Theorem 1.2.22 (Connected Components of Classical Groups) *Let $n \geq 1$.*

1. *The Lie group $\mathrm{GL}(n, \mathbb{R})$ has two connected components $\mathrm{GL}(n, \mathbb{R})_\pm$, determined by the sign of the determinant. The Lie groups $\mathrm{GL}(n, \mathbb{C})$ and $\mathrm{GL}(n, \mathbb{H})$ are connected.*
2. *The special linear groups $\mathrm{SL}(n, \mathbb{K})$ are connected for all $\mathbb{K} = \mathbb{R}, \mathbb{C}, \mathbb{H}$.*

3. *The Lie group* $\mathrm{O}(n)$ *has two connected components* $\mathrm{O}(n)_{\pm}$, *determined by the sign of the determinant. The Lie group* $\mathrm{SO}(n) = \mathrm{O}(n)_{+}$ *is connected.*
4. *The Lie groups* $\mathrm{U}(n)$, $\mathrm{SU}(n)$ *and* $\mathrm{Sp}(n)$ *are connected.*

There are direct proofs for these assertions that the reader can find in the literature. We chose to give a different argument using *homogeneous spaces* in Sect. 3.8.3, which is conceptually clearer and simpler. The proof with homogeneous spaces utilizes certain actions of the classical groups on $\mathbb{K}^n \setminus \{0\}$ and spheres S^m. The assertions then follow by induction over n from the corresponding (trivial) statements for the groups with $n = 1$.

The number of connected components of a Lie group G can be identified with the number of elements of the homotopy group $\pi_0(G)$. In Sect. 2.6 we will discuss higher homotopy groups of Lie groups.

1.3 Homomorphisms of Lie Groups

Lie groups have two structures: the algebraic structure of a group and the smooth structure of a manifold. A homomorphism between Lie groups should be compatible with both structures.

Definition 1.3.1 Let G and H be Lie groups. A map $\phi\colon G \to H$ which is smooth and a group homomorphism, i.e.

$$\phi(g_1 \cdot g_2) = \phi(g_1) \cdot \phi(g_2) \quad \forall g_1, g_2 \in G,$$

is called a **Lie group homomorphism**. The map ϕ is called a **Lie group isomorphism** if it is a diffeomorphism and a homomorphism (hence an isomorphism) of groups. A Lie group isomorphism $\phi\colon G \to G$ is called a **Lie group automorphism** of G.

Remark 1.3.2 We will show in Theorem 1.8.14 as an application of Cartan's Theorem that *continuous* group homomorphisms $\phi\colon G \to H$ between Lie groups are automatically *smooth*, hence Lie group homomorphisms.

Remark 1.3.3 Occasionally we will call a Lie group homomorphism just a homomorphism if the meaning is clear from the context.
We consider some examples of Lie group homomorphisms.

Example 1.3.4 Let G and H be Lie groups. The constant map

$$\begin{aligned} \phi\colon G &\longrightarrow H \\ g &\longmapsto e \end{aligned}$$

is always a Lie group homomorphism, called the **trivial** homomorphism.

Example 1.3.5 Consider $\mathbb{R}$ with addition and the Lie group S^1. Then

$$\begin{aligned} \phi: \mathbb{R} &\longrightarrow S^1 \\ x &\longmapsto e^{ix} \end{aligned}$$

is a surjective homomorphism of Lie groups. The kernel of this map is

$$\phi^{-1}(1) = 2\pi\mathbb{Z}.$$

Taking products, there is a similar surjective homomorphism $\mathbb{R}^n \to T^n$.

Example 1.3.6 It is easy to check that the map

$$\begin{aligned} \phi: S^1 &\longrightarrow \mathrm{SO}(2) \\ e^{i\alpha} &\longmapsto \begin{pmatrix} \cos\alpha & -\sin\alpha \\ \sin\alpha & \cos\alpha \end{pmatrix} \end{aligned}$$

is an isomorphism of Lie groups.

Example 1.3.7 We have

$$\begin{aligned} S^3 &= \{w \in \mathbb{H} \mid ||w||^2 = w\bar{w} = 1\} \\ &= \{x + yi + uj + vk \in \mathbb{H} \mid x^2 + y^2 + u^2 + v^2 = 1\}. \end{aligned}$$

We also have

$$\mathrm{SU}(2) = \left\{ \begin{pmatrix} a & -\bar{b} \\ b & \bar{a} \end{pmatrix} \in \mathrm{Mat}(2 \times 2, \mathbb{C}) \,\middle|\, |a|^2 + |b|^2 = 1 \right\}.$$

It can be checked that the map

$$\begin{aligned} \phi: S^3 &\longrightarrow \mathrm{SU}(2) \\ x + yi + uj + vk &\longmapsto \begin{pmatrix} x + iy & -u - iv \\ u - iv & x - iy \end{pmatrix} \end{aligned}$$

is an isomorphism of Lie groups. This follows from a direct calculation or from Proposition 1.1.26.

Example 1.3.8 (Universal Covering of SO(3)*)* We define a Lie group homomorphism

$$\phi: S^3 \longrightarrow \mathrm{SO}(3)$$

in the following way: Let $w \in \mathbb{H}$ be a unit quaternion, $||w|| = 1$, and consider the map

$$\begin{aligned} \tau_w : \mathbb{H} &\longrightarrow \mathbb{H} \\ z &\longmapsto wzw^{-1}. \end{aligned}$$

This is an $\mathbb{R}$-linear isomorphism of the 4-dimensional real vector space $\mathbb{H}$. Since

$$||wzw^{-1}|| = ||w|| \cdot ||z|| \cdot ||w||^{-1} = ||z||,$$

the map τ_w is orthogonal with respect to the standard Euclidean scalar product on $\mathbb{H} \cong \mathbb{R}^4$. The map τ_w clearly fixes

$$\mathrm{Re}\,\mathbb{H} = \{x \in \mathbb{H} \mid x \in \mathbb{R}\}$$

and therefore restricts to an orthogonal isomorphism

$$\phi(w) = \tau_w|_{\mathrm{Im}\,\mathbb{H}} : \mathrm{Im}\,\mathbb{H} \longrightarrow \mathrm{Im}\,\mathbb{H}$$

on the orthogonal complement

$$\mathrm{Im}\,\mathbb{H} = \mathrm{Re}\,\mathbb{H}^{\perp} = \{yi + uj + vk \in \mathbb{H} \mid y, u, v \in \mathbb{R}\}.$$

This shows that $\phi(w) \in \mathrm{O}(3)$. Since the map $\phi: S^3 \to \mathrm{O}(3)$ is continuous, S^3 is connected and $\phi(1) = I$, it follows that ϕ has image in the connected component $\mathrm{SO}(3)$ and hence defines a map

$$\phi: S^3 \longrightarrow \mathrm{SO}(3).$$

It can be checked that this map is a surjective homomorphism of Lie groups with kernel $\{+1, -1\}$, cf. Exercise 1.9.20. The homomorphism ϕ defines a connected double covering of $\mathrm{SO}(3)$ by S^3 (this is the *universal covering* of $\mathrm{SO}(3)$, since S^3 is simply connected).

1.4 Lie Algebras

Lie algebras are of similar importance for symmetries and gauge theories as Lie groups. We will begin with the general definition of Lie algebras and describe in the next section their relation to Lie groups.

Definition 1.4.1 A vector space V together with a map

$$[\cdot,\cdot]\colon V \times V \longrightarrow V$$

is called a **Lie algebra** if the following hold:

1. $[\cdot,\cdot]$ is bilinear.
2. $[\cdot,\cdot]$ is antisymmetric:

$$[v,w] = -[w,v] \quad \forall v,w \in V.$$

3. $[\cdot,\cdot]$ satisfies the **Jacobi identity**:

$$[u,[v,w]] + [v,[w,u]] + [w,[u,v]] = 0 \quad \forall u,v,w \in V.$$

The map $[\cdot,\cdot]\colon V \times V \to V$ is called the **Lie bracket**. We will only consider Lie algebras defined on real or complex vector spaces. Unless stated otherwise the vector spaces underlying Lie algebras are finite-dimensional.

We collect some examples to show that Lie algebras occur quite naturally (we discuss many more examples in Sect. 1.5.5).

Example 1.4.2 (Abelian Lie Algebras) Every real or complex vector space with the trivial Lie bracket $[\cdot,\cdot] \equiv 0$ is a Lie algebra. Such Lie algebras are called **abelian**. Every 1-dimensional Lie algebra is abelian, because the Lie bracket is antisymmetric.

Example 1.4.3 (Lie Algebra of Matrices) The vector space $V = \mathrm{Mat}(n \times n, \mathbb{K})$ of square matrices with $\mathbb{K} = \mathbb{R}, \mathbb{C}$ is a real or complex Lie algebra with bracket defined by the commutator of matrices A, B:

$$[A,B] = A \cdot B - B \cdot A.$$

The only axiom that has to checked is the Jacobi identity. This example is very important, because the Lie algebras of linear groups have the same Lie bracket, cf. Corollary 1.5.26. It even follows from *Ado's Theorem* 1.5.25 that any finite-dimensional Lie algebra can be embedded into such a matrix Lie algebra.

Example 1.4.4 (Lie Algebra of Endomorphisms) In the same way the vector space $V = \mathrm{End}(W)$ of endomorphisms (linear maps) on a real or complex vector space W is a real or complex Lie algebra with Lie bracket defined by the commutator of endomorphisms f, g:

$$[f,g] = f \circ g - g \circ f.$$

Example 1.4.5 (Lie Algebra Defined by an Associative Algebra) Even more generally, let A be any associative algebra with multiplication $\cdot$. Then the commutator

$$[a, b] = a \cdot b - b \cdot a$$

defines a Lie algebra structure on A.

Example 1.4.6 (Cross Product on $\mathbb{R}^3$) The vector space $\mathbb{R}^3$ is a Lie algebra with the bracket given by the cross product:

$$[v, w] = v \times w.$$

Again, the only axiom that has to be checked is the Jacobi identity. We will identify $(\mathbb{R}^3, \times)$ with a classical Lie algebra in Exercise 1.9.14.

Example 1.4.7 (Lie Algebra of Vector Fields on a Manifold) Let M be a differentiable manifold and $\mathfrak{X}(M)$ the real vector space of smooth vector fields on M. It follows from Theorem A.1.45 that $\mathfrak{X}(M)$ together with the commutator of vector fields is a real Lie algebra, which is infinite-dimensional if the dimension of M is at least one.

As in the case of Lie groups we have two constructions that yield new Lie algebras from given ones.

Definition 1.4.8 Let $(V, [\cdot, \cdot])$ be a Lie algebra. A vector subspace $W \subset V$ is called a **Lie subalgebra** if for all $w, w' \in W$ the Lie bracket $[w, w']$ is an element of W.

Example 1.4.9 Every 1-dimensional vector subspace of a Lie algebra V is an abelian subalgebra.

Example 1.4.10 From the geometric interpretation of the cross product it follows that $(\mathbb{R}^3, \times)$ does not have 2-dimensional Lie subalgebras.

Example 1.4.11 (Intersection of Lie Subalgebras) If $W_1, W_2 \subset V$ are Lie subalgebras, then the **intersection** $W_1 \cap W_2$ is again a Lie subalgebra of V.

Definition 1.4.12 Let $(V, [\cdot, \cdot]_V)$ and $(W, [\cdot, \cdot]_W)$ be Lie algebras over the same field. Then the **direct sum Lie algebra** is the vector space $V \oplus W$ with the Lie bracket

$$\left[v \oplus w, v' \oplus w'\right] = \left[v, v'\right]_V \oplus \left[w, w'\right]_W.$$

Remark 1.4.13 Note that if V, W are Lie subalgebras in a Lie algebra Q which are complementary as vector spaces, so that $Q = V \oplus W$, it does not follow in general that $Q = V \oplus W$ as Lie algebras. For $v, v' \in V, w, w' \in W$ we have

$$\left[v + w, v' + w'\right] = \left[v, v'\right] + \left[w, w'\right] + \left[v, w'\right] + \left[w, v'\right],$$

hence we need in addition

$$[V, W] = 0.$$

We finally want to define homomorphisms between Lie algebras.

Definition 1.4.14 Let $(V, [\cdot, \cdot]_V)$ and $(W, [\cdot, \cdot]_W)$ be Lie algebras. A linear map $\psi: V \to W$ is called **Lie algebra homomorphism** if

$$[\psi(x), \psi(y)]_W = \psi([x, y]_V) \quad \forall x, y \in V.$$

A **Lie algebra isomorphism** is a bijective homomorphism. An **automorphism** of a Lie algebra V is a Lie algebra isomorphism $\psi: V \to V$.

Example 1.4.15 Let V and W be Lie algebras over $\mathbb{K}$. The constant map

$$\psi: V \longrightarrow W$$
$$X \longmapsto 0$$

is always a Lie algebra homomorphism, called the **trivial** homomorphism.

Example 1.4.16 The injection $i: W \hookrightarrow V$ of a Lie subalgebra into a Lie algebra is of course a Lie algebra homomorphism.
The following notion appears, in particular, in physics:

Definition 1.4.17 Let V be a Lie algebra over $\mathbb{K}$ and $T_1, \ldots, T_n$ a vector space basis for V. Then we can write

$$[T_a, T_b] = \sum_{c=1}^{n} f_{abc} T_c,$$

where the coefficients $f_{abc} \in \mathbb{K}$ are called **structure constants** for the given basis $\{T_a\}$.
Because of bilinearity the structure constants determine all commutators between elements of V. The structure constants are antisymmetric in the first two indices

$$f_{abc} = -f_{bac} \quad \forall a, b, c$$

and satisfy the Jacobi identity

$$f_{abd} f_{dce} + f_{bcd} f_{dae} + f_{cad} f_{dbe} = 0 \quad \forall a, b, c, e$$

(here we use the Einstein summation convention and sum over d). Conversely, every set of $n \times n \times n$ numbers $f_{abc} \in \mathbb{K}$ satisfying these two conditions define a Lie algebra structure on $V = \mathbb{K}^n$.

1.5 From Lie Groups to Lie Algebras

So far we have discussed Lie groups and Lie algebras as two independent notions. We now want to turn to a well-known construction that yields for every Lie group an associated Lie algebra, which can be thought of as an infinitesimal or linear description of the Lie group.

Recall that for every smooth manifold M, the set of smooth vector fields $\mathfrak{X}(M)$ on M with the commutator forms a Lie algebra, which is infinite-dimensional if $\dim M \geq 1$. We could associate to a Lie group G the Lie algebra $\mathfrak{X}(G)$ of all vector fields on G. However, as an infinite-dimensional Lie algebra this is somewhat difficult to handle. It turns out that for a Lie group G there exists a canonical *finite-dimensional* Lie subalgebra $\mathfrak{g}$ in $\mathfrak{X}(G)$ which has the same dimension as the Lie group G itself. This will be the Lie algebra associated to the Lie group G.

1.5.1 Vector Fields Invariant Under Diffeomorphisms

We first consider a very general situation. Let M be a smooth manifold and Γ an arbitrary set of diffeomorphisms from M to M.

Definition 1.5.1 We define the set of vector fields on M invariant under Γ by

$$A_\Gamma(M) = \{X \in \mathfrak{X}(M) \mid \phi_*X = X \quad \forall \phi \in \Gamma\}.$$

We have:

Proposition 1.5.2 *For every set Γ of diffeomorphisms of M, the set $A_\Gamma(M)$ is a Lie subalgebra in the Lie algebra $\mathfrak{X}(M)$ with the commutator.*

Proof Suppose $\Gamma = \{\phi\}$ consists of a single diffeomorphism. If $X, Y \in A_{\{\phi\}}(M)$ and $a, b \in \mathbb{R}$, then

$$\begin{aligned}\phi_*(aX + bY) &= a\phi_*X + b\phi_*Y \\ &= aX + bY\end{aligned}$$

and

$$\begin{aligned}\phi_*[X, Y] &= [\phi_*X, \phi_*Y] \\ &= [X, Y]\end{aligned}$$

according to Corollary A.1.51. Hence $A_{\{\phi\}}(M)$ is a Lie subalgebra of $\mathfrak{X}(M)$. The claim for a general set Γ of diffeomorphisms then follows from

$$A_\Gamma(M) = \bigcap_{\phi \in \Gamma} A_{\{\phi\}}(M). \qquad \square$$

1.5.2 Left-Invariant Vector Fields

We now consider the case of a Lie group G. There exist special diffeomorphisms on G that are defined by group elements $g \in G$.

Definition 1.5.3 For $g \in G$ we set:

$$\begin{aligned} L_g: G &\longrightarrow G \\ h &\longmapsto g \cdot h \\ R_g: G &\longrightarrow G \\ h &\longmapsto h \cdot g \\ c_g: G &\longrightarrow G \\ h &\longmapsto g \cdot h \cdot g^{-1}. \end{aligned}$$

These maps are called **left translation**, **right translation** and **conjugation** by g, respectively.

Example 1.5.4 For $G = \mathbb{R}^n$ with vector addition and $a \in \mathbb{R}^n$ we have

$$\begin{aligned} L_a: \mathbb{R}^n &\longrightarrow \mathbb{R}^n \\ x &\longmapsto a + x \\ R_a: \mathbb{R}^n &\longrightarrow \mathbb{R}^n \\ x &\longmapsto x + a. \end{aligned}$$

This explains the names left and right translation.
The following properties are easy to check:

Lemma 1.5.5 (Properties of Translations and Conjugation) *For all $g \in G$ we have:*

1. *The inverses of left and right translations are given by*

$$L_g^{-1} = L_{g^{-1}} \quad R_g^{-1} = R_{g^{-1}}.$$

   The inverse of conjugation is given by

$$c_g^{-1} = c_{g^{-1}}.$$

   In particular, L_g, R_g and c_g are diffeomorphisms of G.
2. *L_g and R_h commute for all $g, h \in G$.*

3. $c_g = L_g \circ R_{g^{-1}} = R_{g^{-1}} \circ L_g$.
4. *The conjugations* c_g *for* $g \in G$ *are Lie group automorphisms of* G*, called* **inner automorphisms**.

Remark 1.5.6 Note that left translation L_g and right translation R_g are not Lie group homomorphisms for $g \neq e$, because

$$L_g(e) = R_g(e) = g \neq e.$$

Example 1.5.7 If G is abelian, then $L_g = R_g$ for all $g \in G$ and $c_g = \mathrm{Id}_G$ for all $g \in G$. Each of these two properties characterizes abelian Lie groups.
We now set:

Definition 1.5.8 A vector field $X \in \mathfrak{X}(G)$ on a Lie group G is called **left-invariant** if $L_{g*}X = X$ for all $g \in G$. In other words, the set of left-invariant vector fields on G is $A_\Gamma(G)$, where Γ is the set of all left translations.
We get with Proposition 1.5.2:

Theorem 1.5.9 (The Lie Algebra of a Lie Group) *The set of left-invariant vector fields together with the commutator* $[\cdot,\cdot]$ *of vector fields on the Lie group* G *forms a Lie subalgebra*

$$L(G) = \mathfrak{g}$$

in the Lie algebra $\mathfrak{X}(G)$ *of all vector fields on* G*. We call* $\mathfrak{g}$ *the* **Lie algebra of** *(or* **associated to***)* G*.*

Remark 1.5.10 We could also define the Lie algebra of a Lie group with right-invariant vector fields. Using left-invariant vector fields is just the standard convention.

Remark 1.5.11 We defined Lie algebras in general on vector spaces over arbitrary fields. The Lie algebra of a Lie group, however, is always a *real* Lie algebra.
As mentioned before, vector fields, their flows and the commutator are only defined on smooth manifolds. This is the reason why only Lie groups have an associated Lie algebra and not other types of groups.

We want to show that there is a vector space isomorphism between the Lie algebra $\mathfrak{g}$ and the tangent space T_eG.

Definition 1.5.12 Let G be a Lie group with neutral element e and associated Lie algebra $\mathfrak{g}$. We define the **evaluation map**

$$\begin{aligned} \mathrm{ev}\colon \mathfrak{g} &\longrightarrow T_eG \\ X &\longmapsto X_e. \end{aligned}$$

Lemma 1.5.13 *The evaluation map is a vector space isomorphism.*

Proof The evaluation map is clearly linear. To construct the inverse of a vector $x \in T_eG$ under the map ev define a vector field X on G by

$$X_h = (D_eL_h)x, \quad h \in G.$$

To show that X is smooth, consider the multiplication map

$$\begin{aligned} \mu: G \times G &\longrightarrow G \\ (h, g) &\longmapsto hg \end{aligned}$$

with differential

$$\begin{aligned} D\mu: TG \times TG &\longrightarrow TG \\ ((h, Y), (g, X)) &\longmapsto (D_gL_h)(X) + (D_hR_g)(Y). \end{aligned}$$

Then the following map is smooth

$$\begin{aligned} G &\longrightarrow TG \\ h &\longmapsto D\mu((h, 0), (e, x)) = (D_eL_h)x \end{aligned}$$

which is just the vector field X.

The vector field X is also left-invariant, because

$$(D_hL_g)X_h = (D_e(L_g \circ L_h))x = (D_eL_{gh})x = X_{gh}$$

for all $g \in G$ and thus $L_{g*}X = X$. The map

$$\begin{aligned} T_eG &\longrightarrow \mathfrak{g} \\ x &\longmapsto X \end{aligned}$$

is the inverse of ev. □

We can therefore think of the tangent space T_eG of the Lie group G at the neutral element e as having the structure of the Lie algebra $\mathfrak{g}$. In particular, we get:

Corollary 1.5.14 *The Lie algebra of a Lie group is finite-dimensional with dimension equal to the dimension of the Lie group. Furthermore, a left-invariant (or right-invariant) vector field on a Lie group is completely determined by its value at one point.*

It is a non-trivial theorem that any abstract real Lie algebra can be realized by the construction above (for a proof, see [77, 83]):

Theorem 1.5.15 (Lie's Third Theorem) *Every finite-dimensional real Lie algebra is isomorphic to the Lie algebra of some connected Lie group.*

Lie's Third Theorem was proved in this form by Élie Cartan. Note that there may be different, non-isomorphic Lie groups with isomorphic Lie algebras: a trivial example is given by the Lie groups $(\mathbb{R}, +)$ and $(S^1, \cdot)$ whose Lie algebras are one-dimensional and hence abelian. The orthogonal and spin groups provide another example, to be discussed in Chap. 6.

1.5.3 Induced Homomorphisms

Just as we get for every Lie group an associated Lie algebra, we get for every homomorphism between Lie groups a homomorphism between the associated Lie algebras.

Definition 1.5.16 Let G, H be Lie groups and $\phi\colon G \to H$ a homomorphism of Lie groups. If X is a left-invariant vector field on G, we can uniquely define a left-invariant vector field $\phi_* X$ on H by

$$\mathrm{ev}(\phi_* X) = (\phi_* X)_e = (D_e\phi)(X_e).$$

This defines a map

$$\phi_*\colon \mathfrak{g} \longrightarrow \mathfrak{h},$$

called the **differential** or **induced homomorphism** of the homomorphism ϕ.

Remark 1.5.17 Here are two remarks concerning this definition:

1. Note that $\phi(e) = e$ for a homomorphism, so that $\phi_* X \in \mathfrak{h}$ is well-defined.
2. As the notation of the theorem indicates, the push-forward on vector fields is defined in the case of Lie groups not only for diffeomorphisms, but also for *Lie group homomorphisms* acting on *left-invariant* vector fields. This definition is possible, because left-invariant vector fields on Lie groups are determined by their value at one point.

Theorem 1.5.18 (The Differential Is a Lie Algebra Homomorphism) *The differential $\phi_*\colon \mathfrak{g} \to \mathfrak{h}$ of a Lie group homomorphism $\phi\colon G \to H$ is a homomorphism of Lie algebras.*

Proof We have to show that

$$[\phi_* X, \phi_* Y] = \phi_*[X, Y] \quad \forall X, Y \in \mathfrak{g}.$$

By Proposition A.1.49 this will follow if we can show that $\phi_* X$ is ϕ-related to X, i.e. that

$$(\phi_* X)_{\phi(g)} = (D_g \phi)(X_g) \quad \forall g \in G.$$

We have

$$\begin{aligned}(\phi_* X)_{\phi(g)} &= (D_e L_{\phi(g)} \circ D_e \phi)(X_e) \\ &= D_e(L_{\phi(g)} \circ \phi)(X_e) \\ &= D_e(\phi \circ L_g)(X_e) \\ &= (D_g \phi)(X_g),\end{aligned}$$

because $\phi_* X$ and X are left-invariant and ϕ is a homomorphism. This proves the claim. □

Note that it is essential for this argument that the map ϕ is a Lie group homomorphism.

Corollary 1.5.19 *Let $H \subset G$ be an immersed or embedded Lie subgroup. Then $\mathfrak{h} \subset \mathfrak{g}$ is a Lie subalgebra.*

Proof The inclusion $i: H \hookrightarrow G$ is a homomorphism of Lie groups and an immersion. Thus the induced inclusion $i_*: \mathfrak{h} \hookrightarrow \mathfrak{g}$ is an injective homomorphism of Lie algebras. □

We can ask whether it is possible to reverse these relations:

- If G is a Lie group with Lie algebra $\mathfrak{g}$ and $\mathfrak{h} \subset \mathfrak{g}$ a Lie subalgebra, does there exist a Lie subgroup H in G whose Lie algebra is $\mathfrak{h}$?
- If $\phi: \mathfrak{g} \rightarrow \mathfrak{h}$ is a Lie algebra homomorphism between the Lie algebras of Lie groups G and H, does there exist a Lie group homomorphism $\psi: G \rightarrow H$ inducing ϕ on Lie algebras?

Both questions are related to the concept of *integration* from a linear object on the level of Lie algebras to a non-linear object on the level of Lie groups. We shall answer the first question in Sect. 1.6 and briefly comment here, without proof, on the second question. The following theorem specifies a sufficient condition for the existence of a Lie group homomorphism inducing a given Lie algebra homomorphism (for a proof, see [77, 142]):

Theorem 1.5.20 (Integrability Theorem for Lie Algebra Homomorphisms) *Let G be a connected and simply connected Lie group, H a Lie group and $\phi: \mathfrak{g} \rightarrow \mathfrak{h}$ a Lie algebra homomorphism. Then there exists a unique Lie group homomorphism $\psi: G \rightarrow H$ such that $\psi_* = \phi$.*

Example 1.5.21 Without the condition that G is simply connected this need not hold: every Lie algebra homomorphism $\phi: \mathfrak{so}(2) \rightarrow \mathfrak{h}$ induces a unique Lie group homomorphism $\psi: \mathbb{R} \rightarrow H$. However, ϕ does not always induce a Lie group homomorphism $\mathrm{SO}(2) \rightarrow H$ (see the discussion after Corollary 2.1.13).

Similarly there are homomorphisms $\mathfrak{so}(n) \to \mathfrak{h}$ for $n \geq 3$ (so-called *spinor representations*, see Sect. 6.5.2) that do not integrate to homomorphisms $\mathrm{SO}(n) \to H$ (it can be shown that $\mathrm{SO}(n)$ has fundamental group $\mathbb{Z}_2$ for $n \geq 3$).

1.5.4 The Lie Algebra of the General Linear Groups

We have defined the Lie algebra associated to a Lie group, but so far we have not seen any explicit examples of this construction. In this and the subsequent subsection we want to study the Lie algebra associated to the linear groups, i.e. closed subgroups of the general linear groups. We can understand the structure of the corresponding Lie algebras by Corollary 1.5.19 once we have understood the structure of the Lie algebra of the general linear groups.

Theorem 1.5.22 (Lie Algebra of General Linear Groups) *The Lie algebra of the general linear group* $\mathrm{GL}(n, \mathbb{R})$ *is* $\mathfrak{gl}(n, \mathbb{R}) = \mathrm{Mat}(n \times n, \mathbb{R})$ *and the Lie bracket on* $\mathfrak{gl}(n, \mathbb{R})$ *is given by the standard commutator of matrices:*

$$[X, Y] = X \cdot Y - Y \cdot X \quad \forall X, Y \in \mathfrak{gl}(n, \mathbb{R}).$$

An analogous result holds for the Lie algebra $\mathfrak{gl}(n, \mathbb{K})$ *of* $\mathrm{GL}(n, \mathbb{K})$ *for* $\mathbb{K} = \mathbb{C}, \mathbb{H}$.

We want to prove this theorem. The Lie group $G = \mathrm{GL}(n, \mathbb{R})$ is an open subset of $\mathbb{R}^{n^2}$, therefore we can canonically identify the tangent space at the unit element I,

$$T_I \mathrm{GL}(n, \mathbb{R}) = \mathfrak{gl}(n, \mathbb{R}) = \mathfrak{g},$$

with the vector space

$$\mathbb{R}^{n^2} = \mathrm{Mat}(n \times n, \mathbb{R}).$$

Lemma 1.5.23 *If* $X \in \mathrm{Mat}(n \times n, \mathbb{R}) = \mathfrak{g}$*, then the associated left-invariant vector field* $\tilde{X}$ *on* G *is given by*

$$\tilde{X}_A = A \cdot X, \quad \forall A \in G,$$

where $\cdot$ *denotes matrix multiplication.*

Proof To show this, let γ_X be an arbitrary curve in G through e and tangent to X. Then

$$\begin{aligned} \tilde{X}_A &= (D_e L_A)(X) \\ &= \left.\frac{d}{dt}\right|_{t=0} L_A(\gamma_X(t)) \end{aligned}$$

$$
\begin{aligned}
&= \frac{d}{dt}\bigg|_{t=0} A \cdot \gamma_X(t) \\
&= A \cdot X.
\end{aligned}
$$

The last equality sign in this calculation can be understood by considering each entry of the time-dependent matrix $A \cdot \gamma_X(t)$ separately. □

Lemma 1.5.24 *Let $\tilde{X}, \tilde{Y}$ be vector fields on an open subset U of a Euclidean space $\mathbb{R}^N$ and $\gamma_{\tilde{X}}, \gamma_{\tilde{Y}}$ curves tangent to $\tilde{X}$ and $\tilde{Y}$ at a point $p \in U$. Then*

$$
[\tilde{X}, \tilde{Y}]_p = \frac{d}{dt}\bigg|_{t=0} \tilde{Y}_{\gamma_{\tilde{X}}(t)} - \frac{d}{dt}\bigg|_{t=0} \tilde{X}_{\gamma_{\tilde{Y}}(t)}.
$$

Proof Let $e_1, \ldots, e_N$ be the standard basis of the Euclidean space and write

$$
\begin{aligned}
\tilde{X} &= \sum_{k=1}^{N} \tilde{X}_k e_k, \\
\tilde{Y} &= \sum_{k=1}^{N} \tilde{Y}_k e_k.
\end{aligned}
$$

Then, because of $[e_k, e_l] = 0$, we get

$$
\begin{aligned}
[\tilde{X}, \tilde{Y}]_p &= \sum_{k,l=1}^{N} \left(\tilde{X}_k \cdot (L_k \tilde{Y}_l) - \tilde{Y}_k \cdot (L_k \tilde{X}_l)\right) e_l \\
&= \sum_{l=1}^{N} \left(L_{\tilde{X}} \tilde{Y}_l - L_{\tilde{Y}} \tilde{X}_l\right)(p) e_l \\
&= \frac{d}{dt}\bigg|_{t=0} \tilde{Y}_{\gamma_{\tilde{X}}(t)} - \frac{d}{dt}\bigg|_{t=0} \tilde{X}_{\gamma_{\tilde{Y}}(t)}.
\end{aligned}
$$

□

We can now prove Theorem 1.5.22.

Proof Since $\mathrm{GL}(n, \mathbb{R})$ is an open subset of a Euclidean space, we can calculate the commutator of the vector fields $\tilde{X}, \tilde{Y}$ at the point I by

$$
\begin{aligned}
[\tilde{X}, \tilde{Y}]_I &= \frac{d}{dt}\bigg|_{t=0} \tilde{Y}_{\gamma_{\tilde{X}}(t)} - \frac{d}{dt}\bigg|_{t=0} \tilde{X}_{\gamma_{\tilde{Y}}(t)} \\
&= \frac{d}{dt}\bigg|_{t=0} (\gamma_{\tilde{X}}(t) \cdot Y) - \frac{d}{dt}\bigg|_{t=0} (\gamma_{\tilde{Y}}(t) \cdot X) \\
&= X \cdot Y - Y \cdot X.
\end{aligned}
$$

This proves the assertion. □

We would like to mention the following conceptually interesting theorem concerning Lie algebras (for the proof in a special case, see Proposition 2.4.4; the general proof can be found in [77, 83]).

Theorem 1.5.25 (Ado's Theorem) *Let $\mathfrak{g}$ be a finite-dimensional Lie algebra over $\mathbb{K} = \mathbb{R}, \mathbb{C}$. Then there exists an injective Lie algebra homomorphism of $\mathfrak{g}$ into $\mathfrak{gl}(n, \mathbb{K})$ for some n.*

As a consequence of Ado's Theorem every Lie algebra $\mathfrak{g}$ over $\mathbb{K}$ is isomorphic to a Lie subalgebra of $\mathfrak{gl}(n, \mathbb{K})$ for some n.

We will later show in Theorem 1.6.4 that for a given Lie group G, every Lie *subalgebra* $\mathfrak{h} \subset \mathfrak{g}$ is the Lie algebra of a connected Lie *subgroup* $H \subset G$. Therefore Lie's Third Theorem 1.5.15 follows from Ado's Theorem 1.5.25 and Theorem 1.6.4, applied to some general linear group $\mathrm{GL}(n, \mathbb{R})$.

1.5.5 The Lie Algebra of the Linear Groups

As a corollary to Theorem 1.5.22 and Corollary 1.5.19 we get:

Corollary 1.5.26 (Lie Algebra of Linear Groups) *If the Lie algebra of an embedded or immersed Lie subgroup of* $\mathrm{GL}(n, \mathbb{K})$ *is identified in the canonical way with a Lie subalgebra of* $\mathrm{Mat}(n \times n, \mathbb{K})$*, then the Lie bracket on the Lie subalgebra is the standard commutator of matrices.*

As simple as this corollary may seem, it is in fact very useful. In general it can be quite difficult to calculate the commutator of two vector fields on a given manifold. Corollary 1.5.26 shows that this is very easy for left-invariant vector fields on Lie subgroups of general linear groups.

Theorem 1.5.27 (Lie Algebras of Classical Groups) *We can identify the Lie algebras of the classical groups with the following real Lie subalgebras of the Lie algebra* $\mathrm{Mat}(n \times n, \mathbb{K})$.

1. The Lie algebras of the special linear groups are:

$$\mathfrak{sl}(n, \mathbb{R}) = \{M \in \mathrm{Mat}(n \times n, \mathbb{R}) \mid \mathrm{tr}(M) = 0\},$$

$$\mathfrak{sl}(n, \mathbb{C}) = \{M \in \mathrm{Mat}(n \times n, \mathbb{C}) \mid \mathrm{tr}(M) = 0\},$$

$$\mathfrak{sl}(n, \mathbb{H}) = \{M \in \mathrm{Mat}(n \times n, \mathbb{H}) \mid \mathrm{Re}(\mathrm{tr}(M)) = 0\}.$$

(continued)

Theorem 1.5.27 (continued)

2. *The Lie algebras of the orthogonal, unitary and symplectic groups are:*

$$\mathfrak{o}(n) = \{M \in \mathrm{Mat}(n \times n, \mathbb{R}) \mid M + M^T = 0\},$$
$$\mathfrak{u}(n) = \{M \in \mathrm{Mat}(n \times n, \mathbb{C}) \mid M + M^\dagger = 0\},$$
$$\mathrm{sp}(n) = \{M \in \mathrm{Mat}(n \times n, \mathbb{H}) \mid M + M^\dagger = 0\}.$$

3. *The Lie algebras of the special orthogonal and special unitary groups are:*

$$\mathfrak{so}(n) = \mathfrak{o}(n),$$
$$\mathfrak{su}(n) = \{M \in \mathrm{Mat}(n \times n, \mathbb{C}) \mid M + M^\dagger = 0, \mathrm{tr}(M) = 0\}.$$

Remark 1.5.28 We can check directly that these subsets of $\mathrm{Mat}(n \times n, \mathbb{K})$ are real vector subspaces and closed under the commutator; see Exercise 1.9.16.

We prove Theorem 1.5.27.

Proof

1. Let $\mathbb{K} = \mathbb{R}, \mathbb{C}$ and suppose that $A \in \mathfrak{sl}(n, \mathbb{K})$. We shall show in Sect. 1.7 (without using the results here) that $e^{tA} \in \mathrm{SL}(n, \mathbb{K})$ and that

$$1 = \det\left(e^{tA}\right) = e^{\mathrm{tr}(A)t}$$

for all $t \in \mathbb{R}$. We get

$$0 = \left.\frac{d}{dt}\right|_{t=0} e^{\mathrm{tr}(A)t} = \mathrm{tr}(A),$$

hence

$$\mathfrak{sl}(n, \mathbb{K}) \subset \{M \in \mathrm{Mat}(n \times n, \mathbb{K}) \mid \mathrm{tr}(M) = 0\}.$$

Since we already know the dimension of $\mathrm{SL}(n, \mathbb{K})$ from Theorem 1.2.17, the assertion follows by calculating the dimension of the subspace on the right of this inclusion.

The claim for $\mathfrak{sl}(n, \mathbb{H})$ follows, because under the adjoint map χ, the group $\mathrm{SL}(n, \mathbb{H})$ gets identified according to Proposition 1.1.25 with the submanifold

$$\left\{X \in \mathrm{Mat}(2n \times 2n, \mathbb{C}) \mid JXJ^{-1} = \overline{X}, \det(X) = 1\right\}.$$

The tangent space to the neutral element I is contained in

$$\left\{A \in \mathrm{Mat}(2n \times 2n, \mathbb{C}) \mid JAJ^{-1} = \overline{A}, \mathrm{tr}(A) = 0\right\}$$

which corresponds under χ to

$$\{M \in \mathrm{Mat}(n \times n, \mathbb{H}) \mid \mathrm{Re}(\mathrm{tr}(M)) = 0\},$$

since

$$\mathrm{tr}(\chi_M) = 2\mathrm{Re}(\mathrm{tr}(M))$$

by Proposition 1.1.26. The claim follows by a similar dimension argument as before.

2. If $A(t)$ is a curve in $\mathrm{O}(n)$ through I with $\dot{A}(0) = M$, then $A(t)A(t)^T = I$, hence

$$0 = \left.\frac{d}{dt}\right|_{t=0} A(t)A(t)^T = M + M^T,$$

i.e. M is skew-symmetric. The claim then follows by comparing the dimensions of $\mathrm{O}(n)$ and the vector space of skew-symmetric matrices. The cases of $\mathrm{u}(n)$ and $\mathrm{sp}(n)$ follow similarly.

3. The case of $\mathrm{su}(n)$ is clear by a similar dimension argument as before. The case of $\mathfrak{so}(n)$ follows, because if $M \in \mathfrak{o}(n)$, then automatically $\mathrm{tr}(M) = 0$.

□

Example 1.5.29 The Lie algebra $\mathfrak{u}(1)$ has dimension 1 and is equal to $\mathrm{Im}\,\mathbb{C}$, spanned by i.

Example 1.5.30 The Lie algebra $\mathfrak{so}(2)$ has dimension 1 and consists of the skew-symmetric 2×2-matrices. A basis is given by the *rotation matrix*

$$r = \begin{pmatrix} 0 & -1 \\ 1 & 0 \end{pmatrix}.$$

The Lie algebra $\mathfrak{so}(2)$ is isomorphic to $\mathfrak{u}(1)$, because both are 1-dimensional and abelian.

Example 1.5.31 The Lie algebra $\mathfrak{so}(3)$ has dimension 3 and consists of skew-symmetric 3×3-matrices. A basis is given by the *rotation matrices*

$$r_1 = \begin{pmatrix} 0 & 0 & 0 \\ 0 & 0 & -1 \\ 0 & 1 & 0 \end{pmatrix} \quad r_2 = \begin{pmatrix} 0 & 0 & 1 \\ 0 & 0 & 0 \\ -1 & 0 & 0 \end{pmatrix} \quad r_3 = \begin{pmatrix} 0 & -1 & 0 \\ 1 & 0 & 0 \\ 0 & 0 & 0 \end{pmatrix}.$$

These matrices satisfy

$$[r_a, r_b] = \epsilon_{abc} r_c,$$

where ϵ_{abc} is totally antisymmetric in a, b, c with $\epsilon_{123} = 1$.

Example 1.5.32 The Lie algebra $\mathfrak{su}(2)$ has dimension 3 and consists of the skew-Hermitian 2×2-matrices of trace zero. We consider the Hermitian **Pauli matrices**:

$$\sigma_1 = \begin{pmatrix} 0 & 1 \\ 1 & 0 \end{pmatrix} \quad \sigma_2 = \begin{pmatrix} 0 & -i \\ i & 0 \end{pmatrix} \quad \sigma_3 = \begin{pmatrix} 1 & 0 \\ 0 & -1 \end{pmatrix}.$$

Then a basis for $\mathfrak{su}(2)$ is given by the matrices

$$\tau_a = -\frac{i}{2}\sigma_a \quad a = 1, 2, 3.$$

The commutators of these matrices are

$$[\tau_a, \tau_b] = \epsilon_{abc}\tau_c.$$

The map

$$\begin{aligned} \mathfrak{so}(3) &\longrightarrow \mathfrak{su}(2) \\ r_a &\longmapsto \tau_a \end{aligned}$$

is a Lie algebra isomorphism.

Example 1.5.33 The Lie algebra $\mathfrak{su}(3)$ has dimension 8 and consists of the skew-Hermitian 3×3-matrices of trace zero. We consider the Hermitian **Gell-Mann matrices**:

$$\lambda_1 = \begin{pmatrix} 0 & 1 & 0 \\ 1 & 0 & 0 \\ 0 & 0 & 0 \end{pmatrix} \quad \lambda_2 = \begin{pmatrix} 0 & -i & 0 \\ i & 0 & 0 \\ 0 & 0 & 0 \end{pmatrix} \quad \lambda_3 = \begin{pmatrix} 1 & 0 & 0 \\ 0 & -1 & 0 \\ 0 & 0 & 0 \end{pmatrix}$$

$$\lambda_4 = \begin{pmatrix} 0 & 0 & 1 \\ 0 & 0 & 0 \\ 1 & 0 & 0 \end{pmatrix} \quad \lambda_5 = \begin{pmatrix} 0 & 0 & -i \\ 0 & 0 & 0 \\ i & 0 & 0 \end{pmatrix}$$

$$\lambda_6 = \begin{pmatrix} 0 & 0 & 0 \\ 0 & 0 & 1 \\ 0 & 1 & 0 \end{pmatrix} \quad \lambda_7 = \begin{pmatrix} 0 & 0 & 0 \\ 0 & 0 & -i \\ 0 & i & 0 \end{pmatrix} \quad \lambda_8 = \frac{1}{\sqrt{3}}\begin{pmatrix} 1 & 0 & 0 \\ 0 & 1 & 0 \\ 0 & 0 & -2 \end{pmatrix}.$$

(continued)

Example 1.5.33 (continued)
Then a basis for $\mathfrak{su}(3)$ is given by the matrices $\frac{i\lambda_a}{2}$ for $a = 1, \ldots, 8$. The matrices $i\lambda_a$ for $a = 1, 2, 3$ span a Lie subalgebra, isomorphic to $\mathfrak{su}(2)$.

Example 1.5.34 The Lie algebra $\mathfrak{sp}(1)$ has dimension 3 and is equal to $\operatorname{Im}\mathbb{H}$, spanned by the imaginary quaternions i, j, k. If we set

$$e_1 = \frac{i}{2}, \quad e_2 = \frac{j}{2}, \quad e_3 = \frac{k}{2},$$

then

$$[e_a, e_b] = \epsilon_{abc} e_c.$$

The map

$$\begin{aligned} \mathfrak{sp}(1) &\longrightarrow \mathfrak{su}(2) \\ e_a &\longmapsto \tau_a \end{aligned}$$

is a Lie algebra isomorphism.

Example 1.5.35 The Lie algebra $\mathfrak{sl}(1, \mathbb{H})$ is equal to $\mathfrak{sp}(1)$.

Example 1.5.36 The Lie algebra $\mathfrak{sl}(2, \mathbb{R})$ has dimension 3 and consists of the real 2×2-matrices of trace zero. A basis is given by the matrices

$$H = \begin{pmatrix} 1 & 0 \\ 0 & -1 \end{pmatrix} \quad X = \begin{pmatrix} 0 & 1 \\ 0 & 0 \end{pmatrix} \quad Y = \begin{pmatrix} 0 & 0 \\ 1 & 0 \end{pmatrix},$$

with commutators

$$\begin{aligned} [H, X] &= 2X, \\ [H, Y] &= -2Y, \\ [X, Y] &= H. \end{aligned}$$

Example 1.5.37 The Lie algebra $\mathfrak{sl}(2, \mathbb{C})$ has dimension 6 and consists of the complex 2×2-matrices of trace zero. It is also a *complex* Lie algebra of complex dimension 3. A complex basis is given by the same matrices H, X, Y as above for $\mathfrak{sl}(2, \mathbb{R})$. In analogy to the *quantum angular momentum* and *quantum harmonic oscillator*, X is sometimes called the **raising operator** and Y the **lowering operator**. According to Exercise 1.9.18, as a complex Lie algebra, $\mathfrak{sl}(2, \mathbb{C})$ is isomorphic to the complex Lie algebra $\mathfrak{su}(2) \otimes_{\mathbb{R}} \mathbb{C}$.

The Lie algebra $\mathfrak{sl}(2,\mathbb{C})$ plays a special role in physics, because as a real Lie algebra it is isomorphic to the Lie algebra of the *Lorentz group* of 4-dimensional spacetime (see Sect. 6.8.2).

Example 1.5.38 (The Heisenberg Lie Algebra) The Lie algebra of the Heisenberg group Nil^3 is

$$\mathfrak{nil}_3 = \left\{ \begin{pmatrix} 0 & a & b \\ 0 & 0 & c \\ 0 & 0 & 0 \end{pmatrix} \in \mathrm{Mat}(3 \times 3, \mathbb{R}) \,\middle|\, a, b, c \in \mathbb{R} \right\}.$$

A basis is given by the matrices

$$q = \begin{pmatrix} 0 & 1 & 0 \\ 0 & 0 & 0 \\ 0 & 0 & 0 \end{pmatrix}, \quad p = \begin{pmatrix} 0 & 0 & 0 \\ 0 & 0 & 1 \\ 0 & 0 & 0 \end{pmatrix}, \quad z = \begin{pmatrix} 0 & 0 & 1 \\ 0 & 0 & 0 \\ 0 & 0 & 0 \end{pmatrix},$$

satisfying

$$\begin{aligned} [q,p] &= z, \\ [q,z] &= 0, \\ [p,z] &= 0. \end{aligned}$$

We see that z commutes with every element in the Lie algebra $\mathfrak{nil}_3$, i.e. z is a *central element*. Furthermore,

$$[\mathfrak{nil}_3, [\mathfrak{nil}_3, \mathfrak{nil}_3]] = 0,$$

so that $\mathfrak{nil}_3$ is an example of a *nilpotent* Lie algebra.

1.6 *From Lie Subalgebras to Lie Subgroups

Let G be a Lie group with Lie algebra $\mathfrak{g}$. In this section we want to show that there exists a 1-to-1 correspondence between Lie subalgebras of $\mathfrak{g}$ and connected (immersed or embedded) Lie subgroups of G (we follow [142]). We need some background on distributions and foliations that can be found in Sect. A.1.12.

Definition 1.6.1 Let $\mathfrak{h} \subset \mathfrak{g}$ be a Lie subalgebra. If we consider $\mathfrak{g}$ as the set of left-invariant vector fields on G, then $\mathfrak{h}$ is a distribution on G, denoted by $\mathscr{H}$. Equivalently, if we think of $\mathfrak{g}$ as the tangent space T_eG and $\mathfrak{h} \subset T_eG$ as a vector subspace, then the distribution $\mathscr{H}$ is defined by

$$\mathscr{H}_p = L_{p*}\mathfrak{h} \quad \forall p \in G.$$

Lemma 1.6.2 *The distribution $\mathscr{H}$ associated to a Lie subalgebra $\mathfrak{h} \subset \mathfrak{g}$ is integrable.*

Proof Let $V_1, \ldots, V_d$ be left-invariant vector fields on G defined by a vector space basis for $\mathfrak{h}$. Then V_i is a section of $\mathscr{H}$ for all $i = 1, \ldots, d$ and since $\mathfrak{h}$ is a subalgebra, the commutators $[V_k, V_l]$ are again sections of $\mathscr{H}$. If X and Y are arbitrary sections of $\mathscr{H}$, then there exist functions f_i, g_i on G such that

$$X = \sum_{k=1}^{d} f_k V_k,$$

$$Y = \sum_{k=1}^{d} g_k V_k.$$

We get

$$[X, Y] = \left[\sum_{k=1}^{d} f_k V_k, \sum_{l=1}^{d} g_l V_l \right]$$

$$= \sum_{k,l=1}^{d} \left(f_k g_l [V_k, V_l] + f_k (L_{V_k} g_l) V_l - g_l (L_{V_l} f_k) V_k \right).$$

This is a section of $\mathscr{H}$. Thus the distribution $\mathscr{H}$ is integrable. □

Definition 1.6.3 For a Lie subalgebra $\mathfrak{h} \subset \mathfrak{g}$, let H denote the maximal connected leaf of the foliation $\mathscr{H}$ through the neutral element $e \in G$.

Theorem 1.6.4 (The Immersed Lie Subgroup Defined by a Lie Subalgebra) *The immersed submanifold H is the unique, connected, immersed Lie subgroup of G with Lie algebra $\mathfrak{h}$.*

Remark 1.6.5 The subgroup H is sometimes called the **integral subgroup** associated to the Lie algebra $\mathfrak{h}$.

Proof We first show that H is a subgroup in the algebraic sense: Let $g \in H$. Since $\mathscr{H}$ is left-invariant, we have

$$L_{g^{-1}*} \mathscr{H} = \mathscr{H},$$

hence $L_{g^{-1}} H$ is a connected leaf of $\mathscr{H}$, containing $g^{-1}g = e$. By maximality of H we have

$$L_{g^{-1}} H \subset H.$$

Hence if $g, h \in H$, then $g^{-1}h \in H$, showing that H is a subgroup.

We want to show that the group operations on H are smooth with respect to the manifold structure. The map

$$\begin{aligned} H \times H &\longrightarrow G \\ (g, h) &\longmapsto gh^{-1} \end{aligned}$$

is smooth and has image in H. Since H is the leaf of a foliation, it follows from Theorem A.1.56 that

$$\begin{aligned} H \times H &\longrightarrow H \\ (g, h) &\longmapsto gh^{-1} \end{aligned}$$

is smooth. We prove the remaining statement on the uniqueness of H below. □

It remains to show that H is the *unique* connected immersed Lie subgroup with Lie algebra $\mathfrak{h}$. We need the following:

Proposition 1.6.6 (Connected Lie Groups Are Generated by Any Open Neighbourhood of the Neutral Element) *Let G be a connected Lie group and $U \subset G$ an open neighbourhood of e. Then*

$$G = \bigcup_{n=1}^{\infty} U^n,$$

where

$$U^n = \underbrace{U \cdot U \cdots U}_{n \text{ factors}}.$$

Proof This follows from Exercise 1.9.4. □

Here is an immediate consequence.

Corollary 1.6.7 *Let K and K' be Lie groups, where K' is connected. Suppose that $\phi\colon K \to K'$ is a Lie group homomorphism, so that $\phi(K)$ contains an open neighbourhood of $e \in K'$. Then ϕ is surjective. In particular, if $K \subset K'$ is an open subgroup, then $K = K'$.*

We now prove the uniqueness part in Theorem 1.6.4.

Proof Let K be another connected, immersed Lie subgroup with Lie algebra $\mathfrak{h}$. Then K must also be a connected leaf of the foliation $\mathscr{H}$ through $e \in G$, hence by maximality of H we get $K \subset H$. The differential of the inclusion $i\colon K \hookrightarrow H$ is an isomorphism at every point, hence $K \subset H$ is an open subgroup. The assertion then follows from Corollary 1.6.7. □

Example 1.6.8 According to Example 1.1.42 the 1-dimensional Lie subalgebras of the Lie algebra of the torus T^2 define embedded Lie subgroups if they have rational slope in $\mathbb{R}^2$ and immersed Lie subgroups if they have irrational slope.

1.7 The Exponential Map

We saw above that the tangent space T_eG of a Lie group G at the neutral element $e \in G$ has the structure of a Lie algebra $\mathfrak{g}$. In this section we want to study the famous exponential map from $\mathfrak{g}$ to G, which is defined using integral curves of left-invariant vector fields (we follow [14, Sect. 1.2] for the construction).

1.7.1 The Exponential Map for General Lie Groups

For the following statements some background on integral curves and flows of vector fields can be found in Sect. A.1.9.

Theorem 1.7.1 (Integral Curves of Left-Invariant Vector Fields) *Let G be a Lie group and $\mathfrak{g}$ its Lie algebra. Let*

$$\begin{aligned}\phi_X\colon \mathbb{R} \supset I &\longrightarrow G\\ t &\longmapsto \phi_X(t)\end{aligned}$$

denote the maximal integral curve of a left-invariant vector field $X \in \mathfrak{g}$ through the neutral element $e \in G$. Then the following holds:

1. *ϕ_X is defined on all of $\mathbb{R}$.*
2. *$\phi_X\colon \mathbb{R} \to G$ is a homomorphism of Lie groups, i.e.*

$$\phi_X(s+t) = \phi_X(s) \cdot \phi_X(t) \quad \forall s, t \in \mathbb{R}.$$

3. *$\phi_{sX}(t) = \phi_X(st)$ for all $s, t \in \mathbb{R}$.*

Definition 1.7.2 The homomorphism $\phi_X\colon \mathbb{R} \to G$ is called the **one-parameter subgroup** of the Lie group G determined by the left-invariant vector field X.
We prove Theorem 1.7.1 in a sequence of steps. Let

$$\phi_X\colon \mathbb{R} \supset I = (t_{min}, t_{max}) \longrightarrow G$$

denote the maximal integral curve of the vector field X through e, satisfying

$$\phi_X(0) = e, \quad \dot{\phi}_X(t) = X_{\phi_X(t)}.$$

Lemma 1.7.3 *For all $s, t \in I$ with $s + t \in I$ the following identity holds:*

$$\phi_X(s+t) = \phi_X(s) \cdot \phi_X(t).$$

Proof Let $g = \phi_X(s) \in G$. Consider the smooth curves

$$\begin{aligned} \eta: I &\longrightarrow G \\ t &\longmapsto g \cdot \phi_X(t) \end{aligned}$$

and

$$\begin{aligned} \tilde{\eta}: (t_{min} - s, t_{max} - s) &\longrightarrow G \\ t &\longmapsto \phi_X(s+t). \end{aligned}$$

It is easy to show that both η and $\tilde{\eta}$ are integral curves of X with $\eta(0) = \tilde{\eta}(0) = g$. Hence by the uniqueness of integral curves (which is a theorem about the uniqueness of solutions to ordinary differential equations) we have

$$\phi_X(s) \cdot \phi_X(t) = \phi_X(s+t) \quad \forall t \in I \cap (t_{min} - s, t_{max} - s).$$

This implies the claim. □

Lemma 1.7.4 *We have $t_{max} = \infty$ and $t_{min} = -\infty$.*

Proof Suppose $t_{max} < \infty$ and set $\alpha = \min\{t_{max}, |t_{min}|\} < \infty$. Consider the curve

$$\begin{aligned} \gamma: \left(-\frac{\alpha}{2}, \frac{3\alpha}{2}\right) &\longrightarrow G \\ t &\longmapsto \phi_X\left(\frac{\alpha}{2}\right) \phi_X\left(t - \frac{\alpha}{2}\right). \end{aligned}$$

It is easy to check that γ is an integral curve of X with $\gamma(0) = e$. However,

$$\frac{3\alpha}{2} > t_{max}$$

by construction, hence γ is an extension of ϕ_X, contradicting the choice of t_{max}. This shows that $t_{max} = \infty$ and similarly $t_{min} = -\infty$. □

Lemma 1.7.5 $\phi_{sX}(t) = \phi_X(st)$ *for all $s, t \in \mathbb{R}$.*

Proof Fix $s \in \mathbb{R}$ and consider the curve

$$\begin{aligned} \delta: \mathbb{R} &\longrightarrow G \\ t &\longmapsto \phi_X(st). \end{aligned}$$

Fig. 1.1 Exponential map

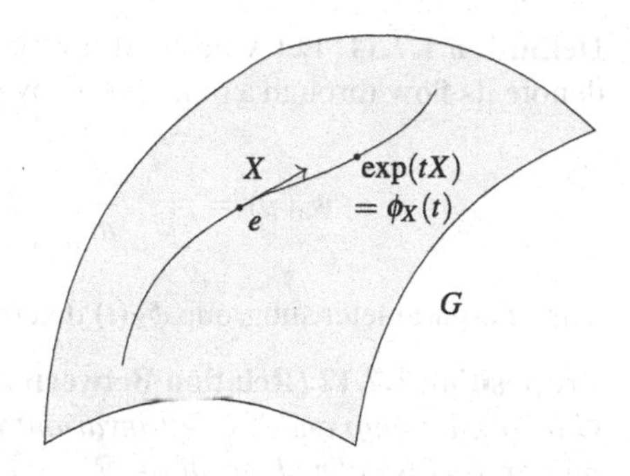

It is easy to show that δ is an integral curve of the vector field sX with $\delta(0) = e$. Hence by uniqueness $\phi_X(st) = \phi_{sX}(t)$. □

Definition 1.7.6 Let $\phi_X : \mathbb{R} \to G$ denote the integral curve through $e \in G$ for an element $X \in \mathfrak{g}$. Then we define the **exponential map**

$$\begin{aligned} \exp : \mathfrak{g} &\longrightarrow G \\ X &\longmapsto \exp(X) = \exp X = \phi_X(1). \end{aligned}$$

See Fig. 1.1.

Remark 1.7.7 The reason for the name *exponential map* will become apparent in Sect. 1.7.3. Note that by definition the exponential map of the Lie group G maps the Lie algebra $\mathfrak{g}$ to the connected component G_e of the neutral element e. Elements in other connected components can never be in the image of the exponential map.

Example 1.7.8 The simplest example is the exponential map of the abelian Lie group $G = \mathbb{R}^n$ with vector addition. We can canonically identify the Lie algebra $\mathfrak{g}$ with $\mathbb{R}^n$. Then the exponential map is the identity map, since the left-invariant vector fields on G are the constant (parallel) vector fields. In this particular case, the exponential map is therefore a diffeomorphism between the Lie algebra and the Lie group.

Proposition 1.7.9 (Properties of the Exponential Map) *The exponential map has the following properties, explaining the name one-parameter subgroup:*

1. $\exp(0) = e$
2. $\exp((s+t)X) = \exp(sX) \cdot \exp(tX)$
3. $\exp(-X) = (\exp X)^{-1}$

for all $X \in \mathfrak{g}$ and $s, t \in \mathbb{R}$.

Proof This is an exercise. □

Remark 1.7.10 One-parameter subgroups are the immersed or embedded Lie subgroups determined as in Theorem 1.6.4 by 1-dimensional (abelian) subalgebras of $\mathfrak{g}$.

Definition 1.7.11 Let X be a left-invariant vector field on a Lie group G. Then we denote its **flow** through a point $p \in G$ by $\phi_t^X(p)$ or $\phi_t(p)$. It is characterized by

$$\phi_0(p) = p, \quad \left.\frac{d}{dt}\right|_{t=s} \phi_t(p) = X_{\phi_s(p)} \quad \forall s \in \mathbb{R}.$$

The one-parameter subgroup $\phi_X(t)$ determined by X is in this notation $\phi_t(e)$.

Proposition 1.7.12 (Relation Between the Flow and the Exponential Map) *Let G be a Lie group and X a left-invariant vector field. Then its flow $\phi_t(p)$ through a point $p \in G$ is defined for all $t \in \mathbb{R}$,*

$$\begin{aligned} \phi\colon \mathbb{R} \times G &\longrightarrow G \\ (t,p) &\longmapsto \phi_t(p), \end{aligned}$$

and given by

$$\phi_t(p) = p \cdot \exp tX = R_{\exp tX}(p) = L_p(\exp tX).$$

Proof Define $\phi_t(p)$ for all $t \in \mathbb{R}$ by the right-hand side. It is clear that

$$\phi_0(p) = p \cdot \exp(0) = p.$$

Furthermore,

$$\begin{aligned} \left.\frac{d}{dt}\right|_{t=s} \phi_t(p) &= \left.\frac{d}{d\tau}\right|_{\tau=0} L_p(\exp sX \cdot \exp \tau X) \\ &= D_e L_{p \exp sX}(X_e) \\ &= X_{p \exp sX} \\ &= X_{\phi_s(p)}, \end{aligned}$$

since X is left-invariant. This implies the claim by uniqueness of solutions of ordinary differential equations. □

Remark 1.7.13 Integral curves of vector fields are defined for all times usually only on *compact* manifolds. It is a special property of Lie groups that integral curves of *left-invariant* vector fields are defined for all times, even on *non-compact* Lie groups.

We want to prove a property of the exponential map that is sometimes useful in applications.

Proposition 1.7.14 (Exponential Map Is a Local Diffeomorphism) *Under the canonical identifications*

$$T_0\mathfrak{g} \cong \mathfrak{g}, \quad T_eG \cong \mathfrak{g},$$

the differential of the exponential map

$$D_0 \exp \colon \mathfrak{g} \longrightarrow \mathfrak{g}$$

is the identity map. In particular, there exist open neighbourhoods V of 0 in $\mathfrak{g}$ and U of e in G such that

$$\exp|_V \colon V \longrightarrow U$$

is a diffeomorphism.

Proof Let $X \in \mathfrak{g}$ and $\gamma(t) = tX$ the associated curve in $\mathfrak{g}$. Then $\dot{\gamma}(0) = X$ and

$$D_0 \exp(X) = \left.\frac{d}{dt}\right|_{t=0} \exp(tX) = X.$$

□

Remark 1.7.15 In general, the exponential map is neither injective nor surjective and hence not a *global* diffeomorphism between the Lie algebra and the Lie group (of course, the exponential map can only be a diffeomorphism if the Lie group itself is not compact and diffeomorphic to a vector space). See Corollary 1.7.20 for a situation when the exponential map is surjective. It is known that the exponential map is a diffeomorphism in the case of simply connected *nilpotent* Lie groups (see [83] for a proof).

Recall from Theorem 1.5.18 that every homomorphism between Lie groups induces a homomorphism between Lie algebras. The exponential map behaves nicely with respect to these homomorphisms.

Theorem 1.7.16 (Induced Homomorphisms on Lie Algebras and the Exponential Map) *Let $\psi \colon G \to H$ be a homomorphism between Lie groups and $\psi_* \colon \mathfrak{g} \to \mathfrak{h}$ the induced homomorphism on Lie algebras. Then*

$$\psi(\exp X) = \exp(\psi_* X) \quad \forall X \in \mathfrak{g},$$

i.e. the following diagram commutes:

$$\begin{array}{ccc} \mathfrak{g} & \xrightarrow{\psi_*} & \mathfrak{h} \\ {\scriptstyle \exp}\downarrow & & \downarrow{\scriptstyle \exp} \\ G & \xrightarrow{\psi} & H \end{array}$$

Proof Consider the curve $\gamma(t) = \psi(\exp tX)$ for $t \in \mathbb{R}$. Then

$$\gamma(0) = \psi(e) = e$$

and

$$
\begin{aligned}
\dot{\gamma}(t) &= D_{\exp tX}\psi\left(\left.\frac{d}{ds}\right|_{s=t}\exp sX\right) \\
&= D_{\exp tX}\psi\left(X_{\exp tX}\right) \\
&= D_e(\psi \circ L_{\exp tX})(X_e) \\
&= D_e(L_{\psi(\exp tX)} \circ \psi)(X_e) \\
&= (\psi_* X)_{\gamma(t)}.
\end{aligned}
$$

Here we used that ψ is a homomorphism and X left-invariant. We conclude that γ is the (unique) integral curve of the left-invariant vector field $\psi_* X$ through $e \in H$ and therefore

$$
\exp(\psi_* X) = \gamma(1) = \psi(\exp X).
$$

□

Corollary 1.7.17 (Exponential Map for Embedded Lie Subgroups) *Let G be a Lie group and $H \subset G$ an embedded Lie subgroup with exponential maps*

$$
\begin{aligned}
\exp^G &: \mathfrak{g} \longrightarrow G, \\
\exp^H &: \mathfrak{h} \longrightarrow H.
\end{aligned}
$$

Then for $X \in \mathfrak{h} \subset \mathfrak{g}$ the following identity holds

$$
\exp^G(X) = \exp^H(X).
$$

Proof This follows from Theorem 1.7.16 with the embedding $i: H \hookrightarrow G$. □

The following generalization of Theorem 1.7.1 to time-dependent vector fields is sometimes useful in applications.

Theorem 1.7.18 (Integral Curves of Time-Dependent Vector Fields on Lie Groups) *Suppose that G is a Lie group and $x: [0,1] \to \mathfrak{g}$ a smooth map. Let $X(t)$ denote the associated left- (or right-)invariant time-dependent vector field on G. Then there exists a unique smooth integral curve $g: [0,1] \to G$ such that*

$$
\begin{aligned}
g(0) &= e, \\
\dot{g}(t) &= X(t) \quad \forall t \in [0,1].
\end{aligned}
$$

Proof We only indicate the idea of the proof in the case of a left-invariant vector field $X(t)$. Details can be found in [14] and are left as an exercise. Let Z be the vector field on $G \times \mathbb{R}$ defined by

$$Z_{(g,s)} = (X_g(s), 1) \in T_{(g,s)}(G \times \mathbb{R}).$$

On the interval $[0, \delta]$, for $\delta > 0$ small enough, there exist integral curves $(g(t), t)$ and $(h(t), t + \delta)$ of Z with $g(0) = h(0) = e$. Then

$$g(t) = g(\delta) \cdot h(t - \delta) \quad \forall t \in [\delta, 2\delta]$$

defines an extension of $g(t)$ to an integral curve of X on $[0, 2\delta]$. □

1.7.2 *The Exponential Map of Tori

It is important to understand the exponential map of the torus T^n, because compact Lie groups always contain embedded Lie subgroups isomorphic to tori (see Exercise 1.9.11 for explicit examples of tori contained in the classical groups).

Proposition 1.7.19 (Exponential Map of Tori) *The exponential map of every torus T^n is surjective.*

Proof This follows, because according to Example 1.7.8 the exponential map of $\mathbb{R}^n$ is surjective. □

Every element of a compact connected Lie group G is contained in some embedded torus subgroup (for a proof, see [24, Sect. IV.1]). Then Corollary 1.7.17 implies:

Corollary 1.7.20 (Exponential Map of Compact Connected Lie Groups) *The exponential map of a compact connected Lie group G is surjective.*

Remark 1.7.21 As Exercise 1.9.27 shows, this is not true in general for non-compact connected Lie groups like $\mathrm{SL}(2, \mathbb{R})$. See Exercise 1.9.28 for a statement which is true in the general case.

Example 1.7.22 Every one-parameter subgroup of SO(3) is given by the subgroup of rotations around a common axis in $\mathbb{R}^3$ (see Example 1.7.32). Corollary 1.7.20 then implies that any rotation of $\mathbb{R}^3$ can be obtained from the identity by rotating around a fixed axis by a certain angle.

The following assertion was discussed in Example 1.1.42.

Proposition 1.7.23 (Embedded and Immersed One-Parameter Subgroups) *A torus of dimension at least two has both embedded and immersed one-parameter subgroups.*

Corollary 1.7.24 *Every Lie group G that contains a torus of dimension at least two has both embedded and immersed one-parameter subgroups.*

Example 1.7.25 The Lie groups SO(4) and SU(3) contain embedded tori of dimension two and thus immersed one-parameter subgroups. The Lie groups SO(3) and SU(2), on the other hand, only contain embedded tori of dimension one. It can be shown that every one-parameter subgroup of SO(3) and SU(2) is closed, hence isomorphic to S^1 (cf. Exercise 1.9.25 for the case of SU(2)).

This is intuitively clear for SO(3), because the one-parameter subgroups are rotations around a fixed axis and thus return to the identity after a rotation by 2π. The result can also be interpreted for SO(4): we can define an embedded Lie subgroup $T^2 = \mathrm{SO}(2)\times\mathrm{SO}(2)$ in SO(4) as those rotations which preserve a splitting of $\mathbb{R}^4$ into two orthogonal planes $\mathbb{R}^2 \oplus \mathbb{R}^2$ and only rotate each plane in itself. If the velocities of the two rotations have an irrational ratio, then both rotations never return at the same time to the identity. This corresponds to an immersed one-parameter subgroup of SO(4).

*1.7.3 *The Matrix Exponential*

In this section we want to determine the exponential map for the linear groups. Let $\mathbb{K} = \mathbb{R}, \mathbb{C}, \mathbb{H}$.

Definition 1.7.26 Let $A \in \mathrm{Mat}(n \times n, \mathbb{K})$ be a square matrix. Then we set

$$e^A = \sum_{k=0}^{\infty} \frac{1}{k!} A^k.$$

Lemma 1.7.27 (Convergence of Exponential Series) *For any square matrix* $A \in \mathrm{Mat}(n \times n, \mathbb{K})$ *the series* $\sum_{k=0}^{\infty} \frac{1}{k!} A^k$ *converges in each entry.*

Proof Let $||\cdot||$ denote the Euclidean norm on $\mathbb{K}^n$. Define the operator norm of a matrix $M \in \mathrm{Mat}(n \times n, \mathbb{K})$ by

$$||M|| = \sup_{||x|| \leq 1} ||Mx||.$$

Then $||\cdot||$ is indeed a norm on the vector space of square matrices and satisfies $||MN|| \leq ||M|| \cdot ||N||$. Since the exponential series for real numbers converges, it follows that the exponential series $\sum_{k=0}^{\infty} \frac{1}{k!} A^k$ is Cauchy and hence converges. □

Lemma 1.7.28 (Exponential of a Sum for Commuting Matrices) *If matrices* $A, B \in \mathrm{Mat}(n \times n, \mathbb{K})$ *commute,* $AB = BA$, *then*

$$e^{A+B} = e^A e^B.$$

In particular, e^{-A} *is the inverse of* e^A, *so that* $e^A \in \mathrm{GL}(n, \mathbb{K})$.

Proof This is Exercise 1.9.23. □

The following is immediate.

Theorem 1.7.29 *For every* $A \in \mathrm{Mat}(n \times n, \mathbb{K})$ *the map*

$$\begin{aligned} \phi_A \colon \mathbb{R} &\longrightarrow \mathrm{GL}(n, \mathbb{K}) \\ t &\longmapsto e^{tA} \end{aligned}$$

is smooth and satisfies

$$\phi_A(0) = I, \quad \left.\frac{d}{dt}\right|_{t=s} \phi_A(t) = \phi_A(s)A \quad \forall s \in \mathbb{R}.$$

We get:

Corollary 1.7.30 (Exponential Map of Linear Group Is Matrix Exponential) *Let*

$$A \in \mathfrak{gl}(n, \mathbb{K}) = \mathrm{Mat}(n \times n, \mathbb{K}).$$

Then

$$\exp(A) = e^A$$

where exp *on the left denotes the canonical exponential map from Lie algebra to Lie group. The same formula holds for the exponential map of any linear group.*

Proof For $A \in \mathrm{Mat}(n \times n, \mathbb{K})$ let $\tilde{A}$ denote the associated left-invariant vector field on $\mathrm{GL}(n, \mathbb{K})$. According to Lemma 1.5.23, $\tilde{A}$ is given at a point $P \in \mathrm{GL}(n, \mathbb{K})$ by

$$\tilde{A}_P = PA.$$

This shows that the map

$$\phi_A \colon \mathbb{R} \longrightarrow \mathrm{GL}(n, \mathbb{K})$$

from Theorem 1.7.29 is the integral curve of the vector field $\tilde{A}$ through I. The first claim now follows by Definition 1.7.6 of the exponential map. The second claim concerning linear groups follows by Corollary 1.7.17. □

Example 1.7.31 The simplest non-trivial case of this theorem is the exponential map

$$\begin{aligned} \exp \colon \mathfrak{u}(1) \cong i\mathbb{R} &\longrightarrow \mathrm{U}(1) \cong S^1 \\ i\alpha &\longmapsto e^{i\alpha}. \end{aligned}$$

Example 1.7.32 A slightly less trivial example is the matrix exponential of tr, where $t \in \mathbb{R}$ and

$$r = \begin{pmatrix} 0 & -1 \\ 1 & 0 \end{pmatrix}$$

is the generator of $\mathfrak{so}(2)$ from Example 1.5.30. It is easy to see that

$$r^{2n} = (-1)^n I$$

and thus

$$r^{2n+1} = (-1)^n r$$

for all $n \geq 0$. Hence

$$\begin{aligned} e^{tr} &= \sum_{n=0}^{\infty} \frac{t^n}{n!} r^n \\ &= \sum_{n=0}^{\infty} \frac{(-1)^n}{(2n)!} t^{2n} \begin{pmatrix} 1 & 0 \\ 0 & 1 \end{pmatrix} + \sum_{n=0}^{\infty} \frac{(-1)^n}{(2n+1)!} t^{2n+1} \begin{pmatrix} 0 & -1 \\ 1 & 0 \end{pmatrix} \\ &= \begin{pmatrix} \cos t & -\sin t \\ \sin t & \cos t \end{pmatrix} \in \mathrm{SO}(2). \end{aligned}$$

This is just the matrix for a rotation in $\mathbb{R}^2$ by an angle t. Similarly the matrix exponential of tr_3, with

$$r_3 = \begin{pmatrix} 0 & -1 & 0 \\ 1 & 0 & 0 \\ 0 & 0 & 0 \end{pmatrix}$$

one of the generators of $\mathfrak{so}(3)$ from Example 1.5.31, is

$$e^{tr_3} = \begin{pmatrix} \cos t & -\sin t & 0 \\ \sin t & \cos t & 0 \\ 0 & 0 & 1 \end{pmatrix} \in \mathrm{SO}(3),$$

which is the matrix for a rotation in $\mathbb{R}^3$ around the z-axis. Rotations around other axes in $\mathbb{R}^3$ are given by one-parameter subgroups conjugate to the one defined by r_3, showing that all one-parameter subgroups of SO(3) are closed.

The proof of the following well-known formula uses that the determinant is multilinear in the columns of a matrix and thus only holds for real and complex matrices.

Theorem 1.7.33 (Determinant of Matrix Exponential) *Let* $A \in \operatorname{Mat}(n \times n, \mathbb{K})$ *where* $\mathbb{K} = \mathbb{R}$ *or* $\mathbb{C}$. *Then*

$$\det\left(e^A\right) = e^{\operatorname{tr}(A)}.$$

Proof We first calculate the differential $D_I \det$. Let $X = (x_1, \ldots, x_n)$ be an arbitrary $n \times n$-matrix with column vectors $x_i \in \mathbb{K}^n$. We have by multilinearity and antisymmetry of the determinant

$$\det(I + (0, \ldots, 0, x_i, 0 \ldots, 0)) = \det \begin{pmatrix} 1 & & & x_{1i} & & & \\ & \ddots & & & & & \\ & & 1 & x_{(i-1)i} & & & \\ & & & (1 + x_{ii}) & & & \\ & & & x_{(i+1)i} & 1 & & \\ & & & & & \ddots & \\ & & & x_{ni} & & & 1 \end{pmatrix}$$
$$= 1 + x_{ii}.$$

Then

$$\begin{aligned} (D_I \det)(X) &= \sum_{i=1}^n (D_I \det)(0, \ldots, 0, x_i, 0 \ldots, 0)) \\ &= \sum_{i=1}^n \frac{d}{dt}\bigg|_{t=0} \det(I + (0, \ldots, 0, tx_i, 0 \ldots, 0)) \\ &= \sum_{i=1}^n x_{ii} \\ &= \operatorname{tr}(X). \end{aligned}$$

Consider the curve

$$\begin{aligned} \gamma \colon \mathbb{R} &\longrightarrow \mathbb{R} \\ t &\longmapsto \det\left(e^{tX}\right). \end{aligned}$$

Then

$$\gamma(0) = 1$$

and for all $s \in \mathbb{R}$

$$\begin{aligned}\left.\frac{d}{dt}\right|_{t=s} \gamma(t) &= \det\left(e^{sX}\right) \left.\frac{d}{d\tau}\right|_{\tau=0} \det\left(e^{\tau X}\right) \\ &= \det\left(e^{sX}\right) (D_I \det)(X) \\ &= \det\left(e^{sX}\right) \operatorname{tr}(X) \\ &= \gamma(s)\operatorname{tr}(X).\end{aligned}$$

The unique solution of this differential equation for $\gamma(t)$ is

$$\gamma(t) = e^{\operatorname{tr}(X)t}.$$

This implies the assertion with $t = 1$. □

Example 1.7.34 Let

$$D = \begin{pmatrix} d_1 & & & \\ & d_2 & & \\ & & \ddots & \\ & & & d_n \end{pmatrix}$$

be a real or complex diagonal matrix. Then

$$e^D = \begin{pmatrix} e^{d_1} & & & \\ & e^{d_2} & & \\ & & \ddots & \\ & & & e^{d_n} \end{pmatrix}$$

and the equation

$$\det\left(e^D\right) = e^{d_1} \cdots e^{d_n} = e^{d_1 + \ldots + d_n} = e^{\operatorname{tr}(D)}$$

is trivially satisfied. The same argument works for upper triangular matrices.
Using Theorem 1.7.16 we can write the statement of Theorem 1.7.33 as follows:

Corollary 1.7.35 *The determinant*

$$\begin{aligned}\det\colon \mathrm{GL}(n, \mathbb{K}) &\longrightarrow \mathbb{K}^* \\ A &\longmapsto \det(A)\end{aligned}$$

is a group homomorphism with differential given by the trace

$$\det_* = \mathrm{tr}\colon \mathrm{Mat}(n \times n, \mathbb{K}) \longrightarrow \mathbb{K}$$
$$X \longmapsto \mathrm{tr}(X).$$

Notice that the trace is indeed a Lie algebra homomorphism to the abelian Lie algebra $\mathbb{K}$.

1.8 *Cartan's Theorem on Closed Subgroups

Our aim in this section is to prove Cartan's Theorem 1.1.44 (we follow [14, 24] and [142]). This theorem is important, because we used it, for example, to show that closed subgroups of the general linear groups are embedded Lie subgroups. We will also employ it later to show that isotropy groups of Lie group actions on manifolds are embedded Lie subgroups. The proof of Cartan's Theorem is one of the more difficult proofs in this book and follows from a sequence of propositions. One direction is quite easy.

Let G be a Lie group.

Definition 1.8.1 Let $H \subset G$ be a subset. A chart $\psi\colon U \to \mathbb{R}^n$ of G such that

$$\psi(U \cap H) = \psi(U) \cap \left(\{0\} \times \mathbb{R}^k\right)$$

for some $k < n$ is called a **submanifold chart** or **flattener** for H around p.

Proposition 1.8.2 *Let $H \subset G$ be an embedded Lie subgroup. Then H is a closed subset in the topology of G.*

Proof Suppose that $H \subset G$ is an embedded Lie subgroup. In a submanifold chart U of G around e for the embedded submanifold H, the set $H \cap U$ is closed in U.

Suppose y is a point in the closure $\bar{H}$ and let x_n be a sequence in H converging to y. Then $x_n^{-1}y$ is a sequence in G converging to e. Hence for sufficiently large index n, the element $x = x_n \in H$ satisfies $x^{-1}y \in U$ and thus $y \in xU$.

Since the group operations on G are continuous, the closure $\bar{H}$ is a subgroup of G. It follows that $y \in \bar{H} \cap xU$ and $x^{-1}y \in \bar{H} \cap U = H \cap U$. Thus $y \in H$ and H is closed. □

The converse statement is more difficult. Assume from now on that $H \subset G$ is a subgroup in the algebraic sense, which is a closed set in the topology of G. To show that H is a k-dimensional embedded Lie subgroup of the n-dimensional Lie group G, we have to find around every point $p \in H$ a chart $\psi\colon U \to \mathbb{R}^n$ of G which is a submanifold chart for H around p. The following argument shows that it suffices to find a submanifold chart for H around e.

Proposition 1.8.3 (Submanifold Charts) *Let $\phi: U \to \mathbb{R}^n$ be a submanifold chart for H around $e \in H$. Suppose $p \in H$. Then*

$$\phi_p = \phi \circ L_{p^{-1}}: L_p(U) \longrightarrow \mathbb{R}^n$$

is a submanifold chart for H around p. Here L_p denotes left translation on G by p.

Proof Note that

$$L_p(H) = H,$$

since H is a subgroup of G, hence

$$L_p(U) \cap H = L_p(U \cap H).$$

This implies

$$\phi_p(L_p(U) \cap H) = \phi(U \cap H)$$

and

$$\phi_p(L_p(U) \cap H) = \phi_p(L_p(U)) \cap \left(\{0\} \times \mathbb{R}^k\right).$$

□

We want to find a submanifold chart for H around e. It turns out that we first have to find a candidate for the Lie algebra of the subgroup H.

Proposition 1.8.4 (The Candidate for the Lie Algebra of H) *Let $H \subset G$ be a subgroup in the algebraic sense which is a closed set in the topology of G. Then*

$$\mathfrak{h} = \{X \in \mathfrak{g} \mid \exp tX \in H \quad \forall t \in \mathbb{R}\}$$

is a vector subspace of $\mathfrak{g}$.

To prove Proposition 1.8.4 note that if $X \in \mathfrak{h}$, then $sX \in \mathfrak{h}$ for all $s \in \mathbb{R}$, since we can choose $t' = ts \in \mathbb{R}$. It remains to show that if $X, Y \in \mathfrak{h}$, then $X + Y \in \mathfrak{h}$. We have to understand terms of the form

$$\exp(t(X + Y)).$$

This is the purpose of the following proposition.

Proposition 1.8.5 (Special Case of the Baker–Campbell–Hausdorff Formula) *Let G be a Lie group with Lie algebra $\mathfrak{g}$. Then for arbitrary vectors $X, Y \in \mathfrak{g}$ we have*

$$\exp(tX) \cdot \exp(tY) = \exp\left(t(X + Y) + O\left(t^2\right)\right) \quad \forall |t| < \epsilon,$$

where $\epsilon > 0$ is small enough and $O(t^2)$ is some function of t such that $O(t^2)/t^2$ stays finite as $t \to 0$.

Proof According to Proposition 1.7.14 the exponential map is a diffeomorphism from an open neighbourhood V of 0 in $\mathfrak{g}$ onto an open neighbourhood W of e in G. We can thus introduce so-called *normal coordinates* on W: Choose a basis $(v_1, \ldots, v_n)$ for the vector space $\mathfrak{g}$. Then there is a unique chart (W, ϕ) of G around e with

$$\phi: G \supset W \longrightarrow \mathbb{R}^n$$

such that

$$\phi(\exp(x_1 v_1 + \ldots + x_n v_n)) = (x_1, \ldots, x_n).$$

Let

$$\phi(\exp(tX)) = tx,$$
$$\phi(\exp(tY)) = ty.$$

Let μ denote group multiplication in G:

$$\mu: G \times G \longrightarrow G$$
$$(g, h) \longmapsto g \cdot h.$$

Utilizing the chart ϕ this induces a map

$$\tilde{\mu}: \mathbb{R}^n \times \mathbb{R}^n \supset \tilde{U} \longrightarrow \mathbb{R}^n,$$

where $\tilde{U}$ is a small open neighbourhood of $(0, 0)$. The map $\tilde{\mu}$ is defined by

$$\tilde{\mu} = \phi \circ \mu \circ \left(\phi^{-1} \times \phi^{-1}\right).$$

We then have to show that

$$\tilde{\mu}(tx, ty) = \phi(\exp(tX) \cdot \exp(tY))$$
$$= t(x + y) + O(t^2).$$

This follows from the Taylor formula for $\tilde{\mu}$ since

$$\mu(e, e) = e$$

hence

$$\tilde{\mu}(0,0) = 0$$

and

$$D_{(e,e)}\mu(u,0) = u = D_{(e,e)}\mu(0,u) \quad \forall u \in T_eG$$

hence

$$D_{(0,0)}\tilde{\mu}(w,0) = w = D_{(0,0)}\tilde{\mu}(0,w) \quad \forall w \in T_0\mathbb{R}^n.$$

□

We will use this proposition in the following special form.

Corollary 1.8.6 (Lie Product Formula) *For arbitrary vectors $X, Y \in \mathfrak{g}$ and all $t \in \mathbb{R}$*

$$\lim_{n\to\infty}\left(\exp\frac{tX}{n}\exp\frac{tY}{n}\right)^n = \exp(t(X+Y)).$$

Proof This follows from Proposition 1.8.5 with the general formula $\exp(Z)^n = \exp(nZ)$ for any $Z \in \mathfrak{g}$. □

We can now finish the proof of Proposition 1.8.4.

Proof If $X, Y \in \mathfrak{h}$, then

$$\left(\exp\frac{tX}{n}\exp\frac{tY}{n}\right)^n \in H \quad \forall n \in \mathbb{N}, t \in \mathbb{R},$$

since H is a subgroup of G. Corollary 1.8.6 together with the assumption that H is a closed subset implies $\exp(t(X+Y)) \in H$ for all $t \in \mathbb{R}$ and thus $X + Y \in \mathfrak{h}$. □

Let $\mathfrak{h}$ be the vector subspace of $\mathfrak{g}$ from Proposition 1.8.4 and $\mathfrak{m}$ an arbitrary complementary vector subspace of $\mathfrak{g}$, so that

$$\mathfrak{g} = \mathfrak{h} \oplus \mathfrak{m}.$$

We now start to construct the submanifold chart for H around e. We fix an arbitrary norm $||\cdot||$ on the vector space $\mathfrak{g}$.

Proposition 1.8.7 (Choice of the Open Subset $V_\mathfrak{m} \subset \mathfrak{m}$) *There exists an open neighbourhood $V_\mathfrak{m} \subset \mathfrak{m}$ of 0 in $\mathfrak{m}$, so that*

$$\exp(V_\mathfrak{m} \setminus \{0\}) \cap H = \emptyset.$$

Proof Suppose that for every open neighbourhood $V_{\mathfrak{m}}$ of 0 in $\mathfrak{m}$ we have

$$\exp(V_{\mathfrak{m}} \setminus \{0\}) \cap H \neq \emptyset.$$

Then there exists a non-zero sequence $(Z_n)_{n\in\mathbb{N}}$ in $\mathfrak{m}$ converging to 0 so that $\exp(Z_n) \in H$. Let K denote the set

$$K = \{Z \in \mathfrak{m} \mid 1 \leq ||Z|| \leq 2\}$$

and choose for every $n \in \mathbb{N}$ a positive integer $c_n \in \mathbb{N}$ such that $c_n Z_n \in K$. Since K is a compact set, we can assume (after passing to a subsequence) that $(c_n Z_n)_{n\in\mathbb{N}}$ converges to some $Z \in K$. Then

$$\frac{Z_n}{||Z_n||} = \frac{c_n Z_n}{||c_n Z_n||} \overset{n\to\infty}{\longrightarrow} \frac{Z}{||Z||}.$$

Since $\exp(Z_n) \in H$ for all $n \in \mathbb{N}$ we get with Lemma 1.8.8 below that $Z \in \mathfrak{h}$. However, $Z \in K \subset \mathfrak{m}$ and $\mathfrak{m}$ is complementary to $\mathfrak{h}$, therefore $Z = 0$. This contradicts that $1 \leq ||Z|| \leq 2$. □

In the proof we used the following lemma.

Lemma 1.8.8 *Let $(Z_n)_{n\in\mathbb{N}}$ be a sequence of non-zero vectors in $\mathfrak{g}$ with $\exp(Z_n) \in H$ and $Z_n \to 0$ as $n \to \infty$. Suppose that the limit*

$$W = \lim_{n\to\infty} \frac{Z_n}{||Z_n||}$$

exists. Then

$$\exp(tW) \in H \quad \forall t \in \mathbb{R}$$

and thus $W \in \mathfrak{h}$.

Proof Let $t \in \mathbb{R}$ be fixed and define

$$c_n = \max\{k \in \mathbb{Z} \mid k||Z_n|| < t\}.$$

We claim that

$$\lim_{n\to\infty} c_n||Z_n|| = t.$$

Note that

$$c_n||Z_n|| < t \leq (c_n + 1)||Z_n|| = c_n||Z_n|| + ||Z_n||.$$

This implies the claim, because $Z_n \to 0$. We get

$$\exp\left(c_n||Z_n|| \cdot \frac{Z_n}{||Z_n||}\right) \longrightarrow \exp(tW).$$

However,

$$\exp\left(c_n||Z_n|| \cdot \frac{Z_n}{||Z_n||}\right) = \exp(c_n Z_n) = \exp(Z_n)^{c_n} \in H.$$

Since H is a closed subset it follows that $\exp(tW) \in H$. □

The following map will be the (inverse of the) submanifold chart.

Lemma 1.8.9 (Definition of the Map F) *The map*

$$\begin{aligned} F\colon \mathfrak{g} = \mathfrak{h} \oplus \mathfrak{m} &\longrightarrow G \\ X + Y &\longmapsto \exp X \cdot \exp Y \end{aligned}$$

is smooth and its differential at 0 *is the identity. Thus F is a local diffeomorphism on a neighbourhood of* 0.

Proof The differential of F maps

$$D_0 F\colon T_0\mathfrak{g} \cong \mathfrak{g} \longrightarrow T_e G \cong \mathfrak{g}.$$

We will show that this map is the identity. For $X \in \mathfrak{h}$ we have

$$\begin{aligned} D_0 F(X) &= \left.\frac{d}{dt}\right|_{t=0} F(tX) \\ &= \left.\frac{d}{dt}\right|_{t=0} \exp(tX) \cdot \exp(0) \\ &= \left.\frac{d}{dt}\right|_{t=0} \exp(tX) \\ &= X. \end{aligned}$$

A similar argument applies for $X \in \mathfrak{m}$. □

Proposition 1.8.10 (The Map F Defines a Submanifold Chart) *Let*

$$\begin{aligned} F\colon \mathfrak{g} = \mathfrak{h} \oplus \mathfrak{m} &\longrightarrow G \\ X + Y &\longmapsto \exp X \cdot \exp Y \end{aligned}$$

be the map from Lemma 1.8.9. Then there exists a small neighbourhood V of 0 in $\mathfrak{g}$ such that $F(V) = U$ is an open neighbourhood of e in G and

$$F^{-1}|_U \colon U \longrightarrow V$$

is a submanifold chart for H around e.

Proof According to Lemma 1.8.9 we can choose an open neighbourhood

$$V = V_{\mathfrak{m}} \times V_{\mathfrak{h}} \subset \mathfrak{m} \oplus \mathfrak{h}$$

of 0 in $\mathfrak{g}$ so that

$$F|_V \colon V \longrightarrow U = F(V)$$

is a diffeomorphism. According to Proposition 1.8.7 we can choose $V_{\mathfrak{m}}$ small enough such that

$$\emptyset = \exp(V_{\mathfrak{m}} \setminus \{0\}) \cap H.$$

Note that $\exp(Y) \in H$ for all $Y \in \mathfrak{h}$, hence

$$\emptyset = F\big((V_{\mathfrak{m}} \setminus \{0\}) \times V_{\mathfrak{h}}\big) \cap H,$$

and

$$\emptyset = \big((V_{\mathfrak{m}} \setminus \{0\}) \times V_{\mathfrak{h}}\big) \cap F^{-1}(H).$$

We conclude that

$$F^{-1}(U \cap H) \subset \{0\} \times V_{\mathfrak{h}}$$

and thus

$$F^{-1}(U \cap H) = \{0\} \times V_{\mathfrak{h}}. \tag{1.1}$$

This implies the claim. □

With Proposition 1.8.3 this finishes the proof of Cartan's Theorem. From the proof we see:

Corollary 1.8.11 (The Lie Algebra of an Embedded Lie Subgroup) *Let $H \subset G$ be as in Cartan's Theorem 1.1.44. Then the Lie algebra of H is given by*

$$\mathfrak{h} = \{X \in \mathfrak{g} \mid \exp(tX) \in H \quad \forall t \in \mathbb{R}\}.$$

Proof We denote the Lie algebra of H for the moment by $L(H)$. It is clear that $L(H) \subset \mathfrak{h}$. We know from Proposition 1.8.4 that $\mathfrak{h}$ is a vector subspace in $\mathfrak{g}$ and from Eq. (1.1) that $\mathfrak{h}$ has the same dimension as H. This implies the claim. □

We collect some consequences of Cartan's Theorem:

Theorem 1.8.12 (Kernel of Lie Group Homomorphism) *Let $\phi: G \to K$ be a Lie group homomorphism. Then $H = \ker \phi$ is an embedded Lie subgroup of G with Lie algebra $\mathfrak{h} = \ker \phi_*$.*

Proof It is clear that H is a subgroup of G and closed, because $H = \phi^{-1}(e)$. By Cartan's Theorem 1.1.44, H is an embedded Lie subgroup of G.

Let X be an element in the Lie algebra $\mathfrak{h}$ of H. Then $\exp tX \in H$ for all $t \in \mathbb{R}$, hence

$$\phi(\exp tX) = e \quad \forall t \in \mathbb{R}.$$

This implies that

$$\phi_* X = \left.\frac{d}{dt}\right|_{t=0} \phi(\exp tX) = 0.$$

Conversely, let $X \in \mathfrak{g}$ with $\phi_* X = 0$. Then

$$\left.\frac{d}{dt}\right|_{t=0} \phi(\exp tX) = 0.$$

This implies for all $s \in \mathbb{R}$

$$\begin{aligned}
\left.\frac{d}{dt}\right|_{t=s} \phi(\exp tX) &= \left.\frac{d}{d\tau}\right|_{\tau=0} \phi(\exp sX)\phi(\exp \tau X) \\
&= D_e L_{\phi(\exp sX)} \left.\frac{d}{d\tau}\right|_{\tau=0} \phi(\exp \tau X) \\
&= 0.
\end{aligned}$$

Therefore, the curve $\phi(\exp tX)$ is constant and equal to $\phi(e) = e$. This implies $\exp tX \in H$ for all $t \in \mathbb{R}$ and thus $X \in \mathfrak{h}$. □

Proposition 1.8.13 (Image of Compact Lie Group Under Homomorphism) *Let $\phi: G \to H$ be a Lie group homomorphism. If G is compact, then the image of ϕ is an embedded Lie subgroup of H.*

Proof This is clear, because the image of ϕ is compact, hence closed. □

Theorem 1.8.14 (Continuous Group Homomorphisms Between Lie Groups Are Smooth) *Let $\phi: G \to K$ be a continuous group homomorphism between Lie groups. Then ϕ is smooth and thus a Lie group homomorphism.*

We use the following lemma from topology, whose proof is left as an exercise.

Lemma 1.8.15 *Let X, Y be topological spaces and $f: X \to Y$ a map. Define the graph of f by*

$$\Gamma_f = \{(x, f(x)) \in X \times Y \mid x \in X\}.$$

If Y is Hausdorff and f continuous, then Γ_f is closed in the product topology of $X \times Y$.

We also need the following.

Lemma 1.8.16 *Let $\psi: K \to G$ be a Lie group homomorphism, which is a homeomorphism. Then ψ is a diffeomorphism, hence a Lie group isomorphism.*

Proof It suffices to show that $\psi_*: \mathfrak{k} \to \mathfrak{g}$ is injective, because K and G have the same dimension. Suppose $\mathfrak{h} \neq 0$ is the kernel of ψ_* and H the kernel of ψ. According to Theorem 1.8.12 the subalgebra $\mathfrak{h}$ is the Lie algebra of H and thus $H \neq \{e\}$. This shows that ψ is not injective and hence not a homeomorphism. □

We now prove Theorem 1.8.14.

Proof The graph $\Gamma_\phi \subset G \times H$ is a closed subgroup of the Lie group $G \times H$, thus an embedded Lie subgroup by Cartan's Theorem 1.1.44. The projection $\mathrm{pr}_1: G \times H \to G$ restricts to a smooth homeomorphism

$$p: \Gamma_\phi \longrightarrow G$$

on the embedded submanifold Γ_ϕ with continuous inverse

$$\begin{aligned} p^{-1}: G &\longrightarrow \Gamma_\phi \\ g &\longmapsto (g, \phi(g)). \end{aligned}$$

It follows by Lemma 1.8.16 that p is a diffeomorphism and thus

$$\phi = \mathrm{pr}_2 \circ p^{-1}$$

is a smooth map. □

Corollary 1.8.17 (Uniqueness of Smooth Lie Group Structure) *Let G be a topological manifold which is a topological group. Then there is at most one smooth structure on G so that G is a Lie group.*

Proof Suppose G' and G'' are smooth Lie group structures on G. The identity map

$$\mathrm{Id}_G: G' \longrightarrow G''$$

is a group isomorphism and a homeomorphism. By Theorem 1.8.14 this map is a diffeomorphism. □

Corollary 1.8.18 (Embeddings of Compact Lie Groups) *Let G, H be Lie groups, G compact, and $\phi: G \to H$ an injective Lie group homomorphism. Then ϕ is a Lie group embedding, i.e. a Lie group isomorphism onto its image, an embedded Lie subgroup of H.*

Proof Since G is compact, the image of ϕ is compact, hence closed in H. This shows that the image of ϕ is an embedded Lie subgroup by Cartan's Theorem 1.1.44. Moreover, $\phi: G \to \phi(G)$ is a closed map, hence a homeomorphism. Lemma 1.8.16 implies that ϕ is an isomorphism onto its image. □

1.9 Exercises for Chap. 1

1.9.1 Let G be a topological group and G_e the connected component containing the neutral element e. Prove that G_e is a normal subgroup of G.

1.9.2 (From [135]) Let G be a topological group which is locally Euclidean. Prove that G is Hausdorff.

1.9.3 Let G be a connected topological group and $H \subset G$ an open subgroup. Prove that $H = G$.

1.9.4 Let G be a connected topological group and $U \subset G$ an open neighbourhood of e. Prove that the set

$$W = \bigcup_{n=1}^{\infty} U^n,$$

where

$$U^n = \underbrace{U \cdot U \cdots U}_{n \text{ factors}},$$

contains an open subgroup of G. Deduce that $W = G$.

1.9.5 (From [129]) Let G be a group which is at the same time a manifold so that the multiplication map

$$\begin{aligned} \mu: G \times G &\longrightarrow G \\ (g, h) &\longmapsto g \cdot h \end{aligned}$$

is smooth.

1. Show that the multiplication map μ is a submersion.
2. Prove that the map

$$\begin{aligned} G &\longrightarrow G \\ g &\longmapsto g^{-1} \end{aligned}$$

is smooth and therefore G is a Lie group.

1.9.6 Prove Proposition 1.1.25 that realizes the space $\mathrm{Mat}(n\times n, \mathbb{H})$ of quaternionic matrices via the adjoint as a subspace of the space $\mathrm{Mat}(2n \times 2n, \mathbb{C})$ of complex matrices.

1.9.7 Prove Proposition 1.1.26 on the properties of the adjoint for quaternionic matrices.

1.9.8 Show that the determinant of a matrix $(a) \in \mathrm{Mat}(1 \times 1, \mathbb{H})$ with $a \in \mathbb{H}$ is equal to

$$\det(a) = ||a||^2.$$

1.9.9 Show that every Lie group homomorphism $\phi\colon S^1 \to \mathbb{R}$ between the Lie groups $(S^1, \cdot)$ and $(\mathbb{R}, +)$ is the constant map to $0 \in \mathbb{R}$.

1.9.10

1. Find an explicit Lie group embedding

$$\mathrm{O}(n) \hookrightarrow \mathrm{SO}(n+1).$$

2. Write $A \in \mathrm{Mat}(n \times n, \mathbb{C})$ as $A = A_1 + iA_2$ with A_1, A_2 real matrices and find Lie group embeddings

$$\mathrm{GL}(n, \mathbb{C}) \hookrightarrow \mathrm{GL}_+(2n, \mathbb{R})$$

and

$$\mathrm{U}(n) \hookrightarrow \mathrm{SO}(2n).$$

3. Identify the image of $\mathrm{Sp}(n)$ in $\mathrm{Mat}(2n \times 2n, \mathbb{C})$ under the adjoint map χ and find a Lie group embedding

$$\mathrm{Sp}(n) \hookrightarrow \mathrm{U}(2n).$$

1.9.11 Let T^n denote the torus of dimension n.

1. Find Lie group embeddings

$$T^n \hookrightarrow \mathrm{U}(n),$$
$$T^{n-1} \hookrightarrow \mathrm{SU}(n),$$
$$T^n \hookrightarrow \mathrm{Sp}(n).$$

2. Find Lie group embeddings

$$T^n \hookrightarrow \mathrm{SO}(2n),$$
$$T^n \hookrightarrow \mathrm{SO}(2n+1).$$

1.9.12 Find an explicit Lie group homomorphism

$$\phi\colon \mathrm{SU}(3) \times \mathrm{SU}(2) \times \mathrm{U}(1) \longrightarrow \mathrm{SU}(5)$$

with discrete kernel.

1.9.13 Show that the Lie group homomorphisms from Example 1.3.7 and Example 1.3.8 together give a Lie group homomorphism $\psi\colon \mathrm{SU}(2) \to \mathrm{SO}(3)$ equal to

$$\psi\begin{pmatrix} x+iy & -u-iv \\ u-iv & x-iy \end{pmatrix} = \begin{pmatrix} x^2+y^2-u^2-v^2 & -2xv+2yu & 2xu+2yv \\ 2xv+2yu & x^2-y^2+u^2-v^2 & -2xy+2uv \\ -2xu+2yv & 2xy+2uv & x^2-y^2-u^2+v^2 \end{pmatrix}.$$

Deduce that

$$\psi\begin{pmatrix} e^{i\alpha/2} & 0 \\ 0 & e^{-i\alpha/2} \end{pmatrix} = \begin{pmatrix} 1 & 0 & 0 \\ 0 & \cos\alpha & -\sin\alpha \\ 0 & \sin\alpha & \cos\alpha \end{pmatrix} \quad \forall \alpha \in \mathbb{R}.$$

1.9.14 Recall that according to Example 1.4.6 the vector space $\mathbb{R}^3$ is a Lie algebra with bracket given by the cross product:

$$[v, w] = v \times w, \quad \forall v, w \in \mathbb{R}^3.$$

Find an explicit isomorphism of $(\mathbb{R}^3, \times)$ with the Lie algebra $\mathfrak{so}(3)$.

1.9.15 Prove that for $n \geq 1$ the sphere S^{2n} does not admit the structure of a Lie group.

1.9.16 (From [24]) Consider the Lie algebras $\mathfrak{g}$ of the classical linear groups G from Theorem 1.5.27.

1. Show directly that the subsets of $\mathrm{Mat}(n \times n, \mathbb{K})$ defined in Theorem 1.5.27 are real vector subspaces and closed under the commutator of matrices.
2. Show also by a direct calculation that these subsets are closed under the following map:

$$X \longmapsto g \cdot X \cdot g^{-1},$$

where $X \in \mathfrak{g}$ is an element of the Lie algebra and $g \in G$ an element of the corresponding linear group (we will identify this map with the adjoint representation in Sect. 2.1.5).

1.9.17

1. Prove that there are, up to isomorphism, only two 2-dimensional real Lie algebras.
2. Show that $\mathfrak{sl}(2, \mathbb{R})$ is not isomorphic to $\mathfrak{su}(2)$.

1.9.18 The Lie algebra $\mathfrak{su}(2)$ is spanned as a real vector space by the matrices τ_1, τ_2, τ_3 from Example 1.5.32. The Lie algebra $\mathfrak{sl}(2, \mathbb{C})$ is spanned as a complex vector space by the matrices H, X, Y from Example 1.5.36.

1. Show that the matrices τ_1, τ_2, τ_3 are a complex basis for $\mathfrak{sl}(2, \mathbb{C})$ and express this basis in terms of H, X, Y.
2. Show that as complex Lie algebras $\mathfrak{sl}(2, \mathbb{C})$ is isomorphic to $\mathfrak{su}(2) \otimes_{\mathbb{R}} \mathbb{C}$, where the Lie bracket on the right is the complex linear extension of the Lie bracket of $\mathfrak{su}(2)$.

1.9.19 Consider the Lie group $\mathrm{U}(n)$ with Lie algebra $\mathfrak{u}(n)$.

1. Find an explicit Lie algebra isomorphism

$$\mathfrak{u}(n) \cong \mathfrak{u}(1) \oplus \mathfrak{su}(n).$$

2. Find an explicit group isomorphism

$$\mathrm{U}(n) \cong (\mathrm{U}(1) \times \mathrm{SU}(n))/\mathbb{Z}_n,$$

where $\mathbb{Z}_n \subset \mathrm{U}(1) \times \mathrm{SU}(n)$ is a normal subgroup.

1.9.20 Recall that

$$\mathrm{SU}(2) = \left\{ \begin{pmatrix} a & -\bar{b} \\ b & \bar{a} \end{pmatrix} \in \mathrm{Mat}(2 \times 2, \mathbb{C}) \,\middle|\, a, b \in \mathbb{C},\ |a|^2 + |b|^2 = 1 \right\}.$$

We identify quaternions $u + jv \in \mathbb{H}$, where $u, v \in \mathbb{C}$, with the following matrices:

$$\mathbb{H} \cong \left\{ \begin{pmatrix} u & -\bar{v} \\ v & \bar{u} \end{pmatrix} \in \mathrm{Mat}(2 \times 2, \mathbb{C}) \,\middle|\, u, v \in \mathbb{C} \right\}.$$

Consider the following isomorphism of real vector spaces:

$$\mathbb{R}^3 \cong i\mathbb{R} \times \mathbb{C} \longrightarrow \mathrm{Im}\,\mathbb{H}$$

$$x = (ic, v) \longmapsto X = \begin{pmatrix} ic & -\bar{v} \\ v & -ic \end{pmatrix}.$$

The Euclidean norm of an element $x \in \mathbb{R}^3$ is given by $||x||^2 = \det X$. Under this identification we set:

$$\mathrm{SU}(2) \times \mathbb{R}^3 \longrightarrow \mathbb{R}^3$$

$$(A, X) \longmapsto AXA^{\dagger}.$$

1. Prove that this map is well-defined and yields a homomorphism

$$\phi\colon \mathrm{SU}(2) \longrightarrow \mathrm{SO}(3)$$

   of Lie groups.
2. Show that ϕ is surjective and that its kernel consists of $\{I, -I\}$.

1.9.21 (From [98]) We identify the Lie group SU(2) and the quaternions $\mathbb{H}$ with subsets of the complex 2×2-matrices as in Exercise 1.9.20. Consider the following isomorphism of real vector spaces:

$$\mathbb{R}^4 \cong \mathbb{C}^2 \longrightarrow \mathbb{H}$$

$$x = (u, v) \longmapsto X = \begin{pmatrix} u & -\bar{v} \\ v & \bar{u} \end{pmatrix}.$$

The Euclidean norm of an element $x \in \mathbb{R}^4 \cong \mathbb{C}^2$ is given by $||x||^2 = \det X$. Under this identification we set:

$$\mathrm{SU}(2) \times \mathrm{SU}(2) \times \mathbb{R}^4 \longrightarrow \mathbb{R}^4$$

$$(A_-, A_+, X) \longmapsto A_- \cdot X \cdot A_+^{\dagger},$$

where $\cdot$ denotes matrix multiplication.

1. Prove that this map is well-defined and yields a homomorphism

$$\psi\colon \mathrm{SU}(2) \times \mathrm{SU}(2) \to \mathrm{SO}(4)$$

   of Lie groups.
2. Show that ψ is surjective with kernel $\{(I, I), (-I, -I)\}$.

1.9.22 Prove that every Lie group homomorphism $\rho\colon S^1 \to S^1$ is of the form

$$\rho(z) = z^k$$

for some $k \in \mathbb{Z}$.

1.9.23 Show that if matrices $A, B \in \mathrm{Mat}(n \times n, \mathbb{K})$ with $\mathbb{K} = \mathbb{R}, \mathbb{C}, \mathbb{H}$ commute, $AB = BA$, then

$$e^{A+B} = e^A e^B.$$

1.9.24 Calculate $\exp(s\tau_a) \in \mathrm{SU}(2)$ for $s \in \mathbb{R}$ and the basis τ_1, τ_2, τ_3 of the Lie algebra $\mathfrak{su}(2)$ from Example 1.5.32.

1.9.25 Consider the Lie group SU(2) with Lie algebra $\mathfrak{su}(2)$.

1. Show that every element $X \in \mathfrak{su}(2)$ can be written as

$$X = -2rA \cdot \tau_3 \cdot A^{-1}$$

with $r \in \mathbb{R}$, $A \in \mathrm{SU}(2)$ and

$$\tau_3 = -\frac{1}{2}\begin{pmatrix} i & 0 \\ 0 & -i \end{pmatrix} \in \mathfrak{su}(2).$$

2. Prove that every one-parameter subgroup of SU(2) is closed, i.e. its image is isomorphic to U(1).

1.9.26 Consider the Lie algebra $\mathfrak{su}(3)$ from Example 1.5.33 with the basis $v_1, \ldots, v_8$, where $v_a = \frac{i\lambda_a}{2}$ and λ_a are the Gell-Mann matrices.

1. Prove that the following three sets of basis vectors

$$\{v_1, v_2, v_3\},$$
$$\{v_4, v_5, \alpha v_3 + \beta v_8\},$$
$$\{v_6, v_7, \gamma v_3 + \delta v_8\},$$

where $\alpha, \beta, \gamma, \delta$ are certain real numbers, span Lie subalgebras of $\mathfrak{su}(3)$ isomorphic to $\mathfrak{su}(2)$. Determine $\alpha, \beta, \gamma, \delta$.
2. Prove that the one-parameter subgroups generated by each of the basis vectors $v_1, \ldots, v_8$ are closed.

1.9.27 (From [24]) Consider a matrix

$$A = \begin{pmatrix} a & b \\ c & -a \end{pmatrix} \in \mathfrak{sl}(2, \mathbb{R}).$$

1. Calculate e^A and $\mathrm{tr}\left(e^A\right)$.
2. Prove that the exponential map $\exp\colon \mathfrak{sl}(2, \mathbb{R}) \to \mathrm{SL}(2, \mathbb{R})$ is not surjective.

1.9.28

1. Let G be a connected Lie group. Show that every group element $g \in G$ is of the form

$$g = \exp(X_1) \cdot \exp(X_2) \cdots \exp(X_n)$$

for finitely many vectors $X_1, \ldots, X_n$ in the Lie algebra $\mathfrak{g}$ of G.
2. Let $\phi\colon G \to H$ be a Lie group homomorphism, where G is connected. Suppose that the induced Lie algebra homomorphism $\phi_*\colon \mathfrak{g} \to \mathfrak{h}$ is trivial. Prove that ϕ is trivial.

1.9.29 (From [77])

1. Calculate the k-th power of the nilpotent matrix

$$N = \begin{pmatrix} 0 & 1 & 0 & 0 & \cdots & 0 \\ \cdot & 0 & 1 & 0 & \cdots & \cdot \\ \cdot & \cdot & \cdot & \cdot & \cdot & \cdot \\ 0 & & \cdots & & \cdots & 1 \\ 0 & & \cdots & & \cdots & 0 \end{pmatrix} \in \mathrm{Mat}(n \times n, \mathbb{C}).$$

2. Calculate the k-th power of a Jordan block matrix $\lambda I_n + N$ with $\lambda \in \mathbb{C}$.
3. Calculate e^{tA} for a matrix A in Jordan normal form and $t \in \mathbb{R}$.

1.9.30 (From [77]) Let $A \in \mathrm{Mat}(n \times n, \mathbb{C})$.

1. Use Exercise 1.9.29 to show that the set

$$\left\{e^{tA} \mid t \in \mathbb{R}\right\}$$

is bounded in $\mathrm{Mat}(n \times n, \mathbb{C})$ if and only if A is diagonalizable with purely imaginary eigenvalues.
2. Show that $e^A = I$ if and only if A is diagonalizable with all eigenvalues contained in $2\pi i\mathbb{Z}$.

Chapter 2
Lie Groups and Lie Algebras: Representations and Structure Theory

At least locally, fields in physics can be described by maps on spacetime with values in vector spaces. Since symmetry groups in field theories act on fields, it is important to understand (linear) actions of Lie groups and Lie algebras on vector spaces, known as *representations*.

For example, we shall see that, in the Standard Model, three *Dirac spinors* for each *quark flavour* are combined and form a vector in a representation space $\mathbb{C}^3$ of the gauge group SU(3) of quantum chromodynamics. Similarly, two *left-handed Weyl spinors*, known as the *left-handed electron* and the *left-handed electron neutrino*, are combined to form a vector in a representation space $\mathbb{C}^2$ of the gauge group $SU(2) \times U(1)$ of the electroweak interaction.

It turns out that every Lie group and Lie algebra has a special representation, known as the *adjoint representation*. The adjoint representation can be used to define the *Killing form*, a canonical symmetric bilinear form on every Lie algebra. Both the adjoint representation and the Killing form are important tools for the classification of Lie algebras. The adjoint representation is also important in physics, because *gauge bosons* correspond to fields on spacetime that transform under the adjoint representation of the gauge group.

The purpose of this chapter is to describe representations of Lie groups and Lie algebras in general as well as the structure of semisimple and compact Lie algebras. We also discuss special scalar products on Lie algebras which will be used in Sect. 7.3.1 to construct Lagrangians for gauge boson fields. We only cover the basics of the representation and structure theory of Lie groups and Lie algebras. Much more details can be found in the references mentioned at the beginning of Chap. 1, which are also the references for this chapter.

M.J.D. Hamilton, *Mathematical Gauge Theory*, Universitext,
https://doi.org/10.1007/978-3-319-68439-0_2

2.1 Representations

2.1.1 Basic Definitions

We begin with the basic concept of representations of Lie groups and Lie algebras.

Definition 2.1.1 Let G be a Lie group and V a vector space over the real or complex numbers. Then a **representation** of G on V is a Lie group homomorphism

$$\rho: G \longrightarrow \mathrm{GL}(V)$$

to the Lie group $\mathrm{GL}(V)$ of linear isomorphisms of V. One sometimes writes $\mathrm{GL}(V) = \mathrm{Aut}(V)$, the Lie group of linear automorphisms of V. The Lie group $\mathrm{GL}(V)$ is by definition isomorphic to a general linear group of the form $\mathrm{GL}(n, \mathbb{K})$, where $\mathbb{K} = \mathbb{R}, \mathbb{C}$ and n is the dimension of V.

If the representation is clear from the context, we sometimes write

$$\rho(g)v = g \cdot v = gv$$

for $g \in G, v \in V$. A representation ρ of a Lie group G is called **faithful** if ρ is injective.
For a Lie group representation ρ the identities

$$\rho(gh) = \rho(g) \circ \rho(h)$$

and

$$\rho\left(g^{-1}\right) = \rho(g)^{-1}$$

hold for all $g, h \in G$. Note that the definition of a representation ρ requires that the map ρ is a homomorphism in the algebraic sense and differentiable (in fact, by Theorem 1.8.14 it suffices to demand that the map ρ is continuous).

Example 2.1.2 By Theorem 1.2.7 any compact Lie group has a faithful representation on some finite-dimensional, complex vector space.

Definition 2.1.3 Let ρ_V, ρ_W be representations of a Lie group G on vector spaces V and W. Then a **morphism** of the representations is a G-**equivariant** linear map $f: V \to W$, so that

$$f(\rho_V(g)v) = \rho_W(g)f(v),$$

i.e.

$$f(gv) = gf(v) \quad \forall g \in G, v \in V.$$

Such a map f is also called an **intertwining map**. An **isomorphism** or **equivalence** of representations is a G-equivariant isomorphism.

Definition 2.1.4 Let $\rho: G \to \mathrm{GL}(V)$ be a representation of a Lie group G. Suppose that $H \subset G$ is an embedded Lie subgroup. Then the restriction

$$\rho|_H: H \longrightarrow \mathrm{GL}(V)$$

of the Lie group homomorphism ρ to H is a representation of H, called a **restricted representation**.

We define representations of Lie algebras in a similar way.

Definition 2.1.5 Let $\mathfrak{g}$ be a (real or complex) Lie algebra and V a vector space over the real or complex numbers. Then a **representation** of $\mathfrak{g}$ on V is a Lie algebra homomorphism

$$\phi: \mathfrak{g} \longrightarrow \mathfrak{gl}(V) = \mathrm{End}(V)$$

to the linear endomorphisms of V (linear maps $V \to V$). If the representation is clear from the context, we sometimes write

$$\phi(X)v = X \cdot v = Xv$$

for $X \in \mathfrak{g}, v \in V$. A representation ϕ of a Lie algebra $\mathfrak{g}$ is called **faithful** if ϕ is injective.

For a Lie algebra representation the following identity holds:

$$\phi([X, Y]) = \phi(X) \circ \phi(Y) - \phi(Y) \circ \phi(X) \quad \forall X, Y \in \mathfrak{g}.$$

Example 2.1.6 By Ado's Theorem 1.5.25 any Lie algebra has a faithful representation on some finite-dimensional vector space.

Remark 2.1.7 Note that if the Lie algebra is complex, then we require the representation $\phi: \mathfrak{g} \to \mathrm{End}(V)$ to be a complex linear map.

Definition 2.1.8 Let ϕ_V, ϕ_W be representations of a Lie algebra $\mathfrak{g}$ on vector spaces V and W. Then a **morphism** of the representations is a $\mathfrak{g}$-**equivariant** linear map $f: V \to W$, so that

$$f(\phi_V(X)v) = \phi_W(X)f(v),$$

i.e.

$$f(Xv) = Xf(v) \quad \forall X \in \mathfrak{g}, v \in V.$$

Such a map f is also called an **intertwining map**. An **isomorphism** or **equivalence** of representations is a $\mathfrak{g}$-equivariant isomorphism.

Definition 2.1.9 Let $\phi: \mathfrak{g} \to \mathrm{End}(V)$ be a representation of a Lie algebra $\mathfrak{g}$. Suppose that $\mathfrak{h} \subset \mathfrak{g}$ is a Lie subalgebra. Then the restriction

$$\phi|_{\mathfrak{h}}: \mathfrak{h} \longrightarrow \mathrm{End}(V)$$

of the Lie algebra homomorphism ϕ to $\mathfrak{h}$ is a representation of $\mathfrak{h}$, called **restricted representation**.

Remark 2.1.10 Unless stated otherwise we only consider representations of Lie groups and Lie algebras on real and complex vector spaces and these vector spaces are finite-dimensional.

Remark 2.1.11 Both types of homomorphisms are called *representations*, because we represent elements in the Lie group or Lie algebra by linear maps on a vector space, i.e. (after a choice of basis for the vector space) by matrices.

Representations of Lie groups and their associated Lie algebras are related:

Proposition 2.1.12 (Induced Representations) *Let $\rho: G \to \mathrm{GL}(V)$ be a representation of a Lie group G on a vector space V. Then the differential $\rho_*: \mathfrak{g} \to \mathrm{End}(V)$ is a representation of the Lie algebra $\mathfrak{g}$.*

Proof The proof follows from Theorem 1.5.18, because the differential of a Lie group homomorphism is a Lie algebra homomorphism. □

With Theorem 1.7.16 we get the following commutative diagram:

$$\begin{array}{ccc} \mathfrak{g} & \xrightarrow{\rho_*} & \mathrm{End}(V) \\ \downarrow{\scriptstyle \exp} & & \downarrow{\scriptstyle \exp} \\ G & \xrightarrow{\rho} & \mathrm{GL}(V) \end{array}$$

Note that the exponential map on the right is just the standard exponential map on endomorphisms (defined in the same way as for matrices, using composition instead of matrix multiplication). We can thus write the commutativity of the diagram as

$$\rho(\exp X) = e^{\rho_* X} \quad \forall X \in \mathfrak{g}.$$

This means: if we know how a Lie algebra element $X \in \mathfrak{g}$ acts in a representation on the vector space V, then we know how the group element $\exp X \in G$ acts on V.

Assuming Theorem 1.5.20 we get the following:

Corollary 2.1.13 (Integrability Theorem for Representations) *Let G be a connected and simply connected Lie group. Suppose $\phi: \mathfrak{g} \to \mathrm{End}(V)$ is a representation of the Lie algebra of G. Then there exists a unique representation $\rho: G \to \mathrm{GL}(V)$ such that $\rho_* = \phi$.*

The discussion in Example 1.5.21 shows that this may not hold if G is not simply connected. In particular, if

$$\phi\colon \mathfrak{so}(2) \cong \mathfrak{u}(1) \longrightarrow \mathrm{End}(V)$$

is a representation and X the generator of $\mathfrak{u}(1)$ with $\exp(2\pi iX) = 1$, then a necessary condition that ϕ comes from a representation

$$\rho\colon \mathrm{U}(1) \longrightarrow \mathrm{GL}(V)$$

is that

$$e^{2\pi i\phi(X)} = \mathrm{Id}_V.$$

Example 2.1.14 For any constant $k \in \mathbb{Z}$ there is a complex 1-dimensional representation

$$\begin{aligned} \rho_k\colon \mathrm{U}(1) &\longrightarrow \mathrm{U}(1) \subset \mathrm{GL}(\mathbb{C}) \\ z &\longmapsto z^k. \end{aligned}$$

We say that these representations have **winding number** k. In the Standard Model these representations appear in connection with the *weak hypercharge gauge group* $\mathrm{U}(1)_Y$.

Example 2.1.15 The Lie groups $\mathrm{GL}(n,\mathbb{R})$ (and $\mathrm{GL}(n,\mathbb{C})$) have canonical representations on $\mathbb{R}^n$ (and $\mathbb{C}^n$) by matrix multiplication on column vectors from the left. These representations induce representations for all linear groups, called **standard, defining** or **fundamental representations** (by a fundamental representation we will always mean the defining representation). There are similar, induced representations of the corresponding Lie algebras.

Definition 2.1.16 A representation of a Lie group G (or Lie algebra $\mathfrak{g}$) on a vector space V is called **irreducible** if there is **no proper invariant** subspace $W \subset V$, i.e. no vector subspace W, different from 0 or V, such that $G\cdot W \subset W$ (or $\mathfrak{g}\cdot W \subset W$). A representation is called **reducible** if it is not irreducible.

Example 2.1.17 The 0-dimensional and every 1-dimensional representation are irreducible, because in these cases there are no proper vector subspaces at all.

Definition 2.1.18 A **singlet representation** is a representation of a Lie group or Lie algebra on a 1-dimensional (real or complex) vector space. Similarly, a **doublet** or **triplet representation** is a representation on a 2- or 3-dimensional vector space. A representation of a Lie group or Lie algebra on an n-dimensional vector space is sometimes denoted by $\mathbf{n}$, in particular, if the dimension uniquely determines the representation.

Example 2.1.19 (Trivial Representations) Let G be a Lie group and V a real or complex vector space. Then

$$\begin{aligned} \rho\colon G &\longrightarrow \mathrm{GL}(V) \\ g &\longmapsto \mathrm{Id}_V, \end{aligned}$$

where every group element gets mapped to the identity, is a representation, called a **trivial representation**. It is irreducible precisely if V is 1-dimensional. Similarly, if $\mathfrak{g}$ is a Lie algebra, then

$$\begin{aligned} \phi\colon \mathfrak{g} &\longrightarrow \mathrm{End}(V) \\ g &\longmapsto 0 \end{aligned}$$

is a trivial representation. Again, it is irreducible precisely if V is 1-dimensional. We will later study a class of Lie algebras where *every* representation is either trivial or faithful, see Exercise 2.7.9.

It is a curious fact that the fundamental and trivial representations of $\mathrm{SU}(3)$ and $\mathrm{SU}(2)$, together with the winding number representations of $\mathrm{U}(1)$ in Example 2.1.14, suffice to describe *all* matter particles (and the Higgs field) in the Standard Model; see Sect. 8.5. The gauge bosons corresponding to these gauge groups are described by the *adjoint representation* that we discuss in Sect. 2.1.5.

Example 2.1.20 The fundamental representation of the Lie algebra $\mathfrak{su}(2)$ is an (irreducible) doublet representation on the vector space $\mathbb{C}^2$. Recall from Example 1.5.32 that there exists an isomorphism $\mathfrak{su}(2) \cong \mathfrak{so}(3)$. The fundamental representation of $\mathfrak{so}(3)$ thus also defines an (irreducible) triplet representation of $\mathfrak{su}(2)$ on $\mathbb{R}^3$ (and $\mathbb{C}^3$). It can be proved that $\mathfrak{su}(2)$ has a unique (up to equivalence) irreducible complex representation V_n of dimension $n+1$ for every natural number $n \geq 0$ (see, e.g. [24]).

Example 2.1.21 (The Heisenberg Lie Algebra and Quantum Mechanics) Recall from Example 1.5.38 that the Heisenberg Lie algebra $\mathfrak{nil}_3$ is a 3-dimensional real Lie algebra spanned by vectors p, q, z with Lie brackets

$$\begin{aligned} [q,p] &= z, \\ [q,z] &= 0, \\ [p,z] &= 0. \end{aligned}$$

Let $\hbar \in \mathbb{R}$ be some real number. A **central representation** of $\mathfrak{nil}_3$ is a representation

$$\mathfrak{nil}_3 \longrightarrow \mathrm{End}(V)$$

on a complex vector space V such that z gets mapped to $i\hbar \cdot \mathrm{Id}_V$. If we denote the images of q and p in $\mathrm{End}(V)$ by $\hat{q}$ and $\hat{p}$, then

$$[\hat{q}, \hat{p}] = i\hbar$$

and the other two commutation relations are satisfied trivially (on the right-hand side we do not write the identity map of V explicitly). This is the **canonical commutation relation** of quantum mechanics.

2.1.2 Linear Algebra Constructions of Representations

There are several well-known constructions that yield new vector spaces from given ones. If the given vector spaces carry a representation, then usually the new vector spaces carry induced representations. We first recall the following notion from complex linear algebra.

Definition 2.1.22 Let V be a complex vector space. Then we define the **complex conjugate** vector space $\bar{V}$ as follows:

1. As a set and abelian group $\bar{V} = V$.
2. Scalar multiplication is defined by

$$\begin{aligned} \mathbb{C} \times \bar{V} &\longrightarrow \bar{V} \\ (\lambda, v) &\longmapsto \bar{\lambda} v. \end{aligned}$$

If $f: V \to V$ is a complex linear map, then the same map (on the set $\bar{V} = V$) is denoted by $\bar{f}: \bar{V} \to \bar{V}$ and is still complex linear. The identity map $V \to \bar{V}$ is complex antilinear.

Definition 2.1.23 Let V and W be real or complex vector spaces with representations

$$\begin{aligned} \rho_V: G &\longrightarrow \mathrm{GL}(V) \\ \rho_W: G &\longrightarrow \mathrm{GL}(W) \end{aligned}$$

of a Lie group G. Then there exist the following representations of G, where $g \in G$ and $v \in V, w \in W$ are arbitrary:

1. The **direct sum** representation $\rho_{V \oplus W}$ on $V \oplus W$, defined by

$$g(v, w) = (gv, gw).$$

2. The **tensor product** representation $\rho_{V \otimes W}$ on $V \otimes W$, defined by

$$g(v \otimes w) = gv \otimes gw.$$

3. The **dual** representation ρ_{V^*} on V^*, defined by

$$(g\lambda)(v) = \lambda\left(g^{-1}v\right), \quad \forall \lambda \in V^*.$$

4. The **exterior power** representation $\rho_{\Lambda^k V}$ on $\Lambda^k V$, defined by

$$g(v_1 \wedge v_2 \wedge \ldots \wedge v_k) = gv_1 \wedge gv_2 \wedge \ldots \wedge gv_k, \quad \forall v_1 \wedge v_2 \wedge \ldots \wedge v_k \in \Lambda^k V.$$

5. The **homomorphism space** representation $\rho_{\mathrm{Hom}(V,W)}$ on $\mathrm{Hom}(V, W)$, defined by

$$(gf)(v) = gf\left(g^{-1}v\right), \quad \forall f \in \mathrm{Hom}(V, W).$$

6. If V is a complex vector space, then the **complex conjugate** representation $\rho_{\bar{V}}$ on $\bar{V}$ is defined by

$$\rho_{\bar{V}}(g)v = \overline{\rho_V(g)}v.$$

Suppose in addition that

$$\tau_W: H \longrightarrow \mathrm{GL}(W)$$

is a representation of a Lie group H. Then there exists the following representation, where $h \in H$ is arbitrary:

7. The **(outer) tensor product** representation $\rho_V \otimes \tau_W$ on $V \otimes W$ of the Lie group $G \times H$, defined by

$$(g, h)(v \otimes w) = gv \otimes hw,$$

for $g \in G, h \in H$.

It is easy to check that each of these maps is indeed a representation.

Remark 2.1.24 The direct sum representation $\rho_V \oplus \tau_W$ on $V \oplus W$ of the Lie group $G \times H$, defined by

$$(g, h)(v, w) = (gv, hw),$$

is less important, because it is can be reduced to the representations ρ_V and τ_W, each tensored with the trivial 1-dimensional representation.

Remark 2.1.25 If V is a complex representation space for a Lie group, we then get in total four complex representations which have the same dimension as V: V, V^*, $\bar{V}$ and $\bar{V}^*$.

The representations of the Lie group

$$G = \mathrm{SU}(3) \times \mathrm{SU}(2) \times \mathrm{U}(1)$$

that appear in the Standard Model of elementary particles are direct sums of outer tensor product representations of the form

$$U \otimes V \otimes W,$$

where U, V, W are certain representations of the factors SU(3), SU(2), U(1) of G. See Sect. 8.5 for details.

Example 2.1.26 We describe these constructions using matrices. Consider the column vector spaces $V = \mathbb{K}^n$, $W = \mathbb{K}^m$ where $\mathbb{K} = \mathbb{R}, \mathbb{C}$. Representations ρ_V and ρ_W of a Lie group G take values in the matrix Lie groups $\mathrm{GL}(n, \mathbb{K})$ and $\mathrm{GL}(m, \mathbb{K})$. We can then identify the canonical representations of G on the vector spaces

$$V \oplus W, \quad V^*, \quad \mathrm{Hom}(V, W), \quad \Lambda^2 V^* \quad \text{and} \quad \bar{V} \text{ (if } V \text{ is complex)}$$

with the following representations:

1. $V \oplus W$ can be identified with $\mathbb{K}^{n+m}$. For a column vector $(x, y)^T \in \mathbb{K}^{n+m}$ the direct sum representation is given by

$$\rho_{V \oplus W}(g) \begin{pmatrix} x \\ y \end{pmatrix} = \begin{pmatrix} \rho_V(g) & 0 \\ 0 & \rho_W(g) \end{pmatrix} \begin{pmatrix} x \\ y \end{pmatrix}.$$

2. V^* can be identified with a row vector space that we here denote by $(\mathbb{K}^n)^*$. For a row vector $s \in (\mathbb{K}^n)^*$ the dual representation is given by

$$\rho_{V^*}(g)s = s \cdot \rho_V(g)^{-1}.$$

3. $\mathrm{Hom}(V, W)$ can be identified with the vector space $\mathrm{Mat}(m \times n, \mathbb{K})$. For a matrix $A \in \mathrm{Mat}(m \times n, \mathbb{K})$ the representation on the homomorphism space is given by

$$\rho_{\mathrm{Hom}(V,W)}(g)A = \rho_W(g) \cdot A \cdot \rho_V(g)^{-1}.$$

4. $\Lambda^2 V^*$ is the space of skew-symmetric, bilinear maps

$$\lambda\colon V \times V \longrightarrow \mathbb{K}$$

and can be identified with $\mathfrak{so}(n, \mathbb{K})$, the space of skew-symmetric $n \times n$-matrices, by sending λ to the matrix A with coefficients $A_{ij} = \lambda(e_i, e_j)$. The representation on $\Lambda^2 V^*$ is then given by

$$\rho_{\Lambda^2 V^*}(g)A = \left(\rho_V(g)^{-1}\right)^T \cdot A \cdot \rho_V(g)^{-1}.$$

5. If $V = \mathbb{C}^n$, then $\bar{V} = \mathbb{C}^n$ as an abelian group and every complex scalar (and hence every complex matrix) acts as the complex conjugate. For a column vector $z \in \mathbb{C}^n$ the complex conjugate representation is given by

$$\rho_{\bar{V}}(g)z = \overline{\rho_V(g)} \cdot z.$$

There are analogous constructions for representations of Lie algebras:

Definition 2.1.27 Let V and W be real or complex vector spaces with representations

$$\phi_V : \mathfrak{g} \longrightarrow \mathrm{End}(V)$$
$$\phi_W : \mathfrak{g} \longrightarrow \mathrm{End}(W)$$

of a Lie algebra $\mathfrak{g}$. Then there exist the following representations of $\mathfrak{g}$, where $X \in \mathfrak{g}$ and $v \in V, w \in W$ are arbitrary:

1. The **direct sum** representation $\phi_{V \oplus W}$ on $V \oplus W$, defined by

$$X(v, w) = (Xv, Xw).$$

2. The **tensor product** representation $\phi_{V \otimes W}$ on $V \otimes W$, defined by

$$X(v \otimes w) = (Xv) \otimes w + v \otimes (Xw).$$

3. The **dual** representation ϕ_{V^*} on V^*, defined by

$$(X\lambda)(v) = \lambda(-Xv), \quad \forall \lambda \in V^*.$$

4. The **exterior power** representation $\phi_{\Lambda^k V}$ on $\Lambda^k V$, defined by

$$X(v_1 \wedge v_2 \wedge \ldots \wedge v_k) = \sum_{i=1}^{k} v_1 \wedge \ldots \wedge Xv_i \wedge \ldots \wedge v_k, \quad \forall v_1 \wedge v_2 \wedge \ldots \wedge v_k \in \Lambda^k V.$$

5. The **homomorphism space** representation $\phi_{\mathrm{Hom}(V,W)}$ on $\mathrm{Hom}(V, W)$, defined by

$$(Xf)(v) = Xf(v) + f(-Xv), \quad \forall f \in \mathrm{Hom}(V, W).$$

6. If V is a complex vector space and $\mathfrak{g}$ a real Lie algebra, then the **complex conjugate** representation $\phi_{\bar{V}}$ on $\bar{V}$ is defined by

$$\phi_{\bar{V}}(X)v = \overline{\phi_V(X)}v.$$

Suppose in addition that

$$\psi_W : \mathfrak{h} \longrightarrow \mathrm{End}(W)$$

is a representation of a Lie algebra $\mathfrak{h}$. Then there exists the following representation, where $Y \in \mathfrak{h}$ is arbitrary:

7. The **(outer) tensor product** representation $\phi_V \otimes \psi_W$ on $V \otimes W$ of the Lie algebra $\mathfrak{g} \oplus \mathfrak{h}$, defined by

$$(X, Y)(v \otimes w) = Xv \otimes w + v \otimes Yw,$$

for $X \in \mathfrak{g}, Y \in \mathfrak{h}$.

Remark 2.1.28 Perhaps the most interesting case in the proof that these maps define representations is the dual representation for both Lie groups and Lie algebras. To check that the formulas here define representations is the purpose of Exercise 2.7.1.

Both constructions are related:

Proposition 2.1.29 *Let G and H be Lie groups with Lie algebras $\mathfrak{g}$ and $\mathfrak{h}$. Let ρ be any of the representations of G on $V \oplus W$, $V \otimes W$, V^*, $\Lambda^k V$,* Hom(V, W) *or $\bar{V}$ (or of $G \times H$ on $V \otimes W$) from Definition 2.1.23. Then the induced representation ρ_* of $\mathfrak{g}$ (or of $\mathfrak{g} \oplus \mathfrak{h}$) is the corresponding one from Definition 2.1.27.*

Proof The proof follows by differentiating the representation of G (or of $G \times H$). □

2.1.3 *The Weyl Spinor Representations of SL(2, ℂ)

We discuss an extended example that is relevant for some theories in physics, like the Standard Model or supersymmetry (see reference [146, Appendix A]). Let $G = \mathrm{SL}(2, \mathbb{C})$. As we will discuss in Sect. 6.8.2 in more detail, the group $\mathrm{SL}(2, \mathbb{C})$ is the *(orthochronous) Lorentz spin group*, i.e. the universal covering of the identity component of the Lorentz group of 4-dimensional spacetime.

We denote by $V = \mathbb{C}^2$ the fundamental $\mathrm{SL}(2, \mathbb{C})$-representation. Then we get the following four complex doublet representations, where $M \in \mathrm{SL}(2, \mathbb{C})$ and $\psi \in \mathbb{C}^2$:

1. The fundamental representation V:

$$\psi \longmapsto M\psi.$$

2. The dual representation V^*:

$$\psi^T \longmapsto \psi^T M^{-1}.$$

3. The complex conjugate representation $\bar{V}$:

$$\bar{\psi} \longmapsto \bar{M}\bar{\psi}.$$

4. The dual of the complex conjugate representation $\bar{V}^*$:

$$\bar{\psi}^T \longmapsto \bar{\psi}^T (\bar{M})^{-1}.$$

Here we denote the elements of the vector spaces V^*, $\bar{V}$ and $\bar{V}^*$ for clarity by ψ^T, $\bar{\psi}$ and $\bar{\psi}^T$.

Remark 2.1.30 In physics the components of the vectors in the spaces V, V^*, $\bar{V}$ and $\bar{V}^*$ are denoted by ψ_α, ψ^α, $\bar{\psi}_{\dot{\alpha}}$ and $\bar{\psi}^{\dot{\alpha}}$. We could denote these representations by **2**, **2***, $\bar{\mathbf{2}}$ and $\bar{\mathbf{2}}^*$.

Definition 2.1.31 In this situation the representation of $\mathrm{SL}(2, \mathbb{C})$ on V is called the **left-handed Weyl spinor** representation and the representation on $\bar{V}^*$ is called the **right-handed Weyl spinor** representation. Both representations are also called **chiral spinor representations**.
We want to show that the remaining two representations are isomorphic to the left- and right-handed Weyl spinor representations.

Definition 2.1.32 We define

$$\epsilon = \begin{pmatrix} 0 & 1 \\ -1 & 0 \end{pmatrix}.$$

Proposition 2.1.33 *We have the following equivalent description of* $\mathrm{SL}(2, \mathbb{C})$*:*

$$\mathrm{SL}(2, \mathbb{C}) = \left\{ M \in \mathrm{Mat}(2 \times 2, \mathbb{C}) \mid M^T \epsilon M = \epsilon \right\}.$$

Proof The proof is an easy calculation; see Exercise 2.7.2. □

Proposition 2.1.34 *The map*

$$\begin{aligned} f: V &\longrightarrow V^* \\ \psi &\longmapsto \psi^T \epsilon \end{aligned}$$

is an isomorphism of representations. Similarly the map

$$\begin{aligned} \bar{f}: \bar{V} &\longrightarrow \bar{V}^* \\ \bar{\psi} &\longmapsto \bar{\psi}^T \epsilon \end{aligned}$$

is an isomorphism of representations.

Proof We only have to show $\mathrm{SL}(2, \mathbb{C})$-equivariance of the maps. This follows by applying Proposition 2.1.33:

$$\begin{aligned} f(M\psi) &= (M\psi)^T \epsilon \\ &= \psi^T M^T \epsilon \\ &= (\psi^T \epsilon) M^{-1} \\ &= f(\psi) M^{-1} \end{aligned}$$

and

$$\begin{aligned} \bar{f}(\bar{M}\bar{\psi}) &= (\bar{M}\bar{\psi})^T \epsilon \\ &= \bar{\psi}^T \bar{M}^T \epsilon \\ &= (\bar{\psi}^T \epsilon) \bar{M}^{-1} \\ &= \bar{f}(\bar{\psi}) \bar{M}^{-1}. \end{aligned}$$

□

See Sect. 6.8 and Lemma 8.5.5 for more details about these isomorphisms.

2.1.4 Orthogonal and Unitary Representations

It is often useful to consider representations compatible with a scalar product on the vector space. Recall that a scalar product on a real vector space is called **Euclidean** if it is bilinear, symmetric and positive definite. A scalar product on a complex vector space is called **Hermitian** if it is sesquilinear (complex linear in the second argument and complex antilinear in the first argument), conjugate symmetric (exchanging the first and second argument changes the scalar product by complex conjugation) and positive definite.

Definition 2.1.35 A representation $\rho\colon G \to \mathrm{GL}(V)$ of a Lie group G on a Euclidean (or Hermitian) vector space $(V, \langle \cdot\,, \cdot \rangle)$ is called **orthogonal** (or **unitary**) if the scalar product is G-invariant, i.e.

$$\langle gv, gw \rangle = \langle \rho(g)v, \rho(g)w \rangle = \langle v, w \rangle,$$

for all $g \in G$, $v, w \in V$. Equivalently, the map ρ has image in the orthogonal subgroup $\mathrm{O}(V)$ (or the unitary subgroup $\mathrm{U}(V)$) of the general linear group $\mathrm{GL}(V)$, determined by the scalar product $\langle \cdot\,, \cdot \rangle$.

In an orthogonal representation the group literally acts through rotations (and possibly reflections) on a Euclidean vector space. There is a similar notion for representations of Lie algebras.

Definition 2.1.36 A representation $\phi\colon \mathfrak{g} \to \mathrm{End}(V)$ of a real Lie algebra $\mathfrak{g}$ on a Euclidean (or Hermitian) vector space $(V, \langle\cdot\,,\cdot\rangle)$ is called **skew-symmetric** (or **skew-Hermitian**) if it satisfies

$$\langle Xv, w\rangle + \langle v, Xw\rangle = \langle \phi(X)v, w\rangle + \langle v, \phi(X)w\rangle = 0,$$

for all $X \in \mathfrak{g}$, $v, w \in V$. Equivalently, the map ϕ has image in the orthogonal Lie subalgebra $\mathfrak{o}(V)$ (or the unitary Lie subalgebra $\mathfrak{u}(V)$) of the general linear algebra $\mathfrak{gl}(V)$, determined by the scalar product $\langle\cdot\,,\cdot\rangle$.
We can similarly define invariance of a form on a vector space under representations of a Lie group or Lie algebra in the case where the form is not non-degenerate or not positive definite.

Invariant scalar products for Lie group and Lie algebra representations are related:

Proposition 2.1.37 (Scalar Products and Induced Representations) *Let $\rho\colon G \to \mathrm{GL}(V)$ be a representation of a Lie group G and $\langle\cdot\,,\cdot\rangle$ a G-invariant Euclidean (or Hermitian) scalar product on V, i.e. the representation ρ is orthogonal (or unitary). Then the induced representation $\rho_*\colon \mathfrak{g} \to \mathrm{End}(V)$ of the Lie algebra $\mathfrak{g}$ is skew-symmetric (or skew-Hermitian).*

Proof We have by Theorem 1.7.16

$$\rho(\exp tX) = \exp(t\rho_* X)$$

and hence by Corollary 1.7.30

$$\begin{aligned}\langle v, w\rangle &= \langle \rho(\exp tX)v, \rho(\exp tX)w\rangle \\ &= \langle \exp(t\rho_* X)v, \exp(t\rho_* X)w\rangle \\ &= \left\langle e^{t\rho_* X}v, e^{t\rho_* X}w\right\rangle \quad \forall t \in \mathbb{R}.\end{aligned}$$

Differentiating both sides by t in $t = 0$ and using the product rule we get:

$$0 = \langle (\rho_* X)v, w\rangle + \langle v, (\rho_* X)w\rangle.$$

This implies the claim. □

Let $\phi\colon \mathfrak{g} \to \mathrm{End}(V)$ be a unitary representation of a real Lie algebra $\mathfrak{g}$ on a complex vector space V. Then $\phi(X)$ is a skew-Hermitian endomorphism for all $X \in \mathfrak{g}$, hence $i\phi(X)$ is Hermitian. This implies that the endomorphism $i\phi(X)$ can be diagonalized with real eigenvalues (and $\phi(X)$ can be diagonalized with imaginary eigenvalues).

Definition 2.1.38 The eigenvalues of $-i\phi(X)$ are called **charges** of $X \in \mathfrak{g}$ in the unitary representation ϕ.
The minus sign in $-i\phi(X)$ is convention: we can write $\phi(X)$ as iA_X, where A_X is a Hermitian operator, and the charges are the eigenvalues of A_X.

If $\mathfrak{h} \subset \mathfrak{g}$ is an abelian subalgebra, then the operators $i\phi(X)$ for all $X \in \mathfrak{h}$ commute and can be diagonalized simultaneously. This idea is related to the notion of *weights* of a representation and used extensively in the classification of representations of Lie algebras and Lie groups (in a certain sense, that can be made precise, irreducible representations are thus determined by their charges).

Existence of Invariant Scalar Products

It is an important fact that representations of *compact* Lie groups always admit *an invariant scalar product.*

Theorem 2.1.39 (Existence of Invariant Scalar Products for Representations of Compact Lie Groups) *Let G be a compact Lie group and $\rho: G \to \mathrm{GL}(V)$ a representation on a real (or complex) vector space. Then we can find a G-invariant Euclidean (or Hermitian) scalar product on V, hence the given representation ρ becomes orthogonal (or unitary) for this scalar product.*
The proof uses the existence of an **integral** over differential forms σ of top degree n on oriented n-manifolds M:

$$\int_M \sigma \in \mathbb{R}, \quad \sigma \in \Omega^n(M).$$

If $\phi: M \to N$ is an orientation preserving diffeomorphism between oriented n-manifolds, then we have the transformation formula

$$\int_N \sigma = \int_M \phi^* \sigma \quad \forall \sigma \in \Omega^n(N).$$

We now prove Theorem 2.1.39.

Proof Suppose G has dimension n and let $X_1, \ldots, X_n$ be a basis of T_eG. We set $\tilde{X}_i$ for the corresponding *right-invariant* vector fields on G, defined by

$$\tilde{X}_i(p) = D_eR_p(X_i) \quad \forall p \in G.$$

This basis has a dual basis of right-invariant 1-forms $\omega^1, \ldots, \omega^n$. Then the wedge product

$$\sigma = \omega^1 \wedge \cdots \wedge \omega^n$$

is a nowhere vanishing, right-invariant differential form on G of top degree. We can assume that the orientation of G coincides with the orientation defined by σ, so that

$$\int_G \sigma > 0,$$

which is finite, because G is compact. Let $\langle\langle\cdot,\cdot\rangle\rangle$ denote an arbitrary Euclidean (or Hermitian) scalar product on V. We construct a new scalar product by *averaging* this scalar product over the action of the group G:

$$\langle v, w\rangle = \int_G \tau_{v,w}\sigma,$$

where $\tau_{v,w}$ is the smooth function

$$\tau_{v,w}: G \longrightarrow \mathbb{R}$$
$$h \longmapsto \langle\langle hv, hw\rangle\rangle$$

(here the representation ρ is implicit and we use that G is compact, so that this integral is finite).

We claim that $\langle\cdot,\cdot\rangle$ is a G-invariant Euclidean (or Hermitian) scalar product on V: It is clear that $\langle\cdot,\cdot\rangle$ is bilinear and symmetric (or sesquilinear and conjugate symmetric in the complex case). For $v \neq 0$ the function $\tau_{v,v}$ is strictly positive on G. As a consequence the integral is

$$\langle v, v\rangle \geq 0 \quad \forall v \in V$$

with equality only if $v = 0$. Therefore $\langle\cdot,\cdot\rangle$ is a positive definite Euclidean (or Hermitian) scalar product on G.

We finally show G-invariance of the new scalar product: Let $g \in G$ be fixed. Then

$$R^*_{g^{-1}}\tau_{gv,gw} = \tau_{v,w} \quad \forall v, w \in V.$$

This follows from a short calculation:

$$\begin{aligned}(R^*_{g^{-1}}\tau_{gv,gw})(h) &= \tau_{gv,gw}(hg^{-1}) \\ &= \langle\langle hg^{-1}(gv), hg^{-1}(gw)\rangle\rangle \\ &= \tau_{v,w}(h),\end{aligned}$$

where we used that ρ (which is implicit) is a representation. This implies

$$R^*_{g^{-1}}(\tau_{gv,gw}\sigma) = \tau_{v,w}\sigma,$$

because σ is right-invariant. Since $R_{g^{-1}}$ is an orientation preserving diffeomorphism from G to G we get:

$$\begin{aligned}\langle gv, gw\rangle &= \int_G \tau_{gv,gw}\sigma \\ &= \int_G R^*_{g^{-1}}(\tau_{gv,gw}\sigma) \\ &= \int_G \tau_{v,w}\sigma \\ &= \langle v, w\rangle\end{aligned}$$

for all $g \in G$ and $v, w \in V$. □

Decomposition of Representations

The existence of an invariant scalar product for every representation of a compact Lie group has an important consequence.

Theorem 2.1.40 (Decomposition of Representations) *Let $\rho: G \to \mathrm{GL}(V)$ be a representation of a Lie group G on a finite-dimensional real (or complex) vector space V. Suppose that there exists a G-invariant Euclidean (or Hermitian) scalar product on V (this is always the case, by Theorem 2.1.39, if G is compact). Then V decomposes as a direct sum*

$$(V, \rho) = (V_1, \rho_1) \oplus \ldots \oplus (V_m, \rho_m)$$

of irreducible G-representations (V_i, ρ_i).

Proof The proof follows, because if $W \subset V$ is a subspace with $\rho(G)W \subset W$, then the orthogonal complement $W^\perp$ with respect to a G-invariant scalar product also satisfies $\rho(G)W^\perp \subset W^\perp$. We have

$$(V, \rho) = (W, \rho_W) \oplus (W^\perp, \rho_{W^\perp}).$$

We can thus continue splitting V until we arrive at irreducible representations (after finitely many steps, since V is finite-dimensional). □

Remark 2.1.41 One of the aims of representation theory for Lie groups G is to understand irreducible representations and to decompose any given representation (at least for compact G) into irreducible ones according to Theorem 2.1.40.

For instance, for $G = \mathrm{SU}(2)$, we can consider the tensor product representation $V_n \otimes V_m$, where V_n, V_m are the irreducible complex representations of dimension $n+1$ and $m+1$ mentioned in Example 2.1.20. The tensor product $V_n \otimes V_m$ is reducible under SU(2) and its decomposition into irreducible summands V_k is determined by the *Clebsch–Gordan formula*. This formula appears in quantum mechanics in the theory of the angular momentum of composite systems.

Remark 2.1.42 One of the basic topics in *Grand Unified Theories* is to study the restriction of representations of a compact Lie group G to embedded Lie subgroups $H \subset G$. If the representation ρ of G is irreducible, it may happen that the representation $\rho|_H$ of H is reducible and decomposes as a direct sum. The actual form of the decomposition of a representation ρ under restriction to a subgroup $H \subset G$ is called the **branching rule**.

For instance, there exist certain 5- and 10-dimensional irreducible representations of the Grand Unification group $G = \mathrm{SU}(5)$ that decompose under restriction to the subgroup $H = \mathrm{SU}(3) \times \mathrm{SU}(2) \times \mathrm{U}(1)$ (more precisely, to a certain $\mathbb{Z}_6$ quotient of this group; see Sect. 8.5.7) into the fermion representations of the Standard Model. Details of this calculation can be found in Sect. 9.5.4.

Remark 2.1.43 Suppose a Lie group G has a unitary representation on a complex vector space V and $e_1, \ldots, e_n$ is some orthonormal basis for V. If we decompose V into invariant, irreducible subspaces according to Theorem 2.1.40, then we can choose an associated orthonormal basis $f_1, \ldots, f_n$, adapted to the decomposition of V (spanning the G-invariant subspaces) and related to the original basis by a unitary matrix. In general, the basis $\{f_i\}$ will be different from $\{e_i\}$.

In the Standard Model where $G = \mathrm{SU}(3) \times \mathrm{SU}(2) \times \mathrm{U}(1)$ this is related to the concept of *quark mixing*. The complex vector space V of fermions, which carries a representation of G, has dimension 45 (plus the same number of corresponding antiparticles) and is the direct sum of two G-invariant subspaces (sectors): a lepton sector of dimension 9 (where we do not include the hypothetical right-handed neutrinos) and a quark sector of dimension 36. Counting in this way, the Standard Model thus contains at the most elementary level 90 fermions (particles and antiparticles).

The quark sector has a natural basis of so-called *mass eigenstates*, given by the quarks of six different flavours u, d, c, s, t, b, each one appearing in three different colours and two chiralities (6 basis vectors for each flavour), yielding in total 36 quarks. However, the basis given by these flavours does not define a splitting into subspaces invariant under SU(2). The SU(2)-invariant subspaces are spanned by a basis of so-called *weak eigenstates* that can be obtained from the mass eigenstates by a certain unitary transformation. The matrix of this unitary transformation is known as the *Cabibbo–Kobayashi–Maskawa (CKM) matrix*, which has to be determined by experiments. The CKM matrix and quark mixing will be explained in more detail in Sect. 8.8.2.

Unitary Representations of Non-Compact Lie Groups

It is an important fact that certain *non-compact* Lie groups do not admit non-trivial *finite-dimensional* unitary representations according to the following theorem (a proof can be found in [12, Chap. 8.1B]):

Theorem 2.1.44 *A connected, simple, non-compact Lie group does not admit finite-dimensional unitary complex representations except for the trivial representation.*
See Definition 2.4.27 for the notion of *simple* Lie groups. For example, the Lie group $G = \mathrm{SL}(2, \mathbb{C})$ is simple and non-compact, hence every non-trivial unitary representation of G is infinite-dimensional. This has important consequences for quantum field theory, see Sect. B.2.4. Of course, $\mathrm{SL}(2, \mathbb{C})$ admits non-trivial finite-dimensional *non-unitary* representations, like the fundamental representation on $\mathbb{C}^2$.

2.1.5 The Adjoint Representation

We want to define a particularly important representation of a Lie group and its Lie algebra. The vector space carrying the representation has the same dimension as the Lie group or Lie algebra (we follow [142] in this subsection).

Recall that for an element g of a Lie group G we defined the inner automorphism (conjugation)

$$\begin{aligned} c_g = L_g \circ R_{g^{-1}} \colon G &\longrightarrow G \\ x &\longmapsto gxg^{-1}. \end{aligned}$$

The differential $(c_g)_* \colon \mathfrak{g} \to \mathfrak{g}$ is an automorphism of the Lie algebra $\mathfrak{g}$, in particular a linear isomorphism.

Theorem 2.1.45 (Adjoint Representation of a Lie Group) *The map*

$$\begin{aligned} \mathrm{Ad} \colon G &\longrightarrow \mathrm{GL}(\mathfrak{g}) \\ g &\longmapsto \mathrm{Ad}(g) = \mathrm{Ad}_g = (c_g)_* \end{aligned}$$

is a Lie group homomorphism, i.e. a representation of the Lie group G on the vector space $\mathfrak{g}$, called the **adjoint representation** *or* **adjoint action of the Lie group** G*. We sometimes write Ad_G instead of* Ad.

Proof Note that

$$c_{gh} = c_g \circ c_h \quad \forall g, h \in G.$$

Hence

$$\mathrm{Ad}_{gh} = (c_{gh})_* = (c_g)_* \circ (c_h)_* = \mathrm{Ad}_g \circ \mathrm{Ad}_h.$$

This shows that Ad is a homomorphism in the algebraic sense. We have to show that Ad is a smooth map. It suffices to show that for every $v \in \mathfrak{g}$ the map

$$\mathrm{Ad}(\cdot)v \colon G \longrightarrow \mathfrak{g}$$

is smooth, because if we choose a basis for the vector space $\mathfrak{g}$, it follows that Ad is a smooth matrix representation. The map $\mathrm{Ad}(\cdot)v$ is equal to the composition of smooth maps

$$G \longrightarrow TG \times TG \longrightarrow T(G \times G) \longrightarrow TG$$

given by

$$g \longmapsto ((g, 0), (e, v)) \longmapsto ((g, e), (0, v)) \longmapsto D_{(g,e)}c(0, v),$$

where we set

$$c \colon G \times G \longrightarrow G$$
$$(g, x) \longmapsto gxg^{-1}.$$

This implies the claim. □

The following identity (whose proof is left as an exercise) is sometimes useful.

Proposition 2.1.46 *Let G be a Lie group with Lie algebra $\mathfrak{g}$ and ρ a representation of G on a vector space V with induced representation ρ_* of $\mathfrak{g}$. Then*

$$\rho_*(\mathrm{Ad}_g X) \circ \rho(g) = \rho(g) \circ \rho_*(X) \quad \forall X \in \mathfrak{g}.$$

Example 2.1.47 The adjoint representation is very simple in the case of abelian Lie groups G: if G is abelian, then $c_g = \mathrm{Id}_G$ for all $g \in G$ and thus $\mathrm{Ad}_g = \mathrm{Id}_{\mathfrak{g}}$ for all $g \in G$, hence the adjoint representation is a trivial representation.

We consider a more general example: Let $G \subset \mathrm{GL}(n, \mathbb{K})$ with $\mathbb{K} = \mathbb{R}, \mathbb{C}, \mathbb{H}$ be a closed subgroup of a general linear group with Lie algebra $\mathfrak{g}$. Fix $Q \in G$.

Proposition 2.1.48 (Adjoint Representation of Linear Groups) *The adjoint action*

$$\mathrm{Ad}_Q \colon \mathfrak{g} \longrightarrow \mathfrak{g}$$

is given by

$$\mathrm{Ad}_Q X = Q \cdot X \cdot Q^{-1},$$

where · denotes matrix multiplication and we identify elements $Q \in G$ and $X \in \mathfrak{g}$ with matrices in the canonical way.

Proof Define a curve $\gamma(t) = e^{tX}$ and take the derivative

$$\mathrm{Ad}_Q X = \left.\frac{d}{dt}\right|_{t=0} Q \cdot \gamma(t) \cdot Q^{-1} = Q \cdot X \cdot Q^{-1}.$$

□

In this situation, the Lie algebra $\mathfrak{g}$ on which the adjoint representation acts is naturally a vector space of *matrices*.

Example 2.1.49 We consider the adjoint representation of the Lie group SU(3). The Lie algebra $\mathfrak{su}(3)$ consists of the skew-Hermitian, tracefree matrices. As a real vector space, $\mathfrak{su}(3)$ has dimension 8 and is spanned by $i\lambda_a$, with $a = 1, \ldots, 8$, where λ_a are the Gell-Mann matrices from Example 1.5.33. We can define an explicit isomorphism

$$\mathbb{R}^8 \longrightarrow \mathfrak{su}(3)$$

$$G \longmapsto X = \sum_{a=1}^{8} iG_a\lambda_a = i \begin{pmatrix} G_3 + \frac{1}{\sqrt{3}}G_8 & G_1 - iG_2 & G_4 - iG_5 \\ G_1 + iG_2 & -G_3 + \frac{1}{\sqrt{3}}G_8 & G_6 - iG_7 \\ G_4 + iG_5 & G_6 + iG_7 & -\frac{2}{\sqrt{3}}G_8 \end{pmatrix}.$$

On such a matrix X the group element $Q \in \mathrm{SU}(3)$ acts as

$$\mathrm{Ad}_Q X = Q \cdot X \cdot Q^{-1}.$$

Using the isomorphism $\mathbb{R}^8 \cong \mathfrak{su}(3)$ we could write this as an explicit representation on $\mathbb{R}^8$.

The following observation is sometimes useful.

Lemma 2.1.50 (Adjoint Representation of Direct Product) *Let $G = H \times K$ be a direct product of Lie groups. Then the adjoint representation of G on $\mathfrak{g} = \mathfrak{h} \oplus \mathfrak{k}$ is the direct sum of the adjoint representations of H on $\mathfrak{h}$ and K on $\mathfrak{k}$:*

$$\mathrm{Ad}_{(h,k)}(X, Y) = (\mathrm{Ad}_h X, \mathrm{Ad}_k Y) \quad \forall (h,k) \in H \times K, (X, Y) \in \mathfrak{h} \oplus \mathfrak{k}.$$

Proof Let γ be a curve in H through e, tangent to $X \in \mathfrak{h}$. Then for $(h,k) \in H \times K$

$$\left.\frac{d}{dt}\right|_{t=0} (h,k)(\gamma(t), e)\left(h^{-1}, k^{-1}\right) = \left.\frac{d}{dt}\right|_{t=0} (h\gamma(t)h^{-1}, e)$$
$$= (\mathrm{Ad}_h X, 0).$$

Similarly for a vector in $\mathfrak{k}$. □

Example 2.1.51 We consider the adjoint representation of the Standard Model Lie group

$$H = \mathrm{SU}(3) \times \mathrm{SU}(2) \times \mathrm{U}(1).$$

We can write a group element $Q \in H$ as a block matrix

$$Q = \begin{pmatrix} Q_{\mathrm{SU}(3)} & & \\ & Q_{\mathrm{SU}(2)} & \\ & & Q_{\mathrm{U}(1)} \end{pmatrix},$$

with $Q_K \in K$ for $K = \mathrm{SU}(3), \mathrm{SU}(2), \mathrm{U}(1)$. We can similarly write the elements of the Lie algebra of H as a block matrix: with the notation from Examples 1.5.29, 1.5.32 and 1.5.33, the Lie algebra $\mathfrak{su}(3)$ is spanned by $i\lambda_a$, where λ_a are the Gell-Mann matrices, the Lie algebra $\mathfrak{su}(2)$ is spanned by $i\sigma_a$, where σ_a are the Pauli matrices, and the Lie algebra $\mathfrak{u}(1)$ is spanned by i. We can then define an isomorphism

$$\mathbb{R}^8 \oplus \mathbb{R}^3 \oplus \mathbb{R} \longrightarrow \mathfrak{su}(3) \oplus \mathfrak{su}(2) \oplus \mathfrak{u}(1)$$

$$(G, W, B) \longmapsto X = \left(\sum_{a=1}^{8} iG_a\lambda_a, \sum_{a=1}^{3} iW_a\sigma_a, iB \right)$$

$$= i \begin{pmatrix} G_3 + \frac{1}{\sqrt{3}}G_8 & G_1 - iG_2 & G_4 - iG_5 & & & \\ G_1 + iG_2 & -G_3 + \frac{1}{\sqrt{3}}G_8 & G_6 - iG_7 & & & \\ G_4 + iG_5 & G_6 + iG_7 & -\frac{2}{\sqrt{3}}G_8 & & & \\ & & & W_3 & W_1 - iW_2 & \\ & & & W_1 + iW_2 & -W_3 & \\ & & & & & B \end{pmatrix}.$$

According to Lemma 2.1.50 the adjoint action is given by multiplication of block matrices:

$$\mathrm{Ad}_Q X = Q \cdot X \cdot Q^{-1}.$$

The representation Ad_H describes the representation of the gauge boson fields in the Standard Model. The coefficients G_a, W_a and B (possibly with a different normalization) are known as the **gluon fields**, **weak gauge fields** and **hypercharge gauge field**; see Sect. 8.5.5 for more details.

Like any other representation of a Lie group, the adjoint representation of G induces a representation of the associated Lie algebra.

Theorem 2.1.52 (Adjoint Representation of a Lie Algebra) *The map*

$$\mathrm{ad}\colon \mathfrak{g} \longrightarrow \mathrm{End}(\mathfrak{g}),$$

given by

$$\mathrm{ad} = \mathrm{Ad}_*,$$

is a Lie algebra homomorphism, i.e. a representation of the Lie algebra $\mathfrak{g}$ *on the vector space* $\mathfrak{g}$*, called the* **adjoint representation of the Lie algebra** $\mathfrak{g}$. *We sometimes write* $\mathrm{ad}_{\mathfrak{g}}$ *instead of* ad. *We have the following commutative diagram according to Theorem 1.7.16:*

$$\begin{array}{ccc} \mathfrak{g} & \xrightarrow{\mathrm{ad}} & \mathrm{End}(\mathfrak{g}) \\ {\scriptstyle \exp}\downarrow & & \downarrow{\scriptstyle \exp} \\ G & \xrightarrow{\mathrm{Ad}} & \mathrm{GL}(\mathfrak{g}) \end{array}$$

The map ad *satisfies the formula*

$$\mathrm{ad}(X)(Y) = \mathrm{ad}_X Y = [X, Y] \quad \forall X, Y \in \mathfrak{g}.$$

Proof We only have to prove the formula $\mathrm{ad}_X Y = [X, Y]$. For left-invariant vector fields X, Y on G, where X has flow ϕ_t, we have according to the commutative diagram

$$\begin{aligned} \mathrm{ad}_X Y &= \left.\frac{d}{dt}\right|_{t=0} \mathrm{Ad}_{\exp tX} Y_e \\ &= \left.\frac{d}{dt}\right|_{t=0} (c_{\exp tX})_* Y_e \\ &= \left.\frac{d}{dt}\right|_{t=0} (R_{\exp -tX})_* (L_{\exp tX})_* Y_e \\ &= \left.\frac{d}{dt}\right|_{t=0} (R_{\exp -tX})_* Y_{\exp tX} \\ &= \left.\frac{d}{dt}\right|_{t=0} (\phi_{-t})_* Y_{\phi_t(e)} \\ &= [X, Y]_e. \end{aligned}$$

Here we used Proposition 1.7.12 and Theorem A.1.46. □

We can write the formula given by the commutative diagram as

$$\mathrm{Ad}_{\exp X} = e^{\mathrm{ad}_X} \quad \forall X \in \mathfrak{g}.$$

A direct consequence of Example 2.1.47 is the following:

Corollary 2.1.53 *If G is an abelian Lie group, then the adjoint representation* ad *is trivial, hence the Lie algebra* $\mathfrak{g}$ *is abelian.*
It can be shown that the converse also holds (for connected Lie groups), cf. Exercise 2.7.7.

Remark 2.1.54 We can *define* for any Lie algebra $\mathfrak{g}$, even if it does not belong *a priori* to a Lie group, the map

$$\mathrm{ad}\colon \mathfrak{g} \longrightarrow \mathrm{End}(\mathfrak{g}),$$

by exactly the same formula

$$\mathrm{ad}_X Y = [X, Y] \quad \forall X, Y \in \mathfrak{g}.$$

Then this map is a representation of $\mathfrak{g}$ (by the Jacobi identity), again called the adjoint representation.

Remark 2.1.55 One should be careful not to confuse the fundamental and the adjoint representation for a linear group. In general, the dimensions are already different. For example, in the case of $\mathrm{SU}(n)$ the dimension of the fundamental representation is n, while the adjoint representation has dimension $n^2 - 1$. For a linear group the fundamental representation acts canonically on a vector space of column vectors, while the adjoint representation acts on a vector space of matrices.

Example 2.1.56 The homomorphism $\phi\colon S^3 \to \mathrm{SO}(3)$ from Example 1.3.8 is the adjoint representation of $S^3 = \mathrm{SU}(2)$.

2.2 Invariant Metrics on Lie Groups

Since a Lie group G is a manifold, we can study metrics (Riemannian or pseudo-Riemannian) on it. We are interested in particular in the following types of metrics.

Definition 2.2.1 Let s be a metric on a Lie group G.

1. The metric s is called
   - **left-invariant** if $L_g^* s = s$ for all $g \in G$
   - **right-invariant** if $R_g^* s = s$ for all $g \in G$.

   Equivalently, either all left translations or all right translations are isometries.
2. The metric s is called **bi-invariant** if it is both left- and right-invariant.

It is clear that every metric induces a scalar product on $\mathfrak{g} \cong T_eG$. On the other hand, given an arbitrary scalar product $\langle \cdot\,, \cdot \rangle$ on $\mathfrak{g}$, it is easy to construct

- a *left-invariant metric* on G by

$$s(X, Y) = \langle L_{g^{-1}*}(X), L_{g^{-1}*}(Y) \rangle$$

- a *right-invariant metric* on G by

$$s(X, Y) = \langle R_{g^{-1}*}(X), R_{g^{-1}*}(Y) \rangle,$$

for all $g \in G$ and $X, Y \in T_g G$.

However, in general we only get a *bi-invariant metric* in this way if G is abelian (if G is not abelian, then $L_g \neq R_g$ for some $g \in G$). Bi-invariant metrics have the following characterization:

Theorem 2.2.2 (Bi-Invariant Metrics and Ad**-Invariance)** *Let s be a left-invariant metric on a Lie group G. Then s is bi-invariant if and only if the scalar product $\langle \cdot, \cdot \rangle$ on $\mathfrak{g}$ defined by the metric s is* Ad*-invariant, i.e.*

$$\langle \mathrm{Ad}_g v, \mathrm{Ad}_g w \rangle = \langle v, w \rangle$$

for all $g \in G$ and $v, w \in \mathfrak{g}$.

Proof Let X and Y be vectors in $T_p G$. Then we can calculate:

$$\begin{aligned}(R_g^* s)_p(X, Y) &= \langle L_{(pg)^{-1}*} R_{g*}(X), L_{(pg)^{-1}*} R_{g*}(Y) \rangle \\ &= \langle \mathrm{Ad}_{g^{-1}} \circ L_{p^{-1}*}(X), \mathrm{Ad}_{g^{-1}} \circ L_{p^{-1}*}(Y) \rangle\end{aligned}$$

and

$$s_p(X, Y) = \langle L_{p^{-1}*}(X), L_{p^{-1}*}(Y) \rangle,$$

where in both equations we used that s is left-invariant. This implies the claim, because $L_{p^{-1}*}$ is an isomorphism of vector spaces. □

Theorem 2.2.3 (Ad**-Invariant Scalar Products for Compact Lie Groups)** *Let G be a compact Lie group. Then there exists a Euclidean (positive definite) scalar product $\langle \cdot, \cdot \rangle$ on the Lie algebra $\mathfrak{g}$ which is* Ad*-invariant. The adjoint representation is orthogonal with respect to this scalar product.*

Proof This follows from Theorem 2.1.39, because Ad is a representation of the compact Lie group G on the vector space $\mathfrak{g}$. □

The existence of positive definite Ad-invariant scalar products on the Lie algebra of compact Lie groups is very important in gauge theory, in particular, for the construction of the gauge-invariant *Yang–Mills Lagrangian*; see Sect. 7.3.1. We will study such scalar products in more detail in Sect. 2.5 after we have discussed the general structure of compact Lie groups. The

(continued)

fact that these scalar products are positive definite is important from a phenomenological point of view, because only then do the kinetic terms in the Yang–Mills Lagrangian have the right sign (the gauge bosons have positive kinetic energy [148]).

Here is a corollary to Theorem 2.2.2 and Theorem 2.2.3:

Corollary 2.2.4 *Every compact Lie group admits a bi-invariant Riemannian metric.*

Remark 2.2.5 It can be shown that the geodesics of a bi-invariant metric on a Lie group G through the neutral element e are of the form $\gamma(t) = \exp(tX)$, with $X \in \mathfrak{g}$. The notions of exponential map for geodesics and Lie groups thus coincide for bi-invariant Riemannian metrics.

2.3 The Killing Form

We want to consider a special Ad-invariant inner product on every Lie algebra $\mathfrak{g}$, which in general is neither non-degenerate nor positive or negative definite. This is the celebrated Killing form.

Theorem 2.3.1 *Let $\mathfrak{g}$ be a Lie algebra over $\mathbb{K} = \mathbb{R}, \mathbb{C}$. The* **Killing form** *$B_{\mathfrak{g}}$ on $\mathfrak{g}$ is defined by*

$$\begin{aligned} B_{\mathfrak{g}} \colon \mathfrak{g} \times \mathfrak{g} &\longrightarrow \mathbb{K} \\ (X, Y) &\longmapsto \mathrm{tr}(\mathrm{ad}_X \circ \mathrm{ad}_Y). \end{aligned}$$

This is a $\mathbb{K}$-bilinear, symmetric form on $\mathfrak{g}$.

Remark 2.3.2 Note that the Killing form for complex Lie algebras is also symmetric and complex bilinear and not Hermitian.

Proof For $Z \in \mathfrak{g}$ we have

$$\mathrm{ad}_X \circ \mathrm{ad}_Y(Z) = [X, [Y, Z]].$$

In particular, $B_{\mathfrak{g}}$ is indeed bilinear. To show that the Killing form is symmetric, recall the definition of the **trace** $\mathrm{tr}(f)$ of a linear endomorphism f of a vector space V: If $v_1, \ldots, v_n$ is a basis of V and we define the representing matrix of f by

$$f(v_j) = \sum_{i=1}^{n} f_{ij} v_i,$$

then

$$\mathrm{tr}(f) = \sum_{i=1}^{n} f_{ii}.$$

This number does not depend on the choice of basis for V: If $\phi: V \to V$ is an arbitrary isomorphism, then

$$\mathrm{tr}\left(\phi \circ f \circ \phi^{-1}\right) = \mathrm{tr}(f).$$

We also have

$$\mathrm{tr}(f \circ g) = \mathrm{tr}(g \circ f)$$

for all endomorphisms $f, g: V \to V$. This shows, in particular, that $B_{\mathfrak{g}}$ is symmetric. □

Theorem 2.3.3 (Invariance of Killing Form Under Automorphisms) *Let $\sigma: \mathfrak{g} \to \mathfrak{g}$ be a Lie algebra automorphism of $\mathfrak{g}$. Then the Killing form $B_{\mathfrak{g}}$ satisfies*

$$B_{\mathfrak{g}}(\sigma X, \sigma Y) = B_{\mathfrak{g}}(X, Y) \quad \forall X, Y \in \mathfrak{g}.$$

If $\mathfrak{g}$ is the Lie algebra of a Lie group G, this holds in particular for the automorphism $\sigma = \mathrm{Ad}_g$ with $g \in G$ arbitrary.

Proof Note that

$$\mathrm{ad}_X Y = [X, Y].$$

Since σ is a Lie algebra automorphism we have

$$\mathrm{ad}_{\sigma X} Y = [\sigma X, Y] = \sigma([X, \sigma^{-1} Y] = \sigma \circ \mathrm{ad}_X(\sigma^{-1} Y).$$

Thus

$$\mathrm{ad}_{\sigma X} = \sigma \circ \mathrm{ad}_X \circ \sigma^{-1}.$$

We get for the Killing form:

$$\begin{aligned} B_{\mathfrak{g}}(\sigma X, \sigma Y) &= \mathrm{tr}(\mathrm{ad}_{\sigma X} \circ \mathrm{ad}_{\sigma Y}) \\ &= \mathrm{tr}\left(\sigma \circ \mathrm{ad}_X \circ \mathrm{ad}_Y \circ \sigma^{-1}\right) \\ &= B_{\mathfrak{g}}(X, Y). \end{aligned}$$

□

Corollary 2.3.4 *The Killing form $B_{\mathfrak{g}}$ defines a bi-invariant symmetric form on any Lie group G.*

Remark 2.3.5 We will determine in Sect. 2.4 when the Killing form is non-degenerate or definite (in the case of a real Lie algebra).

Proposition 2.3.6 (ad Is Skew-Symmetric with Respect to the Killing Form) *Let $\mathfrak{g}$ be a Lie algebra with Killing form $B_{\mathfrak{g}}$. Then*

$$B_{\mathfrak{g}}(\mathrm{ad}_X Y, Z) + B_{\mathfrak{g}}(Y, \mathrm{ad}_X Z) = 0 \quad \forall X, Y, Z \in \mathfrak{g}.$$

Proof This follows from Theorem 2.3.3 and Proposition 2.1.37 if $\mathfrak{g}$ is the Lie algebra of a Lie group G. In the general case we use the formula

$$\mathrm{ad}_{\mathrm{ad}_X Y} = \mathrm{ad}_X \circ \mathrm{ad}_Y - \mathrm{ad}_Y \circ \mathrm{ad}_X \quad \forall X, Y \in \mathfrak{g},$$

which follows from the Jacobi identity. The definition of the Killing form implies

$$\begin{aligned} B_{\mathfrak{g}}(\mathrm{ad}_X Y, Z) + B_{\mathfrak{g}}(Y, \mathrm{ad}_X Z) &= \mathrm{tr}(\mathrm{ad}_X \circ \mathrm{ad}_Y \circ \mathrm{ad}_Z) - \mathrm{tr}(\mathrm{ad}_Y \circ \mathrm{ad}_Z \circ \mathrm{ad}_X) \\ &= 0, \end{aligned}$$

because the trace is invariant under cyclic permutations. □

2.4 *Semisimple and Compact Lie Algebras

In this section we discuss some results concerning the general structure of Lie algebras and Lie groups (we follow [83] and [153]). There are two elements that play a key role in the theory of Lie algebras:

- The adjoint representation $\mathrm{ad}_{\mathfrak{g}}$ of the Lie algebra $\mathfrak{g}$, together with its invariant subspaces, known as *ideals*.
- The Killing form $B_{\mathfrak{g}}$ of $\mathfrak{g}$.

Both notions are related: the definition of the Killing form $B_{\mathfrak{g}}$ involves the adjoint representation $\mathrm{ad}_{\mathfrak{g}}$ and the adjoint representation is skew-symmetric with respect to the Killing form.

The idea is to proceed in a similar way to Theorem 2.1.40 and try to decompose $\mathfrak{g}$ with the adjoint representation into irreducible, pairwise $B_{\mathfrak{g}}$-orthogonal pieces. This works out particularly well for a type of Lie algebra known as a *semisimple* Lie algebra. The next step is to classify the pieces where the adjoint representation is irreducible. These are called *simple* Lie algebras. We will discuss the classification for the simple Lie algebras coming from compact Lie groups, which appear in physics as gauge groups.

2.4.1 Simple and Semisimple Lie Algebras in General

Definition 2.4.1 Let $\mathfrak{g}$ be a Lie algebra. For subsets $\mathfrak{a}, \mathfrak{b} \subset \mathfrak{g}$ we define $[\mathfrak{a}, \mathfrak{b}] \subset \mathfrak{g}$ as the set of all finite sums of elements of the form $[X, Y]$ with $X \in \mathfrak{a}$, $Y \in \mathfrak{b}$.

Definition 2.4.2 Let $\mathfrak{g}$ be a Lie algebra.

1. An **ideal** in $\mathfrak{g}$ is a vector subspace $\mathfrak{a} \subset \mathfrak{g}$ such that $[\mathfrak{g}, \mathfrak{a}] \subset \mathfrak{a}$. Equivalently,

$$\mathrm{ad}_{\mathfrak{g}}\mathfrak{a} \subset \mathfrak{a}.$$

2. The **center** of $\mathfrak{g}$ is defined as

$$\mathfrak{z}(\mathfrak{g}) = \{X \in \mathfrak{g} \mid [X, \mathfrak{g}] \equiv 0\}.$$

3. The **commutator** of $\mathfrak{g}$ is defined as $[\mathfrak{g}, \mathfrak{g}]$.

The following is easy to check.

Lemma 2.4.3 *For any Lie algebra the commutator is an ideal and the center is an abelian ideal.*

Proposition 2.4.4 *The kernel of the adjoint representation of a Lie algebra $\mathfrak{g}$ is the center $\mathfrak{z}(\mathfrak{g})$. The adjoint representation is faithful if and only if $\mathfrak{z}(\mathfrak{g}) = 0$.*

Proof We have $\mathrm{ad}_X \equiv 0$ if and only if $[X, \mathfrak{g}] \equiv 0$. □

This implies Ado's Theorem 1.5.25 for Lie algebras with trivial center.

Definition 2.4.5 Let $\mathfrak{g}$ be a Lie algebra.

1. The Lie algebra $\mathfrak{g}$ is called **simple** if $\mathfrak{g}$ is non-abelian and $\mathfrak{g}$ has no non-trivial ideals (different from 0 and $\mathfrak{g}$).
2. The Lie algebra $\mathfrak{g}$ is called **semisimple** if $\mathfrak{g}$ has no non-zero abelian ideals.

Simple Lie algebras are sometimes defined equivalently as follows:

Lemma 2.4.6 *A Lie algebra $\mathfrak{g}$ is simple if and only if $\mathfrak{g}$ has dimension at least two and $\mathfrak{g}$ has no non-trivial ideals.*

Proof If $\mathfrak{g}$ is non-abelian, then it has dimension at least two. On the other hand, if $\mathfrak{g}$ is abelian and has dimension at least two, then $\mathfrak{g}$ has non-trivial (abelian) ideals. □

It is clear that every simple Lie algebra is semisimple.

Lemma 2.4.7 *If $\mathfrak{g}$ is simple, then $[\mathfrak{g}, \mathfrak{g}] = \mathfrak{g}$.*

Proof The commutator $[\mathfrak{g}, \mathfrak{g}]$ is an ideal, hence equal to $\mathfrak{g}$ or 0. The second possibility is excluded, because $\mathfrak{g}$ is not abelian. □

We can characterize simple Lie algebras as follows:

Proposition 2.4.8 (Criterion for Simplicity) *A Lie algebra $\mathfrak{g}$ is simple if and only if $\mathfrak{g}$ is non-abelian and the adjoint representation $\mathrm{ad}_{\mathfrak{g}}$ of $\mathfrak{g}$ is irreducible.*

Proof The claim follows from the definition of an ideal. □

We can also characterize semisimple Lie algebras (we only prove one direction following [83]; the proof of the converse, which would take us too far afield, can be found in [77, 83]):

Theorem 2.4.9 (Cartan's Criterion for Semisimplicity) *A Lie algebra $\mathfrak{g}$ is semisimple if and only if the Killing form $B_\mathfrak{g}$ is non-degenerate.*

Proof We only prove that the Killing form is degenerate if the Lie algebra is not semisimple. Let $\mathfrak{a}$ be a non-zero abelian ideal in $\mathfrak{g}$. We choose a complementary vector space $\mathfrak{s}$ with

$$\mathfrak{g} = \mathfrak{a} \oplus \mathfrak{s}.$$

Let $X \in \mathfrak{a}$ and $Y \in \mathfrak{g}$ be arbitrary elements. Then

$$[X, \mathfrak{a}] = 0,$$
$$[X, \mathfrak{s}] \subset \mathfrak{a},$$
$$[Y, \mathfrak{a}] \subset \mathfrak{a}.$$

Under the splitting $\mathfrak{g} = \mathfrak{a} \oplus \mathfrak{s}$, the endomorphisms ad_X and ad_Y thus have the form

$$\mathrm{ad}_X = \begin{pmatrix} 0 & * \\ 0 & 0 \end{pmatrix},$$

$$\mathrm{ad}_Y = \begin{pmatrix} * & * \\ 0 & * \end{pmatrix}.$$

It follows that

$$\mathrm{ad}_X \circ \mathrm{ad}_Y = \begin{pmatrix} 0 & * \\ 0 & 0 \end{pmatrix}$$

and

$$B_\mathfrak{g}(X, Y) = \mathrm{tr}(\mathrm{ad}_X \circ \mathrm{ad}_Y) = 0.$$

□

Remark 2.4.10 In general, the Killing form of a semisimple Lie algebra is indefinite, i.e. pseudo-Euclidean.

Assuming Cartan's Criterion we can prove the following.

Theorem 2.4.11 (Structure of Semisimple Lie Algebras) *If a Lie algebra $\mathfrak{g}$ is semisimple, then $\mathfrak{g}$ is the direct sum*

$$\mathfrak{g} = \mathfrak{g}_1 \oplus \ldots \oplus \mathfrak{g}_s$$

of ideals $\mathfrak{g}_i$, each of which is a simple Lie algebra, and which are pairwise orthogonal with respect to the Killing form.

Proof We ultimately would like to apply Theorem 2.1.40 and decompose the adjoint representation on $\mathfrak{g}$ into irreducible summands, orthogonal with respect to the Killing form. There is one problem which requires some work: the Killing form $B = B_{\mathfrak{g}}$ is non-degenerate, but not (positive or negative) definite. Therefore it is not immediately clear that orthogonal complements of invariant subspaces lead to a direct sum decomposition.

Let $\mathfrak{a}$ be an ideal in $\mathfrak{g}$ and

$$\mathfrak{a}^{\perp} = \{X \in \mathfrak{g} \mid B(X, Y) = 0 \quad \forall Y \in \mathfrak{a}\}$$

the orthogonal complement with respect to the Killing form B. Then $\mathfrak{a}^{\perp}$ is also an ideal in $\mathfrak{g}$, because

$$B(\mathrm{ad}_{\mathfrak{g}}\mathfrak{a}^{\perp}, \mathfrak{a}) = -B(\mathfrak{a}^{\perp}, \mathrm{ad}_{\mathfrak{g}}\mathfrak{a}) \subset B(\mathfrak{a}^{\perp}, \mathfrak{a}) = 0,$$

by Proposition 2.3.6. Furthermore, $\mathfrak{b} = \mathfrak{a} \cap \mathfrak{a}^{\perp}$ is an abelian ideal in $\mathfrak{g}$: it is clear that the intersection of two ideals is an ideal and

$$B(\mathrm{ad}_{\mathfrak{b}}\mathfrak{b}, \mathfrak{g}) = -B(\mathfrak{b}, \mathrm{ad}_{\mathfrak{b}}\mathfrak{g}) \subset B(\mathfrak{b}, \mathfrak{b}) = 0.$$

This implies that $\mathfrak{b}$ is abelian, because B is non-degenerate. Since $\mathfrak{g}$ is semisimple, it follows that $\mathfrak{a} \cap \mathfrak{a}^{\perp} = 0$.

This implies

$$\mathfrak{g} = \mathfrak{a} \oplus \mathfrak{a}^{\perp}$$

and the restriction of the Killing form to $\mathfrak{a}$ and $\mathfrak{a}^{\perp}$ (which is just the Killing form on these Lie algebras) is non-degenerate. We can continue splitting the (finite-dimensional) Lie algebra $\mathfrak{g}$ in this fashion until we arrive at irreducible, non-abelian (simple) ideals. □

Remark 2.4.12 In addition to semisimple and abelian Lie algebras there are other classes of Lie algebras, like solvable and nilpotent Lie algebras, which we have not discussed in detail.

2.4.2 Compact Lie Algebras

We are particularly interested in *compact* Lie algebras, including compact simple and compact semisimple Lie algebras.

Definition 2.4.13 A real Lie algebra $\mathfrak{g}$ is called **compact** if it is the Lie algebra of some compact Lie group.

Remark 2.4.14 Even if $\mathfrak{g}$ is compact, there could exist non-compact Lie groups whose Lie algebra is also $\mathfrak{g}$. For example, the abelian Lie algebra $\mathfrak{u}(1)$ is the Lie algebra of the compact Lie group $\mathrm{U}(1) = S^1$ and of the non-compact Lie group $\mathbb{R}$.

Example 2.4.15 Note that the abelian Lie algebra $\mathbb{R}^n = \mathfrak{u}(1)\oplus \ldots \oplus\mathfrak{u}(1)$, for $n \geq 1$, is compact, but neither simple nor semisimple.

Theorem 2.4.16 (Killing Form of Compact Lie Algebras) *Suppose* $\mathfrak{g}$ *is a compact real Lie algebra. Then the Killing form* $B_{\mathfrak{g}}$ *is negative semidefinite: We have*

$$
\begin{aligned}
B_{\mathfrak{g}}(X,X) &= 0 \quad \forall X \in \mathfrak{z}(\mathfrak{g}),\\
B_{\mathfrak{g}}(X,X) &< 0 \quad \forall X \in \mathfrak{g} \setminus \mathfrak{z}(\mathfrak{g}).
\end{aligned}
$$

Proof We follow the proof in [14]. Since $\mathfrak{g}$ is the Lie algebra of a compact Lie group G, according to Theorem 2.2.3 there exists a positive definite scalar product $\langle \cdot, \cdot \rangle$ on $\mathfrak{g}$ which is Ad_G-invariant. Let $e_1, \ldots, e_n$ be an orthonormal basis for $\mathfrak{g}$ with respect to this scalar product. We get

$$
\langle \mathrm{ad}_X \circ \mathrm{ad}_X Y, Y \rangle = -||\mathrm{ad}_X Y||^2 \quad \forall X, Y \in \mathfrak{g}
$$

for the associated norm $||\cdot||$. This implies

$$
\begin{aligned}
B_{\mathfrak{g}}(X,X) &= \mathrm{tr}(\mathrm{ad}_X \circ \mathrm{ad}_X)\\
&= \sum_{i=1}^{n} \langle \mathrm{ad}_X \circ \mathrm{ad}_X e_i, e_i \rangle\\
&= -\sum_{i=1}^{n} ||\mathrm{ad}_X e_i||^2\\
&\leq 0.
\end{aligned}
$$

Equality holds if and only if $\mathrm{ad}_X \equiv 0$ on $\mathfrak{g}$, i.e. $X \in \mathfrak{z}(\mathfrak{g})$. □

Remark 2.4.17 Note as an aside that the notion of a bilinear, symmetric form being (semi-)definite is only meaningful on real and not on complex vector spaces.

Corollary 2.4.18 *Let* $\mathfrak{g}$ *be a compact Lie algebra with trivial center,* $\mathfrak{z}(\mathfrak{g}) = 0$. *Then the Killing form* $B_{\mathfrak{g}}$ *is negative definite.*

Proof This follows from Theorem 2.4.16. □

Remark 2.4.19 The following converse to Corollary 2.4.18 can be proved (see [77]): if the Killing form of a real Lie algebra is negative definite, then it is compact with trivial center. In particular, every Lie subalgebra of a compact Lie algebra is compact.

Corollary 2.4.20 *Let $\mathfrak{g}$ be a compact Lie algebra. Then the Killing form $B_{\mathfrak{g}}$ is negative definite if and only if $\mathfrak{g}$ is semisimple.*

Proof One direction follows from Corollary 2.4.18, because semisimple Lie algebras have trivial center. The other direction follows from Theorem 2.4.9. □

Theorem 2.4.21 (Decomposition of Compact Lie Algebras) *Let $\mathfrak{g}$ be a compact Lie algebra with center $\mathfrak{z}(\mathfrak{g})$. Then there exists an ideal $\mathfrak{h}$ in $\mathfrak{g}$ such that*

$$\mathfrak{g} = \mathfrak{z}(\mathfrak{g}) \oplus \mathfrak{h}.$$

The ideal $\mathfrak{h}$ is a compact semisimple Lie algebra with negative definite Killing form.

Proof Choose a positive definite scalar product $\langle \cdot , \cdot \rangle$ on $\mathfrak{g}$ which is Ad_G-invariant. Let $\mathfrak{h}$ be the orthogonal complement

$$\mathfrak{h} = \mathfrak{z}(\mathfrak{g})^{\perp}$$

with respect to this scalar product. Then $\mathfrak{h}$ is an ideal, because

$$\langle \mathrm{ad}_{\mathfrak{g}}\mathfrak{h}, \mathfrak{z}(\mathfrak{g}) \rangle = -\langle \mathfrak{h}, \mathrm{ad}_{\mathfrak{g}}\mathfrak{z}(\mathfrak{g}) \rangle = 0.$$

It is clear that

$$\mathfrak{g} = \mathfrak{z}(\mathfrak{g}) \oplus \mathfrak{h}.$$

By Theorem 2.4.16 the Killing form is negative definite on $\mathfrak{h}$, which is thus compact by Remark 2.4.19 and semisimple by Theorem 2.4.9. □

Corollary 2.4.22 (Structure of Compact Lie Algebras) *Let $\mathfrak{g}$ be a compact Lie algebra. Then $\mathfrak{g}$ is a direct sum of ideals*

$$\mathfrak{g} = \mathfrak{u}(1) \oplus \ldots \oplus \mathfrak{u}(1) \oplus \mathfrak{g}_1 \oplus \ldots \oplus \mathfrak{g}_s,$$

where the $\mathfrak{g}_i$ are compact simple Lie algebras.

Proof This follows from Theorem 2.4.11. The Lie algebras $\mathfrak{g}_i$ are compact by Remark 2.4.19. □

Using considerable effort it is possible to classify simple Lie algebras, one of the great achievements of 19th and 20th century mathematics. The result for compact simple Lie algebras is the following (see [83] for a proof):

Theorem 2.4.23 (Killing–Cartan Classification of Compact Simple Lie Algebras) *Every compact simple Lie algebra is isomorphic to precisely one of the following Lie algebras:*

1. $\mathfrak{su}(n+1)$ *for* $n \geq 1$.
2. $\mathfrak{so}(2n+1)$ *for* $n \geq 2$.
3. $\mathfrak{sp}(n)$ *for* $n \geq 3$.
4. $\mathfrak{so}(2n)$ *for* $n \geq 4$.
5. *An exceptional Lie algebra of type* $\mathrm{G}_2, \mathrm{F}_4, \mathrm{E}_6, \mathrm{E}_7, \mathrm{E}_8$.

The families in the first four cases are also called $\mathrm{A}_n, \mathrm{B}_n, \mathrm{C}_n, \mathrm{D}_n$ *in this order.*

Remark 2.4.24 The lower index n in the series $\mathrm{A}_n, \mathrm{B}_n, \mathrm{C}_n, \mathrm{D}_n$ as well as in the exceptional cases $\mathrm{G}_2, \mathrm{F}_4, \mathrm{E}_6, \mathrm{E}_7, \mathrm{E}_8$ is the **rank** of the corresponding compact Lie group, i.e. the dimension of a **maximal torus subgroup** embedded in the Lie group.

Remark 2.4.25 The reason for the restrictions on n in the first four cases of the classical Lie algebras is to avoid counting Lie algebras twice, because we have the following isomorphisms (we only proved the first isomorphism in Sect. 1.5.5):

$$\begin{aligned} \mathfrak{so}(3) &\cong \mathfrak{su}(2) \cong \mathfrak{sp}(1), \\ \mathfrak{sp}(2) &\cong \mathfrak{so}(5), \\ \mathfrak{so}(6) &\cong \mathfrak{su}(4). \end{aligned}$$

There is also the abelian Lie algebra

$$\mathfrak{so}(2)$$

and the semisimple Lie algebra

$$\mathfrak{so}(4) \cong \mathfrak{su}(2) \oplus \mathfrak{su}(2),$$

cf. Exercise 1.9.21.
The basic building blocks of all compact Lie algebras are thus

- abelian Lie algebras
- the four families of classical compact non-abelian Lie algebras
- five exceptional compact Lie algebras.

In some sense, most compact Lie algebras are therefore classical or direct sums of classical Lie algebras.

It is sometimes convenient to know that we can choose for a compact semisimple Lie algebra a basis in such a way that the structure constants (see Definition 1.4.17) have a nice form. Let $\mathfrak{g}$ be a compact semisimple Lie algebra. According to Corollary 2.4.20 the Killing form $B_\mathfrak{g}$ is negative definite. Let $T_1, \ldots, T_n$ be an orthonormal basis of $\mathfrak{g}$ with respect to the Killing form:

$$B_\mathfrak{g}(T_a, T_b) = -\delta_{ab} \quad \forall a, b \in \{1, \ldots, n\}.$$

Proposition 2.4.26 *The structure constants f_{abc} for a $B_\mathfrak{g}$-orthonormal basis $\{T_a\}$ of a semisimple Lie algebra $\mathfrak{g}$ are totally antisymmetric:*

$$f_{abc} = -f_{bac} = f_{bca} = -f_{acb} \quad \forall a, b, c \in \{1, \ldots, n\}.$$

Proof This is Exercise 2.7.11. □

2.4.3 Compact Lie Groups

We briefly discuss the structure of compact Lie groups.

Definition 2.4.27 A connected Lie group G is called **simple** (or **semisimple**) if its Lie algebra is simple (or semisimple).

Corollary 2.4.28 *If G is simple, then Ad_G is an irreducible representation.*

Proof The claim follows from Proposition 2.4.8 because an Ad_G-invariant subspace in $\mathfrak{g}$ is also $\mathrm{ad}_\mathfrak{g}$-invariant. □

A proof of the following theorem can be found in [77].

Theorem 2.4.29 (Structure of Compact Lie Groups) *Let G be a compact connected Lie group. Then G is a finite quotient of a product of the form*

$$\tilde{G} \cong \mathrm{U}(1) \times \ldots \times \mathrm{U}(1) \times G_1 \times \ldots \times G_s,$$

where the G_i are compact simple Lie groups.

Compact simple Lie groups and the abelian Lie group U(1) are therefore the building blocks of all compact connected Lie groups.

2.5 *Ad-Invariant Scalar Products on Compact Lie Groups

We know from Theorem 2.2.3 that compact Lie algebras admit scalar products that are invariant under the adjoint action. Such scalar products are important in gauge theory: they are necessary ingredients to construct the gauge-invariant Yang–Mills action and are related to the notion of *coupling constants*. We discuss, in particular, how to fix an Ad-invariant scalar product and how many different ones exist on a given compact Lie algebra.

We first consider Ad-invariant scalar products on compact *simple* Lie algebras. We need the following variant of a famous theorem of Schur.

Theorem 2.5.1 (Schur's Lemma for Scalar Products) *Let $\rho: G \to \mathrm{GL}(V)$ be an irreducible representation of a Lie group G on a real vector space V and $\langle\cdot,\cdot\rangle_1$, $\langle\cdot,\cdot\rangle_2$ two G-invariant symmetric bilinear forms on V, so that $\langle\cdot,\cdot\rangle_2$ is positive definite. Then there exists a real number $a \in \mathbb{R}$ such that*

$$\langle\cdot,\cdot\rangle_1 = a\langle\cdot,\cdot\rangle_2.$$

Remark 2.5.2 The assumption that the group representation is irreducible is important.

Proof We follow the proof in [153]. Let $L: V \to V$ be the unique linear map defined by (using non-degeneracy of the second scalar product)

$$\langle v, w\rangle_1 = \langle v, Lw\rangle_2 \quad \forall v, w \in V.$$

We have

$$\begin{aligned}\langle w, Lv\rangle_2 &= \langle w, v\rangle_1\\ &= \langle v, w\rangle_1\\ &= \langle v, Lw\rangle_2,\end{aligned}$$

hence L is self-adjoint with respect to the second scalar product. We can split V into the eigenspaces of L which are orthogonal with respect to the second scalar product. Since both bilinear forms are G-invariant we have

$$\begin{aligned}\langle gv, gLw\rangle_2 &= \langle v, Lw\rangle_2\\ &= \langle v, w\rangle_1\\ &= \langle gv, gw\rangle_1\\ &= \langle gv, L(gw)\rangle_2.\end{aligned}$$

We conclude that $\rho(g)\circ L = L\circ\rho(g)$ for all $g \in G$ and thus the eigenspaces of L are G-invariant. Since the representation ρ is irreducible, V itself must be an eigenspace and hence $L = a\cdot \mathrm{Id}_V$. This implies the claim. □

Theorem 2.5.3 (Ad-Invariant Scalar Products on Compact Simple Lie Algebras) *Let G be a compact simple Lie group. Then there exists up to a positive factor a unique* Ad*-invariant positive definite scalar product on the Lie algebra $\mathfrak{g}$.*

The negative of the Killing form is an example of such an Ad*-invariant positive definite scalar product.*

Proof Existence follows from 2.2.3. Uniqueness follows from Corollary 2.4.28 and Theorem 2.5.1. The claim about the Killing form follows from Corollary 2.4.18. □

Let $T = \mathrm{U}(1) \times \ldots \times \mathrm{U}(1)$ denote an n-dimensional torus and $\langle \cdot , \cdot \rangle_{\mathfrak{t}}$ a positive definite scalar product on its Lie algebra

$$\mathbb{R}^n = \mathfrak{t} = \mathfrak{u}(1) \oplus \ldots \oplus \mathfrak{u}(1).$$

Since the adjoint representation of an abelian Lie group is trivial, any inner product on an abelian Lie algebra is Ad-invariant. With respect to the standard Euclidean scalar product on $\mathbb{R}^n$, the scalar product $\langle \cdot , \cdot \rangle_{\mathfrak{t}}$ is determined by a positive definite symmetric matrix.

Theorem 2.5.4 (Ad-Invariant Scalar Products on General Compact Lie Algebras) *Let G be a compact connected Lie group of the form*

$$G = \mathrm{U}(1) \times \ldots \times \mathrm{U}(1) \times G_1 \times \ldots \times G_s,$$

where the G_i are compact simple Lie groups. Let $\langle \cdot , \cdot \rangle_{\mathfrak{g}}$ be an Ad_G-invariant positive definite scalar product on the Lie algebra $\mathfrak{g}$ of G. Then $\langle \cdot , \cdot \rangle_{\mathfrak{g}}$ is the orthogonal direct sum of:

1. *a positive definite scalar product $\langle \cdot , \cdot \rangle_{\mathfrak{t}}$ on the center $\mathfrak{t} = \mathfrak{u}(1) \oplus \ldots \oplus \mathfrak{u}(1)$;*
2. *Ad_{G_i}-invariant positive definite scalar products $\langle \cdot , \cdot \rangle_{\mathfrak{g}_i}$ on the Lie algebras $\mathfrak{g}_i$.*

Conversely, the direct sum of any positive definite scalar product $\langle \cdot , \cdot \rangle_{\mathfrak{t}}$ on the abelian Lie algebra $\mathfrak{t}$ and any Ad_{G_i}-invariant positive definite scalar products $\langle \cdot , \cdot \rangle_{\mathfrak{g}_i}$ on the simple Lie algebras $\mathfrak{g}_i$ is an Ad_G-invariant positive definite scalar product on $\mathfrak{g}$.

Proof Let $\langle \cdot , \cdot \rangle_{\mathfrak{g}}$ be an Ad_G-invariant positive definite scalar product on the Lie algebra $\mathfrak{g}$. We have to show that it decomposes as an orthogonal direct sum of scalar products on the summands. For any fixed $i = 1, \ldots , s$ we can write $G = G_i \times H$ with a compact Lie group H. Fix an arbitrary $Y \in \mathfrak{h}$ and let

$$\begin{aligned} f \colon \mathfrak{g}_i &\longrightarrow \mathbb{R} \\ X &\longmapsto \langle X, Y \rangle_{\mathfrak{g}}. \end{aligned}$$

Then f is a linear 1-form on $\mathfrak{g}_i$ and its kernel is a vector subspace of codimension zero or one. Let $g \in G_i$ and $X \in \mathfrak{g}_i$. Then by Lemma 2.1.50

$$\begin{aligned} f(\mathrm{Ad}_g X) &= \langle \mathrm{Ad}_g X, Y \rangle_{\mathfrak{g}} \\ &= \langle \mathrm{Ad}_g X, \mathrm{Ad}_g Y \rangle_{\mathfrak{g}} \\ &= \langle X, Y \rangle_{\mathfrak{g}} \\ &= f(X). \end{aligned}$$

This implies that the kernel of f is Ad_{G_i}-invariant. Since the adjoint representation of G_i is irreducible by Corollary 2.4.28 and since $\dim \mathfrak{g}_i > 1$, the kernel of f cannot have codimension 1. Therefore f must vanish identically.

This proves that the scalar product $\langle\cdot,\cdot\rangle_{\mathfrak{g}}$ on $\mathfrak{g}$ decomposes as an orthogonal direct sum of scalar products $\langle\cdot,\cdot\rangle_{\mathfrak{g}_i}$ on $\mathfrak{g}_i$ and $\langle\cdot,\cdot\rangle_{\mathfrak{h}}$ on $\mathfrak{h}$. The scalar product $\langle\cdot,\cdot\rangle_{\mathfrak{g}}$ is Ad_G-invariant, hence $\langle\cdot,\cdot\rangle_{\mathfrak{g}_i}$ is Ad_{G_i}-invariant and $\langle\cdot,\cdot\rangle_{\mathfrak{h}}$ is Ad_H-invariant. We continue to split the scalar product $\langle\cdot,\cdot\rangle_{\mathfrak{h}}$ on $\mathfrak{h}$ until the remaining Lie algebra is the center.

Conversely, if $\langle\cdot,\cdot\rangle_{\mathfrak{t}}$ is a scalar product on $\mathfrak{t} = \mathfrak{u}(1) \oplus \ldots \oplus \mathfrak{u}(1)$ and $\langle\cdot,\cdot\rangle_{\mathfrak{g}_i}$ are Ad_{G_i}-invariant scalar products on $\mathfrak{g}_i$, then the orthogonal direct sum

$$\langle\cdot,\cdot\rangle_{\mathfrak{g}} = \langle\cdot,\cdot\rangle_{\mathfrak{t}} \oplus \langle\cdot,\cdot\rangle_{\mathfrak{g}_1} \oplus \ldots \oplus \langle\cdot,\cdot\rangle_{\mathfrak{g}_s}$$

is Ad_G-invariant by Lemma 2.1.50. □

In the situation of Theorem 2.5.4 the Ad_G-invariant scalar product $\langle\cdot,\cdot\rangle_{\mathfrak{g}}$ on $\mathfrak{g}$ is determined by certain constants:

1. The scalar product $\langle\cdot,\cdot\rangle_{\mathfrak{t}}$ is determined by a positive definite symmetric matrix with respect to the standard Euclidean scalar product on $\mathbb{R}^n$.
2. The scalar products $\langle\cdot,\cdot\rangle_{\mathfrak{g}_i}$ are determined by positive constants relative to some fixed Ad_{G_i}-invariant positive definite scalar product on the simple Lie algebras $\mathfrak{g}_i$ (like the negative of the Killing form).

Definition 2.5.5 The constants that determine an Ad_G-invariant positive definite scalar product on the compact Lie algebra $\mathfrak{g}$ are called **coupling constants** in physics.

Example 2.5.6

1. In the Standard Model, where $G = \mathrm{SU}(3) \times \mathrm{SU}(2) \times \mathrm{U}(1)$, there are three coupling constants, one for each factor.
2. In GUTs with a simple gauge group, like $G = \mathrm{SU}(5)$ or $G = \mathrm{Spin}(10)$, there is only a single coupling constant.

2.6 *Homotopy Groups of Lie Groups

In this section we collect some results (without proofs) on the homotopy groups of compact Lie groups. The following fact is elementary and can be found in textbooks on topology:

Proposition 2.6.1 (Fundamental Group of Topological Groups) *The fundamental group $\pi_1(G)$ of any connected topological group G is abelian.*

Regarding the order of the fundamental group of Lie groups it can be shown that (for a proof, see [24, Sect. V.7]):

Theorem 2.6.2 (Fundamental Group of Compact Semisimple Lie Groups) *Let G be a compact connected Lie group. Then $\pi_1(G)$ is finite if and only if G is semisimple. In particular, every compact simple Lie group has a finite fundamental group.*

The only-if direction follows from Theorem 2.4.29. As an example, it is possible to calculate the fundamental group of the classical Lie groups (see, for example, [129]).

Proposition 2.6.3 (Fundamental Groups of Classical Compact Groups) *The fundamental groups of the classical compact linear groups are:*

1. *Special orthogonal groups:*

$$\pi_1(\mathrm{SO}(2)) \cong \mathbb{Z},$$

$$\pi_1(\mathrm{SO}(n)) \cong \mathbb{Z}_2 \ \forall n \geq 3.$$

2. *Unitary groups (for all $n \geq 1$):*

$$\pi_1(\mathrm{U}(n)) \cong \mathbb{Z}.$$

3. *Special unitary and symplectic groups (for all $n \geq 1$):*

$$\pi_1(\mathrm{SU}(n)) = 1,$$

$$\pi_1(\mathrm{Sp}(n)) = 1.$$

We have the following result on the second homotopy group (for a proof, see again [24, Sect. V.7]):

Theorem 2.6.4 (Second Homotopy Group of Compact Lie Groups) *Let G be a compact connected Lie group. Then $\pi_2(G) = 0$.*

The next theorem on the third homotopy group was proved by M.R. Bott using Morse theory [19]:

Theorem 2.6.5 (Third Homotopy Group of Compact Lie Groups) *Let G be a compact connected Lie group. Then $\pi_3(G)$ is free abelian, i.e. isomorphic to $\mathbb{Z}^r$ for some integer r. If G is compact, connected and simple, then $\pi_3(G) \cong \mathbb{Z}$.*

Combining Theorem 2.6.2 and Theorem 2.6.5 we get a topological criterion to decide whether a compact Lie group is simple:

Corollary 2.6.6 (Topological Criterion for Simplicity) *Let G be a compact connected Lie group. Then G is simple if and only if $\pi_1(G)$ is finite and $\pi_3(G) \cong \mathbb{Z}$.*

2.7 Exercises for Chap. 2

2.7.1 Verify that the dual representations on V^* defined in Definition 2.1.23 and Definition 2.1.27 are indeed representations of the Lie group G and the Lie algebra $\mathfrak{g}$.

2.7.2 Let

$$\epsilon = \begin{pmatrix} 0 & 1 \\ -1 & 0 \end{pmatrix}.$$

Prove the following equivalent description of $\mathrm{SL}(2, \mathbb{C})$:

$$\mathrm{SL}(2, \mathbb{C}) = \{ M \in \mathrm{Mat}(2 \times 2, \mathbb{C}) \mid M^T \epsilon M = \epsilon \}.$$

2.7.3

1. Let $W \cong \mathbb{C}^2$ denote the fundamental representation of $\mathfrak{su}(2)$ and $\bar{W}$ the complex conjugate representation. Show that there exists a matrix $A \in \mathrm{GL}(2, \mathbb{C})$ such that

$$AMA^{-1} = \bar{M} \quad \forall M \in \mathfrak{su}(2).$$

   Conclude that W and $\bar{W}$ are isomorphic as $\mathfrak{su}(2)$-representations.
2. Let $V_k \cong \mathbb{C}$ denote the representation of $\mathfrak{u}(1)$ with winding number $k \neq 0$. Prove that V_k and $\bar{V}_k$ are not isomorphic as $\mathfrak{u}(1)$-representations.
3. Let $V \cong \mathbb{C}^2$ denote the fundamental representation of the real Lie algebra $\mathfrak{sl}(2, \mathbb{C})$ and $\bar{V}$ the complex conjugate representation. Prove that V and $\bar{V}$ are not isomorphic as $\mathfrak{sl}(2, \mathbb{C})$-representations.
4. Does the complex conjugate representation make sense for complex representations of complex Lie algebras, like the complex Lie algebra $\mathfrak{sl}(2, \mathbb{C})$?

Remark It can be shown that the fundamental representation of $\mathfrak{su}(n)$ for every $n \geq 3$ is not isomorphic to its complex conjugate. The only other compact simple Lie algebras which have complex representations not isomorphic to their conjugate are $\mathfrak{so}(4n + 2)$ for every $n \geq 1$ (Weyl spinor representations) and E_6 (a 27-dimensional representation), see [104]. This is one of the reasons why Lie groups such as SU(5), Spin(10) or E_6 appear as gauge groups of Grand Unified Theories; see Sect. 8.5.3.

2.7.4 Determine the charges of the basis element $\tau_3 \in \mathfrak{su}(2)$ in:

1. the fundamental representation of $\mathfrak{su}(2)$ on $\mathbb{C}^2$;
2. the representation of $\mathfrak{su}(2)$ on $\mathbb{C}^3$ via the isomorphism $\mathfrak{su}(2) \cong \mathfrak{so}(3)$ and the complex fundamental representation of $\mathfrak{so}(3)$.

2.7.5

1. Consider the Lie group SU(2) with the fundamental representation on $\mathbb{C}^2$. Each of the basis vectors τ_1, τ_2, τ_3 of $\mathfrak{su}(2)$ from Example 1.5.32 generates

a one-parameter subgroup isomorphic to U(1). Determine the explicit branching rule for the fundamental representation on $\mathbb{C}^2$ under restriction to these circle subgroups, i.e. determine the corresponding decomposition of $\mathbb{C}^2$ into invariant complex subspaces together with the winding numbers of the induced representations.
2. Do the same exercise with the complex representation of SU(2) on $\mathbb{C}^3$ via the universal covering $\mathrm{SU}(2) \to \mathrm{SO}(3)$ and the complex fundamental representation of SO(3).
3. Do the same exercise for the Lie group SU(3) with the fundamental representation on $\mathbb{C}^3$ and the circle subgroups generated by the basis vectors $v_1, \ldots, v_8$ of $\mathfrak{su}(3)$, where $v_a = \frac{i\lambda_a}{2}$ with the Gell-Mann matrices λ_a from Example 1.5.33 (cf. Exercise 1.9.26).

2.7.6 Consider the embedding

$$\mathrm{U}(n) \hookrightarrow \mathrm{SO}(2n)$$

from Exercise 1.9.10. Let $V = \mathbb{C}^{2n}$ be the complex fundamental representation of $\mathrm{SO}(2n)$. Determine the branching rule of the representation V under restriction to the subgroup $\mathrm{U}(n) \subset \mathrm{SO}(2n)$. It may be helpful to first consider the case $n = 1$.

2.7.7 Let G be a Lie group. The **center** of G is defined as

$$Z(G) = \{g \in G \mid gh = hg \quad \forall h \in G\}.$$

Suppose that G is connected.

1. Prove that the center $Z(G)$ is the kernel of the adjoint representation Ad_G. Conclude that $Z(G)$ is an embedded Lie subgroup in G with Lie algebra given by the center $\mathfrak{z}(\mathfrak{g})$ of $\mathfrak{g}$.
2. Prove that $\mathfrak{g}$ is abelian if and only if G is abelian.
3. Prove that Ad_G is trivial if and only if G is abelian. Conclude that the left-invariant and right-invariant vector fields on a connected Lie group G coincide if and only if G is abelian.

2.7.8 Consider the Lie algebra isomorphism of $\mathfrak{so}(3)$ with $(\mathbb{R}^3, \times)$ from Exercise 1.9.14.

1. Determine the symmetric bilinear form on $\mathbb{R}^3$ corresponding under this isomorphism to the Killing form $B_{\mathfrak{so}(3)}$.
2. Interpret the high school formula

$$z \cdot (x \times y) = -y \cdot (x \times z) \quad \forall x, y, z \in \mathbb{R}^3,$$

where $\cdot$ denotes the scalar product, in light of the first part of this exercise.

2.7.9

1. Let $\mathfrak{g}, \mathfrak{h}$ be Lie algebras and $\phi: \mathfrak{g} \to \mathfrak{h}$ a Lie algebra homomorphism. Suppose that $\mathfrak{g}$ is simple. Show that ϕ is either injective or the trivial homomorphism. In particular, every representation of a simple Lie algebra is either faithful or trivial.
2. Show that every complex 1-dimensional representation of a semisimple Lie algebra is trivial.
3. Show that every homomorphism from a connected semisimple Lie group to $\mathrm{U}(1)$ is trivial. Find a non-trivial homomorphism from $\mathrm{U}(n)$ to $\mathrm{U}(1)$.

2.7.10 Let $\mathfrak{g}$ be a real Lie algebra. The **complexification** of $\mathfrak{g}$ is the complex Lie algebra

$$\mathfrak{g}_{\mathbb{C}} = \mathfrak{g} \otimes_{\mathbb{R}} \mathbb{C} \cong \mathfrak{g} \oplus i\mathfrak{g}$$

with the Lie bracket from $\mathfrak{g}$ extended $\mathbb{C}$-bilinearly. Show that if $\mathfrak{g}_{\mathbb{C}}$ is (semi-)simple, then $\mathfrak{g}$ is (semi-)simple.

Remark The following converses can be shown: If $\mathfrak{g}$ is semisimple, then $\mathfrak{g}_{\mathbb{C}}$ is semisimple (this uses Theorem 2.4.9) and if $\mathfrak{g}$ is compact simple, then $\mathfrak{g}_{\mathbb{C}}$ is simple (see [83]).

2.7.11 Prove Proposition 2.4.26: the structure constants f_{abc} for a $B_{\mathfrak{g}}$-orthonormal basis $\{T_a\}$ of a semisimple Lie algebra $\mathfrak{g}$ are totally antisymmetric:

$$f_{abc} = -f_{bac} = f_{bca} = -f_{acb} \quad \forall a, b, c \in \{1, \ldots, n\}.$$

2.7.12 Let τ_1, τ_2, τ_3 be the basis of the Lie algebra $\mathfrak{su}(2)$ from Example 1.5.32. Fix an arbitrary, positive real number $g > 0$ and let

$$\beta_a = g\tau_a \in \mathfrak{su}(2) \quad (a = 1, 2, 3).$$

Define a unique positive definite scalar product $\langle \cdot, \cdot \rangle_g$ with associated norm $||\cdot||_g$ on $\mathfrak{su}(2)$ so that $\beta_1, \beta_2, \beta_3$ form an orthonormal basis. Determine the relation between $\det(X)$ and the norm $||X||_g$ for $X \in \mathfrak{su}(2)$. Show that the scalar product $\langle \cdot, \cdot \rangle_g$ is $\mathrm{Ad}_{\mathrm{SU}(2)}$-invariant.

2.7.13 Consider the Lie algebra $\mathfrak{su}(2)$.

1. Calculate the Killing form $B_{\mathfrak{su}(2)}$ directly from the definition and determine the constant g so that $-B_{\mathfrak{su}(2)} = \langle \cdot, \cdot \rangle_g$, where $\langle \cdot, \cdot \rangle_g$ is the scalar product from Exercise 2.7.12.
2. Fix an arbitrary, positive, real number $\lambda > 0$ and set

$$\begin{aligned} F_\lambda: \mathfrak{su}(2) \times \mathfrak{su}(2) &\longrightarrow \mathbb{R} \\ (X, Y) &\longmapsto \lambda \mathrm{tr}(X \cdot Y), \end{aligned}$$

where tr denotes the trace and $\cdot$ the matrix product. Show that $-F_\lambda$ is a negative definite $\mathrm{Ad}_{\mathrm{SU}(2)}$-invariant scalar product on $\mathfrak{su}(2)$. Determine the constant λ so that $F_\lambda = B_{\mathfrak{su}(2)}$.

2.7.14 Consider the Lie algebra $\mathfrak{sl}(2,\mathbb{R})$ with Killing form $B_{\mathfrak{sl}(2,\mathbb{R})}$. Show that there exists a constant $\lambda \in \mathbb{R}$ such that

$$B_{\mathfrak{sl}(2,\mathbb{R})}(X,Y) = \lambda \mathrm{tr}(X \cdot Y) \quad \forall X, Y \in \mathfrak{sl}(2,\mathbb{R}),$$

where tr denotes the trace of the matrix and $\cdot$ the matrix product. Determine this constant λ. Is the Killing form $B_{\mathfrak{sl}(2,\mathbb{R})}$ definite? or non-degenerate?

2.7.15 Let $\mathbb{K} = \mathbb{R}, \mathbb{C}$.

1. Show that the Killing form of the Lie algebra $\mathfrak{gl}(n,\mathbb{K})$ can be calculated as

$$B_{\mathfrak{gl}(n,\mathbb{K})}(X,Y) = 2n\mathrm{tr}(X \cdot Y) - 2\mathrm{tr}(X)\mathrm{tr}(Y).$$

   A suitable basis for $\mathfrak{gl}(n,\mathbb{K})$ to evaluate the trace on the left-hand side is given by the elementary matrices E_{ij} with a 1 at the intersection of the i-th row and j-th column and zeros elsewhere.
2. Let $\mathfrak{h}$ be an ideal in a Lie algebra $\mathfrak{g}$. Prove that for all $X, Y \in \mathfrak{h}$

$$B_{\mathfrak{h}}(X,Y) = B_{\mathfrak{g}}(X,Y).$$

3. Show that the Killing form of the Lie algebra $\mathfrak{sl}(n,\mathbb{K})$ is equal to

$$B_{\mathfrak{sl}(n,\mathbb{K})}(X,Y) = 2n\mathrm{tr}(X \cdot Y).$$

   Compare with Exercise 2.7.14.

2.7.16

1. Let $\mathfrak{g}$ be a real Lie algebra and $\mathfrak{g}_\mathbb{C}$ its complexification as in Exercise 2.7.10. Under the canonical inclusion $\mathfrak{g} \subset \mathfrak{g}_\mathbb{C}$ as the real part show that for all $X, Y \in \mathfrak{g}$

$$B_{\mathfrak{g}}(X,Y) = B_{\mathfrak{g}_\mathbb{C}}(X,Y).$$

2. Explain the difference between the results for the Killing form in Exercise 2.7.13 and Exercise 2.7.14, given the isomorphism of complex Lie algebras

$$\mathfrak{su}(2)_\mathbb{C} \cong \mathfrak{sl}(2,\mathbb{C}) \cong \mathfrak{sl}(2,\mathbb{R})_\mathbb{C}$$

   from Exercise 1.9.18.
3. Show that every complex matrix A can be written uniquely as $A = B + iC$ with B, C skew-Hermitian. Conclude that

$$\mathfrak{u}(n)_\mathbb{C} \cong \mathfrak{gl}(n,\mathbb{C}),$$
$$\mathfrak{su}(n)_\mathbb{C} \cong \mathfrak{sl}(n,\mathbb{C}).$$

4. Show that the Killing forms of the Lie algebras $\mathfrak{u}(n)$ and $\mathfrak{su}(n)$ can be calculated as

$$B_{\mathfrak{u}(n)}(X,Y) = 2n\mathrm{tr}(X \cdot Y) - 2\mathrm{tr}(X)\mathrm{tr}(Y),$$
$$B_{\mathfrak{su}(n)}(X,Y) = 2n\mathrm{tr}(X \cdot Y).$$

Compare with Exercise 2.7.13.

2.7.17 Consider the basis of $\mathfrak{su}(3)$ given by the elements $i\lambda_a$, where λ_a are the Gell-Mann matrices from Example 1.5.33, with $a = 1, \ldots, 8$. Show that these basis vectors are orthogonal with respect to the Killing form $B_{\mathfrak{su}(3)}$ and determine $B_{\mathfrak{su}(3)}(i\lambda_a, i\lambda_a)$ for all a.

2.7.18

1. The **rank** of a compact Lie group G is the maximal dimension of an embedded torus subgroup $T \subset G$. Prove that the rank of a product $G \times H$ of compact Lie groups G and H is the sum of the ranks of G and H (you can assume without proof that a connected abelian Lie group is a torus).
2. Classify compact semisimple Lie algebras of rank $r = 1, 2, 3, 4$, assuming Theorem 2.4.23.

Chapter 3
Group Actions

There are different ways in which Lie groups can act as transformation or symmetry groups on geometric objects. One possibility, that we discussed in Chap. 2, is the representation of Lie groups on vector spaces. A second possibility, studied in this chapter, is Lie group actions on manifolds. Both concepts are related: A representation is a linear action of the group where the manifold is a vector space. Conversely, an action on a manifold can be thought of as a *non-linear representation* of the group. More precisely, a linear representation of a group corresponds to a homomorphism into the general linear group of a vector space. A group action then corresponds to a homomorphism of the group into the *diffeomorphism group* of a manifold.

Even though we are most interested in Lie group actions on manifolds, it is useful to consider more general types of actions: actions of groups on sets and actions of topological groups on topological spaces. We will also introduce several standard notions related to group actions, like *orbits* and *isotropy groups*. In the smooth case, if a Lie group G acts on a manifold M, then there is an induced *infinitesimal action* of the Lie algebra $\mathfrak{g}$, defining so-called *fundamental vector fields* on M. This map can be understood as the induced Lie algebra homomorphism from the Lie algebra of G to the Lie algebra of the diffeomorphism group $\mathrm{Diff}(M)$.

In the case of smooth actions of a Lie group G on a manifold M, the interesting question arises under which conditions the *quotient space* M/G again admits the structure of a smooth manifold. The main (and rather difficult) result that we prove in this context is *Godement's Theorem*, which gives a necessary and sufficient condition that quotient spaces under general equivalence relations are smooth manifolds. The smooth structure on the quotient space is defined using so-called *slices* for the equivalence classes.

It turns out that the quotient space of a Lie group action admits the structure of a smooth manifold in particular in the following cases:

- A *compact* Lie group G acting smoothly and *freely* on a manifold M.
- A closed subgroup H of a Lie group G acting on G by right (or left) translations.

M.J.D. Hamilton, *Mathematical Gauge Theory*, Universitext,
https://doi.org/10.1007/978-3-319-68439-0_3

Both cases can be used to construct new and interesting smooth manifolds. In the second case, if the closed subgroup H acts on the *right* on G, then there is an additional *left* action of G on the quotient manifold G/H. This action is *transitive* and G/H is an example of a *homogeneous space*. We will study homogeneous spaces in detail in all of the three cases of group actions on sets, topological spaces and manifolds and prove that any homogeneous space is of the form G/H.

We finally apply the theory of group actions to construct the exceptional compact simple Lie group G_2, which plays an important part in *M-theory*, a conjectured theory of quantum gravity in 11 dimensions, and derive some of its properties.

General references for this chapter are [14, 24] and [142].

3.1 Transformation Groups

In this section we define group actions and study their basic properties. Since many statements in this section are quite elementary, we designate some of the proofs as exercises.

Before we begin with the formal definitions, let us consider some basic examples to get a bird's eye view of group actions. The simplest example is perhaps the canonical left action of the general linear group GL(V) on a vector space V, given by the map

$$\begin{aligned} \Phi\colon \mathrm{GL}(V) \times V &\longrightarrow V \\ (f, v) &\longmapsto \Phi(f, v) = f(v). \end{aligned} \tag{3.1}$$

A representation of a group G on V then corresponds to a group homomorphism

$$\phi\colon G \longrightarrow \mathrm{GL}(V),$$

defining a linear action of G on V.

We would like to extend this idea to other types of actions. Suppose that

- M is a set and $\mathrm{S}(M)$ the symmetric group of all bijections $M \to M$; or
- M is a topological space and Homeo(M) the homeomorphism group of M; or
- M is a manifold and Diff(M) the diffeomorphism group of M.

Replacing V by M and GL(V) by $\mathrm{S}(M)$ (Homeo(M), Diff(M)) in Eq. (3.1) we get canonical actions of these automorphism groups on M. *Actions* of a group G on M are then given by homomorphisms ϕ of G into these groups and thus

(continued)

correspond to *non-linear representations* of G on M (which in the case for $\mathrm{Homeo}(M)$ and $\mathrm{Diff}(M)$ should in some sense be continuous and smooth).

In each of these cases, the images of the group G under the homomorphisms ϕ define subgroups of $\mathrm{GL}(V)$, $\mathrm{S}(M)$, $\mathrm{Homeo}(M)$ and $\mathrm{Diff}(M)$ that are usually easier to handle than the full automorphism groups themselves (which in the case of the diffeomorphism group, for example, are infinite-dimensional if $\dim M \geq 1$).

An explicit example of a Lie group action on a manifold is the famous *Hopf action* of $S^1 = \mathrm{U}(1)$ on S^3 defined by the map

$$\begin{aligned} \Phi: S^3 \times S^1 &\longrightarrow S^3 \\ (v, w, \lambda) &\longmapsto (v, w) \cdot \lambda = (v\lambda, w\lambda), \end{aligned}$$

where S^3 is the unit sphere in $\mathbb{C}^2$ and S^1 the unit circle in $\mathbb{C}$ (this is an example of a *right action*). It is clear that the map is well-defined, i.e. it preserves the 3-sphere, and it is smooth. The map also has the following properties:

1. $(v, w) \cdot (\lambda \cdot \mu) = ((v, w) \cdot \lambda) \cdot \mu$
2. $(v, w) \cdot 1 = (v, w)$

for all $(v, w) \in S^3$ and $\lambda, \mu \in S^1$. We shall see that these are the defining properties of group actions, ensuring that we obtain a homomorphism into the diffeomorphism group. In the case of the Hopf action we can think of it as a homomorphism

$$\phi: S^1 \longrightarrow \mathrm{Diff}(S^3).$$

The Hopf action will also be an important example in subsequent chapters.

We shall later study properties of this and other actions. For example, we can fix a point $(v_0, w_0) \in S^3$ and consider its *orbit* under the action:

$$\begin{aligned} S^1 &\longrightarrow S^3 \\ \lambda &\longmapsto (v_0, w_0) \cdot \lambda. \end{aligned}$$

In this case, the orbit map is injective for all $(v_0, w_0) \in S^3$ and the Hopf action is therefore called *free*.

3.2 Definition and First Properties of Group Actions

We now come to the formal definition of group actions.

Definition 3.2.1 **A left action** of a group G on a set M is a map

$$\begin{aligned}\Phi: G \times M &\longrightarrow M \\ (g,p) &\longmapsto \Phi(g,p) = g \cdot p = gp\end{aligned}$$

satisfying the following properties:

1. $(g \cdot h) \cdot p = g \cdot (h \cdot p)$ for all $p \in M$ and $g, h \in G$.
2. $e \cdot p = p$ for all $p \in M$.

The group G is called a **transformation group** of M.
We can think of a group action as moving a point $p \in M$ around in M as we vary the group element $g \in G$. This is very similar to the concept of a representation of a group on a vector space, where a vector is moved around as we vary the group element.

If G is a topological group, M a topological space and Φ continuous, then Φ is called a **continuous left action**. Similarly, if G is a Lie group, M a smooth manifold and Φ is smooth, then Φ is called a **smooth left action**. Here $G \times M$ carries the canonical product structure as a topological space or smooth manifold.

Similarly **right actions** of a group G on a set M are defined as a map

$$\begin{aligned}\Phi: M \times G &\longrightarrow M \\ (p,g) &\longmapsto \Phi(p,g) = p \cdot g = pg\end{aligned}$$

satisfying the following properties:

1. $p \cdot (g \cdot h) = (p \cdot g) \cdot h$ for all $p \in M$ and $g, h \in G$.
2. $p \cdot e = p$ for all $p \in M$.

There is, of course, also the notion of a continuous or smooth right action (most of the following statements hold for both left and right actions). We can turn every left action into a right action (and vice versa):

Proposition 3.2.2 *Let*

$$\begin{aligned}\Phi: G \times M &\longrightarrow M \\ (g,p) &\longmapsto g \cdot p\end{aligned}$$

be a left action of a group G on a set M. Then

$$\begin{aligned}M \times G &\longrightarrow M \\ (p,g) &\longmapsto p * g = g^{-1} \cdot p\end{aligned}$$

defines a right action of G on M.

Proof This is Exercise 3.12.1. □

A group action Φ is a map with two entries: a group element $g \in G$ and a point $p \in M$. It is useful to consider the maps that we obtain if we fix one of the entries and let only the other one vary.

Definition 3.2.3 Let $\Phi: G \times M \to M$ be a left action. For $g \in G$ we define the **left translation** by

$$\begin{aligned} l_g: M &\longrightarrow M \\ p &\longmapsto g \cdot p. \end{aligned}$$

Similarly, for a right action $\Phi: M \times G \to M$ and $g \in G$ we define the **right translation** by

$$\begin{aligned} r_g: M &\longrightarrow M \\ p &\longmapsto p \cdot g. \end{aligned}$$

For $p \in M$ the **orbit map** is given by

$$\begin{aligned} \phi_p: G &\longrightarrow M \\ g &\longmapsto g \cdot p \end{aligned}$$

for a left action and

$$\begin{aligned} \phi_p: G &\longrightarrow M \\ g &\longmapsto p \cdot g \end{aligned}$$

for a right action.

It is clear that for a continuous (smooth) left action the left translations l_g for all $g \in G$ and the orbit maps ϕ_p for all $p \in M$ are continuous (smooth) maps. The reason is that in the smooth case the map l_g is given by the composition of smooth maps

$$\begin{aligned} M &\longrightarrow G \times M \longrightarrow M \\ p &\longmapsto (g, p) \longmapsto g \cdot p \end{aligned}$$

and ϕ_p is given by the composition

$$\begin{aligned} G &\longrightarrow G \times M \longrightarrow M \\ g &\longmapsto (g, p) \longmapsto g \cdot p. \end{aligned}$$

The continuous case and the case of right actions follow similarly.

We could define left translations as above for any map $\Phi: G \times M \to M$ even if Φ does not satisfy *a priori* the axioms of a left action. It is easy to see that group actions are then characterized by the fact that all left translations l_g for $g \in G$ are bijections of M and

$$\begin{aligned} \phi: G &\longrightarrow \mathrm{S}(M) \\ g &\longmapsto l_g \end{aligned}$$

is a group homomorphism. In the case of a continuous (smooth) left action, the left translations define a group homomorphism

$$\phi: G \longrightarrow \mathrm{Homeo}(M)$$

and

$$\phi: G \longrightarrow \mathrm{Diff}(M),$$

respectively, into the group of homeomorphisms (diffeomorphisms) of M. Note that, as we said before, a continuous (smooth) group action is more than just a group homomorphism into the homeomorphism (diffeomorphism) group, because the group homomorphism has to be in addition continuous (smooth) in the argument $g \in G$ (one could make this precise by defining a topology or smooth structure on the homeomorphism and diffeomorphism groups, which in general are infinite-dimensional).

Here are some additional concepts for group actions (we define them in the general case for group actions on sets, but they apply verbatim for continuous and smooth group actions).

Definition 3.2.4 Let Φ be a left action of a group G on a set M.

1. The **orbit** of G through a point $p \in M$ is

$$\mathcal{O}_p = G \cdot p = \{g \cdot p \mid g \in G\}.$$

The orbit is the image of the orbit map (see Fig. 3.1).

2. The **fixed point set** of a group element $g \in G$ is the set

$$M^g = \{p \in M \mid g \cdot p = p\}.$$

3. The **isotropy group** or **stabilizer** of a point $p \in M$ is

$$G_p = \{g \in G \mid g \cdot p = p\}.$$

In physics, isotropy groups are also called **little groups**. It is an easy exercise to show that the isotropy group G_p is indeed a subgroup of G for all $p \in M$.

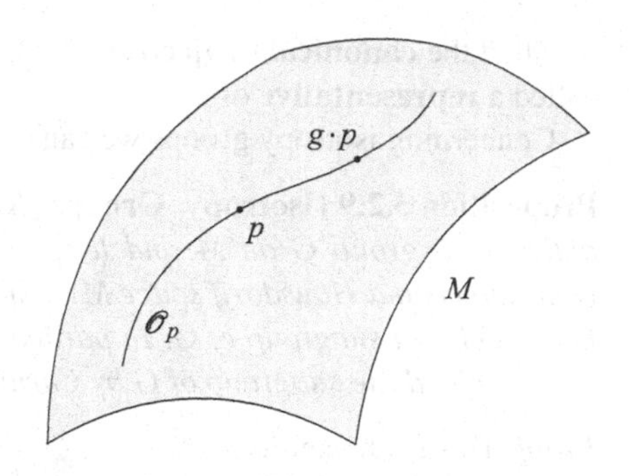

Fig. 3.1 Orbit of a group action

There are analogous definitions for right actions.

Remark 3.2.5 We shall see later in Corollary 3.8.10 that for a smooth action of a Lie group G on a manifold M the orbit $\mathcal{O}_p$ through every point $p \in M$ is an (immersed or embedded) submanifold of M.

Lemma 3.2.6 (Two Orbits Are Either Disjoint or Identical) *Let Φ be an action of a group G on a set M and $p \in M$ an arbitrary point. If $q \in \mathcal{O}_p$, then $\mathcal{O}_q = \mathcal{O}_p$. Hence the orbits of two points in M are either disjoint or identical.*
This means that orbits which intersect in one point are already identical.

Proof Suppose Φ is a left action. Then q is of the form $q = g \cdot p$ for some $g \in G$. We get

$$\mathcal{O}_q = G \cdot q = (G \cdot g) \cdot p = G \cdot p,$$

because the right translation $R_g : G \to G$ is a bijection. □

Remark 3.2.7 We can also phrase this differently: The relation

$$p \sim q \Leftrightarrow \exists g \in G : q = g \cdot p$$

for $p, q \in M$ defines an *equivalence relation* on M and the orbits of G are precisely the equivalence classes. M is therefore the disjoint union of the orbits of G.

Definition 3.2.8 Let Φ be an action of a group G on a set M. Then the following subset of the powerset of M

$$M/G = \{\mathcal{O}_p \subset M \mid p \in M\}$$

is called **the space of orbits** or the **quotient space** of the action.
Note that the *subsets* $\mathcal{O}_p$ of M become *elements* (*points*) in M/G. If we think of the subset $\mathcal{O}_p$ as a point in M/G, we also denote it by $[p]$ or $\bar{p}$. The map

$$\begin{aligned} \pi : M &\longrightarrow M/G \\ p &\longmapsto [p] \end{aligned}$$

is called the **canonical projection**. If $x \in M/G$, then a point $p \in M$ with $[p] = x$ is called a **representative** of x.

Concerning isotropy groups we can say the following.

Proposition 3.2.9 (Isotropy Groups Are (Closed) Subgroups) *Let Φ be an action of a group G on M and let $p \in M$ be any point. If the group action is continuous on a Hausdorff space M or smooth on a manifold M, then the stabilizer G_p is a closed subgroup of G. In particular, in the smooth case the stabilizer G_p is an embedded Lie subgroup of G by Cartan's Theorem 1.1.44.*

Proof This is an exercise. □

Suppose $\phi: G \to \mathrm{S}(M)$ is the group homomorphism induced from a group action. Then

$$\ker\phi = \bigcap_{p\in M} G_p.$$

In particular, for continuous actions on Hausdorff spaces or smooth actions on manifolds, the normal subgroup $\ker\phi$ is closed in G.

We want to compare the isotropy groups of points on the same G-orbit. Suppose p, q are points in M on the same G-orbit. It is easy to check that there exists an element $g \in G$ such that

$$c_g(G_p) = g \cdot G_p \cdot g^{-1} = G_q.$$

In particular, the isotropy groups G_p and G_q are isomorphic.

In the case of a smooth action of a Lie group G we call the Lie algebra $\mathfrak{g}_p$ of the stabilizer G_p of a point $p \in M$ the **isotropy subalgebra**. The following description of the isotropy subalgebra is useful in applications.

Proposition 3.2.10 (The Isotropy Subalgebra and the Orbit Map) *Let Φ be a smooth action of a Lie group G on a manifold M. Fix a point $p \in M$ and let ϕ_p denote the orbit map*

$$\phi_p: G \longrightarrow M$$

as before. Then the kernel of the differential

$$D_e\phi_p: \mathfrak{g} \longrightarrow T_pM$$

is equal to the isotropy subalgebra $\mathfrak{g}_p$.

Proof We assume that the action is on the left, the case of right actions follows similarly. If $X \in \mathfrak{g}_p$, then $\exp(tX) \in G_p$ for all $t \in \mathbb{R}$ and therefore

$$\phi_p(\exp(tX)) = \exp(tX) \cdot p = p \quad \forall t \in \mathbb{R}.$$

This implies that X is in the kernel of the differential $D_e\phi_p$. Conversely, suppose that X is in the kernel of $D_e\phi_p$. Then

$$\left.\frac{d}{d\tau}\right|_{\tau=0} (\exp(\tau X)\cdot p) = 0.$$

This implies

$$\begin{aligned}
\left.\frac{d}{dt}\right|_{t=s} (\exp(tX)\cdot p) &= \left.\frac{d}{d\tau}\right|_{\tau=0} (\exp(sX)\cdot \exp(\tau X)\cdot p) \\
&= D_p l_{\exp(sX)} \left(\left.\frac{d}{d\tau}\right|_{\tau=0} \exp(\tau X)\cdot p \right) \\
&= 0 \quad \forall s \in \mathbb{R}.
\end{aligned}$$

Therefore, the curve $\exp(tX)\cdot p$ is constant and equal to $\exp(0)\cdot p = p$. This implies that $\exp(tX) \in G_p$ for all $t \in \mathbb{R}$ and thus $X \in \mathfrak{g}_p$ by Corollary 1.8.11. □

Definition 3.2.11 Let Φ be an action of a group G on a set M. We distinguish three cases, depending on whether the orbit map is surjective, injective or bijective for every $p \in M$.

1. The action is called **transitive** if the orbit map is surjective for every $p \in M$. In other words, M consists of only one orbit, $M = \mathcal{O}_p$ for every $p \in M$. We then call M a **homogeneous space** for G.
2. The action is called **free** if the orbit map is injective for every $p \in M$.
3. The action is called **simply transitive** if it is both transitive and free, i.e. if the orbit map for every $p \in M$ is a bijection from G onto M.

We leave it as an exercise to show the following properties of G-actions on M:

1. The orbit map is surjective for one $p \in M$ if and only if it is surjective for all $p \in M$.
2. The action is transitive if and only if M/G consists of precisely one point.
3. The orbit map of a point $p \in M$ is injective if and only if the isotropy group of p is trivial, $G_p = \{e\}$.
4. The action is free if and only if $g \cdot p \neq p$ for all $p \in M$, $g \neq e \in G$. Hence the action is free if and only if all points in M have trivial isotropy group or, equivalently, all group elements $g \neq e$ have empty fixed point set.

As a consequence of Proposition 3.2.10 we see:

Corollary 3.2.12 (Orbit Maps of Smooth Free Actions) *If Φ is a smooth free action of a Lie group G on a manifold M, then the orbit maps $\phi_p\colon G \to M$ are injective immersions for every point $p \in M$. If G is compact, then the orbit maps are embeddings.*

In the case of a free action of a compact Lie group G each orbit is therefore an embedded submanifold diffeomorphic to G.

Definition 3.2.13 An action Φ is called **faithful** or **effective** if the induced homomorphism $\phi: G \to \mathrm{S}(M)$ is injective.
It is not difficult to see that if one point in M has trivial isotropy group, then the action is faithful. We can always make a group action faithful by passing to the induced action of the quotient group $G/\ker\phi$.

It is sometimes important to compare actions of a group G on two sets M and N. In particular, we would like to have a notion of isomorphism of group actions.

Definition 3.2.14 Let $\Phi: G\times M \to M$ and $\Psi: G\times N \to N$ be left actions of a group G on sets M and N. Then a G-**equivariant** map $f: M \to N$ is a map such that

$$f(g\cdot p) = g\cdot f(p) \quad \forall p \in M, g \in G.$$

If G is a topological (Lie) group and the actions continuous (smooth), we demand in addition that f is continuous (smooth). A G-equivariant bijection (homeomorphism, diffeomorphism) is called an **isomorphism** of G-actions. There are analogous definitions in the case of right actions.

3.3 Examples of Group Actions

We discuss some common examples of group actions, in particular, smooth actions of Lie groups on manifolds.

It is quite easy to define group actions of *discrete abelian groups* on manifolds: Any diffeomorphism $f: M \to M$ defines a smooth group action

$$\begin{aligned} \mathbb{Z}\times M &\longrightarrow M \\ (k,p) &\longmapsto k\cdot p = f^k(p). \end{aligned}$$

If f happens to be periodic, $f^n = \mathrm{Id}_M$ for some integer n, then this defines a smooth group action of the cyclic group $\mathbb{Z}_n = \mathbb{Z}/n\mathbb{Z}$:

$$\begin{aligned} \mathbb{Z}_n\times M &\longrightarrow M \\ ([k],p) &\longmapsto [k]\cdot p = f^k(p). \end{aligned}$$

If $f_1,\ldots,f_m$ are pairwise commuting diffeomorphisms of M, then

$$\begin{aligned} \mathbb{Z}^m\times M &\longrightarrow M \\ (k_1,\ldots,k_m,p) &\longmapsto (k_1,\ldots,k_m)\cdot p = f_1^{k_1}\circ\ldots\circ f_m^{k_m}(p) \end{aligned}$$

is a smooth group action.

An example of a simply transitive action is to take $M = G$ for an arbitrary group G and let G act on itself by left (and right) translations:

$$\begin{aligned}\Phi: G \times G &\longrightarrow G \\ (g, h) &\longmapsto g \cdot h.\end{aligned}$$

Another type of action that is easy to define is given by *group representations*. Let $\rho: G \rightarrow \mathrm{GL}(V)$ be a representation of a Lie group G on a real (or complex) vector space V. Then

$$\begin{aligned}\Phi: G \times V &\longrightarrow V \\ (g, v) &\longmapsto g \cdot v = \rho(g)v\end{aligned}$$

is a smooth left action on the manifold V (which is diffeomorphic to a Euclidean space). Such an action is called **linear**. We can also define a smooth right action by

$$\begin{aligned}\Phi: V \times G &\longrightarrow V \\ (v, g) &\longmapsto v \cdot g = \rho(g)^{-1}v.\end{aligned}$$

Note that it is important to take the inverse of $\rho(g)^{-1}$, otherwise the first property of a right action is in general not satisfied (see Exercise 3.12.1).

In both cases, the orbit of $0 \in V$ consists only of one point,

$$G \cdot 0 = \{0\}$$

and thus the isotropy group of 0 is all of G,

$$G_0 = G.$$

For a non-zero vector $v \neq 0$ the isotropy group in general will be a proper subgroup of G,

$$G_v \subsetneq G.$$

This is the basic mathematical idea behind **symmetry breaking** (from the full group G to the subgroup G_v), one of the centrepieces of the Standard Model that we discuss in Chap. 8.

For a linear representation, the homomorphism induced by the action has image in $\mathrm{GL}(V)$

$$\phi = \rho: G \longrightarrow \mathrm{GL}(V).$$

The action is faithful if and only if the representation is faithful.

Suppose in addition that $V \cong \mathbb{R}^n$ and the representation ρ is orthogonal. Then ρ has image in $\mathrm{O}(n)$ and the action maps the unit sphere S^{n-1} in $\mathbb{R}^n$ around the origin to itself. We therefore get a smooth left action

$$\begin{aligned} G \times S^{n-1} &\longrightarrow S^{n-1} \\ (g, v) &\longmapsto \rho(g)v. \end{aligned}$$

Similarly, if $V \cong \mathbb{C}^n$ and the representation ρ is unitary, so that it has image in $\mathrm{U}(n)$, then the action preserves the unit sphere S^{2n-1} in $\mathbb{C}^n$. We get a smooth left action

$$\begin{aligned} G \times S^{2n-1} &\longrightarrow S^{2n-1} \\ (g, v) &\longmapsto \rho(g)v. \end{aligned}$$

Finally, assume that $V \cong \mathbb{H}^n$ and the representation ρ is **quaternionic unitary**, by which we mean that ρ has image in $\mathrm{Sp}(n)$. Then the action preserves the standard symplectic scalar product (see Definition 1.2.9) and induces a smooth left action on the unit sphere S^{4n-1} in $\mathbb{H}^n$:

$$\begin{aligned} G \times S^{4n-1} &\longrightarrow S^{4n-1} \\ (g, v) &\longmapsto \rho(g)v. \end{aligned}$$

In each case we can similarly define right actions, using the inverses $\rho(g)^{-1}$. These actions on spheres are again called **linear**.

An important special case of this construction is the following:

Definition 3.3.1 Consider the groups $\mathbb{R}^*$, $\mathbb{C}^*$ and $\mathbb{H}^*$ of non-zero real, complex and quaternionic numbers. We define for $\mathbb{K} = \mathbb{R}, \mathbb{C}, \mathbb{H}$ the following linear right actions by scalar multiplication:

$$\mathbb{K}^{n+1} \setminus \{0\} \times \mathbb{K}^* \longrightarrow \mathbb{K}^{n+1} \setminus \{0\}.$$

These actions are free and induce the following free linear right actions of the groups of real, complex and quaternionic numbers of unit norm on unit spheres:

$$\begin{aligned} S^n \times S^0 &\longrightarrow S^n \\ S^{2n+1} \times S^1 &\longrightarrow S^{2n+1} \\ S^{4n+3} \times S^3 &\longrightarrow S^{4n+3} \\ (x, \lambda) &\longmapsto x\lambda. \end{aligned}$$

These actions are called **Hopf actions**. The most famous example is the action of S^1 on S^3 that we already considered at the beginning of Sect. 3.1.

Note that $S^0 \cong \mathbb{Z}_2$, $S^1 \cong \mathrm{U}(1)$ and $S^3 \cong \mathrm{SU}(2)$. We shall see later in Example 3.7.34 that the *quotient spaces* under these free actions are smooth manifolds

$$\begin{aligned} \mathbb{RP}^n &= S^n/\mathbb{Z}_2 \\ \mathbb{CP}^n &= S^{2n+1}/\mathrm{U}(1) \\ \mathbb{HP}^n &= S^{4n+3}/\mathrm{SU}(2) \end{aligned}$$

of dimension n, $2n$, $4n$, called **real, complex** and **quaternionic projective space**.

We consider another example of linear actions on spheres.

Theorem 3.3.2 (Linear Transitive Actions of Classical Groups) *The defining (fundamental) representations of* $\mathrm{O}(n)$, $\mathrm{SO}(n)$, $\mathrm{U}(n)$, $\mathrm{SU}(n)$ *and* $\mathrm{Sp}(n)$ *define the following linear transitive actions on spheres with associated isotropy groups of the vector* e_1*:*

1. $\mathrm{O}(n)$*-action on* S^{n-1} *with isotropy group*

$$\begin{pmatrix} 1 & 0 \\ 0 & \mathrm{O}(n-1) \end{pmatrix} \cong \mathrm{O}(n-1).$$

Similarly, there is an

2. $\mathrm{SO}(n)$*-action on* S^{n-1} *with isotropy group isomorphic to* $\mathrm{SO}(n-1)$.
3. $\mathrm{U}(n)$*-action on* S^{2n-1} *with isotropy group isomorphic to* $\mathrm{U}(n-1)$.
4. $\mathrm{SU}(n)$*-action on* S^{2n-1} *with isotropy group isomorphic to* $\mathrm{SU}(n-1)$.
5. $\mathrm{Sp}(n)$*-action on* S^{4n-1} *with isotropy group isomorphic to* $\mathrm{Sp}(n-1)$.

For $\mathbb{K} = \mathbb{R}, \mathbb{C}, \mathbb{H}$ *the defining (fundamental) representations of* $\mathrm{GL}(n, \mathbb{K})$ *and* $\mathrm{SL}(n, \mathbb{K})$ *define the following linear transitive actions with associated isotropy groups of the vector* e_1*:*

6. $\mathrm{GL}(n, \mathbb{K})$*-action on* $\mathbb{K}^n \setminus \{0\}$ *with isotropy group*

$$\begin{pmatrix} 1 & \mathbb{K}^{n-1} \\ 0 & \mathrm{GL}(n-1, \mathbb{K}) \end{pmatrix}.$$

7. $\mathrm{GL}(n, \mathbb{R})_+$*-action on* $\mathbb{R}^n \setminus \{0\}$ *with isotropy group*

$$\begin{pmatrix} 1 & \mathbb{R}^{n-1} \\ 0 & \mathrm{GL}(n-1, \mathbb{R})_+ \end{pmatrix}.$$

8. $\mathrm{SL}(n, \mathbb{K})$*-action on* $\mathbb{K}^n \setminus \{0\}$ *with isotropy group*

$$\begin{pmatrix} 1 & \mathbb{K}^{n-1} \\ 0 & \mathrm{SL}(n-1, \mathbb{K}) \end{pmatrix}.$$

Proof This is an exercise. The case of $\mathrm{SL}(n, \mathbb{H})$ uses the following lemma. □

Lemma 3.3.3 *For $A \in \mathrm{Mat}(m \times m, \mathbb{H})$ and $v \in \mathbb{H}^m$ the following equation holds:*

$$\det \begin{pmatrix} 1 & v \\ 0 & A \end{pmatrix} = \det(A).$$

Proof This is an exercise. □

We saw above that from a representation of a group on a vector space, we sometimes get group actions on other manifolds, in particular on spheres. We now show that from smooth actions on manifolds we also get representations on certain vector spaces.

Let $\Phi: G \times M \to M$ be a smooth (left) action of a Lie group G on a manifold M. Let $p \in M$ be a point and G_p its isotropy subgroup. By Proposition 3.2.9 the isotropy group G_p is an embedded Lie subgroup of G. The differential of the left translation l_g is a map

$$l_{g*} = D_p l_g : T_pM \longrightarrow T_pM,$$

for all $g \in G_p$. This is an isomorphism with inverse $l_{g^{-1}*}$.

Theorem 3.3.4 (Isotropy Representation) *The map*

$$\begin{aligned} \rho_p: G_p &\longrightarrow \mathrm{GL}(T_pM) \\ g &\longmapsto l_{g*} \end{aligned}$$

is a representation of the isotropy group G_p on T_pM, called the **isotropy representation**.

Proof We follow [142]. For $g, h \in G$ we calculate

$$\begin{aligned} \rho_p(gh) &= D_p l_{gh} \\ &= D_p(l_g \circ l_h) \\ &= \rho_p(g) \circ \rho_p(h), \end{aligned}$$

where we used the chain rule. Hence ρ_p is a group homomorphism. We want to show that ρ_p is smooth. Let $v \in T_pM$ be arbitrary and fixed. Then the map $\rho_p(\cdot)v$ is the composition of smooth maps

$$G_p \longrightarrow TG_p \times TM \longrightarrow T(G \times M) \longrightarrow TM$$

given by

$$g \longmapsto ((g, 0), (p, v)) \longmapsto ((g, p), (0, v)) \longmapsto D_{(g,p)}\Phi(0, v).$$

It follows that ρ_p is a smooth homomorphism, hence a representation. □
We get an analogous isotropy representation for right actions using the differential of right translations. Here is an almost trivial example of this construction.

Example 3.3.5 Let G be a Lie group, ρ a G-representation on a vector space V and $\Phi: G \times V \to V$ the induced linear action. Then the isotropy group of $0 \in V$ is all of G,

$$G_0 = G,$$

and the isotropy representation on $T_0V \cong V$ can be identified with ρ itself

$$\rho_0 = \rho,$$

because the action is linear.
Here is a more interesting example:

Example 3.3.6 Every Lie group G acts on itself on the left by conjugation:

$$\begin{aligned} G \times G &\longrightarrow G \\ (g, h) &\longmapsto c_g(h) = ghg^{-1}. \end{aligned}$$

The isotropy group of $e \in G$ is the full group G,

$$G_e = G,$$

and the isotropy representation on $T_eG \cong \mathfrak{g}$ is the adjoint representation

$$\rho_e = \mathrm{Ad}_G.$$

The adjoint representation can thus be seen as a special case of the general construction of isotropy representations.

3.4 Fundamental Vector Fields

Suppose a Lie group G acts smoothly on a manifold M. We want to discuss a construction that defines for every vector in the Lie algebra $\mathfrak{g}$ a certain vector field on M. These vector fields correspond to an **infinitesimal action** of $\mathfrak{g}$ on M (the construction only works for smooth Lie group actions on manifolds).

We can think of this from an abstract point of view as follows: if $f: G \to H$ is a Lie group homomorphism, then we saw in Sect. 1.5.3 that there is an induced Lie algebra homomorphism

$$f_*: \mathfrak{g} \longrightarrow \mathfrak{h}.$$

Suppose now that the Lie group G acts smoothly on a manifold M. We know that this action corresponds to a homomorphism

$$\phi: G \longrightarrow \mathrm{Diff}(M),$$

where ϕ is in a certain sense smooth. We can ask whether there is again an induced homomorphism on the level of Lie algebras.

We first have to determine the Lie algebra of the diffeomorphism group Diff(M): note that if Y is a vector field on M, then its *flow* generates a 1-parameter family of diffeomorphisms of M. If we think of the flow of Y as an exponential map applied to Y, it is clear that the Lie algebra of Diff(M) consists of the Lie algebra $\mathfrak{X}(M)$ of vector fields on M with the standard commutator (this is plausible even if we do not formally define Diff(M) as an infinite-dimensional Lie group). Given a Lie group action of G on M we therefore look for an induced Lie algebra homomorphism

$$\phi_*: \mathfrak{g} \longrightarrow \mathfrak{X}(M).$$

For example, in the case of the Hopf action

$$\begin{aligned} \Phi: S^3 \times \mathrm{U}(1) &\longrightarrow S^3 \\ (v, w, \lambda) &\longmapsto (v\lambda, w\lambda) \end{aligned}$$

it follows from the definition below that the induced homomorphism

$$\phi_*: \mathfrak{u}(1) \cong i\mathbb{R} \longrightarrow \mathfrak{X}(S^3)$$

is given by

$$\phi_*(ix)_{(v,w)} = (ivx, iwx)$$

with $x \in \mathbb{R}$. Here $\phi_*(ix)$ is indeed a tangent vector field on S^3.

Definition 3.4.1 Let G be a Lie group and M a manifold. Suppose that $M \times G \rightarrow M$ is a right action. For $X \in \mathfrak{g}$ we define the associated **fundamental vector field** $\tilde{X}$ on M by

$$\tilde{X}_p = \frac{d}{dt}\bigg|_{t=0} (p \cdot \exp(tX)).$$

If we denote by ϕ_p the orbit map for the right action,

$$\begin{aligned} \phi_p : G &\longrightarrow M \\ g &\longmapsto p \cdot g, \end{aligned}$$

then

$$\tilde{X}_p = (D_e\phi_p)(X_e).$$

Similarly, suppose that $G \times M \rightarrow M$ is a left action. Then we define the fundamental vector field by

$$\tilde{X}_p = \frac{d}{dt}\bigg|_{t=0} (\exp(-tX) \cdot p)$$

for $p \in M$. If we denote by ϕ'_p the following orbit map for the left action,

$$\begin{aligned} \phi'_p : G &\longrightarrow M \\ g &\longmapsto g^{-1} \cdot p, \end{aligned}$$

then

$$\tilde{X}_p = (D_e\phi'_p)(X_e).$$

The minus sign in the definition of the fundamental vector field for left actions has a reason that will become clear in Proposition 3.4.4.

The formula for the fundamental vector fields has the following interpretation: recall that vectors X in the Lie algebra define one-parameter subgroups, given by $\exp(tX)$ with $t \in \mathbb{R}$. The action of such a subgroup on a point $p \in M$ defines a curve in M and the fundamental vector field in p is given as the velocity vector of this curve at $t = 0$ (up to the sign in the case of left actions).

Example 3.4.2 Let $\rho: G \rightarrow \mathrm{GL}(V)$ be a representation of a Lie group G on a vector space V. The representation defines a left action

$$\Phi: G \times V \longrightarrow V.$$

Let $\rho_*: \mathfrak{g} \to \text{End}(V)$ be the induced representation of the Lie algebra. For $X \in \mathfrak{g}$, the fundamental vector field $\tilde{X}$ is then given by

$$\begin{aligned}\tilde{X}_v &= \left.\frac{d}{dt}\right|_{t=0} (\exp(-tX) \cdot v) \\ &= -\rho_*(X)(v) \quad \forall v \in V.\end{aligned}$$

Here are some properties of fundamental vector fields.

Proposition 3.4.3 (Fundamental Vector Fields of Free Actions) *Let G be a Lie group acting on a smooth manifold M. If the action is free, then the map*

$$\begin{aligned}\phi_*: \mathfrak{g} &\longrightarrow \mathfrak{X}(M) \\ X &\longmapsto \tilde{X}\end{aligned}$$

is injective.

Proof This follows from Proposition 3.2.10. □

Proposition 3.4.4 (Fundamental Vector Fields Define Lie Algebra Homomorphism) *Let G be a Lie group acting on a manifold M on the right or left. The map*

$$\begin{aligned}\phi_*: \mathfrak{g} &\longrightarrow \mathfrak{X}(M) \\ X &\longmapsto \tilde{X}\end{aligned}$$

that associates to a Lie algebra element the corresponding fundamental vector field on M is a Lie algebra homomorphism, i.e. it is an $\mathbb{R}$-linear map such that

$$\widetilde{[X, Y]} = [\tilde{X}, \tilde{Y}] \quad \forall X, Y \in \mathfrak{g}.$$

In particular, the set of all fundamental vector fields is a Lie subalgebra of the Lie algebra of all vector fields on M.

Proof We prove the claim if G acts on the left on M. The proof for right actions follows similarly. Fix a point $p \in M$ and let ϕ'_p denote the following orbit map

$$\begin{aligned}\phi'_p: G &\longrightarrow M \\ g &\longmapsto g^{-1} \cdot p.\end{aligned}$$

The second definition of $\tilde{X}$,

$$\tilde{X}_p = (D_e \phi'_p)(X_e)$$

shows that the map

$$\phi_*\colon \mathfrak{g} \longrightarrow \mathfrak{X}(M)$$
$$X \longmapsto \tilde{X}$$

is linear.

We want to show that the left-invariant vector field $X \in \mathfrak{g}$ and $\tilde{X} \in \mathfrak{X}(M)$ are ϕ_p'-related. For this we have to show that

$$\tilde{X}_{\phi_p'(a)} = (D_a\phi_p')(X_a)$$

for all $a \in G$. We have, since X is a left-invariant vector field on G,

$$\begin{aligned}(D_a\phi_p')(X_a) &= (D_a\phi_p')(D_eL_a)\left(\frac{d}{dt}\bigg|_{t=0}\exp(tX)\right)\\ &= \frac{d}{dt}\bigg|_{t=0}(\exp(-tX)\,(a^{-1}\cdot p))\\ &= \tilde{X}_{a^{-1}p}\\ &= \tilde{X}_{\phi_p'(a)}.\end{aligned}$$

The claim now follows from Proposition A.1.49. □

Remark 3.4.5 The reason why we defined in Definition 3.4.1 the fundamental vector field for left actions with a minus sign in $\exp(-tX)$ is so that

$$\widetilde{[X,Y]} = [\tilde{X}, \tilde{Y}]$$

holds for all $X, Y \in \mathfrak{g}$. If we defined the fundamental vector field for left actions with $\exp(tX)$ instead (this is sometimes done in the literature), then we would get a minus sign here:

$$\widetilde{[X,Y]} = -[\tilde{X}, \tilde{Y}] \quad \forall X, Y \in \mathfrak{g},$$

because on the left-hand side we have to change the sign once and on the right-hand side twice.

It is sometimes useful to know how fundamental vector fields behave under right or left translations on the manifold. It will turn out that even though fundamental vector fields are defined using the group action, they are in general not invariant under the action.

Proposition 3.4.6 (Action of Right and Left Translations on Fundamental Vector Fields) *Suppose a Lie group G acts on a manifold M. Let $X \in \mathfrak{g}$ and $g \in G$.*

1. *If G acts on the right on M, then*

$$r_{g*}(\tilde{X}) = \tilde{Y},$$

where

$$Y = \mathrm{Ad}_{g^{-1}}X \in \mathfrak{g}.$$

2. *If G acts on the left on M, then*

$$l_{g*}(\tilde{X}) = \tilde{Z},$$

where

$$Z = \mathrm{Ad}_g X \in \mathfrak{g}.$$

Proof We prove the statement for right actions, the statement for left actions follows similarly. At a point $p \in M$ we calculate

$$\begin{aligned}(r_{g*}(\tilde{X}))_p &= (D_{pg^{-1}}r_g)(\tilde{X}_{pg^{-1}}) \\ &= (D_{pg^{-1}}r_g)\left(\frac{d}{dt}\bigg|_{t=0} pg^{-1}\cdot\exp(tX)\right) \\ &= (D_e\phi_p)\left(\frac{d}{dt}\bigg|_{t=0} \alpha_{g^{-1}}(\exp tX)\right),\end{aligned}$$

with the orbit map

$$\begin{aligned}\phi_p\colon G &\longrightarrow M \\ g &\longmapsto p\cdot g.\end{aligned}$$

On the other hand

$$\begin{aligned}Y &= \mathrm{Ad}_{g^{-1}}X \\ &= \frac{d}{dt}\bigg|_{t=0} \alpha_{g^{-1}}(\exp tX).\end{aligned}$$

This implies the claim by the second definition of the fundamental vector field. □

Corollary 3.4.7 (Translations of Fundamental Vector Fields Are Fundamental) *For a right (left) action of a Lie group G on a manifold M the right (left) translations of fundamental vector fields are again fundamental vector fields. If the Lie group G is abelian, then the fundamental vector fields are invariant under all right (left) translations.*

3.5 The Maurer–Cartan Form and the Differential of a Smooth Group Action

3.5.1 *Vector Space-Valued Forms*

Recall from Definition A.2.3 that a k-form on a real vector space V is defined as an alternating multilinear map

$$\lambda\colon \underbrace{V \times \cdots \times V}_{k} \longrightarrow \mathbb{R}.$$

The vector space of all k-forms on V is denoted by $\Lambda^k V^*$.

Suppose W is another real vector space. Then we define a **k-form on V with values in W** as an alternating, multilinear map

$$\lambda\colon \underbrace{V \times \cdots \times V}_{k} \longrightarrow W.$$

The vector space of all k-forms on V with values in W can be identified with the tensor product $\Lambda^k V^* \otimes W$.

Similarly we defined k-forms on a smooth manifold as alternating $\mathscr{C}^\infty(M)$-multilinear maps

$$\lambda\colon \underbrace{\mathfrak{X}(M) \times \cdots \times \mathfrak{X}(M)}_{k} \longrightarrow \mathscr{C}^\infty(M)$$

and we defined $\Omega^k(M)$ as the set of all k-forms on M; see Definition A.2.12.

We now define

$$\mathscr{C}^\infty(M, W)$$

as the set of all smooth maps from M into the vector space W (the vector space W has a canonical structure of a manifold, so that smooth maps into W are defined). A **k-form on M with values in W** is then an alternating $\mathscr{C}^\infty(M)$-multilinear map

$$\lambda\colon \underbrace{\mathfrak{X}(M) \times \cdots \times \mathfrak{X}(M)}_{k} \longrightarrow \mathscr{C}^\infty(M, W).$$

The set of all k-forms on M with values in W can be identified with $\Omega^k(M, W) = \Omega^k(M) \otimes_{\mathbb{R}} W$. One also calls the forms in $\Omega^k(M, W)$ **twisted with** W.

Remark 3.5.1 Note that there is no canonical wedge product of forms on a vector space or a manifold with values in a vector space W, because there is in general no canonical product $W \times W \to W$ (an exception is forms with values in $W = \mathbb{C}$, where there is indeed a canonical wedge product).

3.5.2 The Maurer–Cartan Form

The following notion of a vector space-valued form on a Lie group is useful for studying group actions and principal bundles. Let G be a Lie group with Lie algebra $\mathfrak{g}$.

Definition 3.5.2 The **Maurer–Cartan form** $\mu_G \in \Omega^1(G, \mathfrak{g})$ is the 1-form on G with values in $\mathfrak{g}$ defined by

$$(\mu_G)_g(v) = (D_g L_{g^{-1}})(v) \in T_e G \cong \mathfrak{g}$$

for all $g \in G$ and $v \in T_g G$. The Maurer–Cartan form is also called the **canonical form** or **structure form**.
The Maurer–Cartan form thus associates to a tangent vector v at the point $g \in G$ the unique left-invariant vector field X on G whose value at g is $X_g = v$ (equivalently, the generating vector of this vector field at $e \in G$).

Proposition 3.5.3 (Invariance of Maurer–Cartan Form Under Translations) *The Maurer–Cartan form has the following invariance properties under left and right translations:*

$$L_g^* \mu_G = \mu_G,$$
$$R_g^* \mu_G = \mathrm{Ad}_{g^{-1}} \circ \mu_G,$$

for all $g \in G$.

Proof We calculate for all $h \in G$ and $v \in T_h G$:

$$\begin{aligned} (R_g^* \mu_G)_h(v) &= (\mu_G)_{hg}(D_h R_g)(v) \\ &= (D_{hg} L_{g^{-1}h^{-1}})(D_h R_g)(v) \\ &= (D_e \alpha_{g^{-1}})(D_h L_{h^{-1}})(v) \\ &= \mathrm{Ad}_{g^{-1}}(\mu_G)_h(v). \end{aligned}$$

The statement for L_g follows similarly. □

3.5.3 The Differential of a Smooth Group Action

Recall that a smooth (right) action of a Lie group is a map $\Phi: M \times G \to M$ satisfying certain axioms. It is sometimes useful to determine the differential of this map in a given point $(x, g) \in M \times G$. The formula for this differential involves the Maurer–Cartan form.

Proposition 3.5.4 (The Differential of a Smooth Group Action) *Let G be a Lie group acting smoothly on the right on a manifold M,*

$$\Phi: M \times G \longrightarrow M.$$

Then under the canonical identification

$$T_{(x,g)}M \times G \cong T_xM \oplus T_gG,$$

the differential of the map Φ is given by

$$\begin{aligned} D_{(x,g)}\Phi: T_xM \oplus T_gG &\longrightarrow T_{xg}M \\ (X, Y) &\longmapsto (D_xr_g)(X) + \widetilde{\mu_G(Y)}_{xg}, \end{aligned}$$

where r_g denotes right translation and μ_G denotes the Maurer–Cartan form.

Proof Let $\phi_x: G \to M$ denote the orbit map

$$\phi_x(g) = xg.$$

Let $x(t)$ be a curve in M tangent to X and $g(t)$ a curve in G tangent to Y. Then

$$\begin{aligned} D_{(x,g)}\Phi(X, Y) &= D_{(x,g)}\Phi(X, 0) + D_{(x,g)}\Phi(0, Y) \\ &= D_{(x,g)}\Phi(\dot{x}(0), 0) + D_{(x,g)}\Phi(0, \dot{g}(0)) \\ &= (D_xr_g)(X) + (D_g\phi_x)(Y). \end{aligned}$$

Let $y \in \mathfrak{g}$ denote the left-invariant vector field corresponding to Y. Then $y = \mu_G(Y)$. In the proof of Proposition 3.4.4 we saw that

$$(D_g\phi_x)(Y) = \tilde{y}_{\phi_x(g)}$$

(we proved the statement for left actions, but the corresponding statement also holds for right actions). This proves the claim. □

3.6 Left or Right Actions?

In general there is no difference whether we assume that a group action is a left or right action. However, when we discuss homogeneous spaces in Sect. 3.8, there will be two different actions at the same time, which have to be compatible. We therefore make the following conventions:

- If we are interested in quotient spaces M/G, we take the G-action on M to be a *right action*. In particular, if $H \subset G$ is a subgroup and we want to consider G/H, then H acts on G on the right. When we consider principal bundles in Chap. 4, we will take the G-action on the principal bundle to be a right action as well. For example, the Hopf actions introduced in Definition 3.3.1 are right actions whose quotient spaces are the projective spaces.
- If we are interested in homogeneous spaces, i.e. spaces M with a transitive group action, we will take the G-action on M to be a *left action*. For example, the linear transitive actions on spheres introduced in Theorem 3.3.2 are left actions.

Usually we are not interested in the quotient space of a transitive group action, because it consists only of a single point, so that both cases do not overlap. Occasionally one encounters situations in the literature where we have a right G-action on M with quotient space M/G and a non-transitive left K-action on M/G. Then it makes sense to consider the quotient space $K\backslash M/G$ under the left K-action (we will not consider such quotients in the following).

3.7 *Quotient Spaces

An important objective in the study of group actions is to understand the quotient space of a given action. In this section we are specifically interested in the following question: Suppose that G is a Lie group acting smoothly on a manifold M. Under which circumstances does the quotient set M/G have the structure of a smooth manifold?

This question has many applications, because it is possible to construct new and interesting manifolds as quotients of this form (like projective spaces and lens spaces, to name only two examples). For instance, in the case of the Hopf action

$$\begin{aligned}\Phi: S^3 \times \mathrm{U}(1) &\longrightarrow S^3 \\ (v, w, \lambda) &\longmapsto (v\lambda, w\lambda),\end{aligned}$$

which is a *free* action, it can be shown that the quotient space $S^3/\mathrm{U}(1)$ is a smooth manifold diffeomorphic to $\mathbb{CP}^1 \cong S^2$.

It is useful to study the question of quotients in greater generality: we first consider quotients of manifolds (and topological spaces) under arbitrary equivalence relations and later the case of the equivalence relation defined by group actions.

We follow [130] for smooth manifolds and the excellent exposition in [139] in the general case. An additional reference is [89].

3.7.1 *Quotient Spaces Under Equivalence Relations on Topological Spaces*

Suppose X is a set and $\sim$ an equivalence relation on X. We can describe $\sim$ equivalently by a subset $R \subset X \times X$ so that

$$x \sim y \Leftrightarrow (x, y) \in R.$$

The equivalence class of an element $x \in X$ is the subset

$$[x] = \{y \in X \mid y \sim x\}.$$

As subsets of X, equivalence classes of two elements $x, x' \in X$ are either disjoint or identical. We denote by X/R the space of equivalence classes, called the **quotient space**

$$X/R = \{[x] \mid x \in X\}.$$

We have the canonical projection

$$\begin{aligned} \pi: X &\longrightarrow X/R \\ x &\longmapsto [x]. \end{aligned}$$

We now specialize to the case when X is a topological space. Then we define on X/R the usual **quotient topology** by setting $U \subset X/R$ open if and only if $\pi^{-1}(U) \subset X$ is open. It is easy to check that this indeed defines a topology on X/R. The canonical projection $\pi: X \to X/R$ is continuous. The following is well-known:

Lemma 3.7.1 *A map $f: X/R \to Y$ from a quotient space to another topological space is continuous if and only if $f \circ \pi$ is continuous:*

$$\begin{array}{ccc} X & & \\ {\scriptstyle \pi}\downarrow & \searrow^{f\circ\pi} & \\ X/R & \xrightarrow[f]{} & Y \end{array}$$

We are first interested in the following question: under which conditions is the quotient space X/R Hausdorff? The answer is given by the following lemma.

Lemma 3.7.2 (Hausdorff Property of Quotient Spaces Under Equivalence Relations) *Let X be a topological space.*

1. *If X/R is Hausdorff, then $R \subset X \times X$ is closed.*
2. *If $\pi: X \to X/R$ is open and $R \subset X \times X$ is closed, then X/R is Hausdorff.*

Remark 3.7.3 Note that we do not need to assume that X is Hausdorff.

Proof We use in the proof the following standard fact from point set topology: a topological space Y is Hausdorff if and only if the diagonal

$$\Delta = \{(y, y) \in Y \times Y \mid y \in Y\}$$

is a closed subset in $Y \times Y$. In the following, we denote by Δ the diagonal in the space $X/R \times X/R$.

1. The map

$$\pi \times \pi: X \times X \longrightarrow X/R \times X/R$$

is continuous. Since X/R is Hausdorff, the diagonal Δ is closed, hence the preimage $(\pi \times \pi)^{-1}(\Delta)$ is closed. We have

$$(x, y) \in (\pi \times \pi)^{-1}(\Delta) \Leftrightarrow (x, y) \in R.$$

Hence $R = (\pi \times \pi)^{-1}(\Delta)$ is closed in $X \times X$.

2. The map $\pi \times \pi$ is open and $(X \times X) \setminus R$ is open, hence its image in $X/R \times X/R$ is open. We have

$$\begin{aligned}([x], [y]) \in (\pi \times \pi)((X \times X) \setminus R) &\Leftrightarrow [x] \neq [y] \\ &\Leftrightarrow ([x], [y]) \in (X/R \times X/R) \setminus \Delta.\end{aligned}$$

It follows that Δ is closed and X/R is Hausdorff.

□

3.7.2 Quotient Spaces Under Equivalence Relations on Manifolds

We now consider the case of an equivalence relation R on a smooth manifold M and we would like to determine when the quotient space M/R is a smooth manifold. It is useful to demand that the smooth structure has the additional property that $\pi: M \to M/R$ is a submersion. Consider the following lemma.

Lemma 3.7.4 (Surjective Submersions Admit Local Sections) *Let $p: M \to N$ be a surjective submersion between smooth manifolds. Then p admits* **smooth local**

sections, *i.e. for each* $x \in N$ *there exists an open neighbourhood* $U \subset N$ *of* x *and a smooth map* $s: U \to M$ *such that* $p \circ s = \mathrm{Id}_U$.

Proof This follows from the normal form theorem for submersions (see Theorem A.1.28), because locally submersions are projections. □

The following lemma is very useful in applications.

Lemma 3.7.5 (Smoothness of Maps Out of the Target Space of a Surjective Submersion) *Let* $p: M \to N$ *be a surjective submersion. Then a map* $f: N \to Q$ *is smooth if and only if* $f \circ p: M \to Q$ *is smooth. Moreover,* f *is a submersion if and only if* $f \circ p$ *is a submersion and* f *is surjective if and only if* $f \circ p$ *is surjective.*

$$\begin{array}{ccc} M & & \\ {\scriptstyle p}\downarrow & \searrow^{f\circ p} & \\ N & \xrightarrow[f]{} & Q \end{array}$$

Proof If f is smooth, then $f \circ p$ is smooth. Conversely, assume that $f \circ p$ is smooth. Let $x \in N$ and $U \subset N$ an open neighbourhood of x with a smooth section $s: U \to M$ for p. On U we have $p \circ s = \mathrm{Id}_U$, hence

$$(f \circ p) \circ s = f.$$

Thus f is smooth on U and therefore on all of N.

The claim about submersions and surjectivity is clear, because p and its differential are surjective. □

Corollary 3.7.6 *Let* M *be a manifold and* $p: M \to N$ *a surjective map to a set* N. *Then* N *admits at most one structure of a smooth manifold so that* p *is a submersion.*

Proof Suppose N_1 and N_2 are structures of smooth manifolds on N so that p is a submersion in both cases. By Lemma 3.7.5 the identity map $\mathrm{Id}_N: N_1 \to N_2$ is a diffeomorphism.

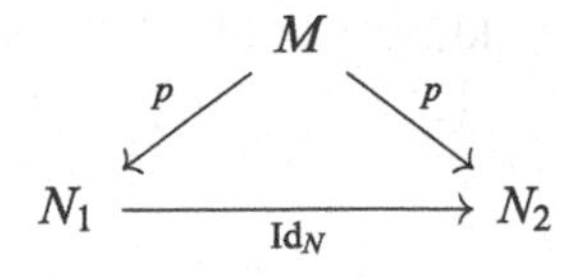

□

Corollary 3.7.7 (Uniqueness of Smooth Manifold Structure on Quotient Spaces) *Let* M *be a smooth manifold and* R *an equivalence relation on* M. *Then* *there exists at most one smooth structure on M/R so that $\pi: M \to M/R$ is a submersion.*

Remark 3.7.8 Lemma 3.7.5 and Corollary 3.7.7 are the reasons why the smooth structure on M/R should have the property that $\pi: M \to M/R$ is a submersion.

We assume from now on that M is a smooth manifold and R an equivalence relation on M. We first derive a necessary condition for M/R to be a smooth manifold such that π is a submersion.

Lemma 3.7.9 *Let M/R have the structure of a smooth manifold so that $\pi: M \to M/R$ is a surjective submersion. Then R is a closed embedded submanifold of $M \times M$ and the restrictions of the projections*

$$\mathrm{pr}_i|_R: R \longrightarrow M$$

are surjective submersions, for $i = 1, 2$.

Proof The graph of the projection

$$\Gamma = \{(x, \pi(x)) \in M \times M/R \mid x \in M\}$$

is a closed embedded submanifold of $M \times M/R$ and

$$F = \mathrm{Id}_M \times \pi: M \times M \longrightarrow M \times M/R$$

is a submersion. Therefore $F^{-1}(\Gamma)$ is a closed embedded submanifold of $M \times M$. We have

$$\begin{aligned} (x, y) \in F^{-1}(\Gamma) &\Leftrightarrow (x, \pi(y)) \in \Gamma \\ &\Leftrightarrow \pi(x) = \pi(y) \\ &\Leftrightarrow (x, y) \in R. \end{aligned}$$

This shows that R is a closed embedded submanifold of $M \times M$.

The map $F|_R: R \to \Gamma$ is a surjective submersion. The projection $\mathrm{pr}_1|_\Gamma: \Gamma \to M$ is also a surjective submersion, because

$$\mathrm{pr}_1|_\Gamma \circ (\mathrm{Id}_M, \pi) = \mathrm{Id}_M: M \longrightarrow M.$$

It follows that

$$\mathrm{pr}_1|_\Gamma \circ F|_R: R \longrightarrow M$$

is a smooth surjective submersion. This map is equal to $\mathrm{pr}_1|_R$. The claim for $\mathrm{pr}_2|_R$ follows by symmetry of the equivalence relation. □

It is a non-trivial fact that the converse also holds.

Theorem 3.7.10 (Godement's Theorem on the Manifold Structure of Quotient Spaces) *Let R be an equivalence relation on a manifold M. Suppose that R is a closed embedded submanifold of $M \times M$ and $\mathrm{pr}_1|_R: R \rightarrow M$ a surjective submersion. Then M/R has a unique structure of a smooth manifold such that the canonical projection $\pi: M \rightarrow M/R$ is a surjective submersion.*

The proof of Godement's Theorem, which is not easy and quite technical, is deferred to Sect. 3.11. We first want to derive some consequences of it.

3.7.3 Quotient Spaces Under Continuous Group Actions

We begin more generally by considering the case of a topological group G acting continuously on the right on a topological space X. The map defining the action is

$$\Phi: X \times G \longrightarrow X.$$

We would like to determine under which conditions the quotient space X/G is Hausdorff (we do not need to assume that X itself is Hausdorff).

Lemma 3.7.11 *The canonical projection $\pi: X \rightarrow X/G$ is open.*

Proof Let U be an open subset of x. We have to show that $\pi^{-1}(\pi(U))$ is open in X. However,

$$\pi^{-1}(\pi(U)) = \bigcup_{g \in G} U \cdot g$$

and each of the sets $U \cdot g$ is open, because right translations are homeomorphisms. □

Corollary 3.7.12 (Hausdorff Property of Quotient Spaces Under Continuous Group Actions) *The quotient space X/G is Hausdorff if and only if the map*

$$\begin{aligned} \Psi: X \times G &\longrightarrow X \times X \\ (x, g) &\longmapsto (x, xg) \end{aligned}$$

has closed image.

Proof According to Lemma 3.7.2 and since $\pi: X \to X/G$ is open, the space X/G is Hausdorff if and only if the equivalence relation $R \subset X \times X$ is closed. We have

$$(x, y) \in R \Leftrightarrow \exists g \in G: y = xg.$$

This shows that R is equal to the image of the map Ψ. □

Let G be a topological group and $H \subset G$ a subgroup with the subspace topology. Then H acts continuously on the right on G via right translations

$$\Phi: G \times H \longrightarrow G$$
$$(g, h) \longmapsto gh.$$

We get a topological quotient space G/H.

Corollary 3.7.13 (Hausdorff Property of G/H) *Let G be a topological group and $H \subset G$ a subgroup. Then G/H is Hausdorff if and only if H is a closed set in the topology of G.*

Proof According to Corollary 3.7.12 we have to show that the image of

$$\Psi: G \times H \longrightarrow G \times G$$
$$(g, h) \longmapsto (g, gh)$$

is closed if and only if H is closed. Consider the map

$$T: G \times G \longrightarrow G \times G$$
$$(g, g') \longmapsto (g, gg').$$

This map is a homeomorphism and we have $\Psi = T|_{G\times H}$. Hence the image of Ψ is closed in $G \times G$ if and only if $G \times H$ is closed in $G \times G$. This happens if and only if H is closed in G. □

As an aside we note the following result, which is useful in applications such as Example 3.8.11 (we follow [34] and [142] in the proof).

Proposition 3.7.14 (Connectedness of G and G/H) *Let G be a topological group and $H \subset G$ a closed subgroup. Suppose that H is connected. Then G/H is connected if and only if G is connected.*

Proof If G is connected, then G/H is connected, because the canonical projection $\pi: G \to G/H$ is surjective and continuous.

Conversely, suppose that G/H is connected and

$$G = U \cup V,$$

where U, V are open non-empty subsets of G. We have to show that U and V cannot be disjoint.

By Lemma 3.7.11 the sets $\pi(U)$ and $\pi(V)$ are open and non-empty in G/H with

$$G/H = \pi(U) \cup \pi(V).$$

Since G/H is connected there exists an element

$$[g] \subset \pi(U) \cap \pi(V).$$

Because of $G = U \cup V$ we get

$$\mathcal{O}_g = gH = (gH \cap U) \cup (gH \cap V).$$

By construction $gH \cap U$, $gH \cap V$ are open and non-empty in gH. Since gH is connected, the claim follows. □

3.7.4 Proper Group Actions

We consider some topological notions that are useful in applications to group actions.

Definition 3.7.15 A topological space X is called **locally compact** if every point in X has a compact neighbourhood.

Lemma 3.7.16 *Let X be a locally compact Hausdorff space. Then a subset $A \subset X$ is closed if and only if the intersection of A with any compact subset of X is compact.*

Proof If A is closed, then the intersection with any compact subset of X is compact. Conversely, assume that $A \cap K$ is compact for every compact subset $K \subset X$. Let $x \in X \setminus A$. Since X is locally compact, there exists an open neighbourhood $U \subset X$ of x contained in a compact subset $K \subset X$. By assumption, $C = A \cap K$ is compact, hence closed in X, since X is Hausdorff. Then $U \setminus C = U \cap (X \setminus C)$ is an open neighbourhood of x contained in $X \setminus A$. This implies the claim. □

Definition 3.7.17 A continuous map $f: X \to Y$ between topological spaces is called **proper** if the preimage $f^{-1}(K)$ of every compact subset $K \subset Y$ is compact in X.

Lemma 3.7.18 *Let $f: X \to Y$ be a continuous proper map between topological spaces, where Y is locally compact Hausdorff. Then f is a closed map.*

Proof Let $A \subset X$ be a closed set. By Lemma 3.7.16 we have to show that $f(A) \cap K$ is compact for every compact subset $K \subset Y$. However,

$$f(A) \cap K = f\left(A \cap f^{-1}(K)\right).$$

Since f is proper, the set $f^{-1}(K)$ is compact and thus $A \cap f^{-1}(K)$ and $f\left(A \cap f^{-1}(K)\right)$ are compact. This implies the claim. □

Lemma 3.7.19 *Let $f: X \to Y$ be a closed continuous map between topological spaces such that $f^{-1}(y)$ is compact for all $y \in Y$. Then f is proper.*

Proof The proof is left as an exercise. □

We consider the following type of group actions.

Definition 3.7.20 A continuous action of a topological group G on a topological space X is called **proper** if the map

$$\Psi: X \times G \longrightarrow X \times X$$
$$(x, g) \longmapsto (x, xg)$$

is proper.

Corollary 3.7.21 (Map Ψ Is Closed If Action Is Proper) *Let $X \times G \to X$ be a continuous, proper action of a topological group G on a topological space X, where X is locally compact Hausdorff. Then the map*

$$\Psi: X \times G \longrightarrow X \times X$$
$$(x, g) \longmapsto (x, xg)$$

is closed. In particular, X/G is Hausdorff.

Proof This follows from Lemma 3.7.18 and Corollary 3.7.12. □

Here is a general situation in which group actions are proper.

Proposition 3.7.22 (Actions of Compact Topological Groups Are Proper) *Let $X \times G \to X$ be a continuous action of a topological group G on a Hausdorff space X. Suppose that G is compact. Then the action is proper.*

Proof Let $K \subset X \times X$ be a compact subset. Then

$$L = \mathrm{pr}_1(K)$$

is a compact subset of X. If $\Psi(x, g) = (x, xg) \in K$, then $x \in L$, hence

$$\Psi^{-1}(K) = \Psi^{-1}(K) \cap (L \times G).$$

However, $\Psi^{-1}(K)$ is closed in $X \times G$ and $L \times G$ is compact, hence $\Psi^{-1}(K)$ is compact. □

Corollary 3.7.23 *Let $X \times G \to X$ be a continuous action of a topological group G on a locally compact Hausdorff space X. Suppose that G is compact. Then X/G is Hausdorff.*

3.7.5 Quotient Spaces Under Smooth Group Actions

We have now arrived at the central topic in this section: to determine under which conditions the quotient of a smooth action of a Lie group on a smooth manifold is again a smooth manifold.

Definition 3.7.24 We call a smooth right action of a Lie group G on a manifold M **principal** if the action is free and the map

$$\Psi: M \times G \longrightarrow M \times M$$
$$(p, g) \longmapsto (p, pg)$$

is closed.

Theorem 3.7.25 (Manifold Structure on Quotient Spaces Under Principal Actions of Lie Groups) *Suppose that Φ is a principal right action of the Lie group G on the manifold M. Then M/G has a unique structure of a smooth manifold such that $\pi: M \to M/G$ is a submersion.*

Proof Since the action of G on M is free, the map Ψ is injective. We want to show that Ψ is an immersion: by Proposition 3.5.4 the differential of Ψ is given by

$$D_{(x,g)}(X, Y) = \left(X, (D_x r_g)(X) + \widetilde{\mu_G(Y)}_{xg}\right).$$

If $D_{(x,g)}(X, Y) = (0, 0)$, then $X = 0$ and $\widetilde{\mu_G(Y)}_{xg} = 0$. From Proposition 3.4.3 we get $\mu_G(Y) = 0$, hence $Y = 0$. This proves that the differential of Ψ is injective.

Since Ψ is a closed injective map, it is a homeomorphism onto its image R and thus an embedding. Hence R is a closed embedded submanifold of $M \times M$. According to Theorem 3.7.10 it remains to show that $\mathrm{pr}_1|_R: R \to M$ is a submersion. However,

$$\mathrm{pr}_1|_R \circ \Psi: M \times G \longrightarrow M$$

is just $\mathrm{pr}_1: M \times G \to M$ and thus a submersion. This implies the claim. □

Corollary 3.7.26 (The Differential of the Projection $\pi: M \to M/G$) *Suppose that Φ is a principal right action of the Lie group G on the manifold M. Then the dimension of the quotient manifold M/G is given by*

$$\dim(M/G) = \dim M - \dim G.$$

In particular, the kernel of the differential

$$D_p\pi: T_pM \longrightarrow T_{[p]}M/G$$

at a point $p \in M$ is equal to the tangent space $T_p\mathcal{O}_p$ of the G-orbit through p.

Proof The claim about the dimension of M/G follows from the proof of Theorem 3.7.10. The second claim then follows from Corollary 3.2.12. □

Remark 3.7.27 For a free action of a Lie group G on a manifold M the preimage $\Psi^{-1}(p,q)$ of every point $(p,q) \in M \times M$ is either empty or consists of a single element (and is thus a compact set). Lemma 3.7.19 implies that principal Lie group actions on manifolds are proper. Together with Corollary 3.7.21 we conclude that principal actions are equivalent to free proper Lie group actions on manifolds.
We formulate this as follows:

Corollary 3.7.28 (Free Proper Actions of Lie Groups Are Equivalent to Principal Actions) *Suppose that $M \times G \to M$ is a smooth free action of a Lie group G on a manifold M. Then the action is principal if and only if it is proper.*
Proposition 3.7.22 then implies:

Corollary 3.7.29 (Free Actions of Compact Lie Groups Are Principal) *Suppose that $M \times G \to M$ is a smooth free action of a compact Lie group G on a manifold M. Then the action is principal.*
We get the following corollary, which is very useful in applications.

Corollary 3.7.30 (Quotients Under Free Actions of Compact Lie Groups) *Let G be a compact Lie group acting smoothly and freely on a manifold M. Then M/G has a unique structure of a smooth manifold such that $\pi: M \to M/G$ is a submersion.*

Proof This follows from Theorem 3.7.25 and Corollary 3.7.29. □

Remark 3.7.31 (Fundamental Groups) If in the situation of Corollary 3.7.30 the manifolds M and G are connected, it follows from Exercise 3.12.6 that π_* maps the fundamental group of M surjectively onto the fundamental group of M/G. In particular, if M is simply connected, then M/G is simply connected.

Example 3.7.32 (Quotients Under Free Actions of Finite Groups) Finite groups with the discrete topology are compact. Hence if a finite group G acts freely and smoothly on a manifold M, then the quotient M/G is a smooth manifold such that the canonical projection is a submersion.

Example 3.7.33 (Lens Spaces) Let $p > 0$ be an integer and $\alpha = e^{2\pi i/p} \in S^1$ the corresponding root of unity. Let $q \neq 0$ be an integer coprime to p. We consider the following smooth action of $\mathbb{Z}_p = \mathbb{Z}/p\mathbb{Z}$ on the unit sphere $S^3 \subset \mathbb{C}^2$:

$$\begin{aligned} S^3 \times \mathbb{Z}_p &\longrightarrow S^3 \\ (z_1, z_2, [k]) &\longmapsto (z_1, z_2) \cdot [k] = \left(z_1 \alpha^k, z_2 \alpha^{kq}\right). \end{aligned}$$

This action is free: if $z_1 \neq 0$ and $z_1\alpha^k = z_1$, then $[k] = 0$. If $z_2 \neq 0$ and $z_2\alpha^{kq} = z_2$, then again $[k] = 0$, because q is coprime to p. According to Example 3.7.32 the quotient

$$L(p,q) = S^3/\mathbb{Z}_p$$

under this action is a smooth 3-dimensional manifold. These manifolds are called **lens spaces**.

Example 3.7.34 (Projective Spaces Are Smooth Manifolds) The projective spaces

$$\begin{aligned} \mathbb{RP}^n &= S^n/\mathbb{Z}_2, \\ \mathbb{CP}^n &= S^{2n+1}/\mathrm{U}(1), \\ \mathbb{HP}^n &= S^{4n+3}/\mathrm{SU}(2) \end{aligned}$$

are quotients of manifolds under smooth free actions of compact Lie groups and therefore smooth manifolds such that the canonical projections are submersions.

Let G be a Lie group and $H \subset G$ a closed subgroup. According to Cartan's Theorem 1.1.44 the subgroup H is an embedded Lie subgroup of G. There is a smooth right action of H on G by right translations

$$\begin{aligned} \Phi\colon G \times H &\longrightarrow G \\ (g,h) &\longmapsto gh. \end{aligned}$$

Corollary 3.7.35 (Manifold Structure on G/H) *Let G be a Lie group and $H \subset G$ a closed subgroup. Then the right action of H on G is principal and G/H has a unique structure of a smooth manifold such that $\pi\colon G \to G/H$ is a submersion.*

Proof It is clear that the orbit maps

$$\begin{aligned} \phi_g\colon H &\longrightarrow G \\ h &\longmapsto gh \end{aligned}$$

are injective, hence the action is free. According to Theorem 3.7.25 it remains to show that the map

$$\begin{aligned} \Psi = \mathrm{Id}_G \times \Phi\colon G \times H &\longrightarrow G \times G \\ (g,h) &\longmapsto (g,gh) \end{aligned}$$

is closed. As in the proof of Corollary 3.7.13 we consider the map

$$T: G \times G \longrightarrow G \times G$$
$$(g, g') \longmapsto (g, gg').$$

This map is a diffeomorphism with $\Psi = T|_{G\times H}$. If $A \subset G \times H$ is closed, then A is closed in $G \times G$, since H is closed in G. This implies that $\Psi(A) = T(A)$ is closed in $G \times G$. □

Corollary 3.7.36 *Let G be a Lie group and $H \subset G$ a closed subgroup. Then the dimension of the quotient manifold G/H is given by*

$$\dim(G/H) = \dim G - \dim H.$$

In particular, the kernel of the differential

$$D_e\pi: T_eG \longrightarrow T_{[e]}G/H$$

is equal to the Lie algebra $\mathfrak{h}$ of H.

Proof This follows from Corollary 3.7.26. □

3.8 *Homogeneous Spaces

Recall that a set M together with a transitive action of a group G is called a *homogeneous space*. An example is the transitive action of SO(n) on the sphere S^{n-1} with isotropy group isomorphic to SO($n-1$), cf. Theorem 3.3.2. In this section we study the structure of homogeneous spaces for actions of groups, topological groups and Lie groups. We are most interested in the case of Lie group actions, but the other two cases are useful as a warm-up. We will show that every homogeneous space is, up to isomorphism of group actions, of the form G/H, where H is a suitable subgroup of G.

3.8.1 Groups and Homogeneous Spaces

Let G be any group and $H \subset G$ a subgroup. Then H acts on the right on G. We get a quotient space G/H of orbits, also called **left cosets**.

Definition 3.8.1 We define a map

$$\Phi: G \times G/H \longrightarrow G/H$$
$$(g, [a]) \longmapsto g \cdot [a] = [ga].$$

Note that G acts on the *left* on the set of *left cosets* G/H.
We want to show that Φ is indeed a group action.

Proposition 3.8.2 (Φ Is a Transitive Left Action of G on G/H)

1. *The map Φ is a well-defined, transitive group action of G on the set G/H.*
2. *The isotropy group of $[e] \in G/H$ is equal to H. Therefore the isotropy group of any point in G/H is isomorphic to H.*

Proof This is an easy exercise. □

We now consider an arbitrary transitive group action of G on a set M. We want to show that up to an equivariant bijection this group action is of the form above.

Proposition 3.8.3 (Structure of Transitive Group Actions on Sets) *Let $G \times M \to M$ be a transitive left action of a group G on a set M. Fix an arbitrary point $p \in M$ and let G_p denote the isotropy group of p. Then $G_p \subset G$ is a subgroup and*

$$\begin{aligned} f: G/G_p &\longrightarrow M \\ [a] &\longmapsto a \cdot p \end{aligned}$$

is a well-defined G-equivariant bijection.

Proof Another easy exercise. □

This implies that every homogeneous G-space is of the form G/H for some subgroup $H \subset G$ (not only as a set, but as the space of an action). We will show in the following subsections that this result essentially still holds in the continuous and smooth category.

3.8.2 Topological Groups and Homogeneous Spaces

Let G be a topological group and $H \subset G$ a subgroup with the subspace topology. Consider the quotient space G/H with the subspace topology. According to Proposition 3.8.2 we get a transitive group action

$$\begin{aligned} \Phi: G \times G/H &\longrightarrow G/H \\ (g, [a]) &\longmapsto g \cdot [a] = [ga]. \end{aligned}$$

Proposition 3.8.4 (Φ Is a Continuous Action for Topological Groups) *Suppose G is a topological group and $H \subset G$ a subgroup. Then the transitive group action*

$$\Phi: G \times G/H \longrightarrow G/H$$

is continuous.

Proof Multiplication in G followed by projection onto G/H is continuous,

$$G \times G \longrightarrow G \longrightarrow G/H.$$

This implies, by the definition of the quotient topology on G/H, that the group action

$$\Phi: G \times G/H \longrightarrow G/H$$

is continuous. □

According to Corollary 3.7.13 the space G/H is Hausdorff if and only if H is a closed subset in G. We now consider the case of an arbitrary transitive continuous group action.

Proposition 3.8.5 (Structure of Transitive Continuous Group Actions on Topological Spaces) *Let $G \times M \rightarrow M$ be a transitive continuous left action of a topological group G on a Hausdorff space M. Fix an arbitrary point $p \in M$ and let G_p denote the isotropy group of p. Then $G_p \subset G$ is a closed subgroup and*

$$f: G/G_p \longrightarrow M$$
$$[a] \longmapsto a \cdot p$$

is a well-defined continuous G-equivariant bijection between Hausdorff spaces. If G is compact, then f is a homeomorphism.

Proof The isotropy group G_p is closed in G by Proposition 3.2.9 (here we need that M is Hausdorff). It is clear by Proposition 3.8.3 that f is a well-defined equivariant bijection. It is also clear from the definition of the quotient topology that f is continuous. The final statement follows, because a continuous bijection from a compact space to a Hausdorff space is a homeomorphism. □

Remark 3.8.6 In general, if G is non-compact, the map f is *not* a homeomorphism.

3.8.3 Lie Groups and Homogeneous Spaces

We now come to the case that we are most interested in: G is a Lie group and $H \subset G$ a closed subgroup. By Corollary 3.7.35 the quotient G/H has a unique structure of a smooth manifold such that $\pi: G \rightarrow G/H$ is a submersion. According to Proposition 3.8.2 we get a transitive group action

$$\Phi: G \times G/H \longrightarrow G/H$$
$$(g, [a]) \longmapsto g \cdot [a] = [ga].$$

Proposition 3.8.7 (Φ Is a Smooth Action for Lie Groups) *Suppose G is a Lie group and $H \subset G$ a closed subgroup. Then the transitive group action*

$$\Phi: G \times G/H \longrightarrow G/H$$

is smooth.

Proof Multiplication in G followed by projection onto G/H is smooth,

$$G \times G \longrightarrow G \longrightarrow G/H.$$

By Lemma 3.7.5 the map

$$\Phi: G \times G/H \longrightarrow G/H$$

is smooth. □

We can now determine the structure of smooth manifolds that are homogeneous under the action of a Lie group.

Theorem 3.8.8 (Structure of Transitive Smooth Group Actions on Manifolds) *Let $G \times M \to M$ be a transitive smooth left action of a Lie group G on a manifold M. Fix an arbitrary point $p \in M$ and let G_p denote the isotropy group of p. Then $G_p \subset G$ is a closed subgroup and*

$$f: G/G_p \longrightarrow M$$
$$[a] \longmapsto a \cdot p$$

is a well-defined G-equivariant diffeomorphism between manifolds.

Proof It follows from Proposition 3.8.5 that f is well-defined, continuous, bijective and G-equivariant. By Corollary 3.7.35 the quotient space G/G_p is a smooth manifold. It remains to show that f is smooth and a diffeomorphism.

The map f is smooth by Lemma 3.7.5, because the orbit map

$$\phi_p: G \longrightarrow M$$
$$a \longmapsto a \cdot p$$

is smooth. To show that f is a diffeomorphism it suffices to show that the differential of f is an isomorphism at every point of G/G_p. By G-equivariance of f we have

$$f([ga]) = g \cdot f([a]).$$

Since left translations are diffeomorphisms of G/G_p and M, the differential of f is an isomorphism at every point of G/G_p if and only if it is an isomorphism at $[e]$.

We first show that the differential of f is injective at $[e]$: let $U \subset G/G_p$ be an open neighbourhood of $[e]$ and $s: U \to G$ a local section with $\pi \circ s = \mathrm{Id}_U$, where $\pi: G \to G/G_p$ is the canonical projection. Without loss of generality $s([e]) = e$. Then $f = \phi_p \circ s$ and

$$D_{[e]}f = D_e\phi_p \circ D_{[e]}s.$$

We also have

$$\mathrm{Id}_{T_{[e]}G/G_p} = D_e\pi \circ D_{[e]}s.$$

This shows that $D_{[e]}s$ is injective and its image is a complementary subspace to the kernel of $D_e\pi$, which is the Lie algebra $\mathfrak{g}_p$ of G_p according to Corollary 3.7.36. The kernel of $D_e\phi_p$ is also equal to $\mathfrak{g}_p$ according to Proposition 3.2.10. This implies that the differential $D_{[e]}f$ is injective.

To show that $D_{[e]}f$ is surjective it suffices to show by G-equivariance that $D_{[a]}f$ is surjective at some point $[a] \in G/G_p$. This follows from the next lemma. □

Lemma 3.8.9 *Let $f: X \to Y$ be a surjective smooth map between manifolds. Then there exists a point $x \in X$ such that $D_x f$ is surjective.*

Proof According to Sard's Theorem A.1.27 there exists a regular value $y \in Y$ of f. Since f is surjective, there exists an $x \in X$ with $f(x) = y$. Then x is a regular point f, i.e. the differential $D_x f$ is surjective. □

Along the way we have shown the following more general result.

Corollary 3.8.10 (The Orbit Map Induces an Injective Immersion of G/G_p into M) *Let $G \times M \to M$ be a smooth left action of a Lie group G on a manifold M, not necessarily transitive. Fix a point $p \in M$. Then*

$$\begin{aligned} f: G/G_p &\longrightarrow M \\ [a] &\longmapsto a \cdot p \end{aligned}$$

is an injective immersion of the manifold G/G_p into M whose image is the orbit $\mathcal{O}_p$ of p. In particular, if the Lie group G is compact, then the orbit $\mathcal{O}_p$ is an embedded submanifold of M, diffeomorphic to G/G_p.

Example 3.8.11 In Theorem 3.3.2 we saw that the standard representation of $\mathrm{O}(n)$ on $\mathbb{R}^n$ induces a transitive action of $\mathrm{O}(n)$ on the unit sphere S^{n-1} with isotropy group of a point $e_1 \in S^{n-1}$ isomorphic to the subgroup $\mathrm{O}(n-1)$. Theorem 3.8.8 then implies that the orbit map descends to a diffeomorphism

$$\mathrm{O}(n)/\mathrm{O}(n-1) \xrightarrow{\cong} S^{n-1}.$$

In a similar way we get diffeomorphisms

$$\mathrm{SO}(n)/\mathrm{SO}(n-1) \xrightarrow{\cong} S^{n-1},$$

$$\mathrm{U}(n)/\mathrm{U}(n-1) \xrightarrow{\cong} S^{2n-1},$$

$$\mathrm{SU}(n)/\mathrm{SU}(n-1) \xrightarrow{\cong} S^{2n-1},$$

$$\mathrm{Sp}(n)/\mathrm{Sp}(n-1) \xrightarrow{\cong} S^{4n-1}.$$

We also get diffeomorphisms

$$\mathrm{GL}(n,\mathbb{K})/\left(\mathrm{GL}(n-1,\mathbb{K}) \times \mathbb{K}^{n-1}\right) \xrightarrow{\cong} \mathbb{K}^n \setminus \{0\},$$

$$\mathrm{GL}(n,\mathbb{R})_+/\left(\mathrm{GL}(n-1,\mathbb{R})_+ \times \mathbb{R}^{n-1}\right) \xrightarrow{\cong} \mathbb{R}^n \setminus \{0\},$$

$$\mathrm{SL}(n,\mathbb{K})/\left(\mathrm{SL}(n-1,\mathbb{K}) \times \mathbb{K}^{n-1}\right) \xrightarrow{\cong} \mathbb{K}^n \setminus \{0\}.$$

Note that the group structure on $\mathrm{GL}(n-1,\mathbb{K}) \times \mathbb{K}^{n-1}$ and $\mathrm{SL}(n-1,\mathbb{K}) \times \mathbb{K}^{n-1}$ is *not* the direct product structure.

We can now prove Theorem 1.2.22 on the connected components of the classical linear groups (the idea for this proof is from [34] and [142]).

Proof Let G be a Lie group and $H \subset G$ a closed connected subgroup. According to Proposition 3.7.14 the quotient manifold G/H is connected if and only if G is connected. We apply this inductively to the homogeneous spaces in Example 3.8.11. We do the case of $\mathrm{SO}(n)$ explicitly, the other cases are left as an exercise. It is clear that $\mathrm{SO}(1) = \{1\}$ is connected. Since S^{n-1} is connected for all $n \geq 2$, the diffeomorphism

$$\mathrm{SO}(n)/\mathrm{SO}(n-1) \xrightarrow{\cong} S^{n-1}$$

shows that $\mathrm{SO}(n)$ is connected for all $n \geq 2$. □

The following fact is sometimes useful:

Corollary 3.8.12 (Smooth Structure on Sets with a Transitive Lie Group Action) *Suppose that M is a set and $G \times M \to M$ a transitive left action of a Lie group G on M with closed isotropy group G_p, for some $p \in M$. Then*

$$f\colon G/G_p \longrightarrow M$$
$$[a] \longmapsto a \cdot p$$

is a bijection. The set M can be given a unique structure of a smooth manifold, so that f becomes a diffeomorphism. If G is compact, then M is compact.

We conclude that in this situation we get the manifold structure on M *for free*, without the (sometimes difficult) task of defining a topology and an atlas of smoothly compatible charts for M.

3.9 *Stiefel and Grassmann Manifolds

We discuss two examples of compact homogeneous spaces where the manifold structure is defined by Corollary 3.8.12.

Example 3.9.1 (Stiefel Manifolds) Let $\mathbb{K} = \mathbb{R}, \mathbb{C}, \mathbb{H}$ and consider positive integers $k \leq n$. The **Stiefel manifold** $V_k(\mathbb{K}^n)$ is defined as the set of ordered k-tuples of orthonormal vectors in $\mathbb{K}^n$ with respect to the standard Euclidean (Hermitian, symplectic) scalar product on $\mathbb{R}^n$ ($\mathbb{C}^n$, $\mathbb{H}^n$) from Definition 1.2.9:

$$V_k(\mathbb{K}^n) = \{(v_1, \ldots, v_k) \mid v_i \in \mathbb{K}^n, \langle v_i, v_j \rangle_{\mathbb{K}^n} = \delta_{ij}\}.$$

We consider the case $\mathbb{K} = \mathbb{R}$ in detail. The group $\mathrm{O}(n)$ acts on the set $V_k(\mathbb{R}^n)$ via

$$A \cdot (v_1, \ldots, v_k) = (Av_1, \ldots, Av_k).$$

Since we can complete the vectors $v_1, \ldots, v_k$ to an orthonormal basis of $\mathbb{R}^n$ and $\mathrm{O}(n)$ acts transitively on orthonormal bases, it follows that the action of $\mathrm{O}(n)$ on $V_k(\mathbb{R}^n)$ is also transitive. The isotropy group of the point

$$p = (e_1, \ldots, e_k) \in V_k(\mathbb{R}^n)$$

is equal to

$$\mathrm{O}(n)_p = \left\{ \begin{pmatrix} E_k & 0 \\ 0 & A \end{pmatrix} \middle| A \in \mathrm{O}(n-k) \right\} \cong \mathrm{O}(n-k).$$

This holds, because if $C \in \mathrm{O}(n)$ satisfies $C \cdot p = p$, then C is of the form

$$C = \begin{pmatrix} E_k & B \\ 0 & A \end{pmatrix}$$

and $CC^T = E$ implies $AA^T = E$ and $BA^T = 0$, hence $A \in \mathrm{O}(n-k)$ and $B = 0$. It follows that the real Stiefel manifold admits the structure of a compact manifold given by

$$V_k(\mathbb{R}^n) = \mathrm{O}(n)/\mathrm{O}(n-k).$$

In particular, $V_k(\mathbb{R}^n)$ has dimension

$$\begin{aligned}\dim V_k(\mathbb{R}^n) &= \dim \mathrm{O}(n) - \dim \mathrm{O}(n-k)\\ &= \dim \mathfrak{o}(n) - \dim \mathfrak{o}(n-k)\\ &= \frac{1}{2}n(n-1) - \frac{1}{2}(n-k)(n-k-1)\\ &= nk - \frac{1}{2}k(k+1).\end{aligned}$$

Similarly, it can be shown that

$$\begin{aligned}V_k(\mathbb{C}^n) &= \mathrm{U}(n)/\mathrm{U}(n-k),\\ V_k(\mathbb{H}^n) &= \mathrm{Sp}(n)/\mathrm{Sp}(n-k).\end{aligned}$$

It follows that the complex and quaternionic Stiefel manifolds are connected for all $k \leq n$. For real Stiefel manifolds and $k < n$ this follows from Exercise 3.12.12.

Example 3.9.2 (Grassmann Manifolds) Let $\mathbb{K} = \mathbb{R}, \mathbb{C}, \mathbb{H}$ and consider non-negative integers $k \leq n$. The **Grassmann manifold** or **Grassmannian** $Gr_k(\mathbb{K}^n)$ is defined as the set of k-dimensional vector subspaces in $\mathbb{K}^n$:

$$Gr_k(\mathbb{K}^n) = \{U \subset \mathbb{K}^n \mid U \text{ is a } k\text{-dimensional vector subspace}\}.$$

We consider the case $\mathbb{K} = \mathbb{R}$. The group $\mathrm{O}(n)$ acts on the set $Gr_k(\mathbb{R}^n)$ via

$$A \cdot U = \{Au \in \mathbb{R}^n \mid u \in U\}.$$

This action is transitive, since we can choose a basis for U and the action of $\mathrm{O}(n)$ on $V_k(\mathbb{R}^n)$ is transitive. The isotropy group of

$$p = \mathrm{span}(e_1, \ldots, e_k) \in Gr_k(\mathbb{R}^n)$$

is equal to

$$\mathrm{O}(n)_p = \left\{ \begin{pmatrix} A & 0 \\ 0 & B \end{pmatrix} \middle| A \in \mathrm{O}(k), B \in \mathrm{O}(n-k) \right\} \cong \mathrm{O}(k) \times \mathrm{O}(n-k).$$

It follows that the real Grassmannian $Gr_k(\mathbb{R}^n)$ admits the structure of a compact manifold given by

$$Gr_k(\mathbb{R}^n) = \mathrm{O}(n)/(\mathrm{O}(k) \times \mathrm{O}(n-k)).$$

Note that there is a diffeomorphism

$$Gr_{n-k}(\mathbb{R}^n) \cong Gr_k(\mathbb{R}^n).$$

The dimension of $Gr_k(\mathbb{R}^n)$ is equal to

$$\begin{aligned}\dim Gr_k(\mathbb{R}^n) &= \dim V_k(\mathbb{R}^n) - \dim \mathrm{O}(k) \\ &= nk - \frac{1}{2}k(k+1) - \frac{1}{2}k(k-1) \\ &= k(n-k).\end{aligned}$$

Similarly, it can be shown that

$$\begin{aligned}Gr_k(\mathbb{C}^n) &= \mathrm{U}(n)/(\mathrm{U}(k) \times \mathrm{U}(n-k)), \\ Gr_k(\mathbb{H}^n) &= \mathrm{Sp}(n)/(\mathrm{Sp}(k) \times \mathrm{Sp}(n-k)).\end{aligned}$$

There are diffeomorphisms

$$Gr_1(\mathbb{K}^{n+1}) \cong \mathbb{KP}^n$$

for $\mathbb{K} = \mathbb{R}, \mathbb{C}, \mathbb{H}$.

3.10 *The Exceptional Lie Group G_2

In this section we discuss the compact simple exceptional Lie group G_2. In particular, we want to show that G_2 has dimension 14. This is a nice application of homogeneous spaces and Stiefel manifolds. We follow the paper [26] by Robert Bryant.

Besides being mathematically interesting, the Lie group G_2 plays an important role in M-theory, a conjectured supersymmetric theory of quantum gravity in 11 dimensions, which is related to the superstring theories in dimension 10. If M-theory is a realistic theory of nature, with 4-dimensional spacetime, 7 of the 11 dimensions have to be very small (*compactified*). The vacuum or background of the theory is thus of the form $\mathbb{R}^4 \times K$, where $\mathbb{R}^4$ is Minkowski spacetime and K is a compact Riemannian 7-manifold. Moreover, for the background to be a solution of the supergravity equations of motion, preserving one supersymmetry in dimension 4 (the most interesting case from a phenomenological point of view), the Riemannian metric on the 7-dimensional compact manifold K has to have *holonomy group* equal to G_2 (assuming that the *flux* is set to zero). The first compact examples of Riemannian manifolds with holonomy equal to G_2 were constructed by Dominic Joyce.

A Riemannian metric has holonomy group G_2 precisely if the 7-manifold admits a certain type of 3-form that is parallel with respect to the Levi-Civita connection. We will introduce the linear model of the 3-form on a vector space of dimension seven and define G_2 as its isotropy group.

3.10.1 *Definition of the 3-Form ϕ and the Lie Group* G_2

We need some preparations: Let $V = \mathbb{R}^7$ with the standard Euclidean scalar product $\langle \cdot, \cdot \rangle$ and standard orthonormal basis $\{e_j\}$. Let $\{\omega^i\}$ denote the dual basis of V^*, defined by

$$\omega^i(e_j) = \delta^i_j.$$

We use a shorthand notation for wedge products of the ω^i. For example,

$$\omega^{123} = \omega^1\omega^{23} = \omega^1 \wedge \omega^2 \wedge \omega^3.$$

Definition 3.10.1 We define a 3-form $\phi \in \Lambda^3 V^*$ by:

$$\phi = \omega^{123} + \omega^1(\omega^{45} + \omega^{67}) + \omega^2(\omega^{46} - \omega^{57}) - \omega^3(\omega^{47} + \omega^{56}).$$

Remark 3.10.2 The peculiar form of ϕ will be justified in Exercise 3.12.15. Other choices, however, are possible and lead to equivalent descriptions of G_2.
The group $\mathrm{GL}(7, \mathbb{R})$ acts on the column vector space V on the left via the standard representation. There is an induced representation on $\Lambda^k V^*$ defined by (cf. Definition 2.1.23):

$$(g\alpha)(v_1, \ldots, v_k) = \alpha\left(g^{-1}v_1, \ldots, g^{-1}v_k\right) \quad g \in \mathrm{GL}(7, \mathbb{R}), v_i \in V.$$

We think of this representation as a left action of $\mathrm{GL}(7, \mathbb{R})$ on $\Lambda^k V^*$.

Definition 3.10.3 We define $G_2 \subset \mathrm{GL}(7, \mathbb{R})$ as the isotropy group of the 3-form ϕ:

$$G_2 = \mathrm{GL}(7, \mathbb{R})_\phi = \{g \in \mathrm{GL}(7, \mathbb{R}) \mid g\phi = \phi\}.$$

This is a closed embedded Lie subgroup of $\mathrm{GL}(7, \mathbb{R})$.

3.10.2 G_2 *as a Compact Subgroup of* $\mathrm{SO}(7)$

Definition 3.10.4 For $x \in V$ we denote by $x \lrcorner \phi$ (contraction of ϕ with x) the 2-form on V defined by

$$(x \lrcorner \phi)(y, z) = \phi(x, y, z) \quad \forall y, z \in V.$$

The following map is very useful in the study of the Lie group G_2.

Definition 3.10.5 We set

$$\begin{aligned} b\colon V \times V &\longrightarrow \Lambda^7 V^* \\ (x, y) &\longmapsto b(x, y) = \frac{1}{6}(x \lrcorner \phi) \wedge (y \lrcorner \phi) \wedge \phi. \end{aligned}$$

Here are some properties of the map b.

Proposition 3.10.6 *The map b is symmetric and bilinear. It is G_2-equivariant, i.e. we have*

$$b(gx, gy) = g(b(x, y)) \quad \forall g \in G_2\, x, y \in V.$$

A calculation shows that

$$b(x, y) = \langle x, y \rangle \cdot \mathrm{vol},$$

where $\mathrm{vol} = \omega^{1234567}$ *is the standard volume form of V.*

Proof It is clear that b is symmetric and bilinear. For $g \in G_2$ and $x, y, z \in V$ we calculate

$$\begin{aligned} ((gx) \lrcorner \phi)(y, z) &= \phi(gx, y, z) \\ &= \phi\left(gx, gg^{-1}y, gg^{-1}z\right) \\ &= \left(g^{-1}\phi\right)\left(x, g^{-1}y, g^{-1}z\right) \\ &= \phi\left(x, g^{-1}y, g^{-1}z\right) \\ &= (x \lrcorner \phi)\left(g^{-1}y, g^{-1}z\right) \\ &= (g(x \lrcorner \phi))(y, z). \end{aligned}$$

Therefore

$$(gx) \lrcorner \phi = g(x \lrcorner \phi)$$

and

$$\begin{aligned} b(gx, gy) &= \frac{1}{6}((gx) \lrcorner \phi) \wedge ((gy) \lrcorner \phi) \wedge \phi \\ &= \frac{1}{6}(g(x \lrcorner \phi)) \wedge (g(y \lrcorner \phi)) \wedge g\phi \\ &= g(b(x, y)). \end{aligned}$$

The final property can be proved by a (tedious) direct calculation using the explicit form of ϕ. Because of symmetry and bilinearity of b it suffices to show that

$$b(e_i, e_j) = \delta_{ij} \cdot \text{vol} \quad \forall i \leq j \in \{1, \ldots, 7\}.$$

We have

$$\begin{aligned}
e_1 \lrcorner \phi &= \omega^{23} + \omega^{45} + \omega^{67}, \\
e_2 \lrcorner \phi &= -\omega^{13} + \omega^{46} - \omega^{57}, \\
e_3 \lrcorner \phi &= \omega^{12} - \omega^{47} - \omega^{56}, \\
e_4 \lrcorner \phi &= -\omega^{15} - \omega^{26} + \omega^{37}, \\
e_5 \lrcorner \phi &= \omega^{14} + \omega^{27} + \omega^{36}, \\
e_6 \lrcorner \phi &= -\omega^{17} + \omega^{24} - \omega^{35}, \\
e_7 \lrcorner \phi &= \omega^{16} - \omega^{25} - \omega^{34}.
\end{aligned}$$

We then calculate all 28 wedge products of the form

$$(e_i \lrcorner \phi) \wedge (e_j \lrcorner \phi) \wedge \phi$$

with $i \leq j$. For example,

$$\begin{aligned}
(e_1 \lrcorner \phi) \wedge (e_1 \lrcorner \phi) \wedge \phi &= 6 \cdot \text{vol}, \\
(e_1 \lrcorner \phi) \wedge (e_2 \lrcorner \phi) \wedge \phi &= 0.
\end{aligned}$$

The claim then follows from these calculations. □

Corollary 3.10.7 (G_2 **Is a Compact Subgroup of** SO(7)) *The following identity holds*

$$\langle gx, gy \rangle = (\det g)^{-1} \cdot \langle x, y \rangle \quad \forall g \in G_2, x, y \in V,$$

and

$$\det g = 1 \quad \forall g \in G_2.$$

In particular, G_2 *preserves the standard scalar product and orientation on* V *and is thus a compact embedded Lie subgroup of* SO(7).

Proof For any $g \in \mathrm{GL}(7, \mathbb{R})$ we have

$$\begin{aligned}(g\mathrm{vol})(e_1, \ldots, e_7) &= \mathrm{vol}\left(g^{-1}e_1, \ldots, g^{-1}e_7\right) \\ &= \det\left(g^{-1}I\right) \\ &= (\det g)^{-1},\end{aligned}$$

hence

$$g\mathrm{vol} = (\det g)^{-1} \cdot \mathrm{vol}.$$

By Proposition 3.10.6 this implies for all $g \in \mathrm{G}_2$

$$\begin{aligned}\langle gx, gy\rangle \mathrm{vol} &= b(gx, gy) \\ &= g(b(x, y)) \\ &= \langle x, y\rangle g\mathrm{vol} \\ &= (\det g)^{-1}\langle x, y\rangle \mathrm{vol}.\end{aligned}$$

Therefore

$$\langle gx, gy\rangle = (\det g)^{-1} \cdot \langle x, y\rangle \quad \forall g \in \mathrm{G}_2, x, y \in V.$$

Consider the matrix $g^T g$. We have

$$(g^T g)_{ij} = \langle ge_i, ge_j\rangle = (\det g)^{-1}\delta_{ij},$$

hence

$$g^T g = (\det g)^{-1} I_7.$$

Calculating the determinant on both sides we get

$$\begin{aligned}(\det g)^2 &= \det\left(g^T g\right) \\ &= (\det g)^{-7},\end{aligned}$$

hence

$$(\det g)^9 = 1$$

and

$$\det g = 1 \quad \forall g \in \mathrm{G}_2.$$

We get

$$\langle gx, gy\rangle = \langle x, y\rangle \quad \forall g \in G_2$$

and with $\det g = 1$ it follows that G_2 is a subgroup of SO(7). Since G_2 is a closed subgroup and SO(7) is compact, it follows that G_2 is compact. □

3.10.3 An SU(2)-Subgroup of G_2

Definition 3.10.8 Let $P\colon V \times V \to V$ be the map defined by

$$\langle P(x, y), z\rangle = \phi(x, y, z) \quad \forall x, y, z \in V.$$

Proposition 3.10.9 *The map P is antisymmetric, bilinear and G_2-equivariant. We have $P(e_1, e_2) = e_3$.*

Proof The first two properties are clear. The third property follows because the standard scalar product on V is G_2-invariant and ϕ is G_2-invariant. The final claim follows immediately from the definition of ϕ. □

Consider the action

$$\begin{aligned} G_2 \times V_2\left(\mathbb{R}^7\right) &\longrightarrow V_2\left(\mathbb{R}^7\right) \\ (g, v_1, v_2) &\longmapsto g \cdot (v_1, v_2) = (gv_1, gv_2). \end{aligned}$$

This is the restriction of the standard action of O(7) on the Stiefel manifold $V_2\left(\mathbb{R}^7\right)$.

Definition 3.10.10 Let $H \subset G_2$ denote the isotropy group of the point $p = (e_1, e_2) \in V_2\left(\mathbb{R}^7\right)$ under this action.

Since P is G_2-equivariant and $P(e_1, e_2) = e_3$ we have $He_3 = e_3$. Therefore H is the subgroup of G_2 defined by

$$He_i = e_i \quad \forall i = 1, 2, 3$$

and the action of H restricts to an action on the orthogonal complement

$$W = \operatorname{span}(e_4, e_5, e_6, e_7).$$

Lemma 3.10.11 *The Lie group H is isomorphic to the subgroup of* SO(4), *acting on W and fixing the 2-forms*

$$\begin{aligned} \beta_1 &= \omega^{45} + \omega^{67}, \\ \beta_2 &= \omega^{46} - \omega^{57}, \\ \beta_3 &= \omega^{47} + \omega^{56}. \end{aligned}$$

Proof This follows, because $H \subset \mathrm{G}_2$ and G_2 fixes the 3-form ϕ. □

Proposition 3.10.12 (*H* Is Isomorphic to SU(2)) *The Lie group H is isomorphic to the subgroup of* SO(4)*, acting on W and fixing the complex structure*

$$Je_4 = e_5,$$
$$Je_6 = e_7$$

and the complex volume form

$$\rho = (\omega^4 + i\omega^5) \wedge (\omega^6 + i\omega^7).$$

Hence H is isomorphic to SU(2).

Proof Since

$$\rho = \beta_2 + i\beta_3,$$

an element $g \in \mathrm{SO}(4)$ fixes ρ if and only if it fixes both β_2 and β_3. For any vector $v \in W$ we have

$$Jv = (v \lrcorner \beta_1)^*,$$

where $*$ denotes the vector dual to the 1-form with respect to the standard scalar product on W. It follows that $g \in \mathrm{SO}(4)$ fixes J if and only if it fixes β_1. □

3.10.4 The Dimension of G$_2$

Corollary 3.10.13 (Upper Bound on the Dimension of G$_2$) *The action of* G_2 *on the Stiefel manifold* $V_2(\mathbb{R}^7)$ *induces an injective immersion of* $\mathrm{G}_2/\mathrm{SU}(2)$ *into* $V_2(\mathbb{R}^7)$. *In particular,*

$$\dim \mathrm{G}_2 \leq 14$$

with equality if and only if the action of G_2 *on* $V_2(\mathbb{R}^7)$ *is transitive.*

Proof The first claim follows from Corollary 3.8.10. We have

$$\dim V_2(\mathbb{R}^7) = 7 \cdot 2 - \frac{1}{2} 2 \cdot 3 = 11,$$

according to the calculation in Example 3.9.1. Since $\dim \mathrm{SU}(2) = 3$ and the map

$$f \colon \mathrm{G}_2/\mathrm{SU}(2) \longrightarrow V_2(\mathbb{R}^7)$$

has injective differential, the second claim follows. The third claim follows since in the case of equality the map f is a submersion, hence has open image, and the image is closed, since G_2 is compact (it can be shown that $V_2(\mathbb{R}^7)$ is connected, cf. Exercise 3.12.12). □

Lemma 3.10.14 (Lower Bound on the Dimension of G_2) *The action of* $GL(7, \mathbb{R})$ *on* $\Lambda^3 V^*$ *induces an injective immersion*

$$h: GL(7, \mathbb{R})/G_2 \longrightarrow \Lambda^3 V^*$$
$$[g] \longmapsto g \cdot \phi.$$

Hence $\dim G_2 \geq 14$, *with equality if and only if the map h has open image.*

Proof The first claim again follows from Corollary 3.8.10. The second claim follows from

$$\dim GL(7, \mathbb{R}) = 7 \cdot 7 = 49,$$
$$\dim \Lambda^3 V^* = \binom{7}{3} = \frac{7 \cdot 6 \cdot 5}{1 \cdot 2 \cdot 3} = 35.$$

□

Collecting our results, we get the following theorem:

Theorem 3.10.15 (G_2 **Has Dimension** 14) *The Lie group* G_2 *has dimension* 14. *It acts transitively on the Stiefel manifold* $V_2(\mathbb{R}^7)$. *In particular, the standard representation of* G_2 *on* $V = \mathbb{R}^7$ *is irreducible. Moreover, the* $GL(7, \mathbb{R})$*-orbit of* ϕ *in* $\Lambda^3 V^*$ *is open.*

Remark 3.10.16 (G_2 Is a Simple Lie Group) A calculation of the homotopy groups of G_2, using the fibration

$$\begin{array}{ccc} SU(2) & \longrightarrow & G_2 \\ & & \downarrow \\ & & V_2(\mathbb{R}^7) \end{array}$$

shows that

$$\pi_0(G_2) = 0, \quad \pi_1(G_2) = 0, \quad \pi_3(G_2) = \mathbb{Z}.$$

Hence G_2 is connected, simply connected and simple, cf. Corollary 2.6.6. The details of this calculation can be found in [26].

3.11 *Godement's Theorem on the Manifold Structure of Quotient Spaces

In this section we want to prove Godement's Theorem 3.7.10. We continue to follow [130] and [139]. Let R be an equivalence relation on a manifold M. Suppose that R is a closed embedded submanifold of $M \times M$ and $\mathrm{pr}_1|_R: R \to M$ a surjective submersion. By symmetry of equivalence relations it follows that $\mathrm{pr}_2|_R: R \to M$ is also a surjective submersion. We endow M/R with the quotient topology.

3.11.1 *Preliminary Facts*

We want to prove two preliminary facts: we first show that the quotient M/R is Hausdorff.

Lemma 3.11.1 (The Quotient Space Is Hausdorff) *The canonical projection $\pi: M \to M/R$ is open and M/R is Hausdorff.*

Proof Suppose $U \subset M$ is open. We claim that

$$\pi^{-1}(\pi(U)) = \mathrm{pr}_1((M \times U) \cap R).$$

This holds because $x \in \pi^{-1}(\pi(U))$ if and only if there exists a $y \in U$ such that $(x, y) \in R$. Since $\mathrm{pr}_1|_R$ is a submersion and $(M \times U) \cap R$ is open in R, the set $\pi^{-1}(\pi(U))$ is an open subset of M, hence $\pi(U)$ is an open subset of M/R by the definition of the quotient topology. This proves that π is an open map. The claim about the Hausdorff property follows from Lemma 3.7.2, because R is by assumption a closed subset of $M \times M$. □

We denote the equivalence class of a point $x \in M$ under the equivalence relation R by $[x]$. We want to show that equivalence classes are embedded submanifolds of M.

Lemma 3.11.2 (Equivalence Classes Are Embedded Submanifolds of M) *Every equivalence of R is a closed embedded submanifold of M of dimension* $\dim R - \dim M$.

Proof We can write

$$[x] = \mathrm{pr}_1\left((\mathrm{pr}_2|_R)^{-1}(\{x\})\right),$$

because

$$(\mathrm{pr}_2|_R)^{-1}(\{x\}) = \{(y, x) \in M \times M \mid y \sim x\}.$$

Since $\mathrm{pr}_2|_R: R \to M$ is a submersion, the subset $K = (\mathrm{pr}_2|_R)^{-1}(\{x\})$ is an embedded submanifold of R of dimension $\dim R - \dim M$. However, K is contained in $M \times \{x\}$

on which pr_1 is a diffeomorphism onto M. Therefore $[x] = \mathrm{pr}_1(K)$ is an embedded submanifold of M of dimension $\dim R - \dim M$. □

3.11.2 The Slice Theorem

Our task is to show that the quotient space M/R has the structure of a smooth manifold. To define charts for M/R we construct so-called *slices* for the equivalence relation on open neighbourhoods for any point of M. In a second step we will then construct slices for *saturated* open neighbourhoods, which are the main tools needed to define the manifold structure on M/R.

Definition 3.11.3 Let $U \subset M$ be an open neighbourhood. Then a **slice** for the intersection of the equivalence classes of R with U is a closed embedded submanifold $S \subset U$ together with a surjective submersion $q: U \to S$ such that for every $x \in U$ the set $[x] \cap U$ intersects S precisely in the single point $q(x)$.

Theorem 3.11.4 (Slice Theorem) *Every point in M has an open neighbourhood $U \subset M$ with a slice (S, q) for the intersection of the equivalence classes of R with U.*

To prove the theorem fix $a \in M$ and let S' be any submanifold of M through a of dimension $\dim M - \dim[a]$ and transverse to the submanifold $[a]$. This means that

$$T_aS \oplus T_a[a] = T_aM.$$

We will show that we can find an open neighbourhood U of a in M such that $S = S' \cap U$ is a slice.

Lemma 3.11.5 *Consider*

$$Z = (\mathrm{pr}_2|_R)^{-1}(S').$$

Then Z is a submanifold of R through (a, a) of dimension $\dim Z = \dim M$ *and* $\mathrm{pr}_1|_Z: Z \to M$ *is a local diffeomorphism around* (a, a).

Proof Since $\mathrm{pr}_2|_R$ is a submersion, it is clear that Z is a submanifold of R with

$$\dim R - \dim Z = \dim M - \dim S' = \dim[a] = \dim R - \dim M.$$

Hence $\dim Z = \dim M$. We have

$$Z = (M \times S') \cap R.$$

Since $a \in S'$ and $a \sim a$, it follows that $(a, a) \in Z$.

It remains to show that the differential of $\mathrm{pr}_1|_Z$ in (a, a) is an isomorphism onto T_aM. We consider the following submanifolds of Z through (a, a):

$$[a] \times \{a\} \text{ and the diagonal } \Delta_{S'} \subset S' \times S'.$$

The tangent spaces to these submanifolds are given by

$$T_a[a] \oplus 0 \text{ and } \Delta_{T_aS'}.$$

These tangent spaces have zero intersection and their dimensions are $\dim[a]$ and $\dim S' = \dim M - \dim[a] = \dim Z - \dim[a]$. Hence

$$T_{(a,a)}Z = (T_a[a] \oplus 0) \oplus \Delta_{T_aS'}.$$

The image of $T_{(a,a)}Z$ under the differential of $\mathrm{pr}_1|_Z$ is

$$T_a[a] + T_aS' = T_aM,$$

hence the differential of $\mathrm{pr}_1|_Z$ is surjective and thus an isomorphism. □

Note that

$$\mathrm{pr}_2|_Z : Z \longrightarrow S'$$

is a submersion. By Lemma 3.11.5 we can choose open neighbourhoods O and U' of $a \in M$ such that

$$\mathrm{pr}_1|_{Z\cap(O\times O)} : Z \cap (O \times O) \longrightarrow U'$$

is a diffeomorphism. Let s denote the inverse of this diffeomorphism and

$$q = \mathrm{pr}_2|_Z \circ s.$$

Then q is a submersion of U' onto an open subset of $S' \cap O$.

Our aim is to shrink U' to U so that $S = S' \cap U$ is a slice together with the restriction of q. Note that

$$s(x) = (x, q(x)) \in Z \cap (O \times O) \quad \forall x \in U'.$$

In particular, $U' \subset O$.

Lemma 3.11.6 *Let $x \in S' \cap U'$. Then $s(x) = (x, x)$ and $q(x) = x$. In particular, if $y \in U'$ and $q(y) \in U'$, then $q(q(y)) = q(y)$.*

Proof We have $\Delta_{S'} \subset R$, hence $\Delta_{S'} \subset Z$. Thus

$$(x, x) \in \Delta_{S'} \cap (U' \times U') \subset Z \cap (O \times O).$$

Moreover,

$$\mathrm{pr}_1(x, x) = x = \mathrm{pr}_1 \circ s(x),$$

since s is the inverse of $\mathrm{pr}_1|_{Z\cap(O\times O)}$. Since $\mathrm{pr}_1|_{Z\cap(O\times O)}$ is injective, this implies $s(x) = (x, x)$ and thus $q(x) = x$.

Finally, if $y \in U'$ and $q(y) \in U'$, then $x = q(y) \in S' \cap U'$ and the claim follows. □

Lemma 3.11.7 *Let*

$$U = U' \cap q^{-1}(U' \cap O),$$
$$S = S' \cap U.$$

Then U and S together with the restriction of q to U satisfy the requirements of Theorem 3.11.4.

Proof Clearly U is an open neighbourhood of a in M, because $a \in U'$ and $a \in S'$, hence $q(a) = a \in U' \cap O$ by Lemma 3.11.6. We also have $S \subset U$. Suppose $x \in U$. Then $x \in U'$ and $q(x) \in U' \cap O$. Thus $q(q(x)) = q(x) \in U' \cap O$ and therefore $q(x) \in U$ by definition of U. But also $q(x) \in S'$ by definition of q, hence $q(x) \in S$. Therefore the restriction of q to U defines a submersion

$$q: U \longrightarrow S.$$

If $x \in S$, then $x \in S' \cap U'$ and $q(x) = x$ by Lemma 3.11.6. This implies that q is surjective.

Finally, suppose that $x \in U$ and $y \in [x] \cap S$. Then

$$(x, y) \in ((M \times S) \cap R) \cap (O \times O) \subset Z \cap (O \times O),$$

because $U' \subset O$. Thus

$$(x, y) = s(x) = (x, q(x)),$$

hence $y = q(x)$. This proves the final requirement for the slice (S, q). □

Definition 3.11.8 If $V \subset M$ is a subset, then we denote the restriction of R to V by R_V. As a subset of $M \times M$ we have $R_V = (V \times V) \cap R$. We denote by $\pi_V: V \to V/R_V$ the canonical projection.

Corollary 3.11.9 (Slice for Open Subset Defines Local Manifold Structure on Quotient) *Every point in M has an open neighbourhood $U \subset M$ such that U/R_U has the structure of a smooth manifold and $\pi_U: U \to U/R_U$ is a surjective submersion.*

Proof Let $U \subset M$ be an open subset with a slice (S, q). Then the map $q: U \to S$ induces a bijection

$$\bar{q}: U/R_U \longrightarrow S.$$

We give U/R_U the structure of a smooth manifold such that $\bar{q}$ is a diffeomorphism. Then $\pi_U = \bar{q}^{-1} \circ q$ is a surjective submersion. □

3.11.3 Slices for Saturated Neighbourhoods and Proof of Godement's Theorem

Definition 3.11.10 A subset $V \subset M$ is called **saturated** if

$$V = \pi^{-1}(\pi(V)).$$

Equivalently, V is a union of equivalence classes. If U is an arbitrary subset of M, then $V = \pi^{-1}(\pi(U))$ is saturated.
We want to show that every point of M is contained in a *saturated* open neighbourhood with a slice. This is the main fact that we need to prove that M/R has the structure of a smooth manifold.

Corollary 3.11.11 (Slices for Saturated Open Subsets) *Let $U \subset M$ be an open subset with a slice (S, q) and V the saturated open subset $V = \pi^{-1}(\pi(U))$. Then there exists a surjective submersion $q': V \to S$ so that (S, q') is a slice for V.*

Proof It is clear that $U \subset V$. Let $j: U \hookrightarrow V$ be the inclusion. We claim that there is a well-defined map

$$\bar{j}: U/R_U \longrightarrow V/R_V$$

and that this map is a bijection. The map is well-defined, because if $x, y \in U$ are equivalent, then they are equivalent in V. The map is also injective. Finally, the map is surjective, because if $x \in V$, then there exists a $y \in U$ with $(x, y) \in R$.

Using the bijection $\bar{q}: U/R_U \to S$ from the proof of Corollary 3.11.9, we get a well-defined map $q': V \to S$:

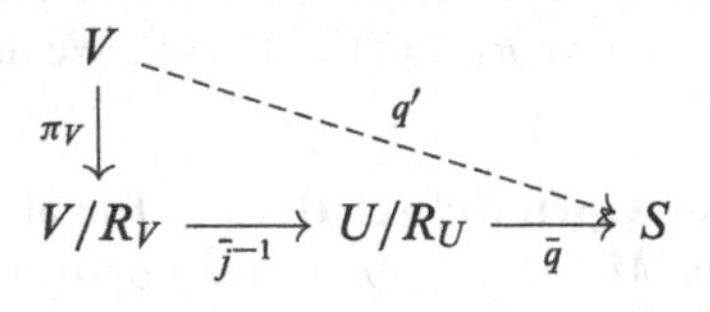

The map q' has the following property: for $x \in V$, there exists a $y \in U$ such that $[x] = [y]$, i.e.

$$\bar{j}^{-1}([x]) = [y].$$

Then

$$\begin{aligned} q'(x) &= \bar{q} \circ \bar{j}^{-1}([x]) \\ &= q(y). \end{aligned}$$

This implies, since $S \subset U$,

$$\begin{aligned} [x] \cap S &= [x] \cap U \cap S \\ &= [y] \cap U \cap S \\ &= \{q(y)\} \\ &= \{q'(x)\}. \end{aligned}$$

Hence $[x]$ intersects S precisely in the point $q'(x)$.

Since $U \subset V$, the map q' is surjective. It remains to show that q' is a submersion. We claim that there is a commutative diagram

$$\begin{array}{ccc} (M \times U) \cap R & \xrightarrow{\mathrm{pr}_2} & U \\ {\scriptstyle \mathrm{pr}_1}\downarrow & & \downarrow{\scriptstyle q} \\ V & \xrightarrow[q']{} & S \end{array}$$

where the arrows on the left, right and top are submersions. The arrow on the right is a submersion, because (S, q) is a slice and the arrows on the top and on the left are submersions, because $\mathrm{pr}_1|_R, \mathrm{pr}_2|_R : R \to M$ are submersions. To show that the diagram is commutative, let $(x, y) \in (M \times U) \cap R$. Then $x \sim y$ and $x \in V$. The statement then is

$$q'(x) = q(y),$$

which we showed above. Lemma 3.7.5 then proves that q' is a submersion. □

Corollary 3.11.12 (Slice for Open Saturated Subset Defines Local Manifold Structure on Quotient) *Let $V \subset M$ be an open subset with a slice (S, q'). Then V/R_V has the structure of a smooth manifold so that $\pi_V : V \to V/R_V$ is a surjective submersion.*

We can now finish the proof of Godement's Theorem 3.7.10.

Proof We have shown that there exists a covering of M by open saturated sets V_i so that the open sets $W_i = V_i/R_{V_i} \subset M/R$ have the structure of a smooth manifold with

$$\pi_i: V_i \longrightarrow W_i$$

being surjective submersions. Suppose $V_i \cap V_j \neq \emptyset$. By Lemma 3.11.13 below we have to show that the manifold structures on $W_i \cap W_j$ induced from W_i and W_j are the same, i.e. the identity map between the open subsets $W_i \cap W_j \subset W_i$ and $W_i \cap W_j \subset W_j$ is a diffeomorphism. Since V_i and V_j are saturated, we have

$$\pi(V_i \cap V_j) = \pi(V_i) \cap \pi(V_j) = W_i \cap W_j.$$

The manifold structure induced from V_i and V_j on $V_i \cap V_j$ are the same. Since π is for each of these structures a submersion from $V_i \cap V_j$ onto $W_i \cap W_j$, it follows from Corollary 3.7.7 that the induced manifold structures on $W_i \cap W_j$ are the same. It is then also clear that

$$\pi: M \longrightarrow M/R$$

is a surjective submersion. □

We used (a slight generalization of) the following lemma, whose proof is clear:

Lemma 3.11.13 *Let X be a topological space, $W_1, W_2 \subset X$ open and*

$$\phi_1: W_1 \longrightarrow U_1$$
$$\phi_2: W_2 \longrightarrow U_2$$

homeomorphisms onto open subsets U_1, U_2 of $\mathbb{R}^n$. Define the unique smooth structure on W_i such that ϕ_i becomes a diffeomorphism, for $i = 1, 2$. Then the change of charts

$$\phi_2 \circ \phi_1^{-1}: \phi_1(W_1 \cap W_2) \longrightarrow \phi_2(W_1 \cap W_2)$$

is a diffeomorphism if and only if

$$\mathrm{Id}: W_1 \supset W_1 \cap W_2 \longrightarrow W_1 \cap W_2 \subset W_2$$

is a diffeomorphism.

3.12 Exercises for Chap. 3

3.12.1 Prove Proposition 3.2.2. Find an example of a left action

$$\begin{aligned} G \times M &\longrightarrow M \\ (g, p) &\longmapsto g \cdot p \end{aligned}$$

so that

$$\begin{aligned} M \times G &\longrightarrow M \\ (p, g) &\longmapsto p * g = g \cdot p \end{aligned}$$

does *not* define a right action of G on M.

3.12.2 Let M be a Hausdorff space with a continuous left action of a topological group G. For a subset $K \subset G$ consider the **fixed point set**

$$M^K = \{p \in M \mid K \cdot p = p\}.$$

Prove the following:

1. If $K = \{k\}$ contains only one element, then M^K is a closed subset of M.
2. M^K is a closed subset of M for arbitrary subsets $K \subset G$.

3.12.3 The Lie group $G = \mathrm{SU}(2) \times \mathrm{U}(1)$ acts on $\mathbb{C}^2$ via

$$\left(A, e^{i\alpha}\right) \cdot v = e^{i\alpha} A v,$$

where Av denotes multiplication of the matrix $A \in \mathrm{SU}(2)$ with the column vector $v \in \mathbb{C}^2$. Let

$$p = \begin{pmatrix} 0 \\ v_0 \end{pmatrix} \in \mathbb{C}^2,$$

where $v_0 \in \mathbb{R}$, $v_0 \neq 0$.

1. Determine the isotropy subalgebra $\mathfrak{g}_p$ and the isotropy subgroup G_p. Which standard Lie group is G_p isomorphic to?
2. Determine the orbit $\mathcal{O}_p$ of p under the action of G. Which standard manifold is $\mathcal{O}_p$ diffeomorphic to?

In the electroweak gauge theory the *Higgs field* takes values in $\mathbb{C}^2$. The vector p is known as a *vacuum vector*. The isotropy group G_p is called the *unbroken subgroup*.

3.12.4 We consider S^3 with the Hopf action:

$$\begin{aligned} S^3 \times S^1 &\longrightarrow S^1 \\ \left(z, e^{i\alpha}\right) &\longmapsto ze^{i\alpha}. \end{aligned}$$

We identify

$$\begin{aligned} \mathbb{R}^4 &\longrightarrow \mathbb{C}^2 \\ (x_1, y_1, x_2, y_2) &\longmapsto (x_1 + iy_1, x_2 + iy_2). \end{aligned}$$

Let s denote the stereographic projection of S^3 through the point $(0, 1) \in S^3$:

$$\begin{aligned} s\colon S^3 \setminus \{(0, 1)\} &\longrightarrow \mathbb{R}^3 \\ (x_1, y_2, x_2, y_2) &\longmapsto \frac{1}{1 - x_2}(x_1, y_1, y_2). \end{aligned}$$

Let $\gamma_i\colon S^1 \to S^3$, for $i = 1, 2, 3$, denote the orbit maps of the points

$$p_1 = (1, 0), \quad p_2 = \frac{1}{\sqrt{2}}(1, 1), \quad p_3 = (0, 1)$$

on S^3 under the Hopf action. Consider the images

$$\begin{aligned} \sigma_i &= s \circ \gamma_i\colon S^1 \longrightarrow \mathbb{R}^3, \quad i = 1, 2 \\ \sigma_3 &= s \circ \gamma_3\colon \mathbb{R} \cong S^1 \setminus \{1\} \longrightarrow \mathbb{R}^3 \end{aligned}$$

of these curves under the stereographic projection. Determine and sketch σ_1, σ_2, σ_3 (for σ_2 it may be helpful to rotate the coordinate system, so that σ_2 lies in a coordinate plane.) Show that σ_1 and σ_2 are circles and σ_3 is a line. The circle σ_1 spans a flat disk in $\mathbb{R}^3$. Show that σ_2 intersects this disk transversely in one point. This means that σ_1, σ_2 and hence γ_1, γ_2 are *linked*.

Remark It is possible to show that all orbits of the Hopf action on S^3 are linked pairwise.

3.12.5 The aim of this exercise is to verify two propositions on fundamental vector fields in a special case with a direct calculation. The standard representation of the Lie group SU(2) on $\mathbb{C}^2$ induces a left-action

$$\mathrm{SU}(2) \times \mathbb{C}^2 \longrightarrow \mathbb{C}^2.$$

We fix the vectors

$$\tau_a = -\frac{i\sigma_a}{2} \in \mathfrak{su}(2), \quad a = 1, 2, 3.$$

1. Determine the fundamental vector fields $\tilde{\tau}_a$ on $\mathbb{C}^2$ and show by direct calculation that

$$[\tilde{\tau}_a, \tilde{\tau}_b] = \widetilde{[\tau_a, \tau_b]} \quad \forall a, b \in \{1, 2, 3\},$$

   without using Proposition 3.4.4.
2. Let

$$A = \begin{pmatrix} r & -\bar{r} \\ r & \bar{r} \end{pmatrix} \in \mathrm{SU}(2), \quad r = \frac{1}{2} - \frac{1}{2}i.$$

   Calculate directly $l_{A*}(\tilde{\tau}_1)$ and compare with $\tilde{Z}$, where $Z = \mathrm{Ad}_A \tau_1$, without using Proposition 3.4.6.

3.12.6 (From [23]) Let G be a compact Lie group acting smoothly and freely on a manifold M. Let $\pi: M \to M/G$ be the canonical projection.

1. Prove that for every smooth curve $\gamma: I \to M/G$, defined on an interval I, there exists a smooth lift $\bar{\gamma}: I \to M$ with $\pi \circ \bar{\gamma} = \gamma$.
2. Suppose that M is connected and at least one of the orbits of G on M is connected (e.g. G is connected). Prove that π_* maps the fundamental group of M surjectively onto the fundamental group of M/G. In particular, if M is simply connected, then M/G is simply connected.

3.12.7 Let G and H be topological groups and M and N topological spaces. Suppose that G acts continuously on the right on M and H acts continuously on the right on N. Let $\phi: G \to H$ be a group homomorphism. Suppose that $f: M \to N$ is ϕ-**equivariant**, i.e.

$$f(p \cdot g) = f(p) \cdot \phi(g) \quad \forall p \in M, g \in G.$$

Prove the following:

1. If f is continuous, then f induces a continuous map $f_\phi: M/G \to N/H$.
2. If ϕ is an isomorphism and f a homeomorphism, then f_ϕ is a homeomorphism.

3.12.8 Use Exercise 3.12.7 to prove the following facts about lens spaces:

1. There exists a homeomorphism $L(p, q) \to L(p, -q)$.
2. If $qr \equiv 1 \bmod p$, then there exists a homeomorphism $L(p, q) \to L(p, r)$.

Remark According to a theorem of Reidemeister there exists a homeomorphism between lens spaces $L(p, q_1)$ and $L(p, q_2)$ only in these two cases, their combination, or in the trivial case $q_1 = q_2$.

3.12.9

1. Show that $\mathbb{CP}^1$ can be covered by two charts diffeomorphic to $\mathbb{C}$ and that $\mathbb{CP}^1$ is diffeomorphic to S^2.
2. Prove that $\mathbb{HP}^1$ is diffeomorphic to S^4.

3.12.10 Consider complex projective space $\mathbb{CP}^n = S^{2n+1}/S^1$. Show that there is a transitive left action of $\mathrm{SU}(n+1)$ on $\mathbb{CP}^n$ with isotropy group isomorphic to $\mathrm{U}(n)$. Deduce that there is a diffeomorphism

$$\mathbb{CP}^n \cong \mathrm{SU}(n+1)/\mathrm{U}(n).$$

3.12.11 Prove that there is a diffeomorphism $\mathbb{RP}^3 \cong \mathrm{SO}(3)$.

3.12.12 Show that for $k < n$ the real and complex Stiefel manifolds can be written as homogeneous spaces

$$V_k(\mathbb{R}^n) = \mathrm{SO}(n)/\mathrm{SO}(n-k),$$
$$V_k(\mathbb{C}^n) = \mathrm{SU}(n)/\mathrm{SU}(n-k).$$

Deduce that for $k < n$ the real Stiefel manifolds $V_k(\mathbb{R}^n)$ are connected.

3.12.13 Consider the half-plane

$$H = \{z \in \mathbb{C} \mid \mathrm{Im}\, z > 0\}.$$

1. Show that the map

$$\mathrm{SL}(2,\mathbb{R}) \times H \longrightarrow H$$
$$(A, z) \longmapsto \frac{az+b}{cz+d},$$

for

$$A = \begin{pmatrix} a & b \\ c & d \end{pmatrix} \in \mathrm{SL}(2,\mathbb{R})$$

is well-defined and defines a left-action of $\mathrm{SL}(2,\mathbb{R})$ on H.

2. Prove that this action is transitive and that the action defines a diffeomorphism between H and $\mathrm{SL}(2,\mathbb{R})/\mathrm{SO}(2)$.

3.12.14 (From [57]) According to Exercise 1.9.10 the group $\mathrm{SO}(2n)$ has a subgroup isomorphic to $\mathrm{U}(n)$. We would like to identify the homogeneous space $\mathrm{SO}(2n)/\mathrm{U}(n)$.

1. Let

$$J_0 = \begin{pmatrix} 0 & -I_n \\ I_n & 0 \end{pmatrix} \in \mathrm{Mat}(2n \times 2n, \mathbb{R}).$$

   Show that the subgroup

$$H = \{A \in \mathrm{SO}(2n) \mid AJ_0 = J_0 A\}$$

   of $\mathrm{SO}(2n)$ is isomorphic to $\mathrm{U}(n)$ (compare with Exercise 1.9.10).
2. Consider the set

$$\mathscr{J}^+\left(\mathbb{R}^{2n}\right) = \left\{J \in \mathrm{SO}(2n) \mid J^2 = -I_{2n}\right\}.$$

   This is the set of *almost complex structures* on $\mathbb{R}^{2n}$, compatible with the scalar product and the orientation. The group $\mathrm{SO}(2n)$ acts on $\mathscr{J}^+\left(\mathbb{R}^{2n}\right)$ by conjugation

$$\begin{aligned} \mathrm{SO}(2n) \times \mathscr{J}^+\left(\mathbb{R}^{2n}\right) &\longrightarrow \mathscr{J}^+\left(\mathbb{R}^{2n}\right) \\ (A, J) &\longmapsto AJA^{-1}. \end{aligned}$$

   Prove that this action is transitive.
3. Conclude that $\mathrm{SO}(2n)/\mathrm{U}(n) \cong \mathscr{J}^+\left(\mathbb{R}^{2n}\right)$.

Remark It can be shown that $\mathrm{SO}(4)/\mathrm{U}(2) \cong S^2$ and $\mathrm{SO}(6)/\mathrm{U}(3) \cong \mathbb{CP}^3$.

3.12.15 Let $V = \mathbb{R}^7$ with standard scalar product $\langle \cdot, \cdot \rangle$ and let $P\colon V \times V \to V$ denote the antisymmetric, bilinear G_2-equivariant map from Definition 3.10.8.

1. Let $x, y \in V$ be arbitrary vectors. Show that there exists an element $g \in \mathrm{G}_2$ such that (at the same time)

$$\begin{aligned} gx &= x_1 e_1, \\ gy &= y_1 e_1 + y_2 e_2, \end{aligned}$$

   with real coefficients x_1, y_1, y_2.
2. Use the first part of this exercise to prove the identity

$$\langle P(x, y), P(x, y) \rangle = \langle x, x \rangle \langle y, y \rangle - \langle x, y \rangle^2 \quad \forall x, y \in V.$$

3. Let

$$\mathbb{O} = \mathbb{R}e_0 \oplus V \cong \mathbb{R}^8$$

and define an $\mathbb{R}$-bilinear multiplication $\cdot$ on $\mathbb{O}$ by

$$e_0 \cdot e_0 = e_0,$$
$$e_0 \cdot x = x = x \cdot e_0,$$
$$x \cdot y = -\langle x, y\rangle e_0 + P(x, y),$$

for all $x, y \in V$. Let $(\cdot\,,\cdot)$ denote the scalar product on $\mathbb{O}$ so that $e_0, e_1, e_2, \ldots, e_7$ are orthonormal, with associated norm $||\cdot||$. Prove that

$$||z \cdot w||^2 = ||z||^2||w||^2 \quad \forall z, w \in \mathbb{O}.$$

Hence $\mathbb{O}$ is a real normed division algebra of dimension 8, known as the **octonions**.

4. Prove that

$$(gx) \cdot (gy) = g(x \cdot y) \quad \forall g \in \mathrm{G}_2, x, y \in V.$$

5. For

$$z = x_0e_0 + x \in \mathbb{O}$$

with $x_0 \in \mathbb{R}$ and $x \in V$ define the conjugate

$$\bar{z} = x_0e_0 - x.$$

Show that

$$z \cdot \bar{z} = \bar{z} \cdot z = ||z||^2 e_0.$$

This implies that every non-zero octonion has a multiplicative inverse.

6. Calculate $(e_1 \cdot e_2) \cdot e_4$ and $e_1 \cdot (e_2 \cdot e_4)$ and show that the octonions are not associative.

3.12.16 (From [27]) We continue with the notation from Exercise 3.12.15.

1. Use the first part of Exercise 3.12.15 to prove the identity

$$P(x, P(x, y)) = -\langle x, x\rangle y + \langle x, y\rangle x \quad \forall x, y \in V.$$

2. Let $x \in V$ be an arbitrary vector of norm 1 and V_x the orthogonal complement of $\mathbb{R}x$ in V. Then V_x is a real 6-dimensional vector subspace of V. Prove that multiplication of octonions defines a linear map

$$J_x \colon V_x \longrightarrow V_x$$
$$v \longmapsto x \cdot v$$

with $J_x^2 = -\mathrm{Id}$, i.e. a complex structure on V_x.

3. Let S^6 be the unit sphere in V. Show that the restriction of the action of SO(7) on S^6 to the subgroup G_2 is transitive with isotropy group isomorphic to SU(3). Conclude that S^6 can be realized as a homogeneous space

$$S^6 \cong G_2/SU(3).$$

Remark Since the rank of the Lie group G_2 is 2, it does not contain Lie subgroups isomorphic to SU(n) for $n \geq 4$.

3.12.17 (From [73]) We continue with the notation from Exercise 3.12.15. Our aim is to show that G_2 contains a certain Lie subgroup isomorphic to SO(4).

1. Consider on $\mathrm{Im}\mathbb{H} \oplus \mathbb{H} \cong \mathbb{R}^7$ the representation of $\mathrm{Sp}(1) \times \mathrm{Sp}(1)$ given by

$$(q_1, q_2) \cdot (a, b) = (\bar{q}_1 a q_1, q_1 b \bar{q}_2).$$

Show that this representation defines an embedding of

$$\mathrm{SO}(4) \cong (\mathrm{Sp}(1) \times \mathrm{Sp}(1))/\mathbb{Z}_2$$

into SO(7) (compare with Exercises 1.9.20 and 1.9.21).

2. Identify V with $\mathrm{Im}\mathbb{H} \oplus \mathbb{H}$ via the embeddings

$$i \mapsto e_1, j \mapsto e_2, k \mapsto e_3 \quad \text{on } \mathrm{Im}\mathbb{H}$$

and

$$1 \mapsto e_4, i \mapsto e_5, j \mapsto e_6, k \mapsto e_7 \quad \text{on } \mathbb{H}.$$

Prove that the embedding $\mathrm{SO}(4) \hookrightarrow \mathrm{SO}(7) = \mathrm{SO}(V)$ above has image in G_2 (for example, by showing that the Lie algebra $\mathfrak{sp}(1) \oplus \mathfrak{sp}(1)$ maps to the Lie algebra $\mathfrak{g}_2$ of G_2).

3.12.18 (From [73]) We continue with the notation from Exercises 3.12.15 and 3.12.17. A 3-dimensional oriented real vector subspace $U \subset V$ is called **associative** if the restriction $\phi|_U$ is positive, i.e. a volume form, where ϕ denotes the 3-form from the definition of the Lie group G_2. Let $G(\phi)$ denote the set of all associative subspaces of V.

1. Show that the action of G_2 on V induces an action of G_2 on $G(\phi)$.
2. Let $U \subset V$ be an associative subspace and $x, y \in U$ orthonormal. Prove that the vectors $x, y, x \cdot y$ span U. Show that the action of G_2 on $G(\phi)$ is transitive.
3. Show that the isotropy group H of $U_0 = \mathrm{span}(e_1, e_2, e_3) \in G(\phi)$ contains the subgroup $\mathrm{SO}(4) \subset G_2$ from Exercise 3.12.17.
4. Let $h \in H$. Show that there exists an element $k \in \mathrm{SO}(4)$ such that

$$g = kh = (\mathrm{Id}, g_2) \in \mathrm{SO}(\mathrm{Im}\mathbb{H}) \times \mathrm{SO}(\mathbb{H})$$

with $g_2(1) = 1$. Show that for $q \in \mathrm{Im}\mathbb{H}$ we can write with multiplication of octonions

$$(0, q) = (q, 0) \cdot (0, 1) \in \mathrm{Im}\mathbb{H} \oplus \mathbb{H} = V.$$

Conclude that $g_2 = \mathrm{Id}$, hence $H = \mathrm{SO}(4)$ and

$$G(\phi) \cong \mathrm{G}_2/\mathrm{SO}(4).$$

Chapter 4
Fibre Bundles

What is gauge theory? It is not an overstatement to say that gauge theory is ultimately the theory of *principal bundles* and *associated vector bundles*. Besides full gauge theories, it also proves beneficial in certain situations to study the theory only involving principal bundles, sometimes called *Yang–Mills theory*. In physics, an example of a full gauge theory would be quantum chromodynamics (QCD), the theory of quarks, gluons and their interactions, while pure Yang–Mills theory would be a theory only of gluons, also called gluodynamics. Even such a simplified theory is very interesting – the Clay Millennium Prize Problem [37] on the mass gap, for instance, is a problem concerning the spectrum of *glueballs* in pure quantum Yang–Mills theory.

With the background knowledge of Lie groups, Lie algebras, representations and group actions, we will now study fibre bundles in general and more specifically principal bundles, vector bundles and associated bundles, which together form the core or the "stage" of gauge theories.

Fibre bundles can be thought of as twisted, non-trivial products between a base manifold and a fibre manifold. Principal and vector bundles are fibre bundles whose fibres are, respectively, Lie groups and vector spaces, so that the bundle admits a special type of *bundle atlas*, preserving some of the additional structure of the fibres.

The fundamental geometric object in a gauge theory is a principal bundle over spacetime with *structure group* given by the gauge group. The fibres of a principal bundle are sometimes thought of as an internal space at every spacetime point, not belonging to spacetime itself. The gauge group acts at every spacetime point on the internal space in a simply transitive way. *Connections* on principal bundles, that we discuss in Chap. 5, correspond to *gauge fields*, whose particle excitations in the associated quantum field theory are the *gauge bosons* that transmit interactions. Matter fields in the Standard Model, like quarks and leptons, or scalar fields, like the Higgs field, correspond to sections of vector bundles *associated* to the principal bundle (and twisted by *spinor bun-*

M.J.D. Hamilton, *Mathematical Gauge Theory*, Universitext,
https://doi.org/10.1007/978-3-319-68439-0_4

dles in the case of fermions). The ultimate reason for the interaction between matter fields and gauge fields is that both are related to the same principal bundle.

Fibre bundles are indispensable in gauge theory and physics in the situation where spacetime, the *base manifold*, has a non-trivial topology. This happens, for example, in string theory where spacetime is typically assumed to be a product $\mathbb{R}^4 \times K$ of Minkowski spacetime with a compact Riemannian manifold K. It also happens if we compactify (Euclidean) spacetime $\mathbb{R}^4$ to the 4-sphere S^4. In these situations, fields on spacetime often cannot be described simply by a map to a fixed vector space, but rather as *sections* of a non-trivial vector bundle.

Even in the case where the fibre bundles are trivial, for example, in the case of principal bundles and vector bundles over contractible manifolds like $\mathbb{R}^n$, there is still a small, but important difference between a trivial fibre bundle and the choice of an actual trivialization. We will see that this is similar to the difference in special relativity between Minkowski spacetime and the choice of an inertial system.

Fibre bundles are not only important in physics, but for a variety of reasons also in differential geometry and differential topology: many non-trivial manifolds can be constructed as (total spaces of) fibre bundles and numerous structures on manifolds, such as vector fields, differential forms and metrics, are defined using bundles. Mathematically, we are especially interested in the construction of *non-trivial* fibre bundles (trivial bundles are just globally products). We discuss the following methods that (potentially) yield non-trivial bundles:

- Mapping tori (Example 4.1.5) and the clutching construction (Sect. 4.6) yield fibre bundles over the circle S^1 and higher-dimensional spheres S^n.
- Principal group actions define principal bundles (Sect. 4.2.2; specific examples are the famous *Hopf fibrations* and principal bundles over homogeneous spaces).
- Actions of the structure group G of a principal bundle $P \to M$ on another manifold F (the general fibre) yield associated fibre bundles (Sect. 4.7) over M. In particular, all vector bundles can be obtained in this way.
- The tangent bundle TM and frame bundle $\mathrm{Fr}(M)$ of smooth manifolds M are specific examples of vector and principal bundles.
- In general, every fibre bundle can be constructed using a cocycle of transition functions (Exercise 4.8.9).

This chapter, like the previous one, contains many definitions and concepts. I hope that there are sufficiently many examples to illustrate the definitions and balance the exposition. References for this chapter for fibre bundles in general are [14, 84, 133] and [136] as well as [5, 25, 39, 74] and [78] for vector bundles in particular.

4.1 General Fibre Bundles

4.1.1 *Definition of Fibre Bundles*

Before we begin with the definition of fibre bundles, we consider two very general notions: suppose $\pi: E \rightarrow M$ is a surjective differentiable map between smooth manifolds (occasionally we will consider the following notions even in the case of a surjective map $\pi: E \rightarrow M$ between sets).

Definition 4.1.1

1. Let $x \in M$ be an arbitrary point. The (non-empty) subset

$$E_x = \pi^{-1}(x) = \pi^{-1}(\{x\}) \subset E$$

is called the **fibre** of π over x.
2. For a subset $U \subset M$ we set

$$E_U = \pi^{-1}(U) \subset E.$$

We can think of E_U as the part of E "above" U. It is clear that E_U is the union of all fibres E_x, where $x \in U$.
3. A differentiable map $s: M \rightarrow E$ such that

$$\pi \circ s = \mathrm{Id}_M$$

is called a **(global) section** of π. A differentiable map $s: U \rightarrow E$, defined on some open subset $U \subset M$, satisfying

$$\pi \circ s = \mathrm{Id}_U$$

is called a **local section.**

Note that a differentiable map $s: U \rightarrow E$ is a (local) section of $\pi: E \rightarrow U$ if and only if $s(x) \in E_x$ for all $x \in U$.

For a general surjective map, the fibres E_x and E_y over points $x \neq y \in M$ can be very complicated and different, in particular, they may not be embedded submanifolds of E and even when they are, they may not be diffeomorphic. The simplest example where these properties *do* hold is a product $E = M \times F$ with π given by the projection onto the first factor.

Fibre bundles are an important generalization of products $E = M \times F$ and can be understood as *twisted* products. The fibres of a fibre bundle are still embedded submanifolds and are all diffeomorphic. However, the fibration in general is only **locally trivial**, i.e. locally a product, and not globally. We shall see later in

Corollary 4.2.9 and Corollary 4.5.12 that if the topology of M is trivial (i.e. M is contractible), then certain types of fibre bundles over M (like principal and vector bundles) are always globally trivial. If M has a non-trivial topology (for example, if M is a sphere S^n), this may not be the case.

Consider, for instance, the Hopf action of $S^1 = \mathrm{U}(1)$ on S^3, introduced in Definition 3.3.1. This is a free action, i.e. the orbit of every point in S^3 is an embedded S^1 and the quotient space $S^3/\mathrm{U}(1)$ of this action is the smooth manifold $\mathbb{CP}^1 \cong S^2$.

However, it is clear (e.g. by considering fundamental groups) that S^3 cannot be diffeomorphic to $S^2 \times S^1$. We will see in Example 4.2.14 that S^3 really is the total space of a *non-trivial* S^1-bundle over S^2. We denote this bundle by

$$S^1 \longrightarrow S^3 \xrightarrow{\pi} S^2$$

or

$$\begin{array}{ccc} S^1 & \longrightarrow & S^3 \\ & & \downarrow \pi \\ & & S^2 \end{array}$$

This is the celebrated *Hopf fibration*. The total space S^3 is simply connected even though the fibres S^1 are not. This is possible, because the fibre bundle is globally non-trivial.

General fibre bundles are defined as follows.

Definition 4.1.2 Let E, F, M be manifolds and $\pi: E \to M$ a surjective differentiable map. Then $(E, \pi, M; F)$ is called a **fibre bundle** (or **locally trivial fibration** or **locally trivial bundle**) if the following holds: For every $x \in M$ there exists an open neighbourhood $U \subset M$ around x such that π restricted to E_U can be **trivialized**, i.e. there exists a diffeomorphism

$$\phi_U: E_U \longrightarrow U \times F$$

such that

$$\mathrm{pr}_1 \circ \phi_U = \pi,$$

hence the following diagram commutes:

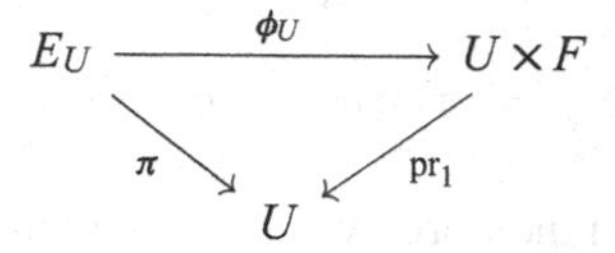

We also write

$$\begin{array}{ccc} F & \longrightarrow & E \\ & & \big\downarrow \pi \\ & & M \end{array}$$

or

$$F \longrightarrow E \xrightarrow{\pi} M$$

to denote a fibre bundle. We call

- E the **total space**
- M the **base manifold**
- F the **general fibre**
- π the **projection**
- (U, ϕ_U) a **local trivialization** or **bundle chart**.

See Fig. 4.1.

Remark 4.1.3 The classic references [133] and [81] use the term *fibre bundle* in a more restrictive sense; see Remark 4.1.15.

It is easy to see, using a local trivialization (U, ϕ_U), that the fibre

$$E_x = \pi^{-1}(x)$$

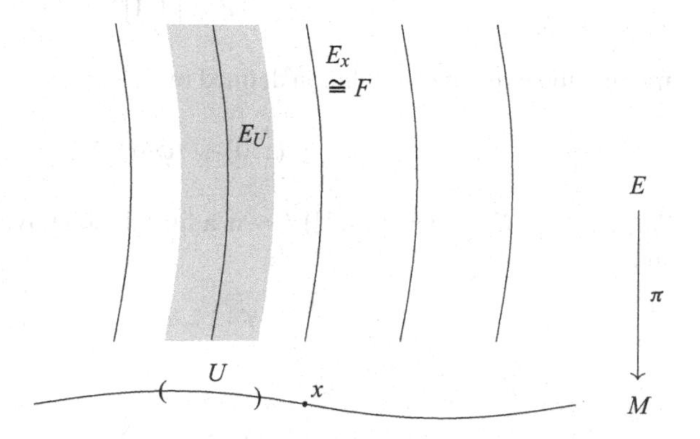

Fig. 4.1 Fibre bundle

is an embedded submanifold of the total space E for every $x \in M$ and the map ϕ_{Ux} defined by

$$\phi_{Ux} = \mathrm{pr}_2 \circ \phi_U|_{E_x}: E_x \longrightarrow F$$

is a diffeomorphism between the fibre over $x \in U$ and the general fibre.

Note that in a local trivialization the map

$$\phi_U: E_U \longrightarrow U \times F$$

is a diffeomorphism and

$$\mathrm{pr}_1: U \times F \longrightarrow U$$

is a submersion (its differential is everywhere surjective). This implies that the projection $\pi: E \to M$ of a fibre bundle is always a submersion. The Regular Value Theorem A.1.32 then shows again that the fibres $E_x = \pi^{-1}(x)$ are embedded submanifolds of E.

Example 4.1.4 (Trivial Bundle) Let M and F be arbitrary smooth manifolds and $E = M \times F$. Then $\pi = \mathrm{pr}_1$ defines a fibre bundle

$$\begin{array}{ccc} F & \longrightarrow & M \times F \\ & & \big\downarrow \mathrm{pr}_1 \\ & & M \end{array}$$

This bundle is called **trivial**.

Example 4.1.5 (Mapping Torus) We discuss an example where the idea of a fibre bundle as a "twisted product" becomes very apparent. Let F be a manifold and $\phi: F \to F$ a diffeomorphism. We construct a fibre bundle E_ϕ as follows: Take

$$F \times [0, 1]$$

modulo the equivalence relation defined by

$$(x, 0) \sim (\phi(x), 1).$$

The quotient $E_\phi = (F \times [0, 1]) / \sim$ is a fibre bundle over the circle S^1 with general fibre F:

$$\begin{array}{ccc} F & \longrightarrow & E_\phi \\ & & \big\downarrow \pi \\ & & S^1 \end{array}$$

The bundle E_ϕ is called the **mapping torus** with general fibre F and **monodromy** ϕ. See Remark 4.6.4 for more details. We can think of the bundle E_ϕ as being obtained by gluing the two ends of $F \times [0, 1]$ together using the diffeomorphism ϕ.

If ϕ is the identity, then the mapping torus is a trivial bundle, but if ϕ is not the identity, the mapping torus may be non-trivial. For example, for the fibre $F = S^1$ we can do the construction with ϕ the identity of S^1, in which case E_ϕ is diffeomorphic to the torus T^2, and with ϕ the reflection $z \mapsto \bar{z}$ on $S^1 \subset \mathbb{C}$, in which case E_ϕ is diffeomorphic to the Klein bottle. Since the Klein bottle is not diffeomorphic to T^2, the second example is a non-trivial S^1-bundle over S^1.

The *clutching construction* that we discuss in Sect. 4.6 is a generalization of the mapping torus construction which yields fibre bundles

$$\begin{array}{ccc} F & \longrightarrow & E_f \\ & & \downarrow \pi \\ & & S^n \end{array}$$

over spheres of arbitrary dimension.

4.1.2 Bundle Maps

Definition 4.1.6 Let $F \to E \xrightarrow{\pi} M$ and $F' \to E' \xrightarrow{\pi'} M$ be fibre bundles over the manifold M. A **bundle map** or **bundle morphism** of these bundles is a smooth map $H: E \to E'$ such that

$$\pi' \circ H = \pi,$$

i.e. such that the following diagram commutes:

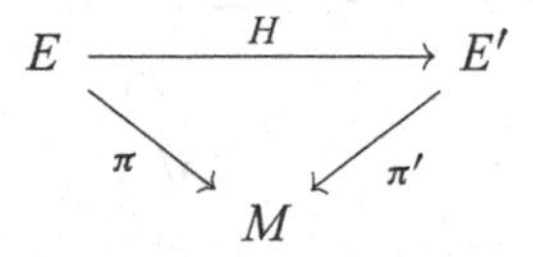

A **bundle isomorphism** is a bundle map which is a diffeomorphism. If such an isomorphism exists, we write $E \cong E'$.

Remark 4.1.7 Note that a morphism $H: E \to E'$ maps a point in the fibre of E over $x \in M$ to a point in the fibre of E' over the same point x. *A bundle map therefore covers the identity of* M. We could consider more general bundle maps between bundles over different manifolds M and N that cover a given smooth map $f: M \to N$. It is clear that a bundle isomorphism induces a diffeomorphism between the fibres of E and E' over any $x \in M$.

Definition 4.1.8 Fibre bundles isomorphic to a trivial bundle as in Example 4.1.4 are also called **trivial.**
It is more difficult to construct non-trivial fibre bundles. The mapping tori defined in Example 4.1.5 are for many choices of (F,ϕ) non-trivial bundles. We will discuss other examples of (potentially) non-trivial bundles in Sect. 4.2.2 and Sect. 4.6.

Remark 4.1.9 Let $F \to E \xrightarrow{\pi} M$ be a fibre bundle. The existence of a local trivialization over $U \subset M$ then means that the restricted bundle

$$\pi|_{E_U}: E_U \longrightarrow U$$

is isomorphic to the trivial bundle

$$\mathrm{pr}_1: U \times F \longrightarrow U.$$

This in hindsight justifies why fibre bundles are called locally trivial.
Isomorphic bundles have diffeomorphic general fibres. The converse is not true in general: There may exist non-isomorphic bundles whose general fibres are diffeomorphic. In particular, as we shall see later in detail, there exist bundles not (globally) isomorphic to a trivial bundle.

We can characterize trivial bundles as follows:

Proposition 4.1.10 (Trivial Bundles and Projections onto the General Fibre) *Let $F \to E \xrightarrow{\pi} M$ be a fibre bundle. Then the bundle is isomorphic to a trivial bundle if and only if there exists a smooth map $\tau: E \to F$ such that the restrictions*

$$\tau|_{E_x}: E_x \longrightarrow F$$

are diffeomorphisms for all $x \in M$.

Proof If the bundle is trivial

$$\begin{array}{ccc} F & \longrightarrow & E = M \times F \\ & & \downarrow \mathrm{pr}_1 \\ & & M \end{array}$$

we can set $\tau = \mathrm{pr}_2$.

Conversely, assume that a map $\tau: E \to F$ exists which restricts to a diffeomorphism on each fibre. Consider the map

$$\begin{aligned} H: E &\longrightarrow M \times F \\ p &\longmapsto (\pi(p), \tau(p)). \end{aligned}$$

Then H is a smooth with

$$\mathrm{pr}_1 \circ H = \pi.$$

The map H is bijective, because it maps E_x bijectively onto F.

We have to show that H is a diffeomorphism. We claim that the differential of H is an isomorphism for every point $p \in E$. Since the dimensions of E and $M \times F$ agree (this follows from the existence of local trivializations for E), it suffices to show that the differential is surjective for every $p \in E$. The details are left as an exercise. □

4.1.3 *Bundle Atlases*

Definition 4.1.11 A **bundle atlas** for a fibre bundle

$$\begin{array}{ccc} F & \longrightarrow & E \\ & & \downarrow \pi \\ & & M \end{array}$$

is an open covering $\{U_i\}_{i \in I}$ of M together with bundle charts

$$\phi_i : E_{U_i} \longrightarrow U_i \times F.$$

Definition 4.1.12 Let $\{(U_i, \phi_i)\}_{i \in I}$ be a bundle atlas for a fibre bundle $F \to E \xrightarrow{\pi} M$. If $U_i \cap U_j \neq \emptyset$, we define the **transition functions** by

$$\phi_j \circ \phi_i^{-1}|_{(U_i \cap U_j) \times F} : (U_i \cap U_j) \times F \longrightarrow (U_i \cap U_j) \times F.$$

The transition functions are diffeomorphisms. These maps have a special structure, because they preserve fibres: For every $x \in U_i \cap U_j$ we get a diffeomorphism

$$\phi_{jx} \circ \phi_{ix}^{-1} : F \longrightarrow F.$$

The maps

$$\begin{aligned} \phi_{ji} : U_i \cap U_j &\longrightarrow \mathrm{Diff}(F) \\ x &\longmapsto \phi_{jx} \circ \phi_{ix}^{-1} \end{aligned}$$

into the group of diffeomorphisms of F are also called transition functions. See Fig. 4.2.

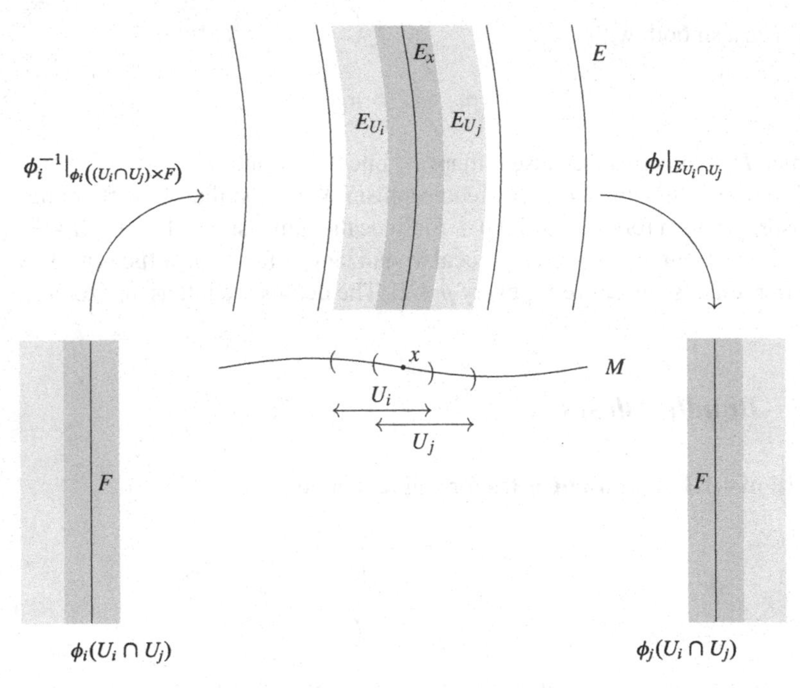

Fig. 4.2 Transition functions

Lemma 4.1.13 (Cocycle Conditions) *The transition functions* $\{\phi_{ji}\}_{i,j\in I}$ *satisfy the following equations:*

$$\begin{aligned} \phi_{ii}(x) &= \mathrm{Id}_F \quad \textit{for}\, x \in U_i, \\ \phi_{ij}(x) \circ \phi_{ji}(x) &= \mathrm{Id}_F \quad \textit{for}\, x \in U_i \cap U_j, \\ \phi_{ik}(x) \circ \phi_{kj}(x) \circ \phi_{ji}(x) &= \mathrm{Id}_F \quad \textit{for}\, x \in U_i \cap U_j \cap U_k. \end{aligned}$$

The third equation is called the **cocycle condition**.

Proof Follows immediately from the definitions. □

Remark 4.1.14 Exercise 4.8.9 shows that a bundle can be (re-)constructed from its transition functions using a suitable quotient space. The three properties of Lemma 4.1.13 ensure the existence of a certain equivalence relation, used in the construction of this quotient space.
A bundle atlas is very similar to an atlas of charts for a manifold. One difference is that in the case of charts for a manifold we demand that the images of the charts are open sets in a Euclidean space $\mathbb{R}^n$. In the case of charts for a bundle the images are of the form $U \times F$. In both cases the transition functions are smooth. In the case of a bundle atlas, the transition functions have an additional special structure, because they preserve fibres.

Table 4.1 Comparison between notions for manifolds and fibre bundles

Manifold	Fibre bundle
Coordinate chart	Bundle chart
Coordinate transformation	Transition functions
Atlas	Bundle atlas
Trivial manifold with only one chart: $\mathbb{R}^n$	Trivial bundle with only one bundle chart: $M \times F$
Non-trivial manifold needs at least two charts (like S^n)	Non-trivial bundle needs at least two bundle charts (like a non-trivial bundle over S^n)

We can compare the definitions of general manifolds and general fibre bundles as in Table 4.1.

Remark 4.1.15 Some references, such as [133] and [81], use the term *fibre bundle* more restrictively. If the topological definition in these books is transferred to a smooth setting, the definition amounts to assuming that the transition functions of a bundle atlas are smooth maps to a Lie group G, acting smoothly as a transformation group on the fibre F, instead of maps to the full diffeomorphism group Diff(F) of the fibre:

$$\begin{aligned} \phi_{ji}\colon U_i \cap U_j &\longrightarrow G \\ x &\longmapsto \phi_{jx} \circ \phi_{ix}^{-1}. \end{aligned}$$

Equivalently, a fibre bundle is with this definition always an associated bundle in the sense of Remark 4.7.8.

4.1.4 *Pullback Bundle

We want to show that we can *pull back* a bundle via a map between the base manifolds. Suppose

$$\begin{array}{ccc} F & \longrightarrow & E \\ & & \downarrow{\scriptstyle \pi_M} \\ & & M \end{array}$$

is a fibre bundle.

Lemma 4.1.16 (Restriction of Bundle over Submanifold in the Base) *If $W \subset M$ is an embedded submanifold, then the restriction*

$$\begin{array}{ccc} F & \longrightarrow & E_W \\ & & \downarrow \pi_M \\ & & W \end{array}$$

is a fibre bundle.

Proof Let $\{(U_i, \phi_i)\}_{i \in I}$ be a bundle atlas for the fibre bundle $F \to E \to M$ with bundle charts

$$\phi_i : E_{U_i} \longrightarrow U_i \times F.$$

Then the sets $V_i = U_i \cap W$ form an open covering of W and

$$\psi_i = \phi_i|_{E_{V_i}} : E_{V_i} \longrightarrow V_i \times F$$

are bundle charts for the restriction of E over W. □

Suppose $f: N \to M$ is a differentiable map from some manifold N to M. We set

$$f^*E = \{(x, e) \in N \times E \mid f(x) = \pi_M(e)\}$$

and

$$\begin{aligned} \pi_N : f^*E &\longrightarrow N \\ (x, e) &\longmapsto x. \end{aligned}$$

Theorem 4.1.17 (Pullback Bundles) *The map π_N with*

$$\begin{array}{ccc} F & \longrightarrow & f^*E \\ & & \downarrow \pi_N \\ & & N \end{array}$$

is a fibre bundle over N, called the **pullback** *of E under f.*

Proof We have an obvious fibre bundle

$$\begin{array}{ccc} F & \longrightarrow & N \times E \\ & & \downarrow \pi_{N \times M} = \mathrm{Id}_N \times \pi_M \\ & & N \times M \end{array}$$

The graph

$$\Gamma_f = \{(x,f(x)) \in N \times M \mid x \in N\}$$

is an embedded submanifold of $N \times M$. Therefore the restriction

$$\begin{array}{ccc} F & \longrightarrow & \pi_{N\times M}^{-1}(\Gamma_f) \subset N \times E \\ & & \downarrow \pi_{N\times M} \\ & & \Gamma_f \end{array}$$

is a fibre bundle by Lemma 4.1.16. Note that

$$(x,e) \in \pi_{N\times M}^{-1}(\Gamma_f) \Leftrightarrow \pi(e) = f(x).$$

Hence as a set

$$\pi_{N\times M}^{-1}(\Gamma_f) = f^*E,$$

which defines a smooth structure on f^*E, and we have a fibre bundle

$$\begin{array}{ccc} F & \longrightarrow & f^*E \\ & & \downarrow \pi_{N\times M} \\ & & \Gamma_f \end{array}$$

There exists a diffeomorphism

$$\begin{aligned} \tau\colon \Gamma_f &\longrightarrow N \\ (x,f(x)) &\longmapsto x. \end{aligned}$$

We can define a bundle over N using the projection $\tau \circ \pi_{N\times M}$:

$$\begin{array}{ccccc} F & \longrightarrow & f^*E & & \\ & & \downarrow \pi_{N\times M} & \searrow & \\ & & \Gamma_f & \xrightarrow{\tau} & N \end{array}$$

But

$$\tau \circ \pi_{N\times M}(x,e) = \tau(x,\pi(e)) = x = \pi_N(x,e),$$

hence

$$\tau \circ \pi_{N\times M} = \pi_N.$$

This shows that

$$\begin{array}{ccc} F & \longrightarrow & f^*E \\ & & \downarrow \pi_N \\ & & N \end{array}$$

is a fibre bundle. □

Remark 4.1.18 Note that the pullback bundle f^*E has the same general fibre F as the bundle E. The fibre of f^*E over a point $x \in N$ is canonically diffeomorphic to the fibre of E over $f(x) \in M$ via the map

$$\begin{aligned} (f^*E)_x &\longrightarrow E_{f(x)} \\ (x, e) &\longmapsto e. \end{aligned}$$

Remark 4.1.19 It is not difficult to show that the pull-back of a trivial bundle is always trivial. The pull-back of a non-trivial bundle may be non-trivial or trivial, depending on the situation; see Exercise 4.8.2.

4.1.5 Sections of Bundles

We want to study sections of fibre bundles. This is particularly simple in the case of trivial bundles.

Definition 4.1.20 Let

$$\begin{array}{ccc} F & \longrightarrow & E \\ & & \downarrow \pi \\ & & M \end{array}$$

be a fibre bundle. We denote the set of smooth global sections $s: M \to E$ by $\Gamma(E)$ and the set of smooth local sections $s: U \to E$, for $U \subset M$ open, by $\Gamma(U, E)$.
It is easy to see that for a trivial bundle E there is a 1-to-1 correspondence between sections of E and maps from the base manifold M to the general fibre F. This implies:

Corollary 4.1.21 (Existence of (Local) Sections)

1. *Every trivial fibre bundle has smooth global sections (for example, under the above correspondence, we could take constant maps from the base M to the fibre F).*
2. *Every fibre bundle has smooth local sections, since every fibre bundle is locally trivial.*

Note that non-trivial fibre bundles can, but do not need to have smooth *global* sections (for example, vector bundles, to be discussed later, always have global sections, but principal bundles in general do not). In particular, for a non-trivial bundle, a map from the base manifold to the general fibre usually does not define a section.

4.2 Principal Fibre Bundles

Principal fibre bundles are a combination of the concepts of fibre bundles and group actions: they are fibre bundles which also have a Lie group action so that both structures are compatible in a certain sense. Principal bundles together with so-called *connections* play an important role in gauge theory. Generally speaking, principal bundles are the primary place where Lie groups appear in gauge theories (Lie groups also appear as global symmetry groups, like the *flavour* or *chiral symmetry* in QCD; see Sect. 9.1).

4.2.1 Definition of Principal Bundles

We consider again the Hopf action of S^1 on S^3, introduced in Definition 3.3.1, with quotient space equal to $\mathbb{CP}^1 \cong S^2$. If we accept for the moment that S^3 is the total space of an S^1-bundle over S^2,

$$\begin{array}{ccc} S^1 & \longrightarrow & S^3 \\ & & \downarrow \pi \\ & & S^2 \end{array}$$

(we will prove this in Example 4.2.14), then we can say the following: there is an action of the Lie group S^1 on the total space S^3 of the bundle which preserves the fibres and is simply transitive on them. In addition we will show that there is a special type of bundle atlas for the Hopf fibration which is compatible with this S^1-action.

This leads us to the following definition:

Definition 4.2.1 Let

$$\begin{array}{ccc} G & \longrightarrow & P \\ & & \downarrow \pi \\ & & M \end{array}$$

be a fibre bundle with general fibre a Lie group G and a smooth action $P \times G \rightarrow P$ on the right. Then P is called a **principal G-bundle** if:

1. The action of G **preserves the fibres of π and is simply transitive on them**, i.e. the action restricts to

$$P_x \times G \longrightarrow P_x$$

and the orbit map

$$\begin{aligned} G &\longrightarrow P_x \\ g &\longmapsto p \cdot g \end{aligned}$$

is a bijection, for all $x \in M, p \in P_x$.
2. There exists a **bundle atlas of G-equivariant bundle charts** $\phi_i \colon P_{U_i} \rightarrow U_i \times G$, satisfying

$$\phi_i(p \cdot g) = \phi_i(p) \cdot g \quad \forall p \in P_{U_i}, g \in G,$$

where on the right-hand side G acts on $(x, a) \in U_i \times G$ via

$$(x, a) \cdot g = (x, ag).$$

We also call such an atlas a **principal bundle atlas** for P and the charts in a principal bundle atlas **principal bundle charts.**

The group G is called the **structure group** of the principal bundle P.

There are two features that distinguish a principal bundle $P \rightarrow M$ from a standard fibre bundle whose general fibre is a Lie group G:

1. there exists a right G-action on P, simply transitive on each fibre P_x, for $x \in M$;
2. the bundle P has a principal bundle atlas.

If $P \rightarrow M$ is a principal G-bundle and $g \in G$, then we denote as before by r_g the right translation

$$\begin{aligned} r_g \colon P &\longrightarrow P \\ p &\longmapsto p \cdot g. \end{aligned}$$

The fibre P_x is a submanifold of the total space P for every $x \in M$ and the orbit map

$$\begin{aligned} G &\longrightarrow P_x \\ g &\longmapsto p \cdot g \end{aligned}$$

is an embedding for all $p \in P_x$, according to Corollary 3.8.10, because the stabilizer $G_p = \{e\}$ is trivial.

Example 4.2.2 The trivial bundle

$$\begin{array}{ccc} G & \longrightarrow & M \times G \\ & & \downarrow \mathrm{pr}_1 \\ & & M \end{array}$$

has the canonical structure of a principal G-bundle with G-action

$$\begin{aligned} (M \times G) \times G &\longrightarrow M \times G \\ (x, h, g) &\longmapsto (x, hg) \end{aligned}$$

and the principal bundle atlas consisting of only one bundle chart

$$\mathrm{Id}: M \times G \longrightarrow M \times G.$$

Example 4.2.3 If $G \to P \xrightarrow{\pi} M$ is a principal bundle and $f: N \to M$ a smooth map, then the pullback f^*P has the canonical structure of a principal G-bundle over N (this is Exercise 4.8.4).

Transition functions for a principal bundle atlas have a special form:

Proposition 4.2.4 (Transition Functions of Principal Bundles) *Let $P \to M$ be a principal G-bundle and $\{(U_i, \phi_i)\}_{i \in I}$ a principal bundle atlas for P. Then the transition functions take values in the subgroup G of* Diff(G),

$$\begin{aligned} \phi_{ji}: U_i \cap U_j &\longrightarrow G \subset \mathrm{Diff}(G) \\ x &\longmapsto \phi_{jx} \circ \phi_{ix}^{-1} \end{aligned}$$

where an element $g \in G$ acts as a diffeomorphism on G through left multiplication,

$$g(h) = g \cdot h.$$

Proof For $x \in U_i \cap U_j$ we have a diffeomorphism

$$\phi_{jx} \circ \phi_{ix}^{-1}: G \longrightarrow G.$$

We set

$$g = \phi_{jx} \circ \phi_{ix}^{-1}(e).$$

Then by equivariance of the bundle charts

$$\phi_{jx} \circ \phi_{ix}^{-1}(h) = g \cdot h.$$

This implies the claim. □

The following criterion sometimes simplifies the task of showing that a group action on a manifold P defines a principal bundle (we follow [14, Theorem 2.4]).

Theorem 4.2.5 (Principal Bundles Defined via Local Sections) *Let G be a Lie group and $\pi: P \to M$ a smooth surjective map between manifolds with a smooth action $P \times G \to P$ on the right. Then P is a principal G-bundle if and only if the following holds:*

1. *The action of G preserves the fibres of π and is simply transitive on them.*
2. *There exists an open covering $\{U_i\}_{i \in I}$ of M together with local sections $s_i: U_i \to P$ of the map π.*

Remark 4.2.6 Recall that we defined in Sect. 4.1.1 the notion of a section for any smooth surjective map, not only for fibre bundles.

Proof Suppose that $\pi: P \to M$ is a principal bundle. Choose a principal bundle atlas $\{(U_i, \phi_i)\}$ for P with

$$\phi_i: P_{U_i} \longrightarrow U_i \times G.$$

Then the following maps are local sections

$$\begin{aligned} s_i: U_i &\longrightarrow P \\ x &\longmapsto \phi_i^{-1}(x, e), \end{aligned}$$

where $e \in G$ is the neutral element.

Conversely, suppose that an open covering $\{U_i\}_{i \in I}$ with sections $s_i: U_i \to P$ is given. According to the following lemma these sections define charts in a principal bundle atlas for P. □

Lemma 4.2.7 *Let G be a Lie group and $\pi: P \to M$ a smooth surjective map between manifolds with a smooth action $P \times G \to P$ on the right. Suppose that the* *action of G preserves the fibres of π and is simply transitive on them. Let $s\colon U \to P$ be a local section for π. Then*

$$
\begin{aligned}
t\colon U \times G &\longrightarrow P_U \\
(x, g) &\longmapsto s(x) \cdot g
\end{aligned}
$$

is a G-equivariant diffeomorphism.

Proof Let $s\colon U \to P$ be a local section of the surjective map $\pi\colon P \to M$. We have to show that

$$
\begin{aligned}
t\colon U \times G &\longrightarrow P_U \\
(x, g) &\longmapsto s(x) \cdot g
\end{aligned}
$$

is a G-equivariant diffeomorphism. It is clear that t is smooth, because the local section s is smooth and the G-action on P is smooth. The map t is also G-equivariant by the definition of group actions and it is bijective: the reason is that the map

$$
\begin{aligned}
t(x, \cdot)\colon G &\longrightarrow P_x \\
g &\longmapsto s(x) \cdot g
\end{aligned}
$$

is bijective for every fixed $x \in U$, since the G-action on P is simply transitive on the fibres. The set $P_U = \pi^{-1}(U)$ is an open subset of P. Since t is smooth and surjective, Sard's Theorem A.1.27 implies that

$$
\dim P = \dim P_U \leq \dim M + \dim G.
$$

It remains to show that the differential of t is injective in each point $(x, g) \in U \times G$. Then t is a diffeomorphism.

The differential

$$
D_{(x,g)}t\colon T_xM \times T_gG \longrightarrow T_{s(x)\cdot g}P
$$

is given according to Proposition 3.5.4 by

$$
D_{(x,g)}t(X, Y) = D_x(r_g \circ s)(X) + \widetilde{\mu_G(Y)}_{s(x)\cdot g}.
$$

We set

$$
s' = r_g \circ s.
$$

The map

$$\begin{aligned} T_gG &\longrightarrow T_{s'(x)}P_x \\ Y &\longmapsto \widetilde{\mu_G(Y)}_{s'(x)} \end{aligned}$$

is an isomorphism, because the action of G is simply transitive on the fibre P_x, cf. Corollary 3.2.12. We consider the map

$$\begin{aligned} T_xM &\longrightarrow T_{s'(x)}P \\ X &\longmapsto D_xs'(X). \end{aligned}$$

Note that s' is also a local section of P over U, since

$$\pi \circ s' = \mathrm{Id}_U.$$

The chain rule shows that

$$D_{s'(x)}\pi \circ D_xs' = \mathrm{Id}_{T_xM}.$$

This implies that D_xs' is injective and the image of D_xs' intersected with $T_{s'(x)}P_x \subset \ker D_{s'(x)}\pi$ is zero. We conclude that $D_{(x,g)}t$ is injective. □

A proof of the following theorem can be found in [81, Chap. 4, Corollary 10.3].

Theorem 4.2.8 (Principal Bundles and Homotopy Equivalences) *Let $f: M \to N$ be a smooth homotopy equivalence between manifolds and G a Lie group. Then the pullback f^* is a bijection between isomorphism classes of principal G-bundles over N and principal G-bundles over M.*

In particular we get:

Corollary 4.2.9 (Principal Bundles over Contractible Manifolds Are Trivial) *If M is a contractible manifold and G a Lie group, then every principal G-bundle over M is trivial. This holds, in particular, if $M = \mathbb{R}^n$ for some n.*

4.2.2 *Principal Bundles Defined by Principal Group Actions

Recall from Definition 3.7.24 that a smooth right action of a Lie group G on a manifold P is called principal if the action is free and the map

$$\begin{aligned} \Psi: P \times G &\longrightarrow P \times P \\ (p, g) &\longmapsto (p, pg) \end{aligned}$$

is closed. We want to show as an application of Theorem 4.2.5 that principal Lie group actions define principal bundles.

Theorem 4.2.10 (Principal Lie Group Actions Define Principal Bundles) *Let Φ be a principal right action of a Lie group G on a manifold P. Then P/G is a smooth manifold and*

$$\begin{array}{ccc} G & \longrightarrow & P \\ & & \downarrow{\scriptstyle \pi} \\ & & P/G \end{array}$$

is a principal bundle with structure group G.

Proof According to Theorem 3.7.25 the topological space P/G has the unique structure of a smooth manifold so that $\pi: P \rightarrow P/G$ is a submersion. In particular, by Lemma 3.7.4, the projection π admits local sections

$$s_i: U_i \longrightarrow P.$$

The claim then follows from Theorem 4.2.5. □

Corollary 4.2.11 (Free Actions by Compact Lie Groups Define Principal Bundles) *Let G be a compact Lie group acting freely on a smooth manifold P. Then P/G is a smooth manifold and*

$$\begin{array}{ccc} G & \longrightarrow & P \\ & & \downarrow{\scriptstyle \pi} \\ & & P/G \end{array}$$

is a principal G-bundle.

Proof This follows from Corollary 3.7.29. □

We can also prove the following converse to Theorem 4.2.10.

Theorem 4.2.12 (Principal Bundles Define Principal Actions) *Let*

$$\begin{array}{ccc} G & \longrightarrow & P \\ & & \downarrow{\scriptstyle \pi} \\ & & M \end{array}$$

be a principal G-bundle. Then the right action of G on P is principal.

Proof The G-action on P is free by the definition of principal bundles. If G is compact, then the claim follows from Corollary 3.7.29. In the general case, consider the map

$$\begin{aligned} \Psi: P \times G &\longrightarrow P \times P \\ (p, g) &\longmapsto (p, p \cdot g). \end{aligned}$$

We have to show that Ψ is closed.

Let $A \subset P \times G$ be a closed subset and $((p_i, q_i))_{i\in\mathbb{N}} \in \Psi(A)$ a sequence converging to $(p, q) \in P \times P$. There exist uniquely determined $g_i \in G$ such that $q_i = p_i \cdot g_i$, where $(p_i, g_i) \in A$ and

$$\Psi(p_i, g_i) = (p_i, q_i).$$

We want to show that the sequence $(g_i)_{i\in\mathbb{N}}$ converges in G.

Let $\pi(p) = x$ and $U \subset M$ be an open neighbourhood of x with a principal bundle chart

$$\phi: P_U \longrightarrow U \times G.$$

There exists an integer N such that for all $i \geq N$ the p_i are contained in P_U. Then we can write

$$\begin{aligned}\phi(p_i) &= (x_i, h_i),\\ \phi(q_i) &= (x_i, h_i g_i),\\ \phi(p) &= (x, h),\end{aligned}$$

with certain $x_i \in U$ and $h_i, h \in G$. Since $q_i \to q$ and $x_i \to x$, it follows that

$$\phi(q) = \left(x, h'\right)$$

for some $h' \in G$. Since $h_i \to h$ and $h_i g_i \to h'$, it follows that the sequence

$$g_i = h_i^{-1}(h_i g_i)$$

converges in G to

$$g = h^{-1}h'.$$

The set A is closed, hence $(p, g) \in A$. We have $q = p \cdot g$ and we conclude that (p, q) is in $\Psi(A)$. □

Corollary 4.2.13 *Principal bundles with structure group G correspond precisely to principal G-actions.*

Example 4.2.14 (Hopf Fibration) Let

$$S^{2n+1} = \left\{ (w_0, \ldots, w_n) \in \mathbb{C}^{n+1} \,\middle|\, \sum_{i=0}^{n} |w_i|^2 = 1 \right\}$$

be a sphere of odd dimension. Consider the Lie group $S^1 = \mathrm{U}(1) \subset \mathbb{C}$ (unit circle). It acts on the sphere S^{2n+1} via

$$\begin{aligned} S^{2n+1} \times S^1 &\longrightarrow S^{2n+1} \\ (w, \lambda) &\longmapsto w\lambda. \end{aligned}$$

This is the Hopf action from Definition 3.3.1. The quotient $S^{2n+1}/\mathrm{U}(1)$ of this action can be identified with the complex projective space $\mathbb{CP}^n$. Corollary 4.2.11 implies that

$$\begin{array}{ccc} S^1 & \longrightarrow & S^{2n+1} \\ & & \downarrow \pi \\ & & \mathbb{CP}^n \end{array}$$

is a principal S^1-bundle, called the **Hopf fibration** or **Hopf bundle**.

To give an alternative proof of this statement, we can also apply Theorem 4.2.5 directly (we follow [14, Example 2.7]). We have to find an open covering of $\mathbb{CP}^n$ together with local sections (the first condition in the theorem is clearly satisfied, because the action of S^1 on S^{2n+1} is free). We set

$$\pi(w_0, \ldots, w_n) = [w_0 : \ldots : w_n] \in \mathbb{CP}^n$$

and define for $i = 0, \ldots, n$

$$U_i = \{[w] = [w_0 : \ldots : w_n] \in \mathbb{CP}^n \mid w_i \neq 0\}.$$

The subset U_i is open in $\mathbb{CP}^n$, since π is an open map by Lemma 3.7.11. We also set

$$v_i([w]) = \left(\frac{w_0}{w_i}, \ldots, \frac{w_{i-1}}{w_i}, 1, \frac{w_{i+1}}{w_i}, \ldots, \frac{w_n}{w_i}\right) \in \mathbb{C}^{n+1} \setminus \{0\}$$

and

$$\begin{aligned} s_i \colon U_i &\longrightarrow S^{2n+1} \\ [w] &\longmapsto \frac{v_i([w])}{||v_i([w])||}. \end{aligned}$$

These are well-defined local sections for the canonical projection π:

$$\pi \circ s_i = \mathrm{Id}_{U_i},$$

since $s_i([w])$ is a complex multiple of w. Therefore we see again that $S^1 \rightarrow S^{2n+1} \xrightarrow{\pi} \mathbb{CP}^n$ is a principal fibre bundle.

It is clear (considering fundamental groups, for example) that S^{2n+1} is not diffeomorphic to $\mathbb{CP}^n \times S^1$. The Hopf fibration is thus an example of a non-trivial (principal) fibre bundle.

Similar arguments for the standard action of the Lie group $S^3 \subset \mathbb{H}$ on $S^{4n+3} \subset \mathbb{H}^{n+1}$ lead to a Hopf fibration

$$\begin{array}{ccc} S^3 & \longrightarrow & S^{4n+3} \\ & & \downarrow \pi \\ & & \mathbb{HP}^n \end{array}$$

over the quaternionic projective space $\mathbb{HP}^n$ (there is also a principal $\mathbb{Z}_2$-bundle $S^n \to \mathbb{RP}^n$ over real projective space). Special cases of this construction are the Hopf fibrations (see Exercise 3.12.9)

$$\begin{array}{ccc} S^1 & \longrightarrow & S^3 \\ & & \downarrow \pi \\ & & S^2 \end{array}$$

and

$$\begin{array}{ccc} S^3 & \longrightarrow & S^7 \\ & & \downarrow \pi \\ & & S^4 \end{array}$$

We consider another class of examples of principal bundles. Let G be a Lie group and $H \subset G$ a closed subgroup, acting smoothly on G by right translations:

$$\Phi: G \times H \longrightarrow G$$
$$(g,h) \longmapsto gh.$$

According to Corollary 3.7.35 there is a (unique) smooth structure on the quotient space G/H, so that $\pi: G \to G/H$ is a submersion.

Theorem 4.2.15 (The Canonical Principal Bundles over Homogeneous Spaces) *If G is a Lie group and $H \subset G$ a closed subgroup, then*

$$\begin{array}{ccc} H & \longrightarrow & G \\ & & \downarrow \pi \\ & & G/H \end{array}$$

is a principal bundle with structure group H.

Proof This follows from Theorem 4.2.10. We can also verify the conditions of Theorem 4.2.5 directly. The first condition is clearly satisfied, because the action of H on G is free. By Lemma 3.7.4 there exist smooth local sections

$$s_i \colon U_i \longrightarrow G$$

for the canonical projection $\pi \colon G \to G/H$, where the open subsets $U_i \subset G/H$ cover G/H. This proves the claim. □

Example 4.2.16 (Principal Bundles over Homogeneous Spheres) From Example 3.8.11 we get the following principal bundles over spheres:

$$\begin{array}{ccc} \mathrm{O}(n-1) & \longrightarrow & \mathrm{O}(n) \\ & & \downarrow \pi \\ & & S^{n-1} \end{array}$$

$$\begin{array}{ccc} \mathrm{SO}(n-1) & \longrightarrow & \mathrm{SO}(n) \\ & & \downarrow \pi \\ & & S^{n-1} \end{array}$$

$$\begin{array}{ccc} \mathrm{U}(n-1) & \longrightarrow & \mathrm{U}(n) \\ & & \downarrow \pi \\ & & S^{2n-1} \end{array}$$

$$\begin{array}{ccc} \mathrm{SU}(n-1) & \longrightarrow & \mathrm{SU}(n) \\ & & \downarrow \pi \\ & & S^{2n-1} \end{array}$$

$$\begin{array}{ccc} \mathrm{Sp}(n-1) & \longrightarrow & \mathrm{Sp}(n) \\ & & \downarrow \pi \\ & & S^{4n-1} \end{array}$$

In particular, we get the following principal sphere bundles over spheres:

$$\begin{array}{ccc} S^1 & \longrightarrow & \mathrm{SO}(3) \\ & & \downarrow \pi \\ & & S^2 \end{array}$$

$$\begin{array}{ccc} S^1 & \longrightarrow & \mathrm{U}(2) \\ & & \downarrow \pi \\ & & S^3 \end{array}$$

$$\begin{array}{ccc} S^3 & \longrightarrow & \mathrm{SU}(3) \\ & & \downarrow \pi \\ & & S^5 \end{array}$$

$$\begin{array}{ccc} S^3 & \longrightarrow & \mathrm{Sp}(2) \\ & & \downarrow \pi \\ & & S^7 \end{array}$$

From the examples in Sect. 3.9 we also get principal bundles over the Stiefel and Grassmann manifolds, such as

$$\begin{array}{ccc} \mathrm{O}(n-k) & \longrightarrow & \mathrm{O}(n) \\ & & \downarrow \pi \\ & & V_k(\mathbb{R}^n) \end{array}$$

and

$$\begin{array}{ccc} \mathrm{O}(k) \times \mathrm{O}(n-k) & \longrightarrow & \mathrm{O}(n) \\ & & \downarrow \pi \\ & & Gr_k(\mathbb{R}^n) \end{array}$$

and similarly for the complex and quaternionic Stiefel and Grassmann manifolds. According to the results in Sect. 3.10.4 there is a principal bundle

$$\begin{array}{ccc} \mathrm{SU}(2) & \longrightarrow & \mathrm{G}_2 \\ & & \downarrow \\ & & V_2(\mathbb{R}^7) \end{array}$$

and according to Exercise 3.12.16 there is a principal bundle

$$\begin{array}{ccc} \mathrm{SU}(3) & \longrightarrow & \mathrm{G}_2 \\ & & \downarrow \\ & & S^6 \end{array}$$

4.2.3 *Bundle Morphisms, Reductions of the Structure Group and Gauges*

We define homomorphisms of principal bundles as follows:

Definition 4.2.17 Suppose $G \to P \xrightarrow{\pi} M$ and $G' \to P' \xrightarrow{\pi'} M$ are principal bundles over the same base manifold M and $f: G \to G'$ is a Lie group homomorphism. Then a **bundle morphism** between P and P' is an f**-equivariant** smooth bundle map $H: P \to P'$, i.e.

$$\pi' \circ H = \pi$$

and

$$H(p \cdot g) = H(p) \cdot f(g) \quad \forall p \in P, g \in G.$$

Given the principal G'-bundle P' and the homomorphism $f: G \to G'$, the principal G-bundle P together with the bundle morphism $H: P \to P'$ is also known as an f**-reduction** of P'.

If $f: G \to G'$ is an embedding, then H is called a G**-reduction** of P' and the image of H is called a **principal** G**-subbundle** of P'. If $G = G', f = \mathrm{Id}_G$ and H is a G-equivariant bundle isomorphism, then H is called a **bundle isomorphism**.

A principal G-bundle isomorphic to the trivial bundle in Example 4.2.2 is also called trivial.

As before in the case of general bundles we could consider morphisms between principal bundles over different base manifolds M and N that cover a smooth map from M to N.

The following notion is especially relevant in gauge theory.

Definition 4.2.18 Let $\pi: P \to M$ be a principal bundle. A **global gauge** for the principal bundle is a global section $s: M \to P$. Similarly, a **local gauge** is a local section $s: U \to P$ defined on an open subset $U \subset M$.

Any local gauge defines a local trivialization of a principal bundle:

Theorem 4.2.19 (Gauges Correspond to Trivializations) *Let*

$$\begin{array}{ccc} G & \longrightarrow & P \\ & & \downarrow{\scriptstyle \pi} \\ & & M \end{array}$$

be a principal G-bundle and $s: U \to P$ *a local gauge. Then*

$$t: U \times G \longrightarrow P_U$$
$$(x, g) \longmapsto s(x) \cdot g$$

is a G-equivariant diffeomorphism. In particular, if $s: M \to G$ *is a global gauge, then the principal bundle is trivial, with trivialization given by the inverse of t:*

$$t^{-1}: P \longrightarrow M \times G.$$

Proof This follows from Lemma 4.2.7. □

Remark 4.2.20 Note that for this construction to work we need the G-action on P. The result would not hold if we just had a fibre bundle with fibre G.

Remark 4.2.21 Theorem 4.2.19 has the following interpretation, see Table 4.2: A local gauge defines a local trivialization of a principal G-bundle, i.e. an identification $\pi^{-1}(U) \cong U \times G$. *A choice of local gauge thus corresponds to the choice of a local coordinate system for a principal bundle in the fibre direction.* This can be compared, in special relativity, to the choice of an inertial system for Minkowski spacetime M, which defines an identification $M \cong \mathbb{R}^4$.

Of course, different choices of gauges are possible, leading to different trivializations of the principal bundle, just as different choices of inertial systems lead to different identifications of spacetime with $\mathbb{R}^4$. The idea of gauge theory is that physics *should be independent of the choice of gauge*. This can be compared to the theory of relativity which says that physics is independent of the choice of inertial system.

Note that, if we consider principal bundles over Minkowski spacetimes $\mathbb{R}^4$, it does not matter for this discussion that principal bundles over Euclidean spaces are always trivial by Corollary 4.2.9. What matters is the independence of the actual choice of *trivialization*, i.e. the choice of (global) gauge. Even on a trivial principal bundle there are non-trivial gauge transformations. This is very similar to special relativity, where spacetime is trivial, i.e. isometric to $\mathbb{R}^4$ with a Minkowski metric, but what matters is the independence of the actual trivialization, i.e. the choice of inertial system. Transformations between inertial systems are called *Lorentz transformations*, transformations between (local) gauges are called *gauge transformations*.

Table 4.2 Comparison between notions for special relativity and gauge theory

	Manifold	Trivialization	Transformations and invariance
Special relativity	Spacetime M	$M \cong \mathbb{R}^4$ via inertial system	Lorentz
Gauge theory	Principal bundle $P \to M$	$P \cong M \times G$ via choice of gauge	Gauge

4.3 *Formal Bundle Atlases

We briefly return to the case of general fibre bundles. We are sometimes in the following situation: We have a manifold M, a set E and a surjective map $\pi: E \to M$. However, we do not *a priori* have a topology or the structure of a smooth manifold on E. Under which circumstances can we define such structures, so that $\pi: E \to M$ becomes a smooth fibre bundle?

Example 4.3.1 Let M be a smooth manifold of dimension n. The tangent space T_pM is an n-dimensional vector space for all $p \in M$. Let TM be the disjoint union

$$TM = \dot{\bigcup}_{p\in M} T_pM$$

with the obvious projection $\pi: TM \to M$. How do we define the structure of a smooth manifold on the set TM, such that TM becomes a fibre bundle over M, with fibres given by T_pM? We can also define for each tangent space T_pM the dual vector space T_p^*M or the exterior algebra $\Lambda^k T_p^*M$. How do we construct smooth fibre bundles that have these vector spaces as fibres?
The following notion is useful in this context (we follow [14, Sect. 2.1]).

Definition 4.3.2 Let M and F be manifolds, E a set and $\pi: E \to M$ a surjective map.

1. Suppose $U \subset M$ is open and

$$\phi_U: E_U \longrightarrow U \times F$$

   is a bijection with

$$\mathrm{pr}_1 \circ \phi_U = \pi|_{E_U}.$$

   Then we call (U, ϕ_U) a **formal bundle chart** for E.
2. A family $\{(U_i, \phi_i)\}_{i\in I}$ of formal bundle charts, where $\{U_i\}_{i\in I}$ is an open covering of M, is called a **formal bundle atlas** for E.
3. We call the charts in a formal bundle atlas $\{(U_i, \phi_i)\}_{i\in I}$ **smoothly compatible** if all transition functions

$$\phi_j \circ \phi_i^{-1}|_{(U_i\cap U_j)\times F}: (U_i \cap U_j) \times F \longrightarrow (U_i \cap U_j) \times F,$$

   for $U_i \cap U_j \neq \emptyset$, are smooth maps (i.e. diffeomorphisms).

We then have:

Theorem 4.3.3 (Formal Bundle Atlases Define Fibre Bundles) *Let M and F be manifolds, E a set and $\pi: E \to M$ a surjective map. Suppose that $\{(U_i, \phi_i)\}_{i\in I}$ is a formal bundle atlas for E of smoothly compatible charts. Then there exists a unique* *topology and a unique structure of a smooth manifold on E such that*

$$\begin{array}{ccc} F & \longrightarrow & E \\ & & \downarrow \pi \\ & & M \end{array}$$

is a smooth fibre bundle with smooth bundle atlas $\{(U_i, \phi_i)\}_{i \in I}$.

The proof consists of several steps. We first define a topology on E: consider the bijections

$$\phi_i : E_{U_i} \longrightarrow U_i \times F.$$

We define a subset $O \subset E$ to be open if and only if

$$\phi_i(O \cap E_{U_i})$$

is open in $U_i \times F$ for all $i \in I$.

Lemma 4.3.4 (The Topology on E Defined by a Formal Bundle Atlas) *This defines a topology on E which is Hausdorff and has a countable base. It is the unique topology on E such that all formal bundle charts* $\phi_i : E_{U_i} \to U_i \times F$ *are homeomorphisms.*

Proof We first show that this defines a topology on E: it is clear that $\emptyset$ and E are open. It is also easy to see that arbitrary unions and finite intersections of open sets are open.

By definition the maps ϕ_i are open. Suppose that $O \subset E_{U_i}$ and $\phi_i(O)$ is open. Then for all $j \in I$

$$\begin{aligned} \phi_j(O \cap E_{U_j}) &= \left(\phi_j \circ \phi_i^{-1}\right)\left(\phi_i(O \cap E_{U_j} \cap E_{U_i})\right) \\ &= \left(\phi_j \circ \phi_i^{-1}\right)\left(\phi_i(O) \cap (U_j \cap U_i) \times F\right). \end{aligned}$$

It follows that O is open in E and that $\phi_i : E_{U_i} \to U_i \times F$ is a homeomorphism.

Since M and F are Hausdorff, it is not difficult to show that the topology on E is Hausdorff, by considering for arbitrary points $p, q \in E$ first the case $\pi(p) \neq \pi(q)$ with $\pi(p) \in U_i$, $\pi(q) \in U_j$ and then the case $\pi(p) = \pi(q) \in U_i$.

To show that the topology on E has a countable base we choose a countable base $\{V_j\}_{j \in J}$ for the topology of M and a countable base $\{W_k\}_{k \in K}$ for the topology of F. Without loss of generality we can assume that the family $\{U_i\}_{i \in I}$ is countable, without changing the topology of E. Let $O \subset E$ be an arbitrary open set and $p \in O$ a point. Then $p \in O \cap E_{U_i}$ for some i and there exist $j \in J$ and $k \in K$ such that

$$p \in \phi_i^{-1}(V_j \times W_k) \subset O \cap E_{U_i}.$$

This shows that the countable family

$$\{\phi_i^{-1}(V_j \times W_k)\}_{i\in I, j\in J, k\in K}$$

of open sets of E forms a base.

The uniqueness statement for the topology of E is clear. □

We can now finish the proof of Theorem 4.3.3.

Proof To define a smooth structure on E, we first define the smooth structure on E_{U_i} such that the homeomorphism

$$\phi_i: E_{U_i} \longrightarrow U_i \times F$$

is a diffeomorphism. Then this defines a smooth structure on E, because the transition functions

$$\phi_j \circ \phi_i^{-1}|_{(U_i\cap U_j)\times F}: (U_i \cap U_j) \times F \longrightarrow (U_i \cap U_j) \times F$$

are diffeomorphisms. This is the unique smooth structure on E so that $\pi: E \to M$ is a smooth fibre bundle with general fibre F and $\{(U_i, \phi_i)\}_{i\in I}$ is a smooth bundle atlas. □

4.4 *Frame Bundles

We want to apply Theorem 4.3.3 to define so-called *frame bundles*. Let M be a smooth, n-dimensional manifold. For a point $p \in M$ we define the set of all bases of T_pM

$$\mathrm{Fr}_{\mathrm{GL}}(M)_p = \{(v_1, \ldots, v_n) \text{ basis of } T_pM\}$$

and define the disjoint union

$$\mathrm{Fr}_{\mathrm{GL}}(M) = \dot{\bigcup}_{p\in M} \mathrm{Fr}_{\mathrm{GL}}(M)_p.$$

There is a natural projection $\pi: \mathrm{Fr}_{\mathrm{GL}}(M) \to M$ and an action

$$\mathrm{Fr}_{\mathrm{GL}}(M) \times \mathrm{GL}(n, \mathbb{R}) \longrightarrow \mathrm{Fr}_{\mathrm{GL}}(M),$$

given by

$$(v_1, \ldots, v_n) \cdot A = \left(\sum_{i=1}^n v_i A_{i1}, \ldots, \sum_{i=1}^n v_i A_{in} \right), \quad \forall (v_1, \ldots, v_n) \in \mathrm{Fr}_{\mathrm{GL}}(M)_p, A \in \mathrm{GL}(n, \mathbb{R}).$$

Theorem 4.4.1 (Frame Bundles) *The projection* π *and the action of* $\mathrm{GL}(n, \mathbb{R})$ *define the structure of a principal* $\mathrm{GL}(n, \mathbb{R})$*-bundle*

$$\begin{array}{ccc} \mathrm{GL}(n,\mathbb{R}) & \longrightarrow & \mathrm{Fr}_{\mathrm{GL}}(M) \\ & & \downarrow \pi \\ & & M \end{array}$$

This bundle is called the **frame bundle** *of the manifold M.*

Proof We defined $\mathrm{Fr}_{\mathrm{GL}}(M)$ so far only as a set. It is clear that the action of $\mathrm{GL}(n, \mathbb{R})$ preserves the fibres of π and is simply transitive on them. Let (U_i, ψ_i) be a local manifold chart for M,

$$\psi_i \colon U_i \longrightarrow \mathbb{R}^n.$$

Then

$$\begin{aligned} s_i \colon U_i &\longrightarrow \mathrm{Fr}_{\mathrm{GL}}(M)_{U_i} \\ p &\longmapsto (\partial_{x_1}, \ldots, \partial_{x_n})(p) \end{aligned}$$

is a local section for π. We have

$$s_i(p) = \left((D_p\psi_i)^{-1}e_1, \ldots, (D_p\psi_i)^{-1}e_n\right).$$

We define the inverse of a formal bundle chart by

$$\begin{aligned} \phi_i^{-1} \colon U_i \times \mathrm{GL}(n,\mathbb{R}) &\longrightarrow \mathrm{Fr}_{\mathrm{GL}}(M)_{U_i} \\ (p, A) &\longmapsto s_i(p) \cdot A. \end{aligned}$$

The transition functions are

$$\phi_j \circ \phi_i^{-1} \colon (U_i \cap U_j) \times \mathrm{GL}(n,\mathbb{R}) \longrightarrow (U_i \cap U_j) \times \mathrm{GL}(n,\mathbb{R})$$

with

$$\phi_j \circ \phi_i^{-1}(p, A) = \left(p, D_{\psi_i(p)}\left(\psi_j \circ \psi_i^{-1}\right) \cdot A\right).$$

These maps are smooth, because the transition functions $\psi_j \circ \psi_i^{-1}$ are smooth. This shows that the maps ϕ_i are smoothly compatible formal bundle charts and by Theorem 4.3.3 there exists a manifold structure on $\mathrm{Fr}_{\mathrm{GL}}(M)$ such that π becomes a fibre bundle with general fibre $\mathrm{GL}(n, \mathbb{R})$.

The $\mathrm{GL}(n,\mathbb{R})$-action is smooth (by considering the action in the bundle charts) and the (inverse) bundle charts ϕ_i^{-1} are $\mathrm{GL}(n,\mathbb{R})$-equivariant:

$$\phi_i^{-1}((p,A)\cdot B) = s_i(p)(A\cdot B) = \phi_i^{-1}(p,A)\cdot B \quad \forall B \in \mathrm{GL}(n,\mathbb{R}).$$

Therefore $\pi\colon \mathrm{Fr}_{\mathrm{GL}}(M) \to M$ is a principal $\mathrm{GL}(n,\mathbb{R})$-bundle over M. □

Remark 4.4.2 (Orthogonal Frame Bundles) If (M,g) is an n-dimensional Riemannian manifold, we can define a principal $\mathrm{O}(n)$-bundle

$$\begin{array}{ccc} \mathrm{O}(n) & \longrightarrow & \mathrm{Fr}_{\mathrm{O}}(M) \\ & & \downarrow \pi \\ & & M \end{array}$$

whose fibre over $p \in M$ consists of the set of orthonormal bases in T_pM. If M is in addition oriented, then there is also a principal $\mathrm{SO}(n)$-bundle

$$\begin{array}{ccc} \mathrm{SO}(n) & \longrightarrow & \mathrm{Fr}_{\mathrm{SO}}(M) \\ & & \downarrow \pi \\ & & M \end{array}$$

defined using oriented orthonormal bases. There are similar constructions of orthonormal frame bundles for pseudo-Riemannian manifolds.

Remark 4.4.3 A frame, i.e. a basis of a tangent space to a manifold, is in physics often called a **vielbein**, in particular in the case of an orthonormal frame to a Lorentz manifold (the word "vielbein" is German and means "many-leg". It is a generalization of the word **tetrad** in the 4-dimensional case.)

Definition 4.4.4 Let G be a Lie group. A principal G-subbundle of the frame bundle $\mathrm{Fr}_{\mathrm{GL}}(M)$ of a smooth manifold M, i.e. a G-reduction of the frame bundle, is called a ***G*-structure** on M.
In particular, a Riemannian metric on M^n defines an $\mathrm{O}(n)$-structure and, together with an orientation, an $\mathrm{SO}(n)$-structure on M.

4.5 Vector Bundles

We consider another class of fibre bundles, called *vector bundles*, that are ubiquitous in differential geometry and gauge theory. The prototype of a vector bundle is the tangent bundle TM of a smooth manifold M. Moreover, in physics, matter fields in gauge theories are described classically by sections of vector bundles. In addition to [14] we follow in this section [25, 74] and [78].

4.5.1 Definitions and Basic Concepts

Let $\mathbb{K}$ be the field $\mathbb{R}$ or $\mathbb{C}$.

Definition 4.5.1 A fibre bundle

$$\begin{array}{ccc} V & \longrightarrow & E \\ & & \downarrow \pi \\ & & M \end{array}$$

is called a **(real or complex) vector bundle of rank** m if:

1. The general fibre V and every fibre E_x, for $x \in M$, are m-dimensional vector spaces over $\mathbb{K}$.
2. There exists a bundle atlas $\{(U_i, \phi_i)\}_{i \in I}$ for E such that the induced maps

$$\phi_{ix}: E_x \longrightarrow V$$

are vector space isomorphisms for all $x \in U_i$. We call such an atlas a **vector bundle atlas** for E and the charts in a vector bundle atlas **vector bundle charts**. See Fig. 4.3.

A vector bundle of rank 1 is called a **line bundle**.

There are two features that distinguish a vector bundle $E \to M$ from a standard fibre bundle whose general fibre is a vector space V:

1. the vector space structure on each fibre E_x, for $x \in M$;
2. the bundle E has a vector bundle atlas.

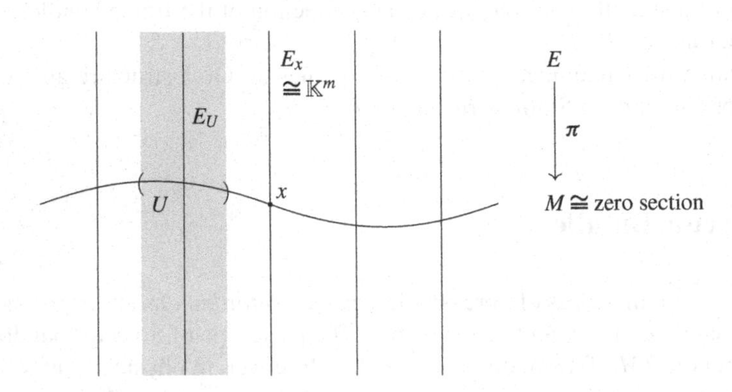

Fig. 4.3 Vector bundle

The vector space structure on each fibre implies that we can add any two sections of a vector bundle E and multiply sections with a scalar or a smooth function on the base manifold M with values in $\mathbb{K}$.

Example 4.5.2 The simplest example of a vector bundle is the trivial bundle $M \times \mathbb{K}^m$, often denoted by $\underline{\mathbb{K}}^m$. It has the canonical vector space structure on each fibre $\{p\} \times \mathbb{K}^m$, for $p \in M$, and the vector bundle atlas consisting of only one bundle chart

$$\mathrm{Id}: M \times \mathbb{K}^m \longrightarrow M \times \mathbb{K}^m.$$

Here is a more interesting example:

Example 4.5.3 (The Tangent Bundle of a Smooth Manifold) We want to show that the tangent bundle of a smooth manifold is canonically a smooth real vector bundle. Let M be a smooth manifold of dimension n. We define the set

$$TM = \dot{\bigcup}_{p \in M} T_pM$$

with the canonical projection $\pi: TM \to M$. We claim that TM has the structure of a smooth real vector bundle of rank n over M: First, the general fibre $\mathbb{R}^n$ and each fibre T_pM are n-dimensional real vector spaces. If

$$\psi_i: U_i \longrightarrow \phi_i(U_i) \subset \mathbb{R}^n$$

is a local manifold chart for M, then

$$\begin{aligned} \Psi_i: TM_{U_i} &\longrightarrow U_i \times \mathbb{R}^n \\ (p, v) &\longmapsto (p, D_p\psi_i(v)) \end{aligned}$$

is a formal bundle chart for TM. These formal bundle charts are smoothly compatible, because

$$\begin{aligned} \Psi_j \circ \Psi_i^{-1}: (U_i \cap U_j) \times \mathbb{R}^n &\longrightarrow (U_i \cap U_j) \times \mathbb{R}^n \\ (p, w) &\longmapsto \left(p, D_p\left(\psi_j \circ \psi_i^{-1}\right) w\right) \end{aligned}$$

is a smooth map. By Theorem 4.3.3, $\pi: TM \to M$ has the structure of a smooth fibre bundle with general fibre diffeomorphic to $\mathbb{R}^n$. Since the bundle charts (U_i, Ψ_i) are linear isomorphisms on each fibre, the bundle TM is a vector bundle of rank n.

Remark 4.5.4 Note that sections of TM are the same as vector fields on M:

$$\Gamma(TM) = \mathfrak{X}(M).$$

Transition functions for a vector bundle atlas have a special form:

Proposition 4.5.5 (Transition Functions of Vector Bundles) *Let $E \to M$ be a $\mathbb{K}$-vector bundle of rank m and $\{(U_i, \phi_i)\}_{i \in I}$ a vector bundle atlas for E. Then the transition functions take values in the subgroup* $\mathrm{GL}(m, \mathbb{K})$ *of* $\mathrm{Diff}\,(\mathbb{K}^m)$,

$$\begin{aligned} \phi_{ji}\colon U_i \cap U_j &\longrightarrow \mathrm{GL}(m, \mathbb{K}) \subset \mathrm{Diff}\,(\mathbb{K}^m) \\ x &\longmapsto \phi_{jx} \circ \phi_{ix}^{-1}. \end{aligned}$$

The following definition applies only to real vector bundles.

Definition 4.5.6 A real vector bundle $E \to M$ of rank m is called **orientable** if it admits a vector bundle atlas $\{(U_i, \phi_i)\}_{i \in I}$ such that all transition functions map to

$$\phi_{ji}\colon U_i \cap U_j \longrightarrow \mathrm{GL}_+(m, \mathbb{R}),$$

where $\mathrm{GL}_+(m, \mathbb{R})$ denotes the subgroup of invertible matrices with positive determinant.
Clearly, if $E \to M$ is a *complex* vector bundle of rank m, then forgetting the complex structure it defines an underlying *real* rank $2m$ vector bundle $E_\mathbb{R} \to M$. The bundle $E_\mathbb{R}$ is always orientable, because the identification $\mathbb{C}^m = \mathbb{R}^{2m}$ as real vector spaces induces an embedding $\mathrm{GL}(m, \mathbb{C}) \subset \mathrm{GL}_+(2m, \mathbb{R})$ by Exercise 1.9.10.

There is a notion of a homomorphism between vector bundles over the same manifold.

Definition 4.5.7 Let $V \to E \stackrel{\pi_E}{\to} M$ and $W \to F \stackrel{\pi_F}{\to} M$ be vector bundles over M over the same field $\mathbb{K}$.

1. A smooth bundle map $L\colon E \to F$, satisfying $\pi_F \circ L = \pi_E$, is called a **vector bundle homomorphism** if the restriction to a fibre

$$L|_{E_x}\colon E_x \longrightarrow F_x$$

is a linear map for all $x \in M$. A vector bundle homomorphism which is injective (surjective) on each fibre is called a **vector bundle monomorphism (epimorphism)**. If $L\colon E \to E$ is a homomorphism, then L is also called a **vector bundle endomorphism**.
2. A **vector bundle isomorphism** is a vector bundle homomorphism which is a diffeomorphism of the total spaces and an isomorphism on each fibre. A vector bundle is called **trivial** if it is isomorphic to the trivial bundle. If the tangent bundle TM of a manifold M is trivial, then M is called **parallelizable**.

Remark 4.5.8 According to Exercise 4.8.5, a vector bundle homomorphism which is an isomorphism on each fibre is a vector bundle isomorphism.
As before, we could consider vector bundle homomorphisms between vector bundles over different base manifolds M and N that cover a smooth map from M

to N. It is not difficult to prove with Remark 4.5.8 that a vector bundle $E \to M$ of rank m is trivial if and only if it has m global sections

$$v_1, \ldots, v_m: M \longrightarrow E,$$

such that $v_1(x), \ldots, v_m(x)$ form a basis of the fibre E_x, for all $x \in M$.

Remark 4.5.9 (Sections of Vector Bundles) Note that (contrary to principal fibre bundles) vector bundles always admit *global* sections: the section that is equal to zero everywhere on M is a trivial example (the fibres of a vector bundle *are* vector spaces, so there is a canonical element, namely 0. The fibres of a principal bundle are *only diffeomorphic* to a Lie group, so the neutral element e is not a canonical element in a fibre.) However, in the case of a vector bundle it is not clear that there are sections without zeros, and even if this is the case, it is not clear that there are m sections which form a basis in each fibre.

Example 4.5.10 (Parallelizable Spheres) We want to show that the spheres S^0, S^1, S^3 and S^7 are parallelizable. This is trivial for S^0, which consists only of two points. We consider S^1 as the unit sphere in $\mathbb{C}$. For $x \in S^1$, the vector $ix \in \mathbb{C}$ is orthogonal to x with respect to the standard Euclidean scalar product:

$$\operatorname{Re}\langle x, ix\rangle = \operatorname{Re} i||x||^2 = 0,$$

where $\langle z, w\rangle = \bar{z}w$ is the standard Hermitian scalar product on $\mathbb{C}$. This implies that

$$\begin{aligned} S^1 \times \mathbb{R} &\longrightarrow TS^1 \subset T\mathbb{C}|_{S^1} \\ (x, t) &\longmapsto (x, tix) \end{aligned}$$

is a trivialization of TS^1.

Similarly, we can consider S^3 as the unit sphere in $\mathbb{H}$. Then

$$\begin{aligned} S^3 \times \mathbb{R}^3 &\longrightarrow TS^3 \subset T\mathbb{H}|_{S^3} \\ (x, t_1, t_2, t_3) &\longmapsto (x, t_1 ix + t_2 jx + t_3 kx) \end{aligned}$$

is a trivialization of TS^3.

Finally, we consider S^7 as the unit sphere in the octonions $\mathbb{O}$. The octonions $\mathbb{O} \cong \mathbb{R}^8$ are spanned by $e_0, e_1, \ldots, e_7$, where

$$e_0^2 = e_0, \quad e_i^2 = -e_0 \quad \forall i = 1, \ldots, 7.$$

The map

$$\begin{aligned} S^7 \times \mathbb{R}^7 &\longrightarrow TS^7 \subset T\mathbb{O}|_{S^7} \\ (x, t_1, t_2, \ldots, t_7) &\longmapsto (x, t_1 e_1 x + t_2 e_2 x + \ldots + t_7 e_7 x) \end{aligned}$$

is a trivialization of TS^7.

It is a deep theorem due to J.F. Adams [2] that S^0, S^1, S^3 and S^7 are the only spheres which are parallelizable. This is related to the fact that division algebras exist only in dimension 1, 2, 4 and 8. A proof using K-theory can be found in [74]. See also Exercise 6.13.8.
A proof of the following theorem can be found in [5, 74] and [81].

Theorem 4.5.11 (Vector Bundles and Homotopy Equivalences) *Let $f: M \to N$ be a smooth homotopy equivalence between manifolds. Then the pullback f^* is a bijection between isomorphism classes of vector bundles over N and vector bundles over M of the same rank and over the same field $\mathbb{K}$.*
In particular, we get:

Corollary 4.5.12 (Vector Bundles over Contractible Manifolds Are Trivial) *If M is a contractible manifold, then every vector bundle over M is trivial. This holds, in particular, if $M = \mathbb{R}^n$ for some n.*

4.5.2 Linear Algebra Constructions for Vector Bundles

A useful fact is that we can construct new vector bundles from given ones by applying linear algebra constructions fibrewise: suppose E, F are vector bundles over M over the same field $\mathbb{K}$. Then there exist canonically defined vector bundles

$$E \oplus F, \quad E \otimes F, \quad E^*, \quad \Lambda^k E, \quad \mathrm{Hom}(E, F)$$

over M. If $\mathbb{K} = \mathbb{C}$ there also exists a complex conjugate vector bundle $\bar{E}$. The fibres of these vector bundles are given by

$$(E \oplus F)_x = E_x \oplus F_x,$$

and similarly in the other cases. This follows from Theorem 4.3.3, because local vector bundle charts for E and F can be combined to yield smoothly compatible formal vector bundle charts for the set $E \oplus F$, defining the structure of a smooth vector bundle on $E \oplus F \to M$. Similarly in the other cases.

Purely linear algebraic constructions, such as the direct sum and tensor product of vector spaces, extend to smooth vector bundles and yield new vector bundles with canonically defined smooth bundle structures.

Example 4.5.13 Consider the tangent bundle $TM \to M$. Then there exist canonically associated vector bundles T^*M and $\Lambda^k T^*M$ over M. Sections of $\Lambda^k T^*M$ are k-forms on M:

$$\Gamma(\Lambda^k T^*M) = \Omega^k(M).$$

More generally, for a vector bundle $E \to M$, sections of the bundle $\Lambda^k T^*M \otimes E$ are k-forms on M with values in E, i.e. elements of $\Omega^k(M,E)$: If $\omega \in \Omega^k(M,E)$, then at a point $x \in M$

$$\omega_x: T_xM \times \ldots \times T_xM \longrightarrow E_x$$

is multilinear and alternating. This generalizes the notion of vector space-valued forms in Sect. 3.5.1 to forms which have values in a vector bundle. One sometimes calls $\Lambda^k T^*M \otimes E$ the **bundle of k-forms over M twisted with E.**
We want to define the concept of vector subbundle (following [25]):

Definition 4.5.14 Let $\pi: E \to M$ be a $\mathbb{K}$-vector bundle of rank m. A subset $F \subset E$ is called a **vector subbundle** of rank k if each point $p \in M$ has an open neighbourhood U together with a vector bundle chart (U, ϕ) of E such that

$$\phi(E_U \cap F) = U \times \mathbb{K}^k \subset U \times \mathbb{K}^m,$$

where $\mathbb{K}^k$ is the vector subspace $\mathbb{K}^k \times \{0\} \subset \mathbb{K}^m$. It follows that F is an embedded submanifold of E and $\pi|_F: F \to M$ has the canonical structure of a $\mathbb{K}$-vector bundle of rank k over M.

Example 4.5.15 (Normal Bundle of Spheres) For $n \neq 0, 1, 3, 7$ the sphere S^n does not have a trivial tangent bundle according to Adams' Theorem mentioned in Example 4.5.10. However, the **normal bundle** $\nu(S^n)$ of S^n in $\mathbb{R}^{n+1}$ is trivial for any $n \geq 0$: The normal bundle is defined as

$$\nu(S^n) = \{(x,u) \in S^n \times \mathbb{R}^{n+1} \mid u \perp T_xS^n\},$$

with projection onto the first factor. It is clear that the normal bundle is a real line bundle. The following map is a trivialization of $\nu(S^n)$:

$$\begin{aligned} S^n \times \mathbb{R} &\longrightarrow \nu(S^n) \\ (x,t) &\longmapsto (x, tx). \end{aligned}$$

Note that

$$TS^n \oplus \nu(S^n) = T\mathbb{R}^{n+1}|_{S^n}.$$

We conclude that the sum of a non-trivial vector bundle (the tangent bundle to the sphere) and a trivial vector bundle (the normal bundle) can be trivial. One says that the tangent bundle of the sphere S^n is **stably trivial**: It becomes trivial after taking the direct sum with a trivial bundle (here a trivial *line* bundle). Both TS^n and $\nu(S^n)$ are vector subbundles of the trivial bundle $T\mathbb{R}^{n+1}|_{S^n}$. Note that this also means that a trivial vector bundle can have non-trivial subbundles.

Definition 4.5.16 Let $E \to M$ be a $\mathbb{K}$-vector bundle over M. A **(Euclidean** or **Hermitian) bundle metric** is a metric on each fibre E_x that varies smoothly with $x \in M$. More precisely, it is a section

$$\langle \cdot , \cdot \rangle \in \Gamma(E^* \otimes E^*) \quad (\mathbb{K} = \mathbb{R})$$

or

$$\langle \cdot , \cdot \rangle \in \Gamma(\bar{E}^* \otimes E^*) \quad (\mathbb{K} = \mathbb{C})$$

which defines in each point $x \in M$ a non-degenerate symmetric ($\mathbb{K} = \mathbb{R}$) or Hermitian ($\mathbb{K} = \mathbb{C}$) form

$$\langle \cdot , \cdot \rangle_x : E_x \times E_x \longrightarrow \mathbb{K}.$$

Proposition 4.5.17 (Existence of Bundle Metrics) *Every $\mathbb{K}$-vector bundle over a manifold M admits a positive definite bundle metric.*

Proof This follows by a partition of unity argument, because a convex combination of positive definite metrics on a vector space is still a positive definite metric. □

For associated vector bundles we will give a more explicit construction of bundle metrics in Proposition 4.7.12.

Example 4.5.18 The tangent bundle TM of any submanifold $M^m \subset \mathbb{R}^n$ has a bundle metric induced from the standard Euclidean scalar product on $\mathbb{R}^n$. In particular, TS^{n-1} has a canonical bundle metric.

Proposition 4.5.19 (Orthogonal Complement of a Vector Subbundle) *Let $E \to M$ be a vector bundle with a positive definite bundle metric and $F \subset E$ a vector subbundle. Then the orthogonal complement $F^\perp$ is a vector subbundle of E and $F \oplus F^\perp$ is isomorphic to E.*

Proof This is Exercise 4.8.16. □

The following is clear.

Proposition 4.5.20 (Transition Functions of Vector Bundles with a Metric) *Let $E \to M$ be a $\mathbb{K}$-vector bundle of rank m with a positive definite bundle metric. Choosing local trivializations given by orthonormal bases it follows that there exists a vector bundle atlas $\{(U_i, \phi_i)\}_{i \in I}$ with transition functions of the form*

$$\phi_{ji} : U_i \cap U_j \longrightarrow \mathrm{O}(m) \quad (\mathbb{K} = \mathbb{R})$$

or

$$\phi_{ji} : U_i \cap U_j \longrightarrow \mathrm{U}(m) \quad (\mathbb{K} = \mathbb{C}).$$

If the bundle is real and orientable, then we can find a vector bundle atlas such that

$$\phi_{ji}: U_i \cap U_j \longrightarrow \mathrm{SO}(m).$$

If $E \to M$ is a real (complex) vector bundle of rank m with a positive definite bundle metric, then the set consisting of all vectors of length 1 in each fibre forms the **unit sphere bundle** $S(E) \to M$, which is a smooth S^{m-1}-bundle (S^{2m-1}-bundle) over M.

4.6 *The Clutching Construction

We want to describe a construction that yields (all) vector bundles over spheres S^n. The idea of this so-called **clutching construction** is to glue together trivial vector bundles over the northern and southern hemisphere of S^n along the equator (we follow [74]).

Let S^n be the unit sphere in $\mathbb{R}^{n+1}$, where $n \geq 1$. We define the north and south pole

$$N_+ = (0, \ldots, 0, +1) \in S^n,$$
$$N_- = (0, \ldots, 0, -1) \in S^n$$

and the open sets

$$U_+ = S^n \setminus \{N_+\},$$
$$U_- = S^n \setminus \{N_-\}.$$

Both U_+ and U_- are diffeomorphic to $\mathbb{R}^n$ via the stereographic projection; cf. Example A.1.8. Let f be any smooth map

$$f: S^{n-1} \longrightarrow \mathrm{GL}(k, \mathbb{K}),$$

where we think of $S^{n-1} \subset S^n$ as the equator of S^n and $\mathbb{K} = \mathbb{R}, \mathbb{C}$. Such a map is called a **clutching function**. We write

$$x = (x_1, \ldots, x_n) \in \mathbb{R}^n,$$
$$z = x_{n+1} \in \mathbb{R}.$$

Let p denote the following retraction of the intersection $U_+ \cap U_-$ onto the equator:

$$p: U_+ \cap U_- \longrightarrow S^{n-1}$$
$$(x, z) \longmapsto \frac{x}{||x||}.$$

Here

$$||x||^2 = x_1^2 + \ldots + x_n^2$$

is the Euclidean norm. Note that this map is well-defined, because $x \neq 0$ on $U_+ \cap U_-$. We use the retraction to extend the clutching function to a smooth map on $U_+ \cap U_-$:

$$\bar{f} = f \circ p \colon U_+ \cap U_- \longrightarrow \mathrm{GL}(k, \mathbb{K}).$$

Definition 4.6.1 Let $E_f = \tilde{E}/\sim$ be the quotient set of the disjoint union

$$\tilde{E} = (U_- \times \mathbb{K}^k) \dot{\bigcup} (U_+ \times \mathbb{K}^k)$$

by identifying

$$(x, z, v) \in (U_- \cap U_+) \times \mathbb{K}^k \subset U_- \times \mathbb{K}^k$$

with

$$(x, z, \bar{f}(x, z) \cdot v) \in (U_+ \cap U_-) \times \mathbb{K}^k \subset U_+ \times \mathbb{K}^k.$$

Theorem 4.6.2 (Vector Bundle over a Sphere Defined by a Clutching Function) *Via the projection*

$$\pi \colon E_f \longrightarrow S^n$$
$$[x, z, v] \longmapsto (x, z)$$

the set E_f has a canonical structure of a $\mathbb{K}$-vector bundle of rank k over the sphere S^n:

$$\begin{array}{ccc} \mathbb{K}^k & \longrightarrow & E_f \\ & & \downarrow \pi \\ & & S^n \end{array}$$

Proof (See also Exercise 4.8.9.) Note that the map π is well-defined on the quotient set E_f and surjective onto S^n. Let σ denote the quotient map

$$\sigma \colon (U_- \times \mathbb{K}^k) \dot{\bigcup} (U_+ \times \mathbb{K}^k) \longrightarrow E_f.$$

The map σ decomposes into injective maps $\sigma_\pm$ on $U_\pm \times \mathbb{K}^k$. For $(x, z, w) \in U_\pm \times \mathbb{K}^k$ define

$$[x, z, w]_\pm = \sigma_\pm(x, z, w).$$

Then

$$[x, z, v]_- = \left[x, z, \bar{f}(x, z) \cdot v\right]_+ .$$

The maps

$$\begin{aligned} \phi_\pm : E_{f\,U_\pm} &\longrightarrow U_\pm \times \mathbb{K}^k \\ [x, z, v] &\longmapsto \sigma_\pm^{-1}([x, z, v]_\pm) \end{aligned}$$

are well-defined formal bundle charts. We want to show that these formal bundle charts are smoothly compatible: We calculate

$$\begin{aligned} \phi_+ \circ \phi_-^{-1}(x, z, v) &= \sigma_+^{-1} \circ \sigma_-(x, z, v) \\ &= \sigma_+^{-1}[x, z, v]_- \\ &= \sigma_+^{-1}\left[x, z, \bar{f}(x, z) \cdot v\right]_+ \\ &= (x, z, \bar{f}(x, z) \cdot v), \end{aligned}$$

which is a smooth map. It follows from Theorem 4.3.3 that $\pi: E_f \to S^n$ has the structure of a fibre bundle. Since the bundle charts ϕ_+, ϕ_- are linear isomorphisms on each fibre, it follows that $\pi: E_f \to S^n$ is a vector bundle with general fibre $\mathbb{K}^k$.

□

The following can be shown, see [5] or [74]:

Theorem 4.6.3 (Vector Bundles over Spheres and Homotopy Classes of Clutching Functions)

1. *Every complex vector bundle over S^n of rank k is isomorphic to a bundle E_f for a certain clutching function $f: S^{n-1} \to \mathrm{U}(k)$, unique up to homotopy.*
2. *Similarly, every orientable real vector bundle over S^n of rank k is isomorphic to a bundle E_f for a certain clutching function $f: S^{n-1} \to \mathrm{SO}(k)$, unique up to homotopy.*

Every vector bundle over a sphere can be constructed using a clutching function, because by Corollary 4.5.12 every vector bundle over S^n is trivial over U_+ and U_-. Given an arbitrary $\mathbb{K}$-vector bundle $E \to S^n$ of rank k we obtain an associated clutching function as follows:

- Let $E_\pm$ denote the restrictions of E to $U_\pm$. Choose vector bundle trivializations

$$h_\pm : E_\pm \longrightarrow U_\pm \times \mathbb{K}^k.$$

- Consider

$$h_+ \circ h_-^{-1}: S^{n-1} \times \mathbb{K}^k \longrightarrow S^{n-1} \times \mathbb{K}^k.$$

This map is a linear isomorphism on each fibre and defines the clutching function

$$f: S^{n-1} \longrightarrow \mathrm{GL}(k, \mathbb{K}).$$

The clutching functions in Theorem 4.6.3 can be taken to have image in $\mathrm{U}(k)$ or $\mathrm{SO}(k)$, because there are deformation retractions

$$\mathrm{GL}(k, \mathbb{C}) \longrightarrow \mathrm{U}(k),$$
$$\mathrm{GL}_+(k, \mathbb{R}) \longrightarrow \mathrm{SO}(k).$$

Complex and real orientable vector bundles over S^n are therefore essentially classified by the homotopy groups $\pi_{n-1}(\mathrm{U}(k))$ and $\pi_{n-1}(\mathrm{SO}(k))$.

Remark 4.6.4 (Clutching Construction for Arbitrary Fibres) Let F be a smooth manifold and

$$f: S^{n-1} \longrightarrow \mathrm{Diff}(F)$$

a "smooth" map, again called a clutching function (the precise formulation in the general case is not completely trivial, because we did not define a smooth structure on the diffeomorphism group $\mathrm{Diff}(F)$). A similar construction to the one above yields a fibre bundle

$$\begin{array}{ccc} F & \longrightarrow & E_f \\ & & \downarrow \pi \\ & & S^n \end{array}$$

over S^n with general fibre F. The mapping torus construction in Example 4.1.5 can be seen as a special case of the clutching construction for $n = 1$: If $\phi: F \to F$ is the monodromy of the mapping torus, we choose

$$\begin{aligned} f: S^0 &\longrightarrow \mathrm{Diff}(F) \\ -1 &\longmapsto \mathrm{Id}_F \\ +1 &\longmapsto \phi. \end{aligned}$$

Example 4.6.5 (Exotic 7-Spheres) Another very nice application appears in Milnor's classic paper [94], where certain *exotic7-spheres* (homeomorphic but not diffeomorphic to S^7) are defined as S^3-bundles over S^4, using the clutching construction with clutching function

$$f_{hj}\colon S^3 \longrightarrow \mathrm{SO}(4) \subset \mathrm{Diff}_+(S^3)$$

given by

$$f_{hj}(u) \cdot v = u^h v u^j.$$

Here $u, v \in S^3 = \mathrm{Sp}(1)$ and $h, j \in \mathbb{Z}$. For certain values of the integers h and j, the S^3-bundle over S^4 determined by the clutching function f_{hj} is an exotic 7-sphere.

Milnor's paper started the field known as *differential topology* and led to an extensive investigation of exotic spheres of arbitrary dimension. The study of the smooth topology of general 4-manifolds, using Donaldson theory and later Seiberg–Witten theory, also belongs to the field of differential topology.

4.7 Associated Vector Bundles

In Chap. 2 we studied the theory of Lie group representations. We now want to combine this theory with the theory of principal bundles from the present chapter.

We said before that principal bundles are the place where Lie groups appear in gauge theories. *Associated vector bundles*, which we discuss in this section, are precisely the place where *representations on vector spaces* are built into gauge theories. We can summarize this in the following diagram:

$$\begin{array}{ccc}
\text{Lie groups (gauge groups)} & \longrightarrow & \text{Representations on vector spaces} \\
\downarrow & & \downarrow \\
\text{Principal bundles} & \longrightarrow & \text{Associated vector bundles (matter fields)} \\
\downarrow & & \downarrow \\
\text{Connections (gauge fields)} & \longrightarrow & \text{Covariant derivatives (interaction/coupling)}
\end{array}$$

The third row will be explained in Chap. 5.

For example, in the Standard Model, one generation of fermions is described by associated complex vector bundles of rank 8 for left-handed fermions and rank 7 for right-handed fermions, associated to representations of the gauge group $\mathrm{SU}(3) \times \mathrm{SU}(2) \times \mathrm{U}(1)$. Taking particles and antiparticles together we get two associated complex vector bundles of rank 15 (right-handed and left-handed) which are related by complex conjugation. The complete fermionic content of the Standard Model is described by the direct sum of three copies of these vector bundles (a complex vector bundle of rank 90), corresponding to the three generations. These constructions will be described in detail in Sect. 8.5.1.

4.7.1 Basic Concepts

As an introduction, consider again the Hopf fibration

$$\begin{array}{ccc} S^1 & \longrightarrow & S^3 \\ & & \big\downarrow \pi \\ & & S^2 \end{array}$$

We are interested in the complex representations of S^1 on $\mathbb{C}$ with winding number $k \in \mathbb{Z}$:

$$\begin{aligned} \rho_k \colon S^1 &\longrightarrow \mathrm{U}(1) \\ z &\longmapsto z^k \end{aligned}$$

Our aim is to define an associated bundle

$$\gamma^k = S^3 \times_{\rho_k} \mathbb{C}.$$

This will be a complex line bundle over S^2

$$\begin{array}{ccc} \mathbb{C} & \longrightarrow & \gamma^k \\ & & \big\downarrow \pi \\ & & S^2 \end{array}$$

whose transition functions are given by the transition functions of the Hopf fibration *composed* with the group homomorphism ρ_k.

The general definition is the following: Let $\mathbb{K}$ denote the field $\mathbb{R}$ or $\mathbb{C}$. We then associate to each principal G-bundle

$$\begin{array}{ccc} G & \longrightarrow & P \\ & & \downarrow \pi_P \\ & & M \end{array}$$

and each representation

$$\rho: G \longrightarrow \mathrm{GL}(V)$$

of the structure group G on a $\mathbb{K}$-vector space V of dimension k a vector bundle $E \rightarrow M$ with fibres isomorphic to V.

Lemma 4.7.1 *Let P be a principal G-bundle and ρ a representation of the Lie group G on a $\mathbb{K}$-vector space V. Then the map*

$$\begin{aligned} (P \times V) \times G &\longrightarrow P \times V \\ (p, v, g) &\longmapsto (p, v) \cdot g = \left(p \cdot g, \rho(g)^{-1} v\right) \end{aligned}$$

defines a free principal right action of the Lie group G on the product manifold $P \times V$. In particular, the quotient space $E = (P \times V)/G$ is a smooth manifold such that the projection $P \times V \rightarrow E$ is a submersion.

Proof It is clear that

$$(P \times V) \times G \longrightarrow P \times V$$

is a right action, which is free since the action of G on P is free. If G is compact, then the claim follows from Corollary 3.7.29. In the general case, the action is principal by an argument very similar to the one in the proof of Theorem 4.2.12. □

Theorem 4.7.2 (Associated Vector Bundle Constructed as a Quotient) *Let P be a principal G-bundle and ρ a representation of the Lie group G on a $\mathbb{K}$-vector space V. Then the quotient space $E = (P \times V)/G$ has the structure of a $\mathbb{K}$-vector bundle over M, with projection*

$$\begin{aligned} \pi_E: E &\longrightarrow M \\ [p, v] &\longmapsto \pi_P(p) \end{aligned}$$

and fibres

$$E_x = (P_x \times V)/G$$

isomorphic to V. The vector space structure on the fibre E_x over $x \in M$ is defined by

$$\lambda[p, v] + \mu[p, w] = [p, \lambda v + \mu w], \quad \forall p \in P, v, w \in V, \lambda, \mu \in \mathbb{K},$$

where $\pi_P(p) = x$.

Proof It is clear that π_E is well-defined and that V is isomorphic to the fibres E_x via $v \mapsto [p_x, v]$ with a fixed $p_x \in P_x$. We need to find a vector bundle atlas for E. Let (U, ϕ_U) be a bundle chart for the principal bundle P:

$$\begin{aligned} \phi_U: P_U &\longrightarrow U \times G \\ p &\longmapsto (\pi_P(p), \beta_U(p)). \end{aligned}$$

We set

$$\begin{aligned} \psi_U: E_U &\longrightarrow U \times V \\ [p, v] &\longmapsto (\pi_P(p), \rho(\beta_U(p))v). \end{aligned}$$

Since $P \times V \to E$ is a submersion, the map ψ_U is smooth. It is a diffeomorphism with smooth inverse

$$\begin{aligned} \psi_U^{-1}: U \times V &\longrightarrow E_U \\ (x, v) &\longmapsto \left[\phi_U^{-1}(x, e), v\right]. \end{aligned}$$

Its restriction to each fibre E_x is a linear isomorphism to the vector space V. Thus ψ_U defines a chart in a vector bundle atlas for E. □

Definition 4.7.3 The vector bundle

$$E = P \times_\rho V = (P \times V)/G$$

is called the **vector bundle associated** to the principal bundle P and the representation ρ on V:

$$\begin{array}{ccc} V & \longrightarrow & P \times_\rho V \\ & & \downarrow \pi_E \\ & & M \end{array}$$

The group G (or its image $\rho(G) \subset \mathrm{GL}(V)$) is known as the **structure group** of E.

Remark 4.7.4 Note that in the definition of the vector space structure on the fibres E_x,

$$\lambda[p, v] + \mu[p, w] = [p, \lambda v + \mu w],$$

we have to choose both representatives with the *same* point p in the fibre of P over $x \in M$.

Example 4.7.5 For every principal G-bundle $P \to M$ and every vector space V, the vector bundle associated to the trivial homomorphism

$$\begin{aligned} \rho: G &\longrightarrow \mathrm{GL}(V) \\ g &\longmapsto \mathrm{Id}_V \end{aligned}$$

is a trivial vector bundle. See Exercise 4.8.20.

It is useful in applications to have a suitable description of local sections of an associated vector bundle.

Proposition 4.7.6 (Local Sections of Associated Vector Bundles) *Let P be a principal bundle and $E = P \times_\rho V$ an associated vector bundle. Let $s: U \to P$ be a local gauge. Then there is a 1-to-1 relation between smooth sections $\tau: U \to E$ and smooth maps $f: U \to V$, given by*

$$\tau(x) = [s(x), f(x)] \quad \forall x \in U.$$

In particular, the local gauge defines a preferred isomorphism between V and every fibre E_x over $x \in U$.

Proof If $f: U \to V$ is a smooth map, then

$$\begin{aligned} U &\longrightarrow P \times V \\ x &\longmapsto (s(x), f(x)) \end{aligned}$$

is smooth and hence $\tau: U \to E$ is smooth. The map τ is a section, because

$$\pi_E \circ \tau(x) = \pi_P \circ s(x) = x.$$

Conversely, let $\tau: U \to E$ be a smooth section. Since $E_x = (P_x \times V)/G$ and the action of G on P_x is simply transitive, there is a unique $f(x) \in V$ such that

$$\tau(x) = [s(x), f(x)].$$

We have to show that $f: U \to V$ is smooth: Define a bundle chart ϕ_U of the principal bundle using the section s:

$$\begin{aligned}\phi_U^{-1}: U \times G &\longrightarrow P_U \\ (x,g) &\longmapsto s(x)\cdot g.\end{aligned}$$

Then with the notation in the proof of Theorem 4.7.2 we have $\beta_U(s(x)) = e$ and

$$\begin{aligned}\psi_U \circ \tau(x) &= \psi_U([s(x), f(x)]) \\ &= (x, \rho(\beta_U(s(x)))f(x)) \\ &= (x, f(x)).\end{aligned}$$

Since ψ_U and τ are smooth, it follows that f is smooth. □

Matter fields in physics are described by smooth sections of vector bundles E associated to principal bundles P via representations of the gauge group G on a vector space V (in the case of fermions the associated bundle E is twisted in addition with a *spinor bundle* S, i.e. the bundle is $S \otimes E$). It follows that, given a local gauge of the gauge bundle P, the section in E corresponds to a unique local map from spacetime into the vector space V.

In particular, since principal bundles on $\mathbb{R}^n$ are trivial by Corollary 4.2.9, we can describe matter fields on a spacetime diffeomorphic to $\mathbb{R}^n$ by unique maps from $\mathbb{R}^n$ into a vector space, *once a global gauge for the principal bundle has been chosen.* A (local) trivialization of the gauge bundle thus determines a unique (local) trivialization of all associated vector bundles.

Definition 4.7.7 Let $E = P \times_\rho V$ be an associated vector bundle. If the representation

$$\rho_*: \mathfrak{g} \longrightarrow \mathrm{End}(V)$$

is non-trivial, then the sections of E are called **charged**.
This term will be explained in more detail in Sect. 5.9.

Remark 4.7.8 (Associated Fibre Bundles with Arbitrary Fibres) Given a principal bundle $P \to M$ with structure group G, a manifold F and a smooth left action

$$\begin{aligned}\Psi: G \times F &\longrightarrow F \\ (g,v) &\longmapsto g\cdot v\end{aligned}$$

a similar construction using the quotient

$$P \times_{\Psi} F = (P \times F)/G$$

under the G-action

$$\begin{aligned} (P \times F) \times G &\longrightarrow P \times F \\ (p, v, g) &\longmapsto \left(p \cdot g, g^{-1} \cdot v\right) \end{aligned}$$

yields an associated fibre bundle

$$\begin{array}{ccc} F & \longrightarrow & P \times_{\Psi} F \\ & & \downarrow \pi \\ & & M \end{array}$$

with structure group given by the image of G in the diffeomorphism group Diff(F), determined by the action Ψ.

Example 4.7.9 (Flat Bundles) Here is an example of the construction in Remark 4.7.8. Let M and F be manifolds and

$$\psi: \pi_1(M) \longrightarrow \mathrm{Diff}(F)$$

a group homomorphism. This defines an action of the (discrete) group $\pi_1(M)$ on F. The universal covering

$$\pi_M: \tilde{M} \longrightarrow M$$

can be considered as a principal bundle with discrete structure group $\pi_1(M)$. The associated fibre bundle

$$\begin{array}{ccc} F & \longrightarrow & \tilde{M} \times_{\psi} F \\ & & \downarrow \\ & & M \end{array}$$

is called the **flat bundle** with **holonomy** ψ. In the case of $M = S^1$ this yields again the mapping torus from Example 4.1.5. More generally, for $M = T^n$, a collection of n pairwise commuting diffeomorphisms

$$f_i: F \longrightarrow F, \quad i = 1, \ldots, n$$

defines a flat bundle

$$\begin{array}{ccc} F & \longrightarrow & E \\ & & \downarrow \\ & & T^n \end{array}$$

4.7.2 Adapted Bundle Atlases for Associated Vector Bundles

We discuss a specific type of bundle atlas for associated vector bundles. Let $P \to M$ be a principal G-bundle and $E = P \times_\rho V$ an associated vector bundle, where $\rho\colon G \to \mathrm{GL}(V)$ is a representation. We choose a principal bundle atlas $\{(U_i, \phi_i)\}_{i\in I}$ for P, determined by local gauges $s_i\colon U_i \to P$.

Definition 4.7.10 The principal bundle atlas for P defines an **adapted bundle atlas** for E with local trivializations

$$\psi_i\colon E_{U_i} \longrightarrow U_i \times V$$

whose inverses are given by

$$\psi_i^{-1}(x, v) = [s_i(x), v].$$

Proposition 4.7.11 (Adapted Bundle Atlases and the Structure Group) *Suppose the transition functions of the principal bundle charts for P are given by*

$$\phi_{ji} = \phi_{j\cdot} \circ \phi_{i\cdot}^{-1}\colon U_i \cap U_j \longrightarrow G.$$

Then the transition functions for the adapted bundle atlas for E are

$$\begin{aligned} \psi_{ji} = \psi_{j\cdot} \circ \psi_{i\cdot}^{-1}\colon U_i \cap U_j &\longrightarrow \mathrm{GL}(V) \\ x &\longmapsto \psi_{ji}(x) = \rho(\phi_{ji}(x)). \end{aligned}$$

The transition functions of E thus have image in the subgroup $\rho(G) \subset \mathrm{GL}(V)$, where G is the structure group of P.

Proof We have

$$s_i(x) = s_j(x) \cdot \phi_{ji}(x).$$

This implies

$$\begin{aligned} \psi_{ix}^{-1}(v) &= [s_j(x) \cdot \phi_{ji}(x), v] \\ &= \psi_{jx}^{-1} \circ \rho(\phi_{ji}(x))(v). \end{aligned}$$

Therefore

$$\psi_{ji}(x) = \psi_{jx} \circ \psi_{ix}^{-1} = \rho(\phi_{ji}(x)).$$

□

4.7.3 Bundle Metrics on Associated Vector Bundles

It is often important to consider bundle metrics on an associated vector bundle. We can construct such metrics as follows: let $\pi_P\colon P \to M$ be a principal bundle with structure group G, $\rho\colon G \to \mathrm{GL}(V)$ a representation and $E \to M$ the associated vector bundle $E = P \times_\rho V$.

Proposition 4.7.12 (Bundle Metrics on Associated Vector Bundles from *G*-Invariant Scalar Products) *Suppose that* $\langle\cdot,\cdot\rangle_V$ *is a G-invariant scalar product on V. Then the bundle metric* $\langle\cdot,\cdot\rangle_E$ *on the associated vector bundle E given by*

$$\langle [p, v], [p, w]\rangle_{E_x} = \langle v, w\rangle_V,$$

for arbitrary $p \in P_x$*, is well-defined.*

Proof This is an easy calculation choosing two different representatives for the vectors in the fibre E_x. □

4.7.4 Examples

Example 4.7.13 (From Vector Bundles to Principal Bundles and Back) We claim that *every* vector bundle has the structure of an associated vector bundle for some principal bundle. We first consider the tangent bundle *TM*: Let *M* be an *n*-dimensional smooth manifold and consider the frame bundle

$$\begin{array}{ccc} \mathrm{GL}(n,\mathbb{R}) & \longrightarrow & \mathrm{Fr}_{\mathrm{GL}}(M) \\ & & \downarrow \\ & & M \end{array}$$

Let

$$\rho_{\mathrm{GL}}\colon \mathrm{GL}(n,\mathbb{R}) \times \mathbb{R}^n \longrightarrow \mathbb{R}^n$$

be the standard representation, given by matrix multiplication from the left on column vectors. Then there exists an isomorphism of vector bundles

$$TM \cong \mathrm{Fr}_{\mathrm{GL}}(M) \times_{\rho_{\mathrm{GL}}} \mathbb{R}^n.$$

An isomorphism is given by

$$H: \mathrm{Fr}_{\mathrm{GL}}(M) \times_{\rho_{\mathrm{GL}}} \mathbb{R}^n \longrightarrow TM$$

$$[(v_1, \ldots, v_n), (x_1, \ldots, x_n)] \longmapsto \sum_{i=1}^{n} v_i x_i.$$

It is easy to check that the map H is well-defined and a bundle isomorphism.

Choosing a Riemannian metric g on M we can define the orthonormal frame bundle $\mathrm{Fr}_{\mathrm{O}}(M)$. Using the standard representation ρ_{O} of $\mathrm{O}(n)$ on $\mathbb{R}^n$ we get another vector bundle isomorphism

$$TM \cong \mathrm{Fr}_{\mathrm{O}}(M) \times_{\rho_{\mathrm{O}}} \mathbb{R}^n.$$

Similarly, every real vector bundle E of rank n is associated to a principal $\mathrm{GL}(n, \mathbb{R})$-bundle (and a principal $\mathrm{O}(n)$-bundle), defined using frames in the fibres of E. If E is orientable, it is associated to a principal $\mathrm{SO}(n)$-bundle. Similar statements hold for complex vector bundles with principal $\mathrm{GL}(n, \mathbb{C})$- and $\mathrm{U}(n)$-bundles.
We get:

Proposition 4.7.14 *Let $E \to M$ be a real or complex vector bundle. Then E is associated to some principal* $\mathrm{O}(n)$*- or* $\mathrm{U}(n)$*-bundle $P \to M$.*
In particular, the vector bundles over spheres that we defined in Sect. 4.6 using the clutching construction are associated vector bundles. Note that the structure as an associated vector bundle is not unique: as we saw above in the case of the frame bundle, the same vector bundle can be associated to principal bundles with different Lie groups.

We can use our constructions of principal bundles over spheres, projective spaces, and Stiefel and Grassmann manifolds to define associated vector bundles over those manifolds.

Example 4.7.15 Recall the principal bundle

$$\begin{array}{ccc} \mathrm{SO}(n-1) & \longrightarrow & \mathrm{SO}(n) \\ & & \downarrow \pi \\ & & S^{n-1} \end{array}$$

from Example 4.2.16. Then any representation of $\mathrm{SO}(n-1)$ on a real or complex vector space defines an associated vector bundle over the sphere S^{n-1}. A similar construction works for any of the other principal bundles over spheres given in Example 4.2.16. Alternatively, these bundles can also be realized (up to isomorphism) by the clutching construction.

The construction also applies to the principal bundles over Stiefel and Grassmann manifolds, like

$$\begin{array}{ccc} \mathrm{O}(n-k) & \longrightarrow & \mathrm{O}(n) \\ & & \downarrow \pi \\ & & V_k(\mathbb{R}^n) \end{array}$$

and

$$\begin{array}{ccc} \mathrm{O}(k) \times \mathrm{O}(n-k) & \longrightarrow & \mathrm{O}(n) \\ & & \downarrow \pi \\ & & Gr_k(\mathbb{R}^n) \end{array}$$

These examples can be generalized: start with any smooth homogeneous space G/H and consider the canonical principal bundle

$$\begin{array}{ccc} H & \longrightarrow & G \\ & & \downarrow \pi \\ & & G/H \end{array}$$

according to Theorem 4.2.15. Then representations of H define associated vector bundles over G/H, known as **homogeneous vector bundles**.

Example 4.7.16 Let

$$\begin{array}{ccc} S^1 & \longrightarrow & S^{2n+1} \\ & & \downarrow \\ & & \mathbb{CP}^n \end{array}$$

be the Hopf fibration. We want to study complex line bundles associated to this principal S^1-bundle. For $k \in \mathbb{Z}$ consider the homomorphism

$$\begin{aligned} \rho_k \colon S^1 &\longrightarrow \mathrm{U}(1) \\ z &\longmapsto z^k \end{aligned}$$

of winding number k. Then the associated bundle

$$\gamma^k = S^{2n+1} \times_{\rho_k} \mathbb{C}$$

is a complex line bundle. The bundle γ^0 is trivial and γ^k is isomorphic to

$$\gamma^k \cong \underbrace{\gamma^1 \otimes \ldots \otimes \gamma^1}_{k \text{ factors}} \quad (k > 0)$$

and

$$\gamma^k \cong \underbrace{\gamma^{1*} \otimes \ldots \otimes \gamma^{1*}}_{|k| \text{ factors}} \quad (k < 0).$$

See Exercise 4.8.21.

Similarly, using representations of $\mathrm{SU}(2) \cong S^3$ we can define vector bundles associated to the quaternionic Hopf fibration

$$\begin{array}{ccc} S^3 & \longrightarrow & S^{4n+3} \\ & & \downarrow \\ & & \mathbb{HP}^n \end{array}$$

Example 4.7.17 (Adjoint Bundle) An important general example of an associated vector bundle is the following: let

$$\begin{array}{ccc} G & \longrightarrow & P \\ & & \downarrow \\ & & M \end{array}$$

be a principal bundle with structure group G. Consider the adjoint representation

$$\mathrm{Ad}\colon G \longrightarrow \mathrm{GL}(\mathfrak{g}).$$

Then the associated vector bundle

$$\mathrm{Ad}(P) = P \times_{\mathrm{Ad}} \mathfrak{g}$$

is called the **adjoint bundle**. Its general fibre is isomorphic to the vector space underlying the Lie algebra $\mathfrak{g}$:

$$\begin{array}{ccc} \mathfrak{g} & \longrightarrow & \mathrm{Ad}(P) \\ & & \downarrow \\ & & M \end{array}$$

4.8 Exercises for Chap. 4

4.8.1 The *Möbius strip* can be defined as the submanifold

$$M = \left\{ \left(e^{i\phi}, re^{i\phi/2}\right) \;\middle|\; \phi \in [0, 2\pi], r \in [-1, 1] \right\} \subset S^1 \times \mathbb{C}.$$

The projection $\pi: M \to S^1$ is defined as $\pi = \mathrm{pr}_1|_M$.

1. Show that $\pi: M \to S^1$ is a fibre bundle with general fibre $[-1, 1]$ (here we consider a small generalization of the notion of a fibre bundle to manifolds with boundary).
2. Prove that the boundary ∂M is connected and that the bundle π is not trivial.
3. Prove that the image of any smooth section $s: S^1 \to M$ intersects the zero section $z: S^1 \to M, z(\alpha) = (\alpha, 0)$.

Hint: Note that the map $S^1 \to S^1, e^{i\phi} \mapsto e^{i\phi/2}$ is not well-defined.

4.8.2 Let $\pi: M \to S^1$ denote the Möbius strip from Exercise 4.8.1 and consider the map $f_n: S^1 \to S^1, f_n(z) = z^n$ for $n \in \mathbb{Z}$.

1. Show that the pull-back bundle $f_n^* M$ is isomorphic to the bundle $M_n \to S^1$ defined by

$$M_n = \left\{ \left(e^{i\psi}, re^{in\psi/2}\right) \;\middle|\; \psi \in [0, 2\pi], r \in [-1, 1] \right\} \subset S^1 \times \mathbb{C}$$

(with projection onto the first factor).

2. Determine those $n \in \mathbb{Z}$ for which $f_n^* M$ is trivial and those for which it is non-trivial.

4.8.3 (Fibre Sum) Suppose that $F \to E \to M$ and $F' \to E' \to M'$ are two fibre bundles over n-dimensional manifolds M and M'. Let D and D' be embedded open n-discs in M and M' together with trivializations $F \times D$ and $F' \times D'$ of the fibrations over D and D'. We assume that F and F' are diffeomorphic and choose a diffeomorphism

$$\phi: F \longrightarrow F'.$$

We write D and D' minus the centre 0 as $S^{n-1} \times (0, 1)$ and fix a diffeomorphism r from $(0, 1)$ to $(0, 1)$ which reverses orientation. Let $\tau: S^{n-1} \to S^{n-1}$ be the diffeomorphism which reverses the sign of one of the coordinates on $S^{n-1} \subset \mathbb{R}^n$. Consider the diffeomorphism

$$\begin{aligned} \psi: F \times (D \setminus 0) &\longrightarrow F' \times (D' \setminus 0) \\ (x, v, t) &\longmapsto (\phi(x), \tau(v), r(t)). \end{aligned}$$

The **fibre sum** $E\#_\psi E'$ is defined by gluing together the manifolds $M \setminus F$ and $M' \setminus F'$ along the diffeomorphism ψ. Prove that $E\#_\psi E'$ is a smooth fibre bundle over the connected sum $M\#M'$ with general fibre F.

4.8.4 Let $G \to P \xrightarrow{\pi} M$ be a principal bundle and $f: N \to M$ a smooth map between manifolds. Prove that the pullback f^*P has the canonical structure of a principal G-bundle over N.

4.8.5

1. Let $F \to E \xrightarrow{\pi} M$ and $F' \to E' \xrightarrow{\pi} M$ be fibre bundles and $H: E \to E'$ a bundle morphism. Suppose that H maps every fibre of E diffeomorphically onto a fibre of E'. Show that H is a diffeomorphism and hence a bundle isomorphism.
2. Let $G \to P \xrightarrow{\pi} M$ and $G' \to P' \xrightarrow{\pi} M$ be principal bundles and $f: G \to G'$ a Lie group isomorphism. Show that every f-equivariant bundle morphism $H: P \to P'$ is a diffeomorphism.

4.8.6 (From [14]) We consider the Hopf bundle

$$\begin{array}{ccc} S^1 & \longrightarrow & S^3 \\ & & \big\downarrow \pi \\ & & S^2 \end{array}$$

The total space S^3 of this bundle admits two different S^1-actions: The standard action

$$\begin{aligned} S^3 \times S^1 &\longrightarrow S^3, \\ (w, \lambda) &\longmapsto w\lambda \end{aligned}$$

and the *reversed action*

$$\begin{aligned} S^3 \times S^1 &\longrightarrow S^3, \\ (w, \lambda) &\longmapsto w\lambda^{-1}. \end{aligned}$$

Both actions endow the same fibre bundle $S^1 \to S^3 \xrightarrow{\pi} S^2$ with the structure of a principal bundle. Prove that these principal bundles are not isomorphic *as principal bundles*.

4.8.7 Recall the definition of lens spaces from Example 3.7.33. Show that the lens space $L(p, 1)$ is the total space of a principal fibre bundle over S^2 with structure group S^1.

4.8.8 Show that there is a canonical free $O(k)$-action on the Stiefel manifold $V_k(\mathbb{R}^n)$ and that this defines a principal $O(k)$-bundle

$$\begin{array}{ccc} O(k) & \longrightarrow & V_k(\mathbb{R}^n) \\ & & \downarrow \\ & & Gr_k(\mathbb{R}^n) \end{array}$$

4.8.9 We want to discuss another way to construct fibre bundles. Let M, F be smooth manifolds and $\{U_i\}_{i\in I}$ an open covering of M together with diffeomorphisms

$$\phi_{ji}\colon (U_i \cap U_j) \times F \longrightarrow (U_i \cap U_j) \times F$$

whenever $U_i \cap U_j \neq \emptyset$, satisfying

$$\mathrm{pr}_1 \circ \phi_{ji} = \mathrm{pr}_1.$$

We also write $\phi_{ji}(x) = \phi_{ji}(x, -)$ for $x \in U_i \cap U_j$. Let $\tilde{E}$ be the disjoint union

$$\tilde{E} = \dot{\bigcup_{i\in I}} U_i \times F.$$

1. Show that

$$(x, v) \sim (x', v') \Leftrightarrow \exists i, j \in I : x = x' \in U_i \cap U_j \text{ and } v' = \phi_{ji}(x)v$$

defines an equivalence relation on $\tilde{E}$ if and only if the ϕ_{ji} satisfy the three conditions of Lemma 4.1.13.
2. Show that if the ϕ_{ji} satisfy the conditions of Lemma 4.1.13, then the quotient set

$$E = \tilde{E}/\sim$$

has the canonical structure of a smooth fibre bundle over M with general fibre F and transition functions ϕ_{ji}.

4.8.10 Prove that the principal bundle

$$\begin{array}{ccc} SO(n-1) & \longrightarrow & SO(n) \\ & & \downarrow \pi \\ & & S^{n-1} \end{array}$$

from Example 4.2.16 is isomorphic to the frame bundle $\mathrm{Fr}_{SO}(S^{n-1})$.

4.8.11 Prove that a subset $F \subset E$ is a subbundle of the vector bundle E if and only if F is the image of a vector bundle monomorphism to E.

4.8.12 Prove that

$$E = \{(U, v) \in Gr_k(\mathbb{K}^n) \times \mathbb{K}^n \mid v \in U\},$$

with projection onto the first factor, defines a $\mathbb{K}$-vector bundle over the Grassmann manifold $Gr_k(\mathbb{K}^n)$ of rank k. This bundle is called the **tautological vector bundle**. Particular examples, for $k = 1$, are the **tautological line bundles** over $\mathbb{RP}^{n-1}$ and $\mathbb{CP}^{n-1}$.

4.8.13 We denote by $L \rightarrow S^1$ the infinite Möbius strip, defined by

$$L = \left\{ \left(e^{i\phi}, re^{i\phi/2}\right) \middle| \phi \in [0, 2\pi], r \in \mathbb{R} \right\} \subset S^1 \times \mathbb{C}.$$

It follows from Exercise 4.8.1 that this is a non-trivial, real line bundle over the circle. Prove that the real vector bundle $L \oplus L \rightarrow S^1$ is trivial.

4.8.14 Let $L \rightarrow S^1$ be the infinite Möbius strip.

1. Show that under the diffeomorphism $S^1 \cong \mathbb{RP}^1$ the infinite Möbius strip is isomorphic to the tautological line bundle over $\mathbb{RP}^1$.
2. Prove that the tautological line bundle over $\mathbb{RP}^n$ is non-trivial for all $n \geq 1$.

4.8.15 Let $E \rightarrow M$ be a real vector bundle of rank m. Show that E is orientable if and only if $\Lambda^m E$ is trivial.

4.8.16 Let $E \rightarrow M$ be a $\mathbb{K}$-vector bundle of rank m with a positive definite (Euclidean or Hermitian) bundle metric. Suppose that $F \subset E$ is a vector subbundle. Prove that the orthogonal complement $F^\perp$ is a vector subbundle of E and that $F \oplus F^\perp$ is isomorphic to E.

4.8.17 Determine the clutching function of the tangent bundle $TS^2 \rightarrow S^2$ geometrically as follows:

1. Draw two disks in the plane and label them N and S. Draw on the boundary circle of disk N four points a, b, c, d counter-clockwise with 90° between consecutive points. Draw on the boundary circle of disk S corresponding points a, b, c, d, such that the disks under identification of the boundary circles yield a sphere S^2.
2. Draw in the center of disk N an orthonormal basis and label the vectors 1 and 2. Parallel transport this basis to the points a, b, c, d. Take these bases and draw the matching bases on the S side in the points a, b, c, d. Call these bases I.
3. Take the basis at the point a on disk S and parallel transport it to the center of disk S. Then parallel transport this basis from the center to the points b, c, d. Call these bases II.

4. Determine how bases I twist against bases II and thus determine the clutching function, i.e. the degree of the map

$$f: S^1 \longrightarrow \mathrm{SO}(2) \cong S^1.$$

To fix the sign of the degree, you probably need at least one more point at 45° between a and b, for example.

What do you get if you do something similar for $TS^3 \to S^3$ by realizing S^3 as two solid cubes identified along their six faces?

4.8.18 Determine the clutching function of the tautological complex line bundle $E \to \mathbb{CP}^1 \cong S^2$. The total space of the line bundle is

$$E = \{([z], wz) \in \mathbb{CP}^1 \times \mathbb{C}^2 \mid z \neq 0, w \in \mathbb{C}\}$$

and $\mathbb{CP}^1$ is covered by

$$U_+ = \{[z : 1] \in \mathbb{CP}^1 \mid z \in \mathbb{C}\},$$
$$U_- = \{[1 : z] \in \mathbb{CP}^1 \mid z \in \mathbb{C}\}.$$

4.8.19 Determine the clutching functions in the sense of Remark 4.6.4 for the Hopf fibrations

$$\begin{array}{ccc} S^1 & \longrightarrow & S^3 \\ & & \downarrow \pi \\ & & S^2 \end{array}$$

and

$$\begin{array}{ccc} S^3 & \longrightarrow & S^7 \\ & & \downarrow \pi \\ & & S^4 \end{array}$$

4.8.20 Let

$$\begin{array}{ccc} G & \longrightarrow & P \\ & & \downarrow \pi \\ & & M \end{array}$$

be a principal G-bundle and $\rho: G \to \mathrm{GL}(V)$, $\rho_i: G \to \mathrm{GL}(V_i)$, for $i = 1, 2$, representations. Let

$$E = P \times_\rho V, \quad E_i = P \times_{\rho_i} V_i$$

be the associated vector bundles. Show that the dual bundle E^*, the direct sum $E_1 \oplus E_2$ and the tensor product $E_1 \otimes E_2$ are isomorphic to vector bundles associated to P. Determine the corresponding representations of G and the vector bundle isomorphisms. Show that the vector bundle associated to the trivial representation is trivial.

4.8.21 (From [14]) Let

$$\begin{array}{ccc} \mathbb{C} & \longrightarrow & \gamma^k \\ & & \downarrow \pi \\ & & \mathbb{CP}^n \end{array}$$

be the complex line bundle defined in Example 4.7.16.

1. Prove that γ^0 is trivial and γ^1 is isomorphic to the tautological line bundle.
2. Prove that $\gamma^{-k} \cong \gamma^{k*}$ for all $k \in \mathbb{Z}$ and

$$\gamma^k \cong \underbrace{\gamma^1 \otimes \ldots \otimes \gamma^1}_{k \text{ factors}} \quad (k > 0).$$

4.8.22 Let $E \to M$ be a complex vector bundle of rank $n \geq 2$. Show that E is associated to a principal $\mathrm{SU}(n)$-bundle over M if and only if $\Lambda^n E$ is a trivial complex line bundle.

4.8.23 Let $E = P \times_\rho V$ be an associated vector bundle and α a section of the adjoint bundle $\mathrm{Ad}(P)$. Prove that α defines a canonical endomorphism of the vector bundle E.

4.8.24 Let $M = G/H$ be a smooth homogeneous space and consider the canonical principal H-bundle

$$\begin{array}{ccc} H & \longrightarrow & G \\ & & \downarrow \pi \\ & & M \end{array}$$

Suppose that $\rho: H \to \mathrm{GL}(V)$ is a representation with associated homogeneous vector bundle

$$\begin{array}{ccc} V & \longrightarrow & E = G \times_\rho V \\ & & \downarrow \pi \\ & & M \end{array}$$

1. Prove that there exists a canonical smooth left action of the Lie group G on the total space E. Show that this action maps fibres of E by linear isomorphisms onto fibres of E and that any given fibre of E can be mapped by a group element onto any other fibre.
2. Identify the space $\Gamma(E)$ of sections of the vector bundle E over the manifold M with a suitable vector subspace $\mathrm{Map}_H(G, V)$ of the vector space $\mathrm{Map}(G, V)$.

Remark The representation of G on $\Gamma(E)$, induced by this construction from the representation of the closed subgroup H on V, is denoted by $\mathrm{Ind}_H^G(V)$.

4.8.25 (From [30]) Let $M = G/H$ be a smooth homogeneous space and consider the canonical principal H-bundle

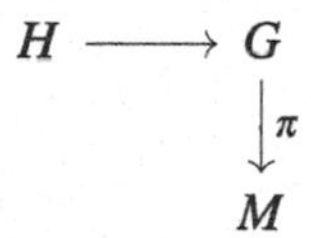

Prove that the tangent bundle TM is isomorphic to the homogeneous vector bundle over M, defined by the representation ρ of H on the vector space $\mathfrak{g}/\mathfrak{h}$, given by

$$\rho(h)[v] = [\mathrm{Ad}_h v] \quad \forall h \in H, [v] \in \mathfrak{g}/\mathfrak{h},$$

where Ad denotes the adjoint representation of G.

Chapter 5
Connections and Curvature

We said in the introduction to Chap. 4 that principal bundles and associated vector bundles are the stage for gauge theories. From a mathematical and physical point of view it is very important that we can define on principal bundles certain fields, known as *connection* 1*-forms*. At least locally (after a choice of local gauge) we can interpret connection 1-forms as fields on spacetime (the base manifold) with values in the Lie algebra of the gauge group. These fields are often called *gauge fields* and correspond in the associated quantum field theory to *gauge bosons*. Every connection 1-form A defines a *curvature* 2*-form* F which can be identified with the field strength of the gauge field. Connection and curvature can be seen as generalizations of the classical potential A_μ and field strength $F_{\mu\nu}$ in electromagnetism, which is a U(1)-gauge theory, to possibly non-abelian gauge groups.

Pure gauge theory, also known as Yang–Mills theory, involves only the gauge field A and its curvature F. Additional matter fields, like fermions or scalars, can be introduced using associated vector bundles. The crucial point is that connections (the gauge fields) define a *covariant derivative* on these associated vector bundles, leading to a *coupling* between gauge fields and matter fields (if the matter fields are *charged*, i.e. the vector bundles are associated to a non-trivial representation of the gauge group). In a gauge-invariant Lagrangian this results in terms of order higher than two in the matter and gauge fields, which are interpreted as interactions between the corresponding particles.

In non-abelian gauge theories, like quantum chromodynamics (QCD), there are also terms in the Lagrangian of order higher than two in the gauge fields themselves, coming from a quadratic term in the curvature that appears in the Yang–Mills Lagrangian. This implies a direct interaction between gauge bosons (the gluons in QCD) that does not occur in abelian gauge theories like quantum electrodynamics (QED). The difficulties that are still present nowadays in trying to understand the quantum version of non-abelian gauge theories, like quantum chromodynamics, can ultimately be traced back to this interaction between gauge bosons.

M.J.D. Hamilton, *Mathematical Gauge Theory*, Universitext,
https://doi.org/10.1007/978-3-319-68439-0_5

Although in this book we are mainly interested in applications of gauge theories to physics, gauge theories are also very influential in pure mathematics, for example, the Donaldson and Seiberg–Witten theories of 4-manifolds and Chern–Simons theory of 3-manifolds (see Exercise 5.15.16 for an introduction to the Chern–Simons action).

References for this chapter are [14, 39] and [84].

5.1 Distributions and Connections

Definition 5.1.1 A **distribution** on a manifold M is a vector subbundle of the tangent bundle TM.

This notion of distributions is not related to the concept of distributions in analysis. **Connections** on principal bundles P, sometimes also called **Ehresmann connections**, are defined as certain distributions on the total space of the principal bundle.

5.1.1 The Vertical Tangent Bundle

We first want to show that on the total space of every principal bundle there is a canonical *vertical bundle*. Let

$$\begin{array}{ccc} G & \longrightarrow & P \\ & & \downarrow \pi \\ & & M \end{array}$$

be a principal G-bundle. For a point $x \in M$ we have the fibre

$$\pi^{-1}(x) = P_x \subset P$$

over x, which is an embedded submanifold of P. Let $p \in P_x$ be a point in the fibre.

Definition 5.1.2 The **vertical tangent space** V_p of the total space P in the point p is the tangent space $T_p(P_x)$ to the fibre.

Proposition 5.1.3 (Vertical Tangent Bundle) *For all $p \in P$ the vertical tangent space has the following properties:*

1. $V_p = \ker D_p\pi$.

2. *The map*

$$\begin{aligned}\phi_*\colon \mathfrak{g} &\longrightarrow V_p\\ X &\longmapsto \tilde{X}_p,\end{aligned}$$

where $\tilde{X}$ is the fundamental vector field associated to $X \in \mathfrak{g}$, determined by the G-action on P, is a vector space isomorphism between $\mathfrak{g}$ and V_p.
3. *The set of all vertical tangent spaces V_p for $p \in P$ forms a smooth distribution on P, called the* **vertical tangent bundle** *V. Its rank is equal to the dimension of G. The distribution V is globally trivial as a vector bundle, with trivialization given by*

$$\begin{aligned}P \times \mathfrak{g} &\longrightarrow V\\ (p,X) &\longmapsto \tilde{X}_p.\end{aligned}$$

4. *The vertical tangent bundle is* **right-invariant**, *i.e.*

$$r_{g*}\left(V_p\right) = V_{p\cdot g} \quad \forall g \in G.$$

Proof

1. We have

$$D_p\pi(Y) = 0 \quad \forall Y \in V_p,$$

because we can write Y as the tangent vector to a curve in P_x, which maps under π to the constant point $x \in M$. Hence

$$V_p \subset \ker D_p\pi.$$

Since $\pi\colon P \to M$ is a submersion, it follows from the Regular Value Theorem A.1.32 that

$$\begin{aligned}\dim\ker D_p\pi &= \dim P - \dim M\\ &= \dim G\\ &= \dim P_x.\end{aligned}$$

This implies the claim.
2. It is clear that this map has image in V_p and considering dimensions it suffices to show that the map is injective. This follows from Proposition 3.4.3.
3. This is clear by 2.
4. This follows because according to Proposition 3.4.6 $r_{g*}(\tilde{X}) = \tilde{Y}$, with $Y = \mathrm{Ad}_{g^{-1}}X \in \mathfrak{g}$.

□

5.1.2 Ehresmann Connections

Let $P \to M$ be a principal G-bundle.

Definition 5.1.4 A **horizontal tangent space** in $p \in P$ is a subspace H_p of T_pP complementary to the vertical tangent space V_p, so that

$$T_pP = V_p \oplus H_p.$$

Note that horizontal tangent spaces are not defined uniquely (if the dimensions of G and M are positive).
The following should be clear:

Proposition 5.1.5 *Let H_p be a horizontal tangent space at $p \in P$, $\pi(p) = x$. Then*

$$D_p\pi : H_p \longrightarrow T_xM$$

is a vector space isomorphism.

Definition 5.1.6 Let H be a distribution on P consisting of horizontal tangent spaces. Then H is called an **Ehresmann connection** or a **connection** on P if it is right-invariant, i.e.

$$r_{g*}(H_p) = H_{p\cdot g} \quad \forall p \in P, g \in G.$$

The distribution H is also called **horizontal tangent bundle** given by the connection.
Right-invariance of an Ehresmann connection means that along a fibre P_x the horizontal subspaces are mutually "parallel" (with respect to right translation along the fibre). In particular, all H_p along a fibre P_x are determined by fixing a single H_{p_0} for some $p_0 \in P_x$, since the G-action is transitive on the fibres of P. Right-invariance of a connection can also be seen as a symmetry property: The right action of the gauge group G on P induces a natural right action on TP and Ehresmann connections are invariant under this action.

Example 5.1.7 (Connections on the Trivial Bundle) Let

$$\begin{array}{ccc} G & \longrightarrow & M \times G \\ & & \downarrow \mathrm{pr}_1 \\ & & M \end{array}$$

be the trivial principal G-bundle. Then the vertical subspaces are given by

$$V_{(x,g)} = T_{(x,g)}(\{x\} \times G) \cong T_gG.$$

We can choose

$$H_{(x,g)} = T_{(x,g)}(M \times \{g\}) \cong T_xM.$$

It is clear that this defines a horizontal subspace complementary to the vertical subspace. Furthermore, the collection of all of these horizontal subspaces are right-invariant and hence define a connection on the trivial bundle, called the **canonical flat connection**.

It can also be shown that every *non-trivial* principal bundle has a connection (see Exercise 5.15.1 for a proof in the case of compact structure groups).

Notice that connections are *not unique* (if $\dim M, \dim G \geq 1$), not even in the case of trivial principal bundles (all connections that appear in the Standard Model over Minkowski spacetime, for example, are defined on trivial principal bundles).

5.2 Connection 1-Forms

In this section we study an equivalent description of connections using differential forms.

5.2.1 Basic Definitions

Recall that we defined in Sect. 3.5.1 the notion of differential forms on a manifold with values in a vector space. We now need this notion to define so-called connection 1-forms.

Definition 5.2.1 A **connection 1-form** or **connection** on a principal G-bundle $\pi: P \rightarrow M$ is a 1-form $A \in \Omega^1(P, \mathfrak{g})$ on the total space P with the following properties:

1. $r_g^* A = \mathrm{Ad}_{g^{-1}} \circ A$ for all $g \in G$.
2. $A(\tilde{X}) = X$ for all $X \in \mathfrak{g}$, where $\tilde{X}$ is the fundamental vector field associated to X.

A connection 1-form is also called a **gauge field** on P.

At a point $p \in P$, a connection 1-form is thus a linear map

$$A_p: T_pP \longrightarrow \mathfrak{g}.$$

Recall that $\mathrm{Ad}_{g^{-1}}$ is a linear isomorphism of $\mathfrak{g}$ onto itself. This shows that the composition $\mathrm{Ad}_{g^{-1}} \circ A$ is well-defined and again an element of $\Omega^1(P, \mathfrak{g})$.

We want to show that the notion of connection 1-forms is completely equivalent to the notion of Ehresmann connections on a principal bundle as defined in Sect. 5.1.2.

Theorem 5.2.2 (Correspondence Between Ehresmann Connections and Connection 1-Forms) *There is a bijective correspondence between Ehresmann connections on a principal G-bundle $\pi: P \to M$ and connection 1-forms:*

1. *Let H be an Ehresmann connection on P. Then*

$$A_p\left(\tilde{X}_p + Y_p\right) = X,$$

for $p \in P$, $X \in \mathfrak{g}$ and $Y_p \in H_p$, defines a connection 1-form A on P.

2. *Let $A \in \Omega^1(P, \mathfrak{g})$ be a connection 1-form on P. Then*

$$H_p = \ker A_p$$

defines an Ehresmann connection H on P.

Proof

1. We have to verify the conditions defining a connection 1-form. It is clear that

$$A\left(\tilde{X}\right) = X \quad \forall X \in \mathfrak{g}.$$

We want to calculate r_g^*A. We have shown in Proposition 3.4.6 that $r_{g*}\left(\tilde{X}\right) = \tilde{Z}$, where $Z = \mathrm{Ad}_{g^{-1}}X$. Note that $r_{g*}Y_p$ is horizontal if Y_p is horizontal by the definition of Ehresmann connections. Therefore

$$\begin{aligned}(r_g^*A)_p\left(\tilde{X}_p + Y_p\right) &= A_{p\cdot g}\left(\tilde{Z}_{p\cdot g} + r_{g*}Y_p\right) \\ &= Z \\ &= \mathrm{Ad}_{g^{-1}} \circ A_p\left(\tilde{X}_p + Y_p\right).\end{aligned}$$

This implies the claim.

2. We have to verify that H is a horizontal right-invariant distribution on P. We first show that H is a subbundle of TP: using a basis $\{T_a\}$ for the Lie algebra $\mathfrak{g}$ we can write $A = \sum_a A_a T_a$, where $A_a \in \Omega^1(P)$ are real-valued 1-forms. Since $A(\tilde{X}) = X$ for all $X \in \mathfrak{g}$, it follows that $A_a(\tilde{T}_b) = \delta_{ab}$. In particular, the 1-forms A_a are linearly independent in each point $p \in P$. Let g be a Riemannian metric

on P and Z_a the vector fields g-dual to the 1-forms A_a. The $\{Z_a\}$ are linearly independent and span a subbundle ζ of TP of rank dim $\mathfrak{g}$. It follows that H is the g-orthogonal complement of ζ in TP and hence a distribution.

To verify that H is horizontal, we first show that $H_p \cap V_p = \{0\}$: Let $Y \in \ker A_p \cap V_p$. Then Y is equal to a fundamental vector, hence $Y = \tilde{X}_p$ for some $X \in \mathfrak{g}$. But then

$$0 = A_p(Y) = X,$$

hence $Y = 0$.

Furthermore, the 1-form A_p is surjective onto $\mathfrak{g}$, hence

$$\dim \ker A_p = \dim T_pP - \dim \mathfrak{g} = \dim T_pP - \dim V_p.$$

Thus $T_pP = \ker A_p \oplus V_p$ and H_p is horizontal. To check that H is right-invariant, let $Y \in H_p$. Then

$$\begin{aligned} A_{p\cdot g}(r_{g*}Y) &= (r_g^*A)_p(Y) \\ &= \mathrm{Ad}_{g^{-1}}(A_p(Y)) \\ &= 0. \end{aligned}$$

This shows $r_{g*}Y \in H_{p\cdot g}$ and hence the claim.

□

Example 5.2.3 Let G be a Lie group and $H \subset G$ a closed subgroup. By Theorem 4.2.15

$$\begin{array}{ccc} H & \longrightarrow & G \\ & & \downarrow \pi \\ & & G/H \end{array}$$

is an H-principal bundle. Suppose there exists a vector subspace $\mathfrak{m} \subset \mathfrak{g}$ such that

$$\mathfrak{g} = \mathfrak{h} \oplus \mathfrak{m}, \quad \mathrm{Ad}(H)\mathfrak{m} \subset \mathfrak{m}.$$

The homogeneous space is then called **reductive**. In this situation we can define a canonical connection 1-form A on the bundle $G \to G/H$. See Exercise 5.15.6 for details.

We describe another explicit example of a connection 1-form in the following subsection.

5.2.2 *A Connection 1-Form on the Hopf Bundle $S^3 \to S^2$

In this subsection we follow reference [14, Example 3.3]. We consider the Hopf bundle

$$\begin{array}{ccc} S^1 & \longrightarrow & S^3 \\ & & \downarrow \pi \\ & & S^2 = \mathbb{CP}^1 \end{array}$$

and think of S^1 as the unit circle in $\mathbb{C}$ with Lie algebra $i\mathbb{R}$. If $Y = iy \in i\mathbb{R}$, then $\exp(Y) = e^{iy} \in S^1$. We also think of S^3 as the unit sphere in $\mathbb{C}^2$ with tangent spaces

$$T_{(z_0,z_1)}S^3 = \{(X_0, X_1) \in \mathbb{C}^2 \mid \bar{z}_0 X_0 + \bar{z}_1 X_1 = 0\}.$$

We define 1-forms $\alpha_j, \bar{\alpha}_j \in \Omega^1(S^3, \mathbb{C})$ by

$$\alpha_j(X_0, X_1) = X_j,$$
$$\bar{\alpha}_j(X_0, X_1) = \bar{X}_j.$$

Proposition 5.2.4 (Connection 1-Form on the Hopf Bundle) *The 1-form A on S^3, given by*

$$A_{(z_0,z_1)} = \frac{1}{2}\left(\bar{z}_0\alpha_0 - z_0\bar{\alpha}_0 + \bar{z}_1\alpha_1 - z_1\bar{\alpha}_1\right),$$

has values in $i\mathbb{R}$ and is a connection 1-form for the Hopf bundle.

Proof It is clear that A has values in $i\mathbb{R}$, since $\bar{A} = -A$. We check the defining properties of connection 1-forms. Since S^1 is abelian, we have $\mathrm{Ad}_{g^{-1}} = \mathrm{Id}$ for all $g \in S^1$. We therefore have to show that

$$r_g^* A = A \quad \forall g \in S^1.$$

We fix a tangent vector $X \in T_{(z_0,z_1)}S^3$, given by the velocity vector

$$X = \left.\frac{d}{dt}\right|_{t=0} (z_0(t), z_1(t)) = (X_0, X_1)$$

of a suitable curve in S^3. Then

$$\begin{aligned} r_{g*}X &= \left.\frac{d}{dt}\right|_{t=0} (z_0(t)g, z_1(t)g) \\ &= (X_0 g, X_1 g). \end{aligned}$$

Hence

$$\begin{aligned}
(r_g^*A)_{(z_0,z_1)}(X) &= A_{(z_0g,z_1g)}(X_0g, X_1g)\\
&= \frac{1}{2}\left(\bar{z}_0\bar{g}X_0g - z_0g\bar{X}_0\bar{g} + \bar{z}_1\bar{g}X_1g - z_1g\bar{X}_1\bar{g}\right)\\
&= \frac{1}{2}\left(\bar{z}_0X_0 - z_0\bar{X}_0 + \bar{z}_1X_1 - z_1\bar{X}_1\right)\\
&= A_{(z_0,z_1)}(X_0, X_1),
\end{aligned}$$

where we used that $g\bar{g} = 1$.

We also have to show that

$$A(\tilde{Y}) = Y$$

for all Y in the Lie algebra of S^1. Let $Y = iy$, with $y \in \mathbb{R}$. Then the associated fundamental vector field is given by

$$\tilde{Y}_{(z_0,z_1)} = \left.\frac{d}{dt}\right|_{t=0} (z_0\exp(ity), z_1\exp(ity)) = (iz_0y, iz_1y).$$

This implies

$$\begin{aligned}
A_{(z_0,z_1)}(\tilde{Y}) &= \frac{1}{2}\left(\bar{z}_0iz_0y + z_0i\bar{z}_0y + \bar{z}_1iz_1y + z_1i\bar{z}_1y\right)\\
&= iy(|z_0|^2 + |z_1|^2)\\
&= Y,
\end{aligned}$$

since $(z_0, z_1) \in S^3$. □

See Exercise 7.9.9 for a generalization of this construction to the Hopf bundle $S^7 \to S^4$ with structure group SU(2).

5.3 Gauge Transformations

Let $\pi: P \to M$ be a principal G-bundle.

Definition 5.3.1 A **(global) gauge transformation** is a **bundle automorphism** of P, i.e. a diffeomorphism $f: P \to P$ which preserves the fibres of P and is G-equivariant:

1. $\pi \circ f = \pi$.
2. $f(p \cdot g) = f(p) \cdot g$ for all $p \in P$ and $g \in G$.

Under composition of diffeomorphisms the set of all gauge transformations forms a group that we denote by $\mathscr{G}(P)$ or $\mathrm{Aut}(P)$. A **local gauge transformation** is a bundle automorphism on the principal G-bundle $\pi: P_U \to U$, where $U \subset M$ is an open set. We sometimes prefer to call gauge transformations in this sense *bundle automorphisms* and leave the name gauge transformations to *physical gauge transformations* that we introduce later. Notice that whether a bundle automorphism f is global or local is not related to the question of whether f is constant or non-constant in some sense. We will later call gauge transformations *rigid* if they are constant in a specific way.

Depending on the context one sometimes calls G or $\mathscr{G}(P) = \mathrm{Aut}(P)$ the **gauge group** of the principal bundle P. The group $\mathscr{G}(P)$ is infinite-dimensional if both the dimensions of M and G are at least 1.

> The group $\mathrm{Aut}(P)$ of bundle automorphisms is one of the places in differential geometry where an infinite-dimensional group appears naturally. Gauge theories, which are field theories invariant under all gauge transformations, in this regard have the huge symmetry group $\mathrm{Aut}(P)$. The diffeomorphism group $\mathrm{Diff}(M)$ of spacetime M plays a comparable role in general relativity.

5.3.1 *Bundle Automorphisms as G-Valued Maps on P*

We would like to give another, equivalent description of bundle automorphisms (we follow [14, Sect. 3.5]).

Definition 5.3.2 We denote by $C^\infty(P,G)^G$ the following set of maps from P to G:

$$C^\infty(P,G)^G = \left\{\sigma: P \to G \text{ smooth} \,\middle|\, \sigma(p \cdot g) = c_{g^{-1}}(\sigma(p)) = g^{-1}\sigma(p)g\right\},$$

where $c_{g^{-1}}$ is conjugation by g^{-1}. This set is a group under pointwise multiplication:

$$(\sigma' \cdot \sigma)(p) = \sigma'(p) \cdot \sigma(p).$$

The neutral element is given by the constant map on P with value $e \in G$.

Proposition 5.3.3 (Correspondence Between Bundle Automorphisms and G-Valued Maps) *The map*

$$\begin{aligned} \mathscr{G}(P) &\longrightarrow \mathscr{C}^\infty(P,G)^G \\ f &\longmapsto \sigma_f \end{aligned}$$

with σ_f *defined by*

$$f(p) = p \cdot \sigma_f(p) \quad \forall p \in P,$$

is a well-defined group isomorphism. We can therefore identify the group of bundle automorphisms $\mathcal{G}(P)$ *with* $\mathcal{C}^\infty(P, G)^G$.

Proof Since $f(p)$ is in the same fibre as p, there exists a unique $g \in G$ such that $f(p) = p \cdot g$. This g we call $\sigma_f(p)$.

We first have to show that σ_f is an element of $\mathcal{C}^\infty(P, G)^G$: It is not difficult to show that σ_f is a smooth map from P to G. We have

$$\begin{aligned}(p \cdot g)\sigma_f(p \cdot g) &= f(p \cdot g)\\ &= f(p) \cdot g\\ &= (p \cdot \sigma_f(p)) \cdot g.\end{aligned}$$

This implies that

$$g \cdot \sigma_f(p \cdot g) = \sigma_f(p) \cdot g \quad \forall g \in G,$$

and thus $\sigma_f \in \mathcal{C}^\infty(P, G)^G$.

The inverse of the map above is given by

$$\begin{aligned}\mathcal{C}^\infty(P, G)^G &\longrightarrow \mathcal{G}(P)\\ \sigma &\longmapsto f_\sigma\end{aligned}$$

with f_σ defined by

$$f_\sigma(p) = p \cdot \sigma(p) \quad \forall p \in P.$$

We only have to show that f_σ is a bundle automorphism. It is clear that $f_\sigma(p)$ is in the same fibre as p, hence f_σ is a bundle map. It is easy to check that f_σ is G-equivariant and that $f_\sigma^{-1} = f_{\sigma^{-1}}$, hence f_σ is a diffeomorphism. Thus $f_\sigma \in \mathcal{G}(P)$.

Finally, we can check that $\sigma_{f' \circ f} = \sigma_{f'} \cdot \sigma_f$, hence the map defines a group isomorphism between $\mathcal{G}(P)$ and $\mathcal{C}^\infty(P, G)^G$. □

In the special case when the structure group G is abelian we have a simpler description.

Proposition 5.3.4 (Bundle Automorphisms for Abelian Structure Groups) *If the Lie group* G *is abelian, then there is a group isomorphism*

$$\begin{aligned}\mathcal{C}^\infty(M, G) &\longrightarrow \mathcal{C}^\infty(P, G)^G\\ \tau &\longmapsto \sigma_\tau\end{aligned}$$

where $\mathscr{C}^\infty(M,G)$ denotes the set of smooth maps from M to G (a group under pointwise multiplication) and σ_τ is defined by

$$\sigma_\tau = \tau \circ \pi,$$

where $\pi: P \to M$ is the projection.

Proof This is Exercise 5.15.2. □

Corollary 5.3.5 *For principal T^n-bundles $P \to M$ there is an isomorphism of the group of bundle automorphisms $\mathscr{G}(P)$ with the group of smooth maps from M to T^n.*

5.3.2 Physical Gauge Transformations

In physics, gauge transformations are often defined as maps on the base manifold M to the structure group G, even for non-abelian Lie groups G. We discuss the relation of this notion to our definition of gauge transformations as bundle automorphisms.

Definition 5.3.6 Let $\pi: P \to M$ be a principal G-bundle. A **physical gauge transformation** is a smooth map $\tau: U \to G$, defined on an open subset $U \subset M$. The set of all physical gauge transformations on U forms a group $\mathscr{C}^\infty(U,G)$ with pointwise multiplication. A **rigid physical gauge transformation** is a constant map $\tau: U \to G$. The rigid physical gauge transformations form a group isomorphic to G.

Proposition 5.3.7 (Physical Gauge Transformations and Bundle Automorphisms) *Let $s: U \to P$ be a local section. Then s defines a group isomorphism*

$$\begin{aligned} \mathscr{C}^\infty(P_U,G)^G &\longrightarrow \mathscr{C}^\infty(U,G) \\ \sigma &\longmapsto \tau_\sigma = \sigma \circ s. \end{aligned}$$

The inverse of this map is given by

$$\begin{aligned} \mathscr{C}^\infty(U,G) &\longrightarrow \mathscr{C}^\infty(P_U,G)^G \\ \tau &\longmapsto \sigma_\tau \end{aligned}$$

where

$$\sigma_\tau(s(x)\cdot g) = g^{-1}\tau(x)g \quad \forall x \in U, g \in G.$$

Proof The proof is left as an exercise. □

The upshot is that *after a choice of local gauge s on U* we can identify local bundle automorphisms on the principal G-bundle $P_U \to U$ with physical gauge transformations on U.

5.3.3 The Action of Bundle Automorphisms on Associated Vector Bundles

Bundle automorphisms on a principal bundle have the important property that they act on every associated vector bundle. Let $\pi_P: P \to M$ be a principal G-bundle and $\pi_E: E = P \times_\rho V \to M$ an associated vector bundle.

Theorem 5.3.8 (Action of Bundle Automorphisms on Associated Bundles) *The group of bundle automorphisms of the principal bundle acts on the associated vector bundle through bundle isomorphisms via*

$$\begin{aligned} \mathscr{G}(P) \times E &\longrightarrow E \\ (f, [p, v]) &\longmapsto f \cdot [p, v] = [f(p), v] = [p \cdot \sigma_f(p), v]. \end{aligned}$$

Proof We only have to show that the action is well-defined: If $[p', v'] = [p, v]$, then $p' = p \cdot g$ and $v' = \rho(g)^{-1} v$ for some $g \in G$, so that

$$\begin{aligned} [f(p'), v'] &= [f(p \cdot g), \rho(g)^{-1} v] \\ &= [f(p) \cdot g, \rho(g)^{-1} v] \\ &= [f(p), v]. \end{aligned}$$

□

We can also describe this action in the language of physics:

Theorem 5.3.9 (Action of Physical Gauge Transformations on Associated Bundles) *Let $s: U \to P$ be a local gauge and $\Phi: U \to E$ a local section. We write the section with respect to the local gauge as*

$$\Phi(x) = [s(x), \phi(x)] \quad \forall x \in U,$$

where $\phi: U \to V$ is a smooth map. Suppose f is a local bundle automorphism of P over U and $\tau_f: U \to G$ the associated physical gauge transformation. Then

$$(f \cdot \Phi)(x) = [s(x), \rho(\tau_f(x))\phi(x)].$$

Proof This is a simple calculation. □

As a consequence the action of a local bundle automorphism on a local section Φ of E is given by the action of the physical gauge transformation on the vector-valued map ϕ. In physics one writes the action of a physical gauge transformations $\tau: U \rightarrow G$ on a field $\phi: U \rightarrow V$ as

$$\phi(x) \longmapsto \tau(x) \cdot \phi(x).$$

The more general notion of bundle automorphism above has the advantage that it also works for non-trivial principal bundles and associated vector bundles, independent of the choice of (local) gauge.

Remark 5.3.10 There is a simple, but profound, difference between gauge theories and general relativity (Edward Witten [150] attributes this insight to Bryce DeWitt). In gauge theories the group of symmetries, the gauge group $\mathscr{G}(P)$, acts through bundle automorphisms, i.e. it preserves all points on the base manifold M. This is related to the fact that gauge theories describe local interactions (the interactions occur in single spacetime points). In general relativity, however, the group of symmetries, the diffeomorphism group Diff(M), acts by moving points around in M. If the diffeomorphism invariance holds in quantum gravity on the level of Green's functions (correlators), then they must be constant, in striking contrast to the behaviour of Green's functions in Poincaré invariant quantum field theories.

It is nowadays thought that gravity cannot be described by a local quantum field theory of point particles and that a theory of quantum gravity must be fundamentally non-local. This leads to alternatives such as string theory, where the graviton and other particles are no longer 0-dimensional point particles, but 1-dimensional strings.

5.4 Local Connection 1-Forms and Gauge Transformations

Let $\pi: P \rightarrow M$ be a principal G-bundle and $A \in \Omega^1(P, \mathfrak{g})$ a connection 1-form. It is very useful to consider the following notion.

Definition 5.4.1 Let $s: U \rightarrow P$ be a local gauge of the principal bundle on an open subset $U \subset M$. Then we define the **local connection 1-form** (or **local gauge field**) $A_s \in \Omega^1(U, \mathfrak{g})$, determined by s, by

$$A_s = A \circ Ds = s^*A.$$

The local connection 1-form is thus defined on an open subset in the base manifold M and can be considered as a "field on spacetime" in the usual sense.

(continued)

Definition 5.4.1 (continued)
If we have a manifold chart on U and $\{\partial_\mu\}_{\mu=1,\ldots,n}$ are the local coordinate basis vector fields on U, we set

$$A_\mu = A_s(\partial_\mu).$$

We can also choose in addition a basis $\{e_a\}$ of the Lie algebra $\mathfrak{g}$ and then expand

$$A_\mu = \sum_{a=1}^{\dim \mathfrak{g}} A_\mu^a e_a.$$

The real-valued fields $A_\mu^a \in \mathscr{C}^\infty(U, \mathbb{R})$ and the corresponding real-valued 1-forms $A_s^a \in \Omega^1(U)$ are called **(local) gauge boson fields**.

A principal bundle can have many local gauges and it is interesting to determine how the local connection 1-forms transform as we change the local gauge. Let $s_i : U_i \to P$ and $s_j : U_j \to P$ be local gauges with $U_i \cap U_j \neq \emptyset$. Recall from the proof of Proposition 4.7.11 that

$$s_i(x) = s_j(x) \cdot g_{ji}(x) \quad \forall x \in U_i \cap U_j,$$

where

$$g_{ji} : U_i \cap U_j \longrightarrow G$$

is the smooth transition function between the associated local trivializations. We can consider g_{ji} as a physical gauge transformation between the local gauges s_i and s_j.

We have local connection 1-forms

$$A_i = A_{s_i} \in \Omega^1(U_i, \mathfrak{g}),$$
$$A_j = A_{s_j} \in \Omega^1(U_j, \mathfrak{g}).$$

We want to calculate the relation between A_i and A_j. Recall that the Maurer–Cartan form $\mu_G \in \Omega^1(G, \mathfrak{g})$ was defined as

$$\mu_G(v) = D_g L_{g^{-1}}(v)$$

for $v \in T_gG$. We set

$$\mu_{ji} = g_{ji}^*\mu_G \in \Omega^1(U_i \cap U_j, \mathfrak{g}).$$

Then we have:

Theorem 5.4.2 (Transformation of Local Gauge Fields Under Changes of Gauge) *With the notation above, the local connection 1-forms transform as*

$$A_i = \mathrm{Ad}_{g_{ji}^{-1}} \circ A_j + \mu_{ji}$$

on $U_i \cap U_j$. If $G \subset \mathrm{GL}(n, \mathbb{K})$ is a matrix Lie group, then

$$A_i = g_{ji}^{-1} \cdot A_j \cdot g_{ji} + g_{ji}^{-1} \cdot dg_{ji},$$

where $\cdot$ denotes matrix multiplication, g_{ji}^{-1} denotes the inverse in G and dg_{ji} is the differential of each component of the function $g_{ji}: U_i \cap U_j \to G \subset \mathbb{K}^{n\times n}$. In particular, if G is abelian, then

$$A_i = A_j + \mu_{ji} = A_j + g_{ji}^{-1} \cdot dg_{ji}.$$

Proof Let $x \in U_i \cap U_j$ and $Z \in T_xM$. We set

$$X = D_xs_j(Z) \in T_{s_j(x)}P,$$

$$Y = D_xg_{ji}(Z) \in T_{g_{ji}(x)}G.$$

With the group action

$$\begin{aligned} \Phi: P \times G &\longrightarrow P \\ (p, g) &\longmapsto p \cdot g \end{aligned}$$

we calculate by Proposition 3.5.4 and the chain rule

$$\begin{aligned} D_xs_i(Z) &= D_x\left(\Phi \circ (s_j, g_{ji})\right)(Z) \\ &= D_{s_j(x)}r_{g_{ji}(x)}(X) + \widetilde{\mu_G(Y)}_{s_i(x)} \\ &= D_{s_j(x)}r_{g_{ji}(x)}(X) + \widetilde{\mu_{ji}(Z)}_{s_i(x)}. \end{aligned}$$

Therefore, by the defining properties of a connection 1-form A,

$$\begin{aligned} A_i(Z) &= A(D_xs_i(Z)) \\ &= A\left(D_{s_j(x)}r_{g_{ji}(x)}(X) + \widetilde{\mu_{ji}(Z)}_{s_i(x)}\right) \end{aligned}$$

$$= \left(r^*_{g_{ji}(x)}A\right)(X) + \mu_{ji}(Z)$$
$$= \mathrm{Ad}_{g_{ji}^{-1}(x)} \circ A_j(Z) + \mu_{ji}(Z).$$

To prove the second claim recall from Proposition 2.1.48 that for a matrix Lie group

$$\mathrm{Ad}_{g^{-1}}a = g^{-1} \cdot a \cdot g,$$

for all $g \in G$ and $a \in \mathfrak{g}$, and $\mu_G(v) = g^{-1} \cdot v$ for $v \in T_gG$, hence

$$\mu_{ji}(Z) = \mu_G\left(D_x g_{ji}(Z)\right) = g_{ji}^{-1} \cdot dg_{ji}(Z).$$

□

Remark 5.4.3 In physics one considers connection 1-forms usually only in the local sense as $\mathfrak{g}$-valued 1-forms A_i on open subset U_i of M together with the transformation rule given by Theorem 5.4.2. The mathematical concept of connections on principal bundles clarifies the invariant geometric object behind this transformation principle.

A very similar argument implies the following global statement:

Theorem 5.4.4 (Transformation of Connections Under Bundle Automorphisms) *Let $P \to M$ be a principal bundle and $A \subset \Omega^1(P, \mathfrak{g})$ a connection 1-form on P. Suppose that $f \in \mathcal{G}(P)$ is a global bundle automorphism. Then f^*A is a connection 1-form on P and*

$$f^*A = \mathrm{Ad}_{\sigma_f^{-1}} \circ A + \sigma_f^* \mu_G.$$

Proof This is Exercise 5.15.3. □

Theorem 5.4.4 corresponds to the "active" point of view for gauge transformations (symmetries are related to the behaviour under certain bundle automorphisms), while Theorem 5.4.2 corresponds to the "passive" point of view (symmetries are implicit in the behaviour under coordinate transformations).

5.5 Curvature

5.5.1 Curvature 2-Forms

Let $\pi: P \to M$ be a principal G-bundle and $A \in \Omega^1(P, \mathfrak{g})$ a connection 1-form on P. Let H be the associated horizontal vector bundle, defined as the kernel of A. We have

$$TP = V \oplus H$$

and set

$$\pi_H: TP \longrightarrow H$$

for the projection onto the horizontal vector bundle.

Definition 5.5.1 The 2-form $F \in \Omega^2(P, \mathfrak{g})$, defined by

$$F(X, Y) = dA(\pi_H(X), \pi_H(Y)) \quad \forall X, Y \in T_pP, p \in P$$

is called the **curvature 2-form** or **curvature** of the connection A. We sometimes write F^A to emphasize the dependence on A.
Here are some simple properties of the curvature.

Proposition 5.5.2 *The following identities hold:*

1. $r_g^* F = \mathrm{Ad}_{g^{-1}} \circ F$ *for all* $g \in G$.
2. $\tilde{X} \lrcorner F = 0$ *for all* $X \in \mathfrak{g}$, *where* $\tilde{X} \lrcorner$ *denotes insertion of the vector field* $\tilde{X}$.

Proof

1. Note that

$$r_{g*}H_p = H_{p \cdot g},$$
$$r_{g*}V_p = V_{p \cdot g}.$$

Hence

$$\pi_H \circ r_{g*} = r_{g*} \circ \pi_H$$

on T_pP, since both sides evaluated on $X = X_h + X_v \in T_pP$, where X_h is horizontal and X_v is vertical, are equal to $r_{g*}(X_h)$. We now calculate for vectors $X, Y \in T_pP$:

$$\begin{aligned}(r_g^* F)_p(X, Y) &= dA(\pi_H \circ r_{g*}(X), \pi_H \circ r_{g*}(Y)) \\ &= (r_g^* dA)(\pi_H(X), \pi_H(Y)) \\ &= d\left(\mathrm{Ad}_{g^{-1}} \circ A\right)(\pi_H(X), \pi_H(Y)) \\ &= \mathrm{Ad}_{g^{-1}} \circ F(X, Y).\end{aligned}$$

2. This is clear, because $\pi_H(\tilde{X}) = 0$.

□

5.5.2 The Structure Equation

Definition 5.5.3 Let P be a manifold and $\mathfrak{g}$ a Lie algebra. For $\eta \in \Omega^k(P, \mathfrak{g})$ and $\phi \in \Omega^l(P, \mathfrak{g})$ we define $[\eta, \phi] \in \Omega^{k+l}(P, \mathfrak{g})$ by

$$[\eta, \phi](X_1, \ldots, X_{k+l}) = \frac{1}{k!l!} \sum_{\sigma \in S_{k+l}} \text{sgn}(\sigma)[\eta(X_{\sigma(1)}, \ldots, X_{\sigma(k)}), \phi(X_{\sigma(k+1)}, \ldots, X_{\sigma(n)}],$$

where the commutators on the right are the commutators in the Lie algebra $\mathfrak{g}$. In the literature one also finds the notation $\eta \wedge \phi$ or $[\eta \wedge \phi]$ for $[\eta, \phi]$.
If we expand in a vector space basis $\{T_a\}$ for the Lie algebra $\mathfrak{g}$

$$\eta = \sum_{a=1}^{\dim \mathfrak{g}} \eta^a \otimes T_a,$$

$$\phi = \sum_{a=1}^{\dim \mathfrak{g}} \phi^a \otimes T_a,$$

with η^a, ϕ^a standard real-valued k- and l-forms, then the definition is equivalent to

$$[\eta, \phi] = \sum_{a,b=1}^{\dim \mathfrak{g}} \eta^a \wedge \phi^b \otimes [T_a, T_b].$$

Most of the time we need the definition only for 1-forms $\eta, \phi \in \Omega^1(P, \mathfrak{g})$, where we have

$$[\eta, \phi](X, Y) = [\eta(X), \phi(Y)] - [\eta(Y), \phi(X)],$$

and

$$[\eta, \eta](X, Y) = 2[\eta(X), \eta(Y)].$$

We can now state the following important formula for the curvature 2-form.

Theorem 5.5.4 (Structure Equation) *The curvature form F of a connection form A satisfies*

$$F = dA + \frac{1}{2}[A, A].$$

We need the following lemma:

Lemma 5.5.5 *Let $X = \tilde{V}$ be a fundamental vector field and Y a horizontal vector field on P. Then the commutator $[X, Y]$ is horizontal.*

Proof The flow of X is given by $\phi_t = r_{\exp(tV)}$. This implies by Theorem A.1.46

$$[X, Y]_p = \frac{d}{dt}\bigg|_{t=0} \phi_{-t*} Y_{p\cdot\exp(tV)} \in H_p,$$

since $Y_{p\cdot\exp(tV)} \in H_{p\cdot\exp(tV)}$ and ϕ_{-t*} preserves the horizontal tangent bundle. □

We can now prove Theorem 5.5.4.

Proof We check the formula by inserting $X, Y \in T_pP$ on both sides of the equation, where we distinguish the following three cases:

1. Both X and Y are vertical: Then X and Y are fundamental vectors,

$$X = \tilde{V}_p,$$
$$Y = \tilde{W}_p,$$

for certain elements $V, W \in \mathfrak{g}$. We get

$$F(X, Y) = dA(\pi_H(X), \pi_H(Y)) = 0.$$

On the other hand we have

$$\frac{1}{2}[A, A](X, Y) = [A(X), A(Y)] = [V, W].$$

The differential dA of a 1-form A is given according to Proposition A.2.22 by

$$dA(X, Y) = L_X(A(Y)) - L_Y(A(X)) - A([X, Y]),$$

where we extend the vectors X and Y to vector fields in a neighbourhood of p. If we choose the extension by the fundamental vector fields $\tilde{V}$ and $\tilde{W}$, then

$$\begin{aligned} dA(X, Y) &= L_X(W) - L_Y(V) - [V, W] \\ &= -[V, W], \end{aligned}$$

since V and W are constant maps from P to $\mathfrak{g}$ and we used that $[\tilde{V}, \tilde{W}] = \widetilde{[V, W]}$ according to Proposition 3.4.4. This implies the claim.

2. Both X and Y are horizontal: Then

$$F(X, Y) = dA(X, Y)$$

and

$$\frac{1}{2}[A,A](X,Y) = [A(X),A(Y)] = [0,0] = 0.$$

This implies the claim.

3. X is vertical and Y is horizontal: Then $X = \tilde{V}_p$ for some $V \in \mathfrak{g}$. We have

$$F(X,Y) = dA(\pi_H(X), \pi_H(Y)) = dA(0,Y) = 0$$

and

$$\frac{1}{2}[A,A](X,Y) = [A(X),A(Y)] = [V,0] = 0.$$

Furthermore,

$$\begin{aligned} dA(X,Y) &= L_{\tilde{V}}(A(Y)) - L_Y(V) - A([\tilde{V},Y]) \\ &= -A([\tilde{V},Y]) \\ &= 0 \end{aligned}$$

since $[\tilde{V},Y]$ is horizontal by Lemma 5.5.5. This implies the claim.

□

The structure equation is very useful when we want to calculate the curvature of a given connection.

5.5.3 The Bianchi Identity

Let F be the curvature 2-form of a connection A. Then dF is a 3-form on P with values in the Lie algebra $\mathfrak{g}$. We want to consider the situation where we insert in all three arguments of dF a vector in the horizontal subbundle H defined by A.

Theorem 5.5.6 (Bianchi Identity (First Form)) *The differential dF of the curvature 2-form vanishes on $H \times H \times H$.*

Proof We use the following formula for the differential of a 2-form η on P, see Proposition A.2.22:

$$\begin{aligned} d\eta(X,Y,Z) = &\, L_X(\eta(Y,Z)) + L_Y(\eta(Z,X)) + L_Z(\eta(X,Y)) \\ &- \eta([X,Y],Z) - \eta([Y,Z],X) - \eta([Z,X],Y) \end{aligned}$$

for all vector fields X, Y, Z on P. By the structure equation we have $F = dA + \frac{1}{2}[A, A]$ so that

$$dF = \frac{1}{2}d[A, A].$$

We set $\eta = \frac{1}{2}[A, A]$. Then

$$dF(X, Y, Z) = d\eta(X, Y, Z)$$

for all $X, Y, Z \in T_pP$. We have $V \lrcorner \eta \equiv 0$ if V is a horizontal vector field, since

$$\eta(V, W) = [A(V), A(W)] = [0, A(W)] = 0$$

for an arbitrary vector field W on P. This implies the claim, because we can assume that X, Y, Z are horizontal in the neighbourhood of $p \in P$. □

5.6 Local Curvature 2-Forms

Let A be a connection 1-form on the principal bundle P and $s: U \to P$ a local section (local gauge), defined on an open subset $U \subset M$. We then defined the local connection 1-form (or local gauge field) $A_s \in \Omega^1(U, \mathfrak{g})$ by

$$A_s = A \circ Ds = s^*A.$$

Similarly we define:

Definition 5.6.1 The **local curvature 2-form** (or **local field strength**) $F_s \in \Omega^2(U, \mathfrak{g})$, determined by s, is defined by

$$F_s = F \circ (Ds, Ds) = s^*F.$$

If we have a manifold chart on U and $\{\partial_\mu\}$ are local coordinate basis vector fields on U, we set

$$F_{\mu\nu} = F_s(\partial_\mu, \partial_\nu).$$

(continued)

Definition 5.6.1 (continued)
If we choose in addition a basis $\{e_a\}$ of the Lie algebra $\mathfrak{g}$, we can expand the local field strength as

$$F_{\mu\nu} = \sum_{a=1}^{\dim \mathfrak{g}} F^a_{\mu\nu} e_a.$$

Proposition 5.6.2 (Local Structure Equation) *The local field strength can be calculated as*

$$F_s = dA_s + \frac{1}{2}[A_s, A_s]$$

and

$$F_{\mu\nu} = \partial_\mu A_\nu - \partial_\nu A_\mu + [A_\mu, A_\nu].$$

If the structure group G is abelian, then $F_s = dA_s$ and

$$F_{\mu\nu} = \partial_\mu A_\nu - \partial_\nu A_\mu.$$

Proof We calculate

$$\begin{aligned} s^*F &= s^*dA + \frac{1}{2}s^*[A, A] \\ &= ds^*A + \frac{1}{2}[s^*A, s^*A] \\ &= dA_s + \frac{1}{2}[A_s, A_s]. \end{aligned}$$

Here we used that

$$(s^*[A, A])(X, Y) = [s^*A, s^*A](X, Y),$$

which is easy to verify. This implies the first formula. The second formula follows from

$$\begin{aligned} F_{\mu\nu} &= dA_s(\partial_\mu, \partial_\nu) + \frac{1}{2}[A_s, A_s](\partial_\mu, \partial_\nu) \\ &= \partial_\mu(A_s(\partial_\nu)) - \partial_\nu(A_s(\partial_\mu)) - A_s([\partial_\mu, \partial_\nu]) + [A_s(\partial_\mu), A_s(\partial_\nu)] \\ &= \partial_\mu A_\nu - \partial_\nu A_\mu + [A_\mu, A_\nu]. \end{aligned}$$

Here we used that $[\partial_\mu, \partial_\nu] = 0$, because the basis vector fields $\{\partial_\mu\}$ come from a chart on U. □

In physics, the quadratic term $[A_\mu, A_\nu]$ in the expression for $F_{\mu\nu}$ (leading to cubic and quartic terms in the *Yang–Mills Lagrangian*, see Definition 7.3.1 and the corresponding local formula in Eq. (7.1)) is interpreted as a direct interaction between gauge bosons described by the gauge field A_μ. The quadratic term in the curvature is only present if the gauge group G is non-abelian, like $G = \mathrm{SU}(3)$ in quantum chromodynamics (QCD), but not if G is abelian, like $G = \mathrm{U}(1)$ in quantum electrodynamics (QED).

This explains why gluons, the gauge bosons of QCD, interact directly with each other, while photons, the gauge bosons of QED, do not. It is also the reason for phenomena in QCD such as *colour confinement* (at low energies) and *asymptotic freedom* (at high energies).

We would like to determine how the local field strength transforms under local gauge transformations. Let $s_i: U_i \to P$ and $s_j: U_j \to P$ be local gauges with $U_i \cap U_j \neq \emptyset$ and associated local curvature 2-forms F_i, F_j. The local gauge transformation

$$g_{ij}: U_i \cap U_j \longrightarrow G$$

is defined by

$$s_i(x) = s_j(x) \cdot g_{ji}(x) \quad \forall x \in U_i \cap U_j.$$

We then have:

Theorem 5.6.3 *The local curvature 2-forms transform as*

$$F_i = \mathrm{Ad}_{g_{ji}^{-1}} \circ F_j$$

on $U_i \cap U_j$. If G is a matrix Lie group, then

$$F_i = g_{ji}^{-1} \cdot F_j \cdot g_{ji}.$$

Proof Recall from the proof of Theorem 5.4.2 that

$$D_x s_i(V) = D_{s_j(x)} r_{g_{ji}(x)} \circ D_x s_j(V) + \widetilde{\mu_{ji}(V)}_{s_i(x)}$$

for a vector $V \in T_xM$. Hence we get for $V, W \in T_xM$, since F vanishes if a vertical vector is inserted:

$$\begin{aligned} F_i(V,W) &= F(D_{s_j(x)}r_{g_{ji}(x)} \circ D_x s_j(V), D_{s_j(x)}r_{g_{ji}(x)} \circ D_x s_j(W)) \\ &= \left(r^*_{g_{ji}(x)}F\right)(D_x s_j(V), D_x s_j(W)) \\ &= \mathrm{Ad}_{g_{ji}^{-1}(x)} \circ F_j(V,W). \end{aligned}$$

□

Corollary 5.6.4 *If G is abelian, then F_s is independent of the choice of local gauge and hence determines a well-defined, global, closed 2-form $F_M \in \Omega^2(M, \mathfrak{g})$.*

Proof In the abelian case the curvature defines a global form on M by Theorem 5.6.3. It remains to check that F_M is closed. In a local gauge s we have according to the local structure equation

$$F_s = dA_s + \frac{1}{2}[A_s, A_s].$$

Since G is abelian, we have $[A_s, A_s] = 0$. We conclude that

$$dF_M = dF_s = ddA_s = 0.$$

□

Remark 5.6.5 Note one important point about this corollary: Locally we have $F_s = dA_s$ if G is abelian, hence F_s is locally exact. However, the 2-form F_M in general is *not globally exact*, because A_s does not define a global 1-form on M (there is a change of A_s under changes of local gauge even if the structure group is abelian, see Theorem 5.4.2).

5.6.1 *The Curvature 2-Form of the Connection on the Hopf Bundle $S^3 \to S^2$

We consider the connection 1-form A on the Hopf bundle from Sect. 5.2.2 and continue to use the same notation (we follow [14, Example 3.10]).

Proposition 5.6.6 *The curvature of the connection 1-form A on the Hopf bundle is given by*

$$F^A = dA = -(\alpha_0 \wedge \bar{\alpha}_0 + \alpha_1 \wedge \bar{\alpha}_1).$$

Proof Since S^1 is abelian, we have $[A,A] = 0$, hence $F^A = dA$. The claim follows once we have shown that

$$dz_i = \alpha_i,$$
$$d\bar{z}_i = \bar{\alpha}_i,$$

since then also $d\alpha_i = 0 = d\bar{\alpha}_i$. This is clear from the definition of α_i and $\bar{\alpha}_i$: for example, if

$$(X_0, X_1) = (\dot{\gamma}_0(0), \dot{\gamma}_1(0))$$

with curves γ_0, γ_1, then

$$dz_j(X_0, X_1) = \left.\frac{d}{dt}\right|_{t=0} z_j(\gamma_0(t), \gamma_1(t)) = \left.\frac{d}{dt}\right|_{t=0} \gamma_j(t) = X_j.$$

□

According to Corollary 5.6.4 the curvature F^A determines a well-defined, global, closed 2-form F_{S^2} on S^2, where

$$F_{S^2}|_U = s^* F^A$$

for any local gauge $s: U \to S^3$ on an open subset $U \subset S^2$. We want to determine the 2-form F_{S^2}. Consider the open subset

$$U_1 = \{[z] = [z_0 : z_1] \in \mathbb{CP}^1 \mid z_1 \neq 0\}$$

together with the chart map

$$\psi_1: U_1 \longrightarrow \mathbb{C}$$
$$[z_0 : z_1] \longmapsto \frac{z_0}{z_1}.$$

We consider a 2-form $\tilde{F}$ on $\mathbb{C}$, defined by

$$\tilde{F}_w = -\frac{1}{(1+|w|^2)^2} dw \wedge d\bar{w}.$$

Proposition 5.6.7 *The 2-form F_{S^2} is given on $U_1 \subset S^2$ by*

$$F_{S^2}|_{U_1} = \psi_1^* \tilde{F}.$$

Proof It suffices to show that

$$\pi^*\psi_1^*\tilde{F} = F^A,$$

because then

$$s^*F^A = (\pi \circ s)^*\psi_1^*\tilde{F} = \psi_1^*\tilde{F}$$

for all local gauges $s: U_1 \to S^3$. To prove the formula note that

$$\psi_1 \circ \pi(z_0, z_1) = \psi_1([z_0 : z_1]) = \frac{z_0}{z_1}.$$

This implies for $(z_0, z_1) \in S^3$

$$\begin{aligned}
\pi^*\psi_1^*\tilde{F} &= (\psi_1 \circ \pi)^*\tilde{F} \\
&= -\frac{1}{(1+|\frac{z_0}{z_1}|^2)^2} d\left(\frac{z_0}{z_1}\right) \wedge d\left(\frac{\bar{z}_0}{\bar{z}_1}\right) \\
&= -z_1^2\left(\frac{1}{z_1}dz_0 - \frac{z_0}{z_1^2}dz_1\right) \wedge \bar{z}_1^2\left(\frac{1}{\bar{z}_1}d\bar{z}_0 - \frac{\bar{z}_0}{\bar{z}_1^2}d\bar{z}_1\right) \\
&= -(z_1 dz_0 - z_0 dz_1) \wedge (\bar{z}_1 d\bar{z}_0 - \bar{z}_0 d\bar{z}_1) \\
&= -|z_1|^2\alpha_0 \wedge \bar{\alpha}_0 - |z_0|^2\alpha_1 \wedge \bar{\alpha}_1 \\
&\quad + z_1\bar{z}_0\alpha_0 \wedge \bar{\alpha}_1 + z_0\bar{z}_1\alpha_1 \wedge \bar{\alpha}_0.
\end{aligned}$$

We have $\bar{z}_0 z_0 + \bar{z}_1 z_1 = 1$, hence

$$\bar{z}_0\alpha_0 + z_0\bar{\alpha}_0 + \bar{z}_1\alpha_1 + z_1\bar{\alpha}_1 = 0.$$

This implies

$$\begin{aligned}
z_1\bar{z}_0\alpha_0 \wedge \bar{\alpha}_1 + z_0\bar{z}_1\alpha_1 \wedge \bar{\alpha}_0 &= -z_1(z_0\bar{\alpha}_0 + \bar{z}_1\alpha_1 + z_1\bar{\alpha}_1) \wedge \bar{\alpha}_1 \\
&\quad - z_0(\bar{z}_0\alpha_0 + z_0\bar{\alpha}_0 + z_1\bar{\alpha}_1) \wedge \bar{\alpha}_0 \\
&= -|z_1|^2\alpha_1 \wedge \bar{\alpha}_1 - |z_0|^2\alpha_0 \wedge \bar{\alpha}_0.
\end{aligned}$$

Therefore

$$\begin{aligned}
\pi^*\psi_1^*\tilde{F} &= -\left(|z_1|^2 + |z_0|^2\right)(\alpha_0 \wedge \bar{\alpha}_0 + \alpha_1 \wedge \bar{\alpha}_1) \\
&= -(\alpha_0 \wedge \bar{\alpha}_0 + \alpha_1 \wedge \bar{\alpha}_1) \\
&= F^A.
\end{aligned}$$

□

Proposition 5.6.8 *For the connection on the Hopf bundle the following equation holds:*

$$\frac{1}{2\pi i}\int_{S^2} F_{S^2} = 1.$$

Proof Since

$$\mathbb{CP}^1 = U_1 \cup \{[1:0]\}$$

(i.e. $S^2 = \mathbb{CP}^1$ is the one-point compactification of $\mathbb{C}$) we can calculate

$$\begin{aligned}
\int_{S^2} F_{S^2} &= \int_{U_1} F_{S^2}|_{U_1} = \int_{U_1} \psi_1^* \tilde{F} = \int_{\mathbb{C}} \tilde{F} \\
&= -\int_{\mathbb{C}} \frac{1}{(1+|w|^2)^2} dw \wedge d\bar{w} \\
&= 2i \int_{\mathbb{R}^2} \frac{1}{(1+x^2+y^2)^2} dx \wedge dy \\
&= 2i \int_0^{2\pi}\int_0^{\infty} \frac{1}{(1+r^2)^2} r dr d\phi \\
&= -i \int_0^{2\pi} \left[\frac{1}{1+r^2}\right]_0^{\infty} d\phi \\
&= i \int_0^{2\pi} d\phi \\
&= 2\pi i.
\end{aligned}$$

□

The form $\psi_1^*\tilde{F}$ extends to a well-defined 2-form on all of $\mathbb{CP}^1$, which is equal to F_{S^2}. The 2-form

$$\omega_{FS} = \frac{1}{2i} F_{S^2}$$

is known as the **Fubini–Study form** of $\mathbb{CP}^1$. It is related to the standard volume form ω_{std} on S^2 of area 4π by

$$\omega_{FS} = \frac{1}{4}\omega_{std}.$$

Remark 5.6.9 We can define for any principal S^1-bundle $P \to M$ over a manifold M the **first Chern class** or **Euler class** as

$$c_1(P) = e(P) = \left[-\frac{1}{2\pi i} F_M \right].$$

This is a real cohomology class in $H^2_{dR}(M)$. It turns out that this class does not depend on the choice of connection 1-form on P (even though the 2-form F_M does). In the case of the Hopf bundle we have

$$c_1(\text{Hopf}) = -\frac{1}{4\pi}[\omega_{std}].$$

5.7 *Generalized Electric and Magnetic Fields on Minkowski Spacetime of Dimension 4

For the following notion from physics see, for example, [100]. Suppose $\pi: P \to M$ is a principal G-bundle and M is $\mathbb{R}^4$ with Minkowski metric η of signature $(+,-,-,-)$ (a similar construction works locally on any four-dimensional Lorentz manifold). Let x_0, x_1, x_2, x_3 be global coordinates in an inertial frame with coordinate vector fields satisfying

$$\begin{aligned} \eta(\partial_0, \partial_0) &= +1, \\ \eta(\partial_i, \partial_i) &= -1 \quad i = 1, 2, 3, \\ \eta(\partial_\mu, \partial_\nu) &= 0 \quad \mu \neq \nu. \end{aligned}$$

We also write for the coordinates

$$x_0 = t, \quad x_1 = x, \quad x_2 = y, \quad x_3 = z.$$

Suppose A is a connection 1-form on P with curvature F. Let $s: M \to P$ be a global gauge and

$$A_s = s^*A, \quad F_s = s^*F$$

as above. We write

$$\begin{aligned} A_\mu &= A_s(\partial_\mu), \\ F_{\mu\nu} &= F_s(\partial_\mu, \partial_\nu) \end{aligned}$$

and we have the local structure equation

$$F_{\mu\nu} = \partial_\mu A_\nu - \partial_\nu A_\mu + [A_\mu, A_\nu].$$

The 4×4-matrix $(F_{\mu\nu})$ comes from a 2-form on M and is skew-symmetric.

Definition 5.7.1 The **generalized electric** and **magnetic field**, determined by the connection, the choice of gauge and the inertial frame, are the $\mathfrak{g}$-valued functions

$$E_i, B_i \in \mathscr{C}^\infty(M, \mathfrak{g}), \quad i = x, y, z$$

defined by

$$(F_{\mu\nu}) = \begin{pmatrix} 0 & E_x & E_y & E_z \\ -E_x & 0 & -B_z & B_y \\ -E_y & B_z & 0 & -B_x \\ -E_z & -B_y & B_x & 0 \end{pmatrix}.$$

Equivalently,

$$E_i = F_{0i},$$
$$\epsilon_{ijk} B_k = -F_{ij},$$

where ϵ_{ijk} is totally antisymmetric with $\epsilon_{123} = 1$. We could expand the generalized electric and magnetic fields further in a basis for the Lie algebra $\mathfrak{g}$.

For quantum electrodynamics (QED) with $G = \mathrm{U}(1)$ these are the standard real-valued electric and magnetic fields (after choosing a basis for $\mathfrak{u}(1) \cong \mathbb{R}$). In this situation the electric and magnetic fields do not depend on the choice of gauge according to Corollary 5.6.4, because G is abelian (the gauge field A_s *does* depend on the choice of gauge).

For $G = \mathrm{SU}(n)$, in particular $G = \mathrm{SU}(3)$ corresponding to quantum chromodynamics (QCD), these $\mathfrak{g}$-valued fields are also called **chromo-electric** and **chromo-magnetic fields** (or colour-electric and colour-magnetic fields). They describe the field strength of the gluon field corresponding to the connection 1-form A_μ.

5.8 Parallel Transport

Connections define an important concept: *parallel transport* in principal and associated vector bundles. The notion of parallel transport also leads to the concept of *covariant derivative* on associated vector bundles.

Let $\pi: P \to M$ be a principal G-bundle with a connection A. We want to lift curves in M to horizontal curves in P, which are defined in the following way (by a **curve** we always mean in this and the following sections a smooth curve).

Definition 5.8.1 A curve $\gamma^*: I \rightarrow P$ is called a **horizontal lift** of a curve $\gamma: I \rightarrow M$, defined on an interval I, if:

1. $\pi \circ \gamma^* = \gamma$
2. the velocity vectors $\dot{\gamma}^*(t)$ are horizontal, i.e. elements of $H_{\gamma^*(t)}$, for all $t \in I$.

The following theorem says that a horizontal lift of a curve in the base manifold always exists and is unique once the starting point has been given.

Theorem 5.8.2 (Existence and Uniqueness of Horizontal Lifts of Curves) *Let* $\gamma: [a, b] \rightarrow M$ *be a curve with* $\gamma(a) = x$. *Let* p *be a point in the fibre* P_x. *Then there exists a unique horizontal lift* γ_p^* *of* γ *with* $\gamma_p^*(a) = p$.

Proof Since P is locally trivial, there exists some lift δ of γ with $\delta(a) = p$ (one could also argue that the pullback of the bundle P under the map γ is trivial, because $[a, b]$ is contractible). We want to find a map $g: [a, b] \rightarrow G$ such that

$$\gamma^*(t) = \delta(t) \cdot g(t)$$

is horizontal. We will determine $g(t)$ as the solution of a differential equation.

The curve $\gamma^*(t)$ will be horizontal if

$$A\left(\dot{\gamma}^*(t)\right) = 0 \quad \forall t \in [a, b].$$

We can calculate $\dot{\gamma}^*(t)$ with Proposition 3.5.4:

$$\dot{\gamma}^*(t) = r_{g(t)*}\dot{\delta}(t) + \widetilde{\mu_G(\dot{g}(t))}_{\gamma^*(t)}.$$

Hence

$$\begin{aligned} A(\dot{\gamma}^*(t)) &= \mathrm{Ad}(g(t)^{-1}) \circ A(\dot{\delta}(t)) + \mu_G(\dot{g}(t)) \\ &= L_{g(t)^{-1}*}\left(R_{g(t)*}A(\dot{\delta}(t)) + \dot{g}(t)\right). \end{aligned}$$

We conclude that $g(t)$ has to be the solution of the differential equation

$$\dot{g}(t) = -R_{g(t)*}A(\dot{\delta}(t))$$

with $g(0) = e$. This is the integral curve in the Lie group G through e of the time-dependent right-invariant vector field on G, corresponding to the Lie algebra element $-A(\dot{\delta}(t)) \in \mathfrak{g}$. Such an integral curve on the interval $[a, b]$ exists by Theorem 1.7.18. An explicit solution for $g(t)$ in the case of a linear Lie group G can also be found in Proposition 5.10.4. □

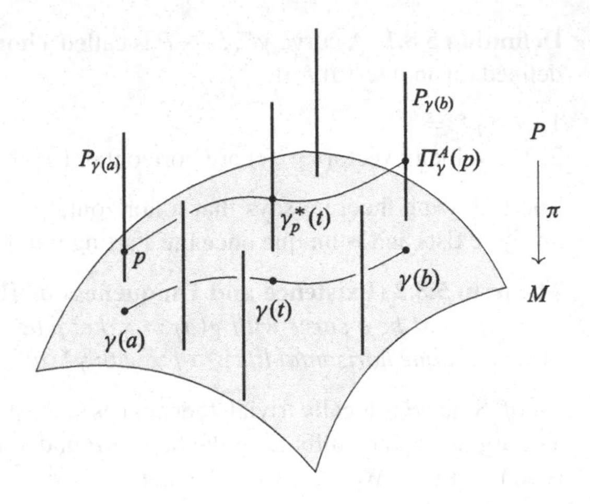

Fig. 5.1 Parallel transport

Definition 5.8.3 Let $\gamma\colon [a,b] \to M$ be a curve in M. The map

$$\begin{aligned} \Pi^A_\gamma\colon P_{\gamma(a)} &\longrightarrow P_{\gamma(b)} \\ p &\longmapsto \gamma^*_p(b) \end{aligned}$$

is called **parallel transport in the principal bundle** P along γ with respect to the connection A. See Fig. 5.1.

Theorem 5.8.4 (Properties of Parallel Transport) *Let P be a principal bundle with connection A.*

1. *Parallel transport Π^A_γ is a smooth map between the fibres $P_{\gamma(a)}$ and $P_{\gamma(b)}$ and does not depend on the parametrization of the curve γ.*
2. *Let γ be a curve in M from x to y and γ' a curve from y to z. Denote the concatenation by $\gamma * \gamma'$, where γ comes first. Then*

$$\Pi^A_{\gamma*\gamma'} = \Pi^A_{\gamma'} \circ \Pi^A_\gamma.$$

3. *If γ^- denotes the curve γ traversed backwards, then*

$$\Pi^A_{\gamma^-} = \left(\Pi^A_\gamma\right)^{-1}.$$

In particular, parallel transport is a diffeomorphism between the fibres.

4. *Parallel transport is G-equivariant: The following identity holds:*

$$\Pi^A_\gamma \circ r_g = r_g \circ \Pi^A_\gamma \quad \forall g \in G.$$

Proof Properties 1–3 follow from the theory of ordinary differential equations. We only prove 4: let γ be a curve from x to y in M and $p \in P_x$. Let γ^*_p be the horizontal lift of γ to p. For $g \in G$ consider the curve $r_g \circ \gamma_p^*$. This curve starts at $p \cdot g$ and projects to γ. Furthermore, it is horizontal, because r_{g*} maps horizontal vectors to horizontal vectors by the definition of connections. It follows that $r_g \circ \gamma_p^*$ is equal to $\gamma_{p \cdot g}^*$. We get

$$\begin{aligned} \Pi_\gamma^A \circ r_g(p) &= \Pi_\gamma^A(p \cdot g) \\ &= \gamma_{p \cdot g}^*(b) \\ &= r_g \circ \gamma_p^*(b) \\ &= r_g \circ \Pi_\gamma^A(p). \end{aligned}$$

□

Since parallel transport does not depend on the parametrization of the curve γ, we will often assume that γ is defined on the interval $[0, 1]$.

5.9 The Covariant Derivative on Associated Vector Bundles

So far we have considered connections on principal bundles. Associated vector bundles play an important role in gauge theory, because matter fields are sections of such bundles. It turns out that connections on principal bundles define so-called *covariant derivatives* on all associated vector bundles (this will explain the third row in the diagram at the beginning of Sect. 4.7). These covariant derivatives appear in physics, in particular, in the Lagrangians and field equations defining gauge theories.

We first want to define the notion of parallel transport in associated vector bundles. Let $P \to M$ be a principal G-bundle with a connection A, $\rho\colon G \to \mathrm{GL}(V)$ a representation on a $\mathbb{K}$-vector space V ($\mathbb{K} = \mathbb{R}, \mathbb{C}$) and $E = P \times_\rho V$ the associated vector bundle.

Theorem 5.9.1 *For a curve* $\gamma\colon [0, 1] \to M$ *the map*

$$\begin{aligned} \Pi_\gamma^{E,A}\colon E_{\gamma(0)} &\longrightarrow E_{\gamma(1)} \\ [p, v] &\longmapsto [\Pi_\gamma^A(p), v] \end{aligned}$$

is well-defined and a linear isomorphism. This map is called **parallel transport in the associated vector bundle** *E along the curve* γ *with respect to the connection A.*

Proof We first show that $\Pi_\gamma^{E,A}$ is well-defined, independent of the choice of representative $[p, v]$. Suppose that

$$[p, v] = \left[p', v'\right] \in E_{\gamma(0)}.$$

Then there exists an element $g \in G$ such that

$$(p', v') = \left(p \cdot g, \rho(g)^{-1} v\right).$$

Part 4 of Theorem 5.8.4 then implies

$$\begin{aligned}\left[\Pi_\gamma^A(p'), v'\right] &= \left[\Pi_\gamma^A(p \cdot g), \rho(g)^{-1} v\right]\\ &= \left[\Pi_\gamma^A(p) \cdot g, \rho(g)^{-1}(v)\right]\\ &= \left[\Pi_\gamma^A(p), v\right].\end{aligned}$$

Hence the map $\Pi_\gamma^{E,A}$ is well-defined. It is then also clear that $\Pi_\gamma^{E,A}$ is a linear isomorphism. □

Let Φ be a section of E, $x \in M$ a point and $X \in T_xM$ a tangent vector. We want to define a **covariant derivative** as follows: choose an arbitrary curve $\gamma\colon (-\epsilon, \epsilon) \to M$ with

$$\begin{aligned}\gamma(0) &= x,\\ \dot{\gamma}(0) &= X.\end{aligned}$$

For each $t \in (-\epsilon, \epsilon)$ parallel transport the vector $\Phi(\gamma(t)) \in E_{\gamma(t)}$ back to E_x along γ. Then take the derivative in $t = 0$ of the resulting curve in the fibre E_x, giving an element in E_x. More formally, we set

$$D(\Phi, \gamma, x, A) = \left.\frac{d}{dt}\right|_{t=0} \left(\Pi_{\gamma_t}^{E,A}\right)^{-1} (\Phi(\gamma(t)) \in E_x.$$

Here γ_t denotes the restriction of the curve γ starting at time 0 and ending at time t, for $t \in (-\epsilon, \epsilon)$.

We want to prove the following formula.

Theorem 5.9.2 *Let $s\colon U \to P$ be a local gauge, $A_s = s^*A$ and $\phi\colon U \to V$ the map with $\Phi = [s, \phi]$. Then the vector $D(\Phi, \gamma, x, A) \in E_x$ is given by*

$$D(\Phi, \gamma, x, A) = [s(x), d\phi(X) + \rho_*(A_s(X))\phi(x)].$$

Proof We have

$$\left(\Pi_{\gamma_t}^{E,A}\right)^{-1} (\Phi(\gamma(t)) = \left[\left(\Pi_{\gamma_t}^{A}\right)^{-1} (s(\gamma(t)), \phi(\gamma(t))\right].$$

Let $q(t)$ be the uniquely determined smooth curve in the fibre P_x such that

$$\Pi^A_{\gamma_t}(q(t)) = s(\gamma(t)).$$

Write

$$q(t) = s(x) \cdot g(t)$$

with a uniquely determined smooth curve $g(t)$ in G. Then

$$\begin{aligned}\left(\Pi^{E,A}_{\gamma_t}\right)^{-1} (\Phi(\gamma(t)) &= [q(t), \phi(\gamma(t))] \\ &= [s(x), \rho(g(t))\phi(\gamma(t))].\end{aligned}$$

For $t = 0$ we have

$$s(x) = s(\gamma(0)) = \Pi^A_{\gamma_0}(q(0)) = q(0),$$

hence

$$g(0) = e \in G.$$

This implies

$$\dot{g}(0) \in \mathfrak{g}.$$

It follows that

$$\begin{aligned}D(\Phi, \gamma, x, A) &= \left.\frac{d}{dt}\right|_{t=0} [s(x), \rho(g(t))\phi(\gamma(t))] \\ &= [s(x), \rho_*(\dot{g}(0))\phi(x) + d\phi(X)].\end{aligned}$$

It remains to calculate $\rho_*(\dot{g}(0))$. We have

$$\left.\frac{d}{dt}\right|_{t=0} s(\gamma(t)) = ds(X)$$

and

$$\left.\frac{d}{dt}\right|_{t=0} \Pi^A_{\gamma_t}(q(t)) = \dot{q}(0) + \left.\frac{d}{dt}\right|_{t=0} \Pi^A_{\gamma_t}(s(x)).$$

Since the curve $\Pi^A_{\gamma_t}(s(x))$ is horizontal with respect to A, we get

$$A_s(x) = A(ds(X)) = A(\dot{q}(0)).$$

However,

$$\dot{q}(0) = \widetilde{\dot{g}(0)}_{s(x)},$$

hence

$$A(\dot{q}(0)) = \dot{g}(0)$$

by the definition of connection 1-form. It follows that

$$\rho_*(\dot{g}(0)) = \rho_*(A_s(X))$$

and thus the claim. □

The theorem implies that $D(\Phi, \gamma, x, A)$ depends only on the tangent vector X and not on the curve γ itself. We can therefore set:

Definition 5.9.3 Let Φ be a section of an associated vector bundle E and $X \in \mathfrak{X}(M)$ a vector field on M. Then the **covariant derivative** $\nabla^A_X \Phi$ is the section of E defined by

$$(\nabla^A_X \Phi)(x) = D(\Phi, \gamma, x, A),$$

where γ is any curve through x and tangent to X_x. The covariant derivative is a map

$$\nabla^A : \Gamma(E) \longrightarrow \Omega^1(M, E).$$

The fact that $\nabla^A \Phi$ is a smooth 1-form in $\Omega^1(M, E)$ for every $\Phi \in \Gamma(E)$ is clear from the local formula.

We often write in a local gauge $s: U \to P$, with $\Phi = [s, \phi]$, the covariant derivative as

$$\nabla^A_X \Phi = \left[s, \nabla^A_X \phi\right],$$

(continued)

Definition 5.9.3 (continued)
where

$$\nabla_X^A \phi = d\phi(X) + \rho_*(A_s(X))\phi,$$

i.e.

$$(\nabla_X^A \phi)(x) = d\phi(X_x) + \rho_*(A_s(X_x))\phi(x) \in V.$$

Here are some properties of the covariant derivative.

Proposition 5.9.4 (Properties of Covariant Derivative) *The map ∇^A is $\mathbb{K}$-linear in both entries and satisfies*

$$\nabla_{fX}^A \Phi = f\nabla_X^A \Phi$$

for all smooth functions $f \in \mathscr{C}^\infty(M, \mathbb{R})$ and the **Leibniz rule**

$$\nabla_X^A(\lambda\Phi) = (L_X\lambda)\Phi + \lambda\nabla_X^A\Phi$$

for all smooth functions $\lambda \in \mathscr{C}^\infty(M, \mathbb{K})$.

Proof $\mathbb{K}$-linearity of ∇^A and function linearity in X is clear. Let $\lambda: U \rightarrow \mathbb{K}$ be a smooth function. Then

$$\begin{aligned}
\nabla_X^A(\lambda\Phi)(x) &= [s(x), d(\lambda\phi)(X_x) + \rho_*(A_s(X_x))(\lambda\phi)(x)] \\
&= [s(x), d\lambda(X_x)\phi(x) + \lambda(x)d\phi(X_x) + \lambda(x)\rho_*(A_s(X_x))\phi(x)] \\
&= d\lambda(X_x)\,[s(x), \phi(x)] + \lambda(x)(\nabla_X^A\Phi)(x) \\
&= (L_{X_x}\lambda)\Phi(x) + \lambda(x)(\nabla_X^A\Phi)(x).
\end{aligned}$$

Here we used the product rule for functions to the vector space V multiplied by functions to the scalars $\mathbb{K}$. □

Remark 5.9.5 If $\{\partial_\mu\}$ are local basis vector fields on U, we get

$$\nabla_\mu^A \phi = \nabla_{\partial_\mu}^A \phi = \partial_\mu\phi + A_\mu\phi,$$

(continued)

Remark 5.9.5 (continued)
where

$$A_\mu \phi = \rho_*(A_\mu)\phi.$$

In physics the covariant derivative is typically written in this form and acts on functions ϕ on U with values in the vector space V, determined by sections Φ in E and the local gauge s. In mathematics the covariant derivative acts directly on the sections of the vector bundle. We denote both operators by ∇^A (it will be clear from the context which operator is meant).

From a physics point of view it is important that the second summand $A_\mu \phi$ in the covariant derivative is non-linear (quadratic) in the fields A_μ and ϕ. This non-linearity, called *minimal coupling*, leads to non-quadratic terms in the Lagrangian (see Definition 7.5.5 and Definition 7.6.2 as well as the local formulas in Eqs. (7.3) and (7.4)), which are interpreted as an interaction between gauge bosons described by A_μ and the particles described by the field ϕ.

Notice the crucial role played by the representation ρ: It is not only needed to define the associated vector bundle E, but also to define the covariant derivative ∇^A. The gauge field A can act on maps with values in V (or sections of E) only if V carries a representation ρ of the gauge group G. If the representation

$$\rho_*: \mathfrak{g} \longrightarrow \mathrm{End}(V)$$

is non-trivial and hence the coupling between the gauge field A_μ and the field ϕ is (potentially) non-trivial, then the particles corresponding to ϕ are called **charged** (charged particles are affected by the gauge field). In Chaps. 8 and 9 we will discuss in some detail the representations that appear in the description of matter particles in the Standard Model and in Grand Unified Theories.

Figure 5.2 shows the Feynman diagrams for the cubic and quartic terms which appear in the Klein–Gordon Lagrangian in Eq. (7.3), representing the interaction between a gauge field A and a charged scalar field described locally by a map ϕ with values in V.

Remark 5.9.6 In physics the covariant derivative is often *defined* (without referring to parallel transport) by the local formula

$$\nabla_X^A \Phi = \left[s, \nabla_X^A \phi\right],$$

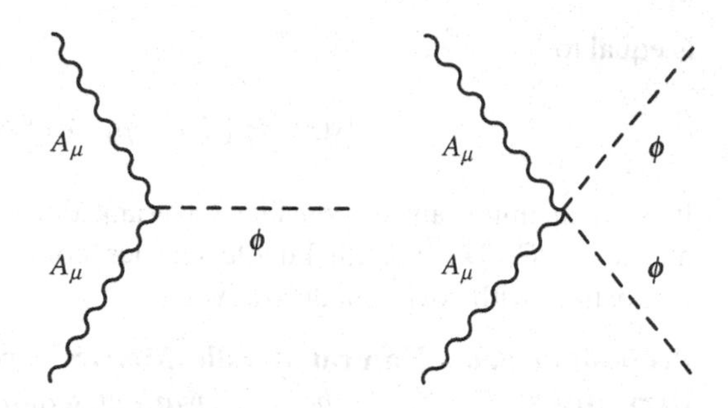

Fig. 5.2 Feynman diagrams for interaction between gauge field and charged scalar

where

$$\nabla_X^A \phi = d\phi(X) + \rho_*(A_s(X))\phi.$$

One then has to show that this definition is independent of the choice of local gauge: Suppose $s' : U \rightarrow P$ is another local gauge. Then there exists a smooth physical gauge transformation $g: U \rightarrow G$ such that

$$s' = s \cdot g.$$

We have

$$\Phi|_U = [s, \phi] = [s', \phi'],$$

with

$$\phi' = \rho(g)^{-1}\phi.$$

Furthermore,

$$A_{s'} = \mathrm{Ad}_{g^{-1}} \circ A_s + \mu,$$

where

$$\mu = g^* \mu_G.$$

It follows that

$$d\phi'(X_x) = \rho(g(x))^{-1} d\phi(X_x) + (D_x \rho(g)^{-1}(X_x))\phi.$$

A lengthy calculation (if done in this abstract setting) then shows that

$$\left[s'(x), d\phi'(X_x) + \rho_*(A_{s'}(X_x))\phi'(x)\right]$$

is equal to

$$[s(x), d\phi(X_x) + \rho_*(A_s(X_x))\phi(x)] .$$

It is often important to consider covariant derivatives compatible with a bundle metric on E. The natural bundle metrics constructed in Proposition 4.7.12 are compatible with covariant derivatives.

Proposition 5.9.7 (Natural Bundle Metrics Are Compatible with Covariant Derivatives) *Let $\langle\cdot,\cdot\rangle_V$ be a G-invariant scalar product on the vector space V and $\langle\cdot,\cdot\rangle_E$ the induced bundle metric on the associated vector bundle $E = P \times_\rho V$. Then the covariant derivative associated to a connection A is compatible with the bundle metric in the sense that*

$$L_X \langle \Phi, \Phi' \rangle_E = \langle \nabla^A_X \Phi, \Phi' \rangle_E + \langle \Phi, \nabla^A_X \Phi' \rangle_E$$

for all sections Φ, Φ' of E and all vector fields X on M.

Proof Since the scalar product on V is G-invariant, the map ρ_* induced by the representation satisfies

$$\langle \rho_*(\alpha)\phi, \phi' \rangle_V + \langle \phi, \rho_*(\alpha)\phi' \rangle_V = 0$$

for all $\alpha \in \mathfrak{g}$ and $\phi, \phi' \in V$; see Proposition 2.1.37. This implies:

$$\begin{aligned}
\langle \nabla^A_X \Phi, \Phi' \rangle_E + \langle \Phi, \nabla^A_X \Phi' \rangle_E &= \langle \nabla^A_X \phi, \phi' \rangle_V + \langle \phi, \nabla^A_X \phi' \rangle_V \\
&= \langle d\phi(X) + \rho_*(A_s(X))\phi, \phi' \rangle_V \\
&\quad + \langle \phi, d\phi'(X) + \rho_*(A_s(X))\phi' \rangle_V \\
&= \langle d\phi(X), \phi' \rangle_V + \langle \phi, d\phi'(X) \rangle_V \\
&= L_X \langle \Phi, \Phi' \rangle_E .
\end{aligned}$$

□

5.10 *Parallel Transport and Path-Ordered Exponentials

We derive in this section a formula that is used in physics to calculate the parallel transport on principal bundles. The following arguments are outlined in [103]. Recall that for the proof of Theorem 5.8.2 concerning the existence of a horizontal lift γ^* of a curve $\gamma: [0, 1] \rightarrow M$, where

$$\gamma^*(0) = p \in P_{\gamma(0)},$$

we had to solve the differential equation

$$\dot{g}(t) = -R_{g(t)*}A(\dot{\delta}(t)),$$

with $g(0) = e$, where δ is some lift of γ and $g\colon [0,1] \to G$ is a map with

$$\gamma^*(t) = \delta(t) \cdot g(t).$$

There is a nice way to write the solution $g(t)$ explicitly, at least if G is a matrix Lie group and γ is contained in an open set over which the principal bundle is trivial.

Suppose that the curve γ is contained in an open set $U \subset M$, so that P_U is trivial over U. Let $s\colon U \to P$ be a local gauge with $s(\gamma(0)) = p$. We can choose

$$\delta = s \circ \gamma.$$

We then have to solve

$$\begin{aligned} \dot{g}(t) &= -R_{g(t)*}A(\dot{\delta}(t)) \\ &= -R_{g(t)*}A(s_*\dot{\gamma}(t)) \\ &= -R_{g(t)*}A_s(\dot{\gamma}(t)). \end{aligned}$$

Suppose that $G \subset \mathrm{GL}(n,\mathbb{K})$ is a linear group, i.e. a closed Lie subgroup. Then the differential equation can be written as

$$\frac{dg(t)}{dt} = -A_s(\dot{\gamma}(t)) \cdot g(t).$$

We write this as

$$\frac{dg(t)}{dt} = f(t) \cdot g(t),$$

where $f\colon [0,1] \to \mathfrak{g}$ is a smooth map determined by $\gamma(t)$, independent of g.

5.10.1 Path-Ordered Exponentials

Let G be a linear group with Lie algebra $\mathfrak{g}$.

Definition 5.10.1 For a smooth map $f\colon [0,1] \to \mathfrak{g}$ we define for all $t \in [0,1]$ the following matrices in $\mathrm{Mat}(n \times n, \mathbb{K})$:

$$P_0(f,t) = I_n,$$
$$P_1(f,t) = \int_0^t f(s_0)\, ds_0,$$
$$P_n(f,t) = \int_0^t \int_0^{s_0} \int_0^{s_1} \ldots \int_0^{s_{n-2}} f(s_0)f(s_1)\ldots f(s_{n-1})\, ds_{n-1} \ldots ds_1 ds_0 \quad \forall n \geq 2,$$

where in the definition of $P_n(f,t)$

$$1 \geq t \geq s_0 \geq s_1 \geq \ldots \geq s_{n-2} \geq 0.$$

The following is easy to show:

Lemma 5.10.2 *For all $n \geq 2$ the integral*

$$Q_n(t) = \int_0^t \int_0^{s_0} \int_0^{s_1} \ldots \int_0^{s_{n-2}} ds_{n-1} \ldots ds_1 ds_0,$$

where

$$1 \geq t \geq s_0 \geq s_1 \geq \ldots \geq s_{n-2} \geq 0,$$

evaluates to

$$Q_n(t) = \frac{1}{n!} t^n.$$

Considering

$$||f|| = \max_{s \in [0,1]} ||f(s)||$$

with respect to a matrix norm on $\mathrm{Mat}(n \times n, \mathbb{K})$ it follows that:

Proposition 5.10.3 *The series*

$$P(f,t) = \sum_{n=0}^{\infty} P_n(f,t)$$

converges for every $t \in [0,1]$ and defines a smooth map

$$P(f,\cdot)\colon [0,1] \longrightarrow \mathrm{Mat}(n \times n, \mathbb{K}).$$

We write

$$\mathscr{P}\exp\left(\int_0^t f(s)\,ds\right) = P(f,t)$$

and call this the **path-ordered exponential** *of the function f.*

Path-ordered exponentials are useful, because they define solutions to the ordinary differential equation we are interested in.

Proposition 5.10.4 (Path-Ordered Exponential Defines Solution of ODE) *Consider a smooth map $f\colon [0,1] \to \mathfrak{g}$ and define a smooth map*

$$g\colon [0,1] \longrightarrow \mathrm{Mat}(n \times n, \mathbb{K})$$

by

$$g(t) = \mathscr{P}\exp\left(\int_0^t f(s)\,ds\right).$$

Then

$$g(0) = I_n,$$
$$\frac{dg(t)}{dt} = f(t)\cdot g(t) \quad \forall t \in [0,1].$$

In particular, g is a map

$$g\colon [0,1] \longrightarrow G.$$

Proof It is clear that $g(0) = I_n$. Calculating the derivative with respect to t we get

$$\begin{aligned}\frac{d}{dt}\mathscr{P}\exp\left(\int_0^t f(s)\,ds\right) &= f(t) + f(t)\int_0^t f(s_1)\,ds_1 + f(t)\int_0^t\int_0^{s_1} f(s_1)f(s_2)\,ds_2 ds_1 \ldots \\ &= f(t)\cdot\mathscr{P}\exp\left(\int_0^t f(s)\,ds\right).\end{aligned}$$

This implies the first claim.

To prove the claim that g takes values in G, note that

$$X_A(t) = f(t)\cdot A \quad \forall A \in \mathrm{Mat}(n\times n,\mathbb{K}), t \in [0,1]$$

defines a time-dependent vector field X on $\mathrm{Mat}(n \times n, \mathbb{K})$, which is right-invariant in the sense that

$$X_{A\cdot B}(t) = X_A(t)\cdot B \quad \forall B \in \mathrm{Mat}(n\times n,\mathbb{K}).$$

The smooth map

$$g\colon [0,1] \longrightarrow \mathrm{Mat}(n \times n, \mathbb{K})$$

is an integral curve of the vector field X through the unit matrix I_n. Let

$$h\colon [0,1] \longrightarrow G$$

be the integral curve through I_n of the restriction of the right-invariant vector field X to G, given by Theorem 1.7.18 (the vector field X is tangent to G, because f takes values in the Lie algebra $\mathfrak{g}$). Then uniqueness of the solution to ordinary differential equations shows that $g \equiv h$, hence g takes values in G. □

5.10.2 Explicit Formula for Parallel Transport

Returning to the situation before Sect. 5.10.1, we can write the curve $\gamma(t)$ in a chart on U with coordinates x^μ as $\gamma(t) = x^\mu(t)$. Then

$$\frac{dg(t)}{dt} = -\sum_{\mu=1}^{n} A_{s\mu}(\gamma(t))\frac{dx^\mu}{dt} \cdot g(t).$$

The solution to this differential equation is

$$\begin{aligned} g(t) &= \mathscr{P}\exp\left(-\int_0^t \sum_{\mu=1}^{n} A_{s\mu}(\gamma(s))\frac{dx^\mu}{ds}ds\right) \\ &= \mathscr{P}\exp\left(-\int_{\gamma(0)}^{\gamma(t)} \sum_{\mu=1}^{n} A_{s\mu}(x^\mu)dx^\mu\right) \\ &= \mathscr{P}\exp\left(-\int_{\gamma_t} A_s\right), \end{aligned}$$

where γ_t denotes the restriction of the curve γ to $[0,t]$. In particular,

$$g(1) = \mathscr{P}\exp\left(-\int_{\gamma} A_s\right).$$

We therefore get:

Theorem 5.10.5 (Parallel Transport Expressed with Path-Ordered Exponential) *Let $P \to M$ be a principal bundle with matrix structure group G. Suppose that $\gamma\colon [0,1] \to M$ is a curve inside an open set $U \subset M$ over which P_U is trivial. Let*

$p \in P_{\gamma(0)}$ be a point and $s: U \to P$ a local gauge, such that $s(\gamma(0)) = p$. Suppose $s(\gamma(1)) = q$. Then the parallel transport of p can be written as

$$\Pi_\gamma^A(p) = q \cdot \mathscr{P}\exp\left(-\int_\gamma A_s\right).$$

5.11 *Holonomy and Wilson Loops

We saw that the induced parallel transport on associated vector bundles can be used to define covariant derivatives. We want to explain another concept where parallel transport on associated vector bundles appears in physics (we follow the definition in [47]).

Suppose $\gamma: [0,1] \to M$ is a closed curve in M (a *loop*) with $\gamma(0) = \gamma(1) = x$. Then parallel transport $\Pi_\gamma^{E,A}$ is a linear isomorphism of the fibre E_x to itself.

Definition 5.11.1 We call the isomorphism $\Pi_\gamma^{E,A}$ of E_x the **holonomy** $\mathrm{Hol}_{\gamma,x}^E(A)$ of the loop γ in the basepoint x with respect to the connection A.
We can express the holonomy using path-ordered exponentials.

Proposition 5.11.2 (Holonomy Expressed with Path-Ordered Exponential) *Suppose that G is a matrix Lie group and the loop γ is contained in an open set $U \subset M$ over which P is trivial and $s: U \to P$ is a local gauge. Then s determines an isomorphism*

$$\begin{aligned} V &\longrightarrow E_x \\ v &\longmapsto [s(x), v] \end{aligned}$$

and with respect to this isomorphism

$$\mathrm{Hol}_{\gamma,x}^E(A) = \rho\left(\mathscr{P}\exp\left(-\oint_\gamma A_s\right)\right) = \mathscr{P}\exp\left(-\oint_\gamma \rho_* A_s\right).$$

Proof This is Exercise 5.15.8. □

We want to understand how the holonomy changes if we choose a different base point on the curve γ.

Lemma 5.11.3 *Let y be another point on the closed curve γ and σ the part of γ from x to y. Then*

$$\mathrm{Hol}_{\gamma,y}^E(A) = \Pi_\sigma \circ \mathrm{Hol}_{\gamma,x}^E(A) \circ (\Pi_\sigma)^{-1},$$

where we abbreviate $\Pi_\sigma = \Pi_\sigma^{E,A}$.

Proof Let σ' be the remaining part of γ from y to x. Then $\gamma = \sigma * \sigma'$ and

$$\mathrm{Hol}^E_{\gamma,x}(A) = \Pi_{\sigma'} \circ \Pi_\sigma,$$

$$\mathrm{Hol}^E_{\gamma,y}(A) = \Pi_\sigma \circ \Pi_{\sigma'}.$$

This implies the claim. □

Therefore the following map is well-defined.

Definition 5.11.4 The **Wilson operator** or **Wilson loop** is the map W^E_γ that associates to a connection A and a loop γ the number

$$\begin{aligned} W^E_\gamma(A) &= \mathrm{tr}\left(\mathrm{Hol}^E_{\gamma,x}(A)\right) \\ &= \mathrm{tr}\left(\mathscr{P}\exp\left(-\oint_\gamma \rho_* A_s\right)\right), \end{aligned}$$

where tr denotes trace, x is any point on γ, and the second formula holds if G is a matrix Lie group and γ is inside a trivializing open set U for P.

Proposition 5.11.5 (Wilson Loops Are Gauge Invariant) *The Wilson loop is invariant under all bundle automorphisms of P:*

$$W^E_\gamma(f^*A) = W^E_\gamma(A) \quad \forall f \in \mathscr{G}(P).$$

Proof This is Exercise 5.15.9. □

In quantum field theory, the gauge field A_μ is a function on spacetime with values in the operators on the Hilbert state space V (if we ignore for the moment questions of whether this operator is well-defined and issues of regularization). The formula

$$W^E_\gamma(A) = \mathrm{tr}\left(\mathscr{P}\exp\left(-\oint_\gamma \rho_* A_s\right)\right)$$

shows that the Wilson loop $W^E_\gamma(A)$ is a gauge invariant operator on this Hilbert space.

5.12 The Exterior Covariant Derivative

In Sect. 5.9 we defined a covariant derivative

$$\nabla^A : \Gamma(E) \longrightarrow \Omega^1(M, E).$$

We can think of this map as a generalization of the differential

$$d\colon \mathscr{C}^\infty(M) \longrightarrow \Omega^1(M).$$

In fact, the differential d can be identified with the covariant derivative on the trivial line bundle over M induced from the trivial connection. The differential d can be uniquely extended in the standard way to an exterior derivative

$$d\colon \Omega^k(M) \longrightarrow \Omega^{k+1}(M)$$

by demanding that $ddf = 0$ for all functions $f \in \mathscr{C}^\infty(M)$ and

$$d(\alpha \wedge \beta) = d\alpha \wedge \beta + (-1)^k \alpha \wedge d\beta$$

for all $\alpha \in \Omega^k(M)$ and $\beta \in \Omega^l(M)$. This differential satisfies $d \circ d = 0$ on all forms, see Exercise 5.15.11. Because of this property, the *de Rham cohomology*

$$H^k_{dR}(M) = \frac{\ker\left(d\colon \Omega^k(M) \to \Omega^{k+1}(M)\right)}{\mathrm{im}\left(d\colon \Omega^{k-1}(M) \to \Omega^k(M)\right)}$$

is well-defined for all k.

We want to show that we can extend the covariant derivative in a similar way to an *exterior covariant derivative*

$$d_A\colon \Omega^k(M,E) \longrightarrow \Omega^{k+1}(M,E).$$

This exterior covariant derivative, however, in general does *not* satisfy $d_A \circ d_A = 0$. The non-vanishing of $d_A \circ d_A$ is precisely measured by the curvature of A, see Exercise 5.15.12.

The constructions in this section work for general covariant derivatives on vector bundles, which are defined as follows.

Definition 5.12.1 Let $E \to M$ be a $\mathbb{K}$-vector bundle. Then a **covariant derivative** ∇ on E is a $\mathbb{K}$-linear map

$$\nabla\colon \Gamma(E) \longrightarrow \Omega^1(M,E)$$

such that

$$\nabla_{fX} e = f \nabla_X e$$

for all smooth functions $f \in \mathscr{C}^\infty(M,\mathbb{R})$ and the Leibniz rule

$$\nabla_X(\lambda e) = (L_X \lambda) e + \lambda \nabla_X e$$

holds for all smooth functions $\lambda \in \mathscr{C}^\infty(M, \mathbb{K})$ and sections $e \in \Gamma(E)$.
We need the following wedge product.

Definition 5.12.2 There is a well-defined wedge product

$$\wedge: \Omega^k(M) \times \Omega^l(M, E) \longrightarrow \Omega^{k+l}(M, E)$$

between standard differential forms (with values in $\mathbb{K}$) and differential forms with values in E.
To explain this definition, we only have to see that the standard definition of the wedge product works in this case. The standard definition involves the sum over products of the two differential forms after we inserted a permutation of the vectors (cf. Definition A.2.5). In the standard case we get the product between two scalars in $\mathbb{K}$, while here we get the product between a scalar in $\mathbb{K}$ and a vector in E, which is still well defined.

To define the exterior covariant derivative, let ω be an element of $\Omega^k(M, E)$. We choose a local basis $e_1, \ldots e_r$ of E over an open set $U \subset M$. Then ω can be written as

$$\omega = \sum_{i=1}^{r} \omega_i \otimes e_i$$

with uniquely defined k-forms $\omega_i \in \Omega^k(U)$ (with values in $\mathbb{K}$).

Definition 5.12.3 Let ∇ be a covariant derivative on a vector bundle E. Then we define the **exterior covariant derivative** or **covariant differential**

$$d_\nabla: \Omega^k(M, E) \longrightarrow \Omega^{k+1}(M, E)$$

by

$$d_\nabla \omega = \sum_{i=1}^{r} \left(d\omega_i \otimes e_i + (-1)^k \omega_i \wedge \nabla e_i \right).$$

If $\nabla = \nabla^A$ is the covariant derivative on an associated vector bundle determined by a connection A on a principal bundle, we write $d_A = d_\nabla$.

Lemma 5.12.4 *The definition of d_∇ is independent of the choice of local basis $\{e_i\}$ for E.*

Proof Let $\{e_i'\}$ be another local basis of E over U. Then there exist unique functions $C_{ji} \in \mathscr{C}^\infty(U, \mathbb{K})$ with

$$e_j' = \sum_{i=1}^{r} C_{ji} e_i.$$

The matrix C with entries C_{ji} is invertible. Let C^{-1} be the inverse matrix with entries C_{lj}^{-1} and define

$$\omega_j' = \sum_{l=1}^{r} C_{lj}^{-1} \omega_l.$$

Then

$$\omega = \sum_{i=1}^{r} \omega_i \otimes e_i = \sum_{j=1}^{r} \omega_j' \otimes e_j'.$$

We calculate

$$\begin{aligned}
\sum_{j=1}^{r} (d\omega_j' \otimes e_j' + (-1)^k \omega_j' \wedge \nabla e_j') &= \sum_{i,j,l=1}^{r} \left(d\left(C_{lj}^{-1}\right) \wedge C_{ji}\omega_l \otimes e_i + C_{lj}^{-1} C_{ji} d\omega_l \otimes e_i \right.\\
&\quad \left. + (-1)^k C_{lj}^{-1} \omega_l \wedge dC_{ji} \otimes e_i + (-1)^k C_{lj}^{-1} C_{ji} \omega_l \wedge \nabla e_i\right)\\
&= \sum_{i=1}^{r} (d\omega_i \otimes e_i + (-1)^k \omega_i \wedge \nabla e_i)\\
&\quad + \sum_{i,j,l=1}^{r} \left(d\left(C_{lj}^{-1}\right) C_{ji} + C_{lj}^{-1} dC_{ji}\right) \wedge \omega_l \otimes e_i.
\end{aligned}$$

But

$$0 = d\delta_{li} = d\left(\sum_{j=1}^{r} C_{lj}^{-1} C_{ji}\right) = \sum_{j=1}^{r} \left(d\left(C_{lj}^{-1}\right) C_{ji} + C_{lj}^{-1} dC_{ji}\right).$$

This implies the claim. □

The first part of the next proposition follows immediately from the definition by considering a local basis $\{e_i\}$ for E. The second part is clear.

Proposition 5.12.5 *The exterior covariant derivative d_∇ satisfies*

$$d_\nabla(\omega + \omega') = d_\nabla \omega + d_\nabla \omega',$$

$$d_\nabla(\sigma \otimes e) = d\sigma \otimes e + (-1)^k \sigma \wedge \nabla e,$$

for all $\omega, \omega' \in \Omega^k(M, E)$, $\sigma \in \Omega^k(M)$ and $e \in \Gamma(E)$. Furthermore, we have on $\Gamma(E) = \Omega^0(M, E)$

$$d_\nabla|_{\Gamma(E)} = \nabla,$$

so that the exterior covariant derivative d_∇ *is an extension of the covariant derivative* ∇.

We want to show the following formula:

Proposition 5.12.6 (Leibniz Formula for Exterior Covariant Derivative) *The exterior covariant derivative* d_∇ *satisfies*

$$d_\nabla(\sigma \wedge \omega) = d\sigma \wedge \omega + (-1)^k \sigma \wedge d_\nabla \omega$$

for all $\sigma \in \Omega^k(M)$ *and* $\omega \in \Omega^l(M, E)$.

Note that this reduces in the case of a 0-form ω with values in E to the second formula in Proposition 5.12.5, because d_∇ on $\Omega^0(M, E) = \Gamma(E)$ is equal to ∇.

Proof We write

$$\omega = \sum_{i=1}^{r} \omega_i \otimes e_i$$

with a local basis $\{e_i\}$ of E over U and $\omega_i \in \Omega^l(U)$. Then

$$\sigma \wedge \omega = \sum_{i=1}^{r} (\sigma \wedge \omega_i) \otimes e_i$$

and

$$\begin{aligned} d_\nabla(\sigma \wedge \omega) &= \sum_{i=1}^{r} \left(d\sigma \wedge \omega_i \otimes e_i + (-1)^k \sigma \wedge d\omega_i \otimes e_i + (-1)^{k+l} \sigma \wedge \omega_i \wedge \nabla e_i \right) \\ &= d\sigma \wedge \omega + (-1)^k \sigma \wedge \sum_{i=1}^{r} \left(d\omega_i \otimes e_i + (-1)^l \omega_i \wedge \nabla e_i \right) \\ &= d\sigma \wedge \omega + (-1)^k \sigma \wedge d_A \omega. \end{aligned}$$

□

Remark 5.12.7 Contrary to the case of the standard exterior derivative d, it can be shown that d_∇ in general has square

$$d_\nabla \circ d_\nabla \neq 0.$$

The non-vanishing of d_∇^2 is related to the curvature F^∇ of the covariant derivative ∇ (see Exercise 5.15.12).

We finally want to derive a local formula for the exterior covariant derivative d_A in the case of an associated vector bundle. Let $P \to M$ be a principal G-bundle,

$\rho: G \to \mathrm{GL}(V)$ a representation and $E = P \times_\rho V$ the associated vector bundle. Let A be a connection 1-form on P.

Definition 5.12.8 We define the wedge product

$$\begin{aligned}\wedge: \Omega^k(M, \mathfrak{g}) \times \Omega^l(M, V) &\longrightarrow \Omega^{k+l}(M, V)\\ (\alpha, \omega) &\longmapsto \alpha \wedge \omega\end{aligned}$$

by expanding $\omega = \sum_{i=1}^n \omega_i \otimes v_i$ in an arbitrary basis $\{v_i\}$ for V and setting

$$\alpha \wedge \omega = \sum_{i=1}^n (\rho_*(\alpha) v_i) \wedge \omega_i.$$

This is independent of the choice of basis $\{v_i\}$ for V.

Let $s: U \to P$ be a local gauge. With respect to the local gauge a form $\sigma \in \Omega^l(M, E)$ defines a form $\sigma_s \in \Omega^l(M, V)$. We get:

Theorem 5.12.9 *With respect to a local gauge $s: U \to P$ we can write*

$$(d_A\omega)_s = d\omega_s + A_s \wedge \omega_s$$

for all $\omega \in \Omega^k(M, E)$.

Proof Choose a basis $v_1, \ldots, v_n$ for V. This determines a local frame $e_1, \ldots, e_n$ for E by setting $e_i = [s, v_i]$. If we expand a form $\sigma \in \Omega^l(M, E)$ as

$$\sigma = \sum_{i=1}^n \sigma_i \otimes e_i,$$

then

$$\sigma_s = \sum_{i=1}^n \sigma_i \otimes v_i.$$

We write

$$\omega = \sum_{i=1}^n \omega_i \otimes e_i$$

and calculate

$$d_A\omega = \sum_{i=1}^n \left(d\omega_i \otimes e_i + (-1)^k \omega_i \wedge \nabla^A e_i\right),$$

which implies

$$\begin{aligned}(d_A\omega)_s &= \sum_{i=1}^{n} \left(d\omega_i \otimes v_i + (-1)^k \omega_i \wedge (\rho_*(A_s)v_i)\right)\\ &= d\omega_s + A_s \wedge \omega_s.\end{aligned}$$

□

5.13 Forms with Values in Ad(P)

Recall that connections are 1-forms on the total space of a principal bundle P with values in the Lie algebra $\mathfrak{g}$. We now want to show that the difference between two connections can be understood as a field on the *base manifold* M with values in the *vector bundle* $\mathrm{Ad}(P)$. We then get a better understanding of why gauge bosons in physics are said to transform under the adjoint representation.

Let $\pi_P: P \to M$ be a principal G-bundle. We then have the vector space $\Omega^k(P, \mathfrak{g})$ of k-forms on P with values in the Lie algebra $\mathfrak{g}$. We want to consider a certain vector subspace of this vector space (we follow [14, Chap. 3]).

Definition 5.13.1 Let $\omega \in \Omega^k(P, \mathfrak{g})$ be a k-form on P with values in the Lie algebra $\mathfrak{g}$. We call ω

1. **horizontal** if for all $p \in P$

$$\omega_p(X_1, \ldots, X_k) = 0$$

whenever at least one of the vectors $X_i \in T_pP$ is vertical.

2. **of type** Ad if

$$r_g^*\omega = \mathrm{Ad}_{g^{-1}} \circ \omega$$

for all $g \in G$.

We denote the set of horizontal k-forms of type Ad on P with values in $\mathfrak{g}$ by

$$\Omega^k_{\mathrm{hor}}(P, \mathfrak{g})^{\mathrm{Ad}},$$

which is clearly a real vector space (usually infinite-dimensional).

This notion is useful for the following reason.

Proposition 5.13.2 *Let $P \to M$ be a principal G-bundle.*

1. Suppose $A, A' \in \Omega^1(P, \mathfrak{g})$ are connection 1-forms on P. Then

$$A' - A \in \Omega^1_{\mathrm{hor}}(P, \mathfrak{g})^{\mathrm{Ad}(P)}.$$

Moreover, if ω is an arbitrary element in $\Omega^1_{hor}(P, \mathfrak{g})^{\mathrm{Ad}(P)}$, then $A + \omega$ is a connection on P.

2. *The curvature F of a connection A on P is an element of $\Omega^2_{hor}(P, \mathfrak{g})^{\mathrm{Ad}}$.*

Proof This follows immediately from the defining properties of connections and the curvature. □

Corollary 5.13.3 *The set of connection 1-forms on P is an affine space over the vector space $\Omega^1_{hor}(P, \mathfrak{g})^{\mathrm{Ad}(P)}$. A base point is given by any connection 1-form on P.*
It is sometimes useful to have a different description of the vector space of horizontal k-forms of type Ad on P. Recall that we defined in Example 4.7.17 the adjoint bundle

$$\mathrm{Ad}(P) = P \times_{\mathrm{Ad}} \mathfrak{g},$$

which is the real vector bundle associated to the principal bundle P via the adjoint representation $\mathrm{Ad}\colon G \to \mathrm{GL}(\mathfrak{g})$.

Theorem 5.13.4 *The vector space $\Omega^k_{hor}(P, \mathfrak{g})^{\mathrm{Ad}}$ is canonically isomorphic to the vector space $\Omega^k(M, \mathrm{Ad}(P))$.*

Proof We define a map

$$\Lambda\colon \Omega^k_{hor}(P, \mathfrak{g})^{\mathrm{Ad}} \longrightarrow \Omega^k(M, \mathrm{Ad}(P))$$

as follows: Let $\bar{\omega}$ be an element of $\Omega^k_{hor}(P, \mathfrak{g})^{\mathrm{Ad}}$. Then we define $\omega = \Lambda(\bar{\omega})$ by

$$\omega_x(X_1, \ldots, X_k) = [p, \bar{\omega}_p(Y_1, \ldots, Y_k)] \in \mathrm{Ad}(P)_x = (P_x \times \mathfrak{g})/G,$$

where

- $x \in M$ and $p \in P$ are arbitrary with $\pi_P(p) = x$.
- $X_i \in T_xM$ and $Y_i \in T_pP$ are arbitrary with $\pi_{P*}(Y_i) = X_i$.

We first show that ω is well-defined. For fixed $p \in P$ the definition is independent of the choice of vectors Y_i: If Y_i' are a different set of vectors with $\pi_{P*}(Y_i') = X_i$, then

$$\pi_{P*}\left(Y_i' - Y_i\right) = 0,$$

hence $Y_i' - Y_i$ is vertical. Since $\bar{\omega}$ is horizontal, we get

$$\begin{aligned}\bar{\omega}_p\left(Y_1', \ldots, Y_k'\right) &= \bar{\omega}_p\left(Y_1 + \left(Y_1' - Y_1\right), \ldots, Y_k + \left(Y_k' - Y_k\right)\right)\\ &= \bar{\omega}_p\left(Y_1, \ldots, Y_k\right).\end{aligned}$$

We now show independence of the choice of p in the fibre P_x: Let p' be another point in P with $\pi_P(p') = x$. Then $p' = p \cdot g^{-1}$ for some $g \in G$. Let $Y_1, \ldots, Y_k$ be vectors in $T_{p'}P$. We calculate

$$\begin{aligned}
[p', \bar{\omega}_{p'}(Y_1, \dots, Y_k)] &= [p \cdot g^{-1}, \bar{\omega}_{p \cdot g^{-1}}(Y_1, \dots, Y_k)] \\
&= [p, \mathrm{Ad}_{g^{-1}} \bar{\omega}_{p \cdot g^{-1}}(Y_1, \dots, Y_k)] \\
&= [p, (r_g^* \bar{\omega})_{p \cdot g^{-1}}(Y_1, \dots, Y_k)] \\
&= [p, \bar{\omega}_p(r_{g*} Y_1, \dots, r_{g*} Y_k)] \\
&= [p, \bar{\omega}_p(Z_1, \dots, Z_k)],
\end{aligned}$$

where we set $Z_i = r_{g*} Y_i$. We have

$$\pi_{P*}(Z_i) = (\pi_P \circ r_g)_*(Y_i) = \pi_{P*}(Y_i).$$

This proves independence of the choice of p.

We prove that the form ω is smooth: Let $s: U \to P$ be a local gauge and $X_1, \dots, X_k$ smooth vector fields on U. Then

$$\omega(X_1, \dots, X_k)|_U = [s, \bar{\omega}(s_* X_1, \dots, s_* X_k)] \in \Gamma(U, \mathrm{Ad}(P)).$$

Hence $\omega \in \Omega^k(M, \mathrm{Ad}(P))$.

This shows that the map Λ is well-defined and it is clearly linear. It remains to show that Λ is bijective: Let $\omega \in \Omega^k(M, \mathrm{Ad}(P))$ and define $\bar{\omega} \in \Omega^k(P, \mathfrak{g})$ by

$$[p, \bar{\omega}_p(Y_1, \dots, Y_k)] = \omega_x(\pi_{P*} Y_1, \dots, \pi_{P*} Y_k).$$

Then $\bar{\omega} \in \Omega^k_{\mathrm{hor}}(P, \mathfrak{g})^{\mathrm{Ad}}$ and $\Lambda(\bar{\omega}) = \omega$. □

As a consequence, we get the following statement about connection 1-forms and curvature 2-forms.

Corollary 5.13.5 (Connections, Curvature and Forms with Values in $\mathrm{Ad}(P)$**)** *Let $P \to M$ be a principal G-bundle.*

1. *The difference of two connection* 1*-forms on P can be identified with an element of $\Omega^1(M, \mathrm{Ad}(P))$. The set of all connections on P is an affine space over $\Omega^1(M, \mathrm{Ad}(P))$.*
2. *The curvature F^A of a connection A on P can be identified with an element F^A_M of $\Omega^2(M, \mathrm{Ad}(P))$.*

The notation F^A_M generalizes the notation in Corollary 5.6.4, because for an abelian structure group G the adjoint bundle $\mathrm{Ad}(P)$ is trivial and F^A_M has values in $\mathfrak{g}$.

In quantum field theory, particles in general are described as excitations of a given vacuum field. In the case of a gauge field the vacuum field is a certain specific connection 1-form A^0 on the principal bundle (the form $A^0 \equiv 0$ is not a connection). Strictly speaking, gauge bosons, the excitations of the gauge field, should then be described classically by the difference $A - A^0$, where A is some other connection 1-form, and not by the field A itself. By Corollary 5.13.5 this difference can be identified with a 1-form on spacetime M with values in $\mathrm{Ad}(P)$. In physics this fact is expressed by saying that gauge bosons, the differences $A_\mu - A^0_\mu$, are fields on spacetime that transform in the adjoint representation of G under gauge transformations.

5.14 *A Second and Third Version of the Bianchi Identity

Let $P \to M$ be a principal G-bundle and A a connection 1-form with curvature F^A. According to Exercise 5.15.14 we can state the Bianchi identity in the following equivalent form:

Theorem 5.14.1 (Bianchi Identity (Second Form)) *The curvature $F^A \in \Omega^2(P, \mathfrak{g})$ satisfies*

$$dF^A + [A, F^A] = 0$$

for any connection A on P.

In Sect. 5.13 we saw that the curvature F^A can be identified with an element F^A_M in $\Omega^2(M, \mathrm{Ad}(P))$. On the other hand the connection A defines an exterior covariant derivative d_A on the associated bundle $\mathrm{Ad}(P)$:

$$d_A \colon \Omega^k(M, \mathrm{Ad}(P)) \longrightarrow \Omega^{k+1}(M, \mathrm{Ad}(P)).$$

We can then write the Bianchi identity in a third equivalent form:

Theorem 5.14.2 (Bianchi Identity (Third Form)) *The curvature $F^A_M \in \Omega^2(M, \mathrm{Ad}(P))$ satisfies*

$$d_A F^A_M = 0$$

for any connection A on P.

Proof This is an immediate consequence of Theorem 5.12.9 and Theorem 5.14.1. □

5.15 Exercises for Chap. 5

5.15.1 Let G be a compact Lie group.

1. Suppose $P \times G \to P$ is a right-action of G on a manifold P. Prove that P has a G-invariant Riemannian metric.
2. Prove that every principal G-bundle $\pi: P \to M$ has a connection.

5.15.2 Suppose that $\pi: P \to M$ is a principal G-bundle where the Lie group G is abelian. Show that the following map is a group isomorphism

$$\mathscr{C}^\infty(M, G) \longrightarrow \mathscr{C}^\infty(P, G)^G$$
$$\tau \longmapsto \sigma_\tau,$$

where $\mathscr{C}^\infty(M, G)$ denotes the set of smooth maps from M to G (a group under pointwise multiplication) and σ_τ is defined by

$$\sigma_\tau = \tau \circ \pi.$$

5.15.3 Let $P \to M$ be a principal bundle and $A \in \Omega^1(P, \mathfrak{g})$ a connection 1-form on P. Suppose that $f \in \mathscr{G}(P)$ is a global bundle automorphism. Prove that f^*A is a connection 1-form on P and

$$f^*A = \mathrm{Ad}_{\sigma_f^{-1}} \circ A + \sigma_f^* \mu_G.$$

5.15.4 Let $P \to M$ be a principal G-bundle with a connection 1-form $A \in \Omega^1(P, \mathfrak{g})$ and curvature 2-form $F \in \Omega^2(P, \mathfrak{g})$. Let X and Y be horizontal vector fields on P with respect to the Ehresmann connection H defined by A.

1. Show that $F(X, Y) = -A([X, Y])$.
2. Prove that the curvature F vanishes identically if and only if the distribution H is integrable, i.e. $[X, Y]$ is a horizontal vector field for all horizontal vector fields X, Y on P.
3. Suppose that M is connected and simply connected ($\pi_1(M) = 1$) and the curvature F vanishes identically. Prove that P is trivial and there exists a global gauge $s: M \to P$ such that $A_s = s^*A \equiv 0$.

5.15.5 Let G be a Lie group. Then G acts by right-multiplication on the right of G:

$$G \times G \longrightarrow G$$
$$(p, g) \longmapsto pg.$$

Since the action is simply transitive, it follows that this defines a principal G-bundle over the manifold consisting of one point:

$$\begin{array}{ccc} G & \longrightarrow & G \\ & & \downarrow \pi \\ & & \{*\} \end{array}$$

1. Show that the Maurer–Cartan form $\mu_G \in \Omega^1(G, \mathfrak{g})$ is a connection on this principal bundle and that it is the only one.
2. Determine the curvature of the connection μ_G. What is the interpretation of the structure equation?

5.15.6 Let G be a Lie group and $H \subset G$ a closed subgroup. By Theorem 4.2.15

$$\begin{array}{ccc} H & \longrightarrow & G \\ & & \downarrow \pi \\ & & G/H \end{array}$$

is an H-principal bundle. We assume that there exists a vector subspace $\mathfrak{m} \subset \mathfrak{g}$ such that

$$\mathfrak{g} = \mathfrak{h} \oplus \mathfrak{m}, \quad \mathrm{Ad}(H)\mathfrak{m} \subset \mathfrak{m},$$

i.e. the homogeneous space is reductive.

1. Consider $A = \mathrm{pr}_{\mathfrak{h}} \circ \mu_G \in \Omega^1(G, \mathfrak{h})$. Prove that A is a connection 1-form on $G \to G/H$.
2. Show that the vertical and horizontal subspaces (defined by the connection A) at a point $g \in G$ are given by $L_{g*}\mathfrak{h}$ and $L_{g*}\mathfrak{m}$.
3. Prove that the curvature of the connection A is given by

$$F = -\frac{1}{2}\mathrm{pr}_{\mathfrak{h}} \circ [\mathrm{pr}_{\mathfrak{m}} \circ \mu_G, \mathrm{pr}_{\mathfrak{m}} \circ \mu_G] \in \Omega^2(G, \mathfrak{h})$$

(the commutator on the right is taken in $\mathfrak{g}$).

5.15.7 Recall from Exercise 4.8.7 that the Hopf right action of S^1 on S^3 induces a right action of $S^1/\mathbb{Z}_p \cong S^1$ on $S^3/\mathbb{Z}_p = L(p,1)$ and the lens space $L(p,1)$ thus has the structure of a principal circle bundle over S^2:

$$\begin{array}{ccc} S^1 & \longrightarrow & L(p,1) \\ & & \downarrow \pi \\ & & S^2 \end{array}$$

Prove that the connection A on the Hopf bundle defined in Sect. 5.2.2 induces a connection A' on $L(p, 1) \to S^2$. Determine the relation between the global curvature 2-form F'_{S^2} of this connection and the curvature 2-form F_{S^2} of the Hopf connection, as well as the integral

$$\frac{1}{2\pi i} \int_{S^2} F'_{S^2}.$$

5.15.8 Suppose that $P \to M$ is a principal G-bundle with a matrix Lie group G, E an associated vector bundle, A a connection on P and γ a loop in M. Suppose that the loop γ is inside an open set $U \subset M$ over which P is trivial and $s\colon U \to P$ is a local gauge. Then s determines an isomorphism

$$\begin{aligned} V &\longrightarrow E_x \\ v &\longmapsto [s(x), v]. \end{aligned}$$

Prove that with respect to this isomorphism

$$\mathrm{Hol}^E_{\gamma,x}(A) = \rho\left(\mathscr{P}\exp\left(-\oint_\gamma A_s\right)\right) = \mathscr{P}\exp\left(-\oint_\gamma \rho_* A_s\right).$$

5.15.9 Let $P \to M$ be a principal G-bundle with a connection A and $f \in \mathscr{G}$ a bundle automorphism. Suppose $\gamma\colon [0, 1] \to M$ is a curve in M.

1. Show that parallel transport with respect to the connection f^*A is given by

$$\Pi^{f^*A}_\gamma = f^{-1} \circ \Pi^A_\gamma \circ f.$$

2. Let $E \to M$ be a vector bundle associated to P and suppose that γ is a closed curve in M. Show that the Wilson loop is invariant under bundle automorphisms of P:

$$W^E_\gamma(f^*A) = W^E_\gamma(A) \quad \forall f \in \mathscr{G}(P).$$

5.15.10 Let

$$\begin{array}{ccc} S^1 & \longrightarrow & S^3 \\ & & \downarrow \\ & & \mathbb{CP}^1 \end{array}$$

be the Hopf bundle with the connection A defined in Sect. 5.2.2. Let σ denote the equator in $S^2 \cong \mathbb{CP}^1$, starting and ending at the point $[1:1] \in \mathbb{CP}^1$.

1. Show that σ can be parametrized as

$$\begin{aligned}\sigma\colon [0,2\pi] &\longrightarrow \mathbb{CP}^1\\ t &\longmapsto [1:e^{it}]\end{aligned}$$

for a suitable identification of S^2 with $\mathbb{CP}^1$.
2. Determine the horizontal lift $\sigma^*\colon [0,2\pi] \to S^3$ of σ with respect to the connection A, starting at $\frac{1}{\sqrt{2}}(1,1)$.
3. Let $\gamma^k \to S^2$ be the complex line bundle associated to the Hopf bundle via the representation

$$\begin{aligned}\rho_k\colon S^1 &\longrightarrow \mathrm{U}(1)\\ z &\longmapsto z^k\end{aligned}$$

as in Example 4.7.16. Determine the Wilson loop $W_\sigma^{\gamma^k}(A)$.

5.15.11 Define the differential

$$d\colon \Omega^k(M) \longrightarrow \Omega^{k+1}(M)$$

by demanding that

- $d\colon \mathscr{C}^\infty(M) \to \Omega^1(M)$ is the standard differential of functions
- $ddf = 0$ for all $f \in \mathscr{C}^\infty(M)$
- $d(\alpha \wedge \beta) = d\alpha \wedge \beta + (-1)^k \alpha \wedge d\beta$ for all $\alpha \in \Omega^k(M)$ and $\beta \in \Omega^l(M)$.

Prove that $dd\omega = 0$ for all $\omega \in \Omega^k(M)$ and all k.

5.15.12 Let $E \to M$ be a vector bundle with a covariant derivative ∇. We define the curvature of ∇ by

$$F^\nabla(X,Y)\Phi = \nabla_X\nabla_Y\Phi - \nabla_Y\nabla_X\Phi - \nabla_{[X,Y]}\Phi,$$

where $X, Y \in \mathfrak{X}(M)$ and $\Phi \in \Gamma(E)$.

1. Show that $F^\nabla(X,Y)\Phi$ is function linear in each argument X, Y, Φ. The curvature thus defines an element

$$F^\nabla \in \Omega^2(M, \mathrm{End}(E)),$$

where $\mathrm{End}(E)$ denotes the endomorphism bundle of E over M, whose fibre $\mathrm{End}(E)_x$ over $x \in M$ is given by $\mathrm{End}(E_x)$.
2. Show that, as 2-forms with value in E,

$$d_\nabla d_\nabla \Phi = F^\nabla \Phi$$

for all sections $\Phi \in \Gamma(E)$.

3. Define a wedge product

$$\wedge: \Omega^k(M, \mathrm{End}(E)) \times \Omega^l(M, E) \longrightarrow \Omega^{k+l}(M, E)$$

by writing $\omega \in \Omega^l(M, E)$ with a local frame $\{e_i\}$ for E over U as $\omega = \sum_{i=1}^n \omega_i \otimes e_i$ with $\omega_i \in \Omega^l(U)$ and setting for $\alpha \in \Omega^k(M, \mathrm{End}(E))$

$$\alpha \wedge \omega = \sum_{i=1}^n \alpha(e_i) \wedge \omega_i$$

(this definition is independent of choices). Prove that

$$d_\nabla d_\nabla \omega = F^\nabla \wedge \omega$$

for all $\omega \in \Omega^l(M, E)$.

5.15.13 Let $P \to M$ be a principal G-bundle and $E = P \times_\rho V \to M$ an associated vector bundle. We fix a connection A on P with induced covariant derivative ∇^A on E. Let F^∇ be the curvature of ∇^A defined in Exercise 5.15.12. Suppose that $s: U \to P$ is a local gauge and write a section Φ of E locally as $\Phi = [s, \phi]$ with $v: U \to V$. Show that

$$F^\nabla(X, Y)\Phi = [s, \rho_*(F_s(X, Y))\phi],$$

where $F_s = s^*F$ and F is the curvature of A on P.

5.15.14 Suppose that P is a manifold and $\mathfrak{g}$ a Lie algebra. Consider forms $\phi \in \Omega^1(P, \mathfrak{g})$, $\omega \in \Omega^k(P, \mathfrak{g})$ and $\tau \in \Omega^l(P, \mathfrak{g})$. Prove the following identities:

$$[\omega, \tau] = -(-1)^{kl}[\tau, \omega],$$

$$[\phi, [\phi, \phi]] = 0,$$

$$d[\omega, \tau] = [d\omega, \tau] + (-1)^k[\omega, d\tau].$$

Derive as an application the following **second form of the Bianchi identity**

$$dF^A + \left[A, F^A\right] = 0,$$

where F^A is the curvature of a connection 1-form A on a principal bundle $P \to M$.

5.15.15 This exercise is a preparation for Exercise 5.15.16. Suppose that P is a manifold and $\mathfrak{g}$ a Lie algebra with a scalar product $\langle \cdot, \cdot \rangle$ which is skew-symmetric with respect to the adjoint representation ad. Let $\omega \in \Omega^k(P, \mathfrak{g})$ and $\tau \in \Omega^l(P, \mathfrak{g})$. We define a real-valued form $\langle \omega, \tau \rangle \in \Omega^{k+l}(P)$ by

$$\langle \omega, \tau \rangle (X_1, \ldots, X_{k+l}) = \frac{1}{k!l!} \sum_{\sigma \in S_{k+l}} \operatorname{sgn}(\sigma) \langle \eta(X_{\sigma(1)}, \ldots, X_{\sigma(k)}), \phi(X_{\sigma(k+1)}, \ldots, X_{\sigma(n)}) \rangle.$$

Consider also a 1-form $\phi \in \Omega^1(P, \mathfrak{g})$. Prove the following identities:

$$d\langle \omega, \tau \rangle = \langle d\omega, \tau \rangle + (-1)^k \langle \omega, d\tau \rangle$$
$$\langle [\phi, \omega], \tau \rangle = -(-1)^k \langle \omega, [\phi, \tau] \rangle.$$

5.15.16 (From [52]) We use the notation from Exercise 5.15.15. Suppose that $P \to M$ is a principal G-bundle over a manifold M and $\langle \cdot, \cdot \rangle$ an Ad-invariant scalar product on the Lie algebra $\mathfrak{g}$. Let A be a connection 1-form on P with curvature F. We define the **Chern–Simons form** $\alpha(A) \in \Omega^3(P)$ by

$$\alpha(A) = \alpha = \langle A, F \rangle - \frac{1}{6} \langle A, [A, A] \rangle.$$

1. Prove that $d\alpha = \langle F, F \rangle$.
2. Let $f \in \mathcal{G}(P)$ be a bundle automorphism with induced map $\sigma_f : P \to G$ and set $\phi = \sigma_f^* \mu_G$. Prove that the Chern–Simons form changes under bundle automorphisms as

$$f^* \alpha = \alpha + d \langle \mathrm{Ad}_{\sigma_f^{-1}} A, \phi \rangle - \frac{1}{6} \langle \phi, [\phi, \phi] \rangle.$$

3. Show that the form

$$\tau_G = -\frac{1}{6} \langle \mu_G, [\mu_G, \mu_G] \rangle \in \Omega^3(G)$$

is closed.
4. Suppose that M is a closed oriented 3-manifold, $P \to M$ a trivial G-bundle and τ_G represents an integral class in $H^3(G; \mathbb{R})$. If $s: M \to P$ is a global gauge, define the **Chern–Simons action** by

$$S_M(s, A) = \int_M s^* \alpha(A).$$

Prove that modulo $\mathbb{Z}$ the number $S_M(A) = S_M(s, A) \in \mathbb{R}/\mathbb{Z}$ is independent of the choice of global gauge s.

Remark Notice that the Chern–Simons action is purely *topological*, i.e. does not depend on the choice of a metric on M. This leads to the concept of *topological quantum field theories (TQFT)*. Similar Chern–Simons terms appear in many places in physics, for example, in the actions of supergravity and in the actions for D-branes in string theory.

Chapter 6
Spinors

In Chap. 5 we studied gauge fields that mediate interactions between particles in gauge theories. Gauge fields correspond to gauge bosons (spin 1 particles) and are described by 1-forms or, dually, vector fields. In physics, of course, there also exist matter particles, like electrons, quarks and neutrinos. These particles are fermions (spin $\frac{1}{2}$ particles) and are described by *spinor fields (spinors)*. Like vector fields or tensor fields, spinors have a specific transformation behaviour under rotations. However, spinors do not transform directly under the orthogonal group, but under a certain double covering, called the *(orthochronous) spin group*. In the case of Minkowski spacetime, rotations correspond to *Lorentz transformations*. The corresponding spin group is the Lorentz spin group.

In many mathematical expositions the discussion of spinors is restricted to the Riemannian case, because in most situations, manifolds in differential geometry carry a Riemannian metric. The pseudo-Riemannian case, like the case of Minkowski spacetime, is discussed less often, even though it is very important for physics (a notable exception is the thorough discussion in Helga Baum's book [13]). Since we are ultimately interested in applications of differential geometry and gauge theory to physics, it seems worthwhile to study orthogonal groups, Clifford algebras, spin groups and spinors from a mathematical point of view also in the Lorentzian and general pseudo-Riemannian case.

The discussion in the present chapter is far more extensive than strictly necessary to understand particle physics and the Standard Model in 4-dimensional flat spacetime, but may be useful for further studies leading to theories such as supersymmetry, supergravity and superstrings in higher dimensions. We sometimes only sketch the arguments that are more or less standard and can be found in other references, and focus instead on topics which are perhaps less well-known in the mathematics literature, like Majorana spinors and scalar products. Even though spinors are elementary objects, some of their properties (like the periodicity modulo 8, real and quaternionic structures, or bilinear and Hermitian scalar products) are

M.J.D. Hamilton, *Mathematical Gauge Theory*, Universitext,
https://doi.org/10.1007/978-3-319-68439-0_6

not at all obvious, already on the level of linear algebra, and do not have a direct analogue in the bosonic world of vectors and tensors.

General references for this chapter are [13, 15, 20, 28, 35, 55, 88] and [93].

6.1 The Pseudo-Orthogonal Group $O(s,t)$ of Indefinite Scalar Products

We begin by describing the structure of the pseudo-orthogonal group for a general indefinite scalar product. This includes the particular case of the Lorentz group of Minkowski spacetime. In Sect. 6.5 we will then introduce the spin group, which is a certain double covering of a (pseudo-)orthogonal group.

In this chapter $\mathbb{K}$ denotes the field $\mathbb{R}$ or $\mathbb{C}$. Let V be a finite-dimensional $\mathbb{K}$-vector space.

Definition 6.1.1 A **symmetric bilinear form** is a map

$$Q: V \times V \longrightarrow \mathbb{K}$$

which is symmetric and $\mathbb{K}$-bilinear. The form Q is called **non-degenerate** if for each $v \neq 0 \in V$ there exists a vector $w \in V$ with $Q(v,w) \neq 0$.
Notice that in the complex case we also consider*complex bilinear* and not Hermitian forms.

Example 6.1.2 We denote by $\mathbb{R}^{s,t}$ the vector space $\mathbb{R}^{s+t}$ with standard basis $e_1, \ldots, e_{s+t}$ and the **standard** symmetric bilinear form η defined by

$$\begin{aligned} \eta(e_i, e_i) &= +1 \quad \forall 1 \leq i \leq s, \\ \eta(e_i, e_i) &= -1 \quad \forall s+1 \leq i \leq s+t, \\ \eta(e_i, e_j) &= 0 \quad \forall i \neq j. \end{aligned}$$

The bilinear form η has **signature** (s,t), also written as

$$(\underbrace{+,\ldots,+}_{s},\underbrace{-,\ldots,-}_{t}),$$

and is non-degenerate. We set $n = s + t$.

The space

- $\mathbb{R}^s = \mathbb{R}^{s,0}$ is **Euclidean space**, and
- $\mathbb{R}^{s,1}$ and $\mathbb{R}^{1,t}$ are the two versions of **Minkowski spacetime** (both versions are used in physics).

In general, if both s and t are non-zero, we call η a **pseudo-** or **semi-Euclidean** scalar product.

A set of vectors $v_1, \ldots, v_s, w_1, \ldots, w_t$ in $\mathbb{R}^{s,t}$ with

$$\eta(v_i.v_j) = \delta_{ij},$$
$$\eta(w_i, w_j) = -\delta_{ij},$$
$$\eta(v_i, w_j) = 0$$

is called an **orthonormal basis**. Occasionally we write for $v \in \mathbb{R}^{s,t}$

$$||v||_\eta^2 = \eta(v, v),$$

called the (η)**-norm squared**. We usually try to avoid this notation, because it suggests that this norm is non-negative, which may not be the case.

Example 6.1.3 On the vector space $\mathbb{C}^d$ we consider the non-degenerate **standard** symmetric complex bilinear form q, given in the standard basis $e_1, \ldots, e_d$ by

$$q(e_i, e_i) = +1 \quad \forall 1 \leq i \leq d,$$
$$q(e_i, e_j) = 0 \quad \forall i \neq j.$$

The following is a well-known result from linear algebra.

Proposition 6.1.4 *Every non-degenerate symmetric bilinear form Q on an $\mathbb{R}$- or $\mathbb{C}$-vector space V is isomorphic to precisely one of these examples. In particular, every non-degenerate symmetric bilinear form on an $\mathbb{R}$-vector space V has a well-defined signature (s, t).*

Remark 6.1.5 Since we can multiply a vector in a $\mathbb{C}$-vector space by i and the form Q is complex bilinear, we can change $Q(v, v) = +1$ to $Q(iv, iv) = -1$. This explains why there is no signature for symmetric bilinear forms on complex vector spaces.

Definition 6.1.6 Let V be a $\mathbb{K}$-vector space with a non-degenerate symmetric bilinear form Q. Then the **(pseudo-)orthogonal group** of (V, Q) is defined as the automorphism group of Q:

$$\mathrm{O}(V, Q) = \{f \in \mathrm{GL}(V) \mid Q(fv, fw) = Q(v, w) \quad \forall v, w \in V\}.$$

In the following we will only consider the case of the standard symmetric bilinear form η on $\mathbb{R}^{s+t}$ of signature (s, t), where we write

$$\mathrm{O}(s, t) = \left\{A \in \mathrm{GL}(s + t, \mathbb{R}) \mid \eta(Av, Aw) = \eta(v, w) \quad \forall v, w \in \mathbb{R}^{s+t}\right\}.$$

The groups $\mathrm{O}(1, t)$ and $\mathrm{O}(s, 1)$ are called **Lorentz groups**.

If we write a matrix $A \in \mathrm{Mat}((s+t) \times (s+t), \mathbb{R})$ in the block form

$$A = \begin{pmatrix} A_{11} & A_{12} \\ A_{21} & A_{22} \end{pmatrix} \tag{6.1}$$

with

$$\begin{aligned} A_{11} &\in \mathrm{Mat}(s \times s, \mathbb{R}), \\ A_{12} &\in \mathrm{Mat}(s \times t, \mathbb{R}), \\ A_{21} &\in \mathrm{Mat}(t \times s, \mathbb{R}), \\ A_{22} &\in \mathrm{Mat}(t \times t, \mathbb{R}), \end{aligned}$$

then $A \in \mathrm{O}(s,t)$ if and only if

$$A^T \begin{pmatrix} I_s & 0 \\ 0 & -I_t \end{pmatrix} A = \begin{pmatrix} I_s & 0 \\ 0 & -I_t \end{pmatrix}.$$

Taking the determinant on both sides of this equation shows that:

Lemma 6.1.7 *Matrices* $A \in \mathrm{O}(s,t)$ *satisfy* $\det A = \pm 1$.
If we write a matrix $A \in \mathrm{Mat}((s+t) \times (s+t), \mathbb{R})$ in the form

$$A = (v_1, \ldots, v_s, w_1, \ldots, w_t)$$

with $v_i, w_j \in \mathbb{R}^{s+t}$, then $A \in \mathrm{O}(s,t)$ if and only if

$$\begin{aligned} \eta(v_i, v_j) &= \delta_{ij}, \\ \eta(w_i, w_j) &= -\delta_{ij}, \\ \eta(v_i, w_j) &= 0, \end{aligned}$$

i.e. the vectors $v_1, \ldots, v_s, w_1, \ldots, w_t$ form an orthonormal basis for $\mathbb{R}^{s,t}$, generalizing a well-known property of orthogonal matrices.

Lemma 6.1.8 *There exists a canonical isomorphism* $\mathrm{O}(s,t) \cong \mathrm{O}(t,s)$.

Proof This is Exercise 6.13.1. □
We note the following facts:

Proposition 6.1.9 (Properties of Pseudo-Orthogonal Groups)

1. *The group* $\mathrm{O}(s,t)$ *is a linear Lie group.*
2. *If both* $s, t \neq 0$, *then* $\mathrm{O}(s,t)$ *is not compact.*
3. *Let* $\mathfrak{o}(s,t)$ *denote the Lie algebra of* $\mathrm{O}(s,t)$. *Then as complex Lie algebras*

$$\mathfrak{o}(s,t) \otimes_{\mathbb{R}} \mathbb{C} \cong \mathfrak{o}(s+t) \otimes_{\mathbb{R}} \mathbb{C}.$$

In particular,

$$\dim \mathrm{O}(s,t) = \dim \mathrm{O}(s+t) = \frac{1}{2}(s+t)(s+t-1).$$

Proof

1. A similar argument to the one in Theorem 1.2.17 shows that the group $\mathrm{O}(s,t)$ is a closed subgroup of $\mathrm{GL}(s+t,\mathbb{R})$, hence a linear Lie group by Cartan's Theorem 1.1.44.
2. This is Exercise 6.13.2.
3. This follows for a reason similar to the one explained in Remark 6.1.5.

□

It is also clear that $\mathrm{O}(0,n) \cong \mathrm{O}(n,0) = \mathrm{O}(n)$, the standard orthogonal group. This group has two connected components and the connected component of the identity is $\mathrm{SO}(n)$, the group of orthogonal matrices of determinant $+1$.

We would like to determine the number of connected components of the pseudo-orthogonal group $\mathrm{O}(s,t)$ for general values of s and t. Let V_+ and V_- be the vector subspaces of $V = \mathbb{R}^{s,t}$ defined by

$$V_+ = \mathrm{span}\{e_1,\ldots,e_s\},$$
$$V_- = \mathrm{span}\{e_{s+1},\ldots,e_{s+t}\}.$$

Then η is positive definite on V_+ and negative definite on V_-. Let $\pi\colon V \to V_+$ denote the projection along V_-.

Suppose that $W \subset V$ is any maximally η-positive definite vector subspace of dimension s. Then

$$\pi|_W\colon W \longrightarrow V_+$$

is an isomorphism, because any non-zero element in the kernel must have negative η-norm squared and hence cannot be an element of W.

Definition 6.1.10 Fix an orientation on the vector subspace V_+. Then there exists for every maximally positive definite vector subspace $W \subset V$ a unique orientation so that the isomorphism $\pi|_W$ is orientation preserving.
Suppose $A \in \mathrm{O}(s,t)$. Since A preserves η, it maps (maximally) positive definite subspaces of V to (maximally) positive definite subspaces.

Definition 6.1.11 We define the **time-orientability** of $A \in \mathrm{O}(s,t)$ to be $+1$ or -1 depending on whether

$$A|_{V_+}\colon V_+ \longrightarrow A(V_+)$$

preserves or does not preserve orientation, where the orientation on $A(V_+)$ is chosen via the projection π as above.

Remark 6.1.12 It might seem more natural to call this notion *space-orientability*, because often a positive definite subspace of V is called spacelike. We could then define time-orientability as the corresponding notion with V_+ replaced by V_-. However, as mentioned above, depending on the convention, 4-dimensional Minkowski spacetime in quantum field theory can have signature $(+,-,-,-)$, so that time caries the plus sign. Furthermore, it follows from Lemma 6.1.18 that if $\det A = 1$ (the only situation in which we are going to consider time-orientability) time-orientability and space-orientability are equivalent. Since the term time-orientability is much more common in physics, we will continue to use it.

Lemma 6.1.13 *Let $A \in \mathrm{O}(s,t)$ and $W \subset V$ an arbitrary maximally positive definite subspace. Then A has time-orientability $+1$ if and only if*

$$A|_W : W \longrightarrow A(W)$$

preserves orientation, with orientation on W and $A(W)$ chosen via the projection π as above.

Proof The proof is a deformation argument. Suppose that $A \in \mathrm{O}(s,t)$ and $W \subset V$ is a maximally positive definite subspace. Since $\pi|_W : W \to V_+$ is an isomorphism, we can find a unique basis $w_1, \ldots, w_s$ for W of the form

$$w_i = e_i + v_i^-,$$

where v_i^- are elements of V_-. The fact that W is positive definite means that

$$\sum_{i=1}^{s} a_i^2 + \langle \alpha, \alpha \rangle > 0 \quad \forall (a_1, \ldots, a_s) \in \mathbb{R}^s \setminus \{0\},$$

where

$$\alpha = \sum_{i=1}^{s} a_i v_i^-.$$

By construction, $\langle \alpha, \alpha \rangle < 0$.

For $\tau \in [0,1]$ consider the following subspace of V:

$$W_\tau = \mathrm{span}\{e_1 + \tau v_1^-, \ldots, e_s + \tau v_s^-\}.$$

It is easy to see that W_τ is maximally positive definite for all $\tau \in [0,1]$ and $W_0 = V_+$, $W_1 = W$. Let

$$A_\tau = A|_{W_\tau}$$

and consider the following commutative diagram:

$$\begin{array}{ccc} W_\tau & \xrightarrow{A_\tau} & A(W_\tau) \\ \pi \downarrow & & \downarrow \pi \\ V_+ & \xrightarrow[B_\tau]{} & V_+ \end{array}$$

We know that π restricted to W_τ and $A(W_\tau)$ is an orientation preserving isomorphism to V_+ for all $\tau \in [0, 1]$, because these subspaces are maximally positive definite. Since A_τ is an isomorphism, it follows that B_τ is an isomorphism for all $\tau \in [0, 1]$, i.e. an element of $\mathrm{GL}(V_+)$. Moreover, A has time-orientability $+1$ if and only if B_0 has positive determinant. Since the determinant is continuous and B_τ is a continuous curve in $\mathrm{GL}(V_+)$, it follows that B_0 has positive determinant if and only if B_1 has positive determinant. This proves the claim. □

The construction of the family W_τ in the proof shows the following:

Lemma 6.1.14 *The set of all maximally positive definite subspaces $W \subset V$ forms a contractible subset of the Grassmannian $Gr_s(V)$.*

Lemma 6.1.13 implies:

Proposition 6.1.15 *If $A, B \in \mathrm{O}(s, t)$ both have time-orientability $+1$, then the same holds for AB and A^{-1}.*

Definition 6.1.16 We call

$$\begin{aligned} \mathrm{SO}(s,t) &= \{A \in \mathrm{O}(s,t) \mid \det A = 1\}, \\ \mathrm{O}^+(s,t) &= \{A \in \mathrm{O}(s,t) \mid A \text{ has time-orientability } +1\}, \\ \mathrm{SO}^+(s,t) &= \mathrm{SO}(s,t) \cap \mathrm{O}^+(s,t), \end{aligned}$$

respectively,

- the **proper** or **special** pseudo-orthogonal group,
- the **orthochronous** pseudo-orthogonal group, and
- the **proper orthochronous** pseudo-orthogonal group.

These subsets are indeed subgroups of $\mathrm{O}(s, t)$: this is clear for $\mathrm{SO}(s, t)$ and follows for $\mathrm{O}^+(s, t)$ from Proposition 6.1.15.

In particular, for $s = 1$ or $t = 1$ these groups are called

- the **proper** Lorentz group,
- the **orthochronous** Lorentz group, and
- the **proper orthochronous** Lorentz group.

If we write a matrix $A \in \mathrm{O}(s,t)$ as in Eq. (6.1) in the block form

$$A = \begin{pmatrix} A_{11} & A_{12} \\ A_{21} & A_{22} \end{pmatrix},$$

then

$$\mathrm{O}^+(s,t) = \{A \in \mathrm{O}(s,t) \mid \det A_{11} > 0\},$$

$$\mathrm{SO}^+(s,t) = \{A \in \mathrm{O}(s,t) \mid \det A = 1, \det A_{11} > 0\}.$$

The definition of time-orientability also works for $\mathrm{O}(n)$, in which case the time-orientability of A is equal to the determinant of A and

$$\mathrm{O}^+(n) = \mathrm{SO}(n) = \mathrm{SO}^+(n).$$

If η is indefinite, then the three corresponding groups are not identical.

A similar argument to the proof of Theorem 1.2.22 in Sect. 3.8.3 shows (see Exercise 6.13.3):

Proposition 6.1.17 (Connected Component of the Identity of Pseudo-Orthogonal Groups) *The subgroup* $\mathrm{SO}^+(s,t)$ *is the connected component of the identity in* $\mathrm{O}(s,t)$.

If both s and t are non-zero, this implies that $\mathrm{O}(s,t)$ has precisely *four* connected components.

Replacing in Definition 6.1.11 V_+ by V_- we can define a new time-orientability using an orientation on V_-. Since the connected component of the identity has to be the same whether we define the time-orientability via orientations on V_+ or V_-, we conclude that:

Lemma 6.1.18 *The subgroup* $\mathrm{SO}^+(s,t)$ *can also be characterized by*

$$\mathrm{SO}^+(s,t) = \{A \in \mathrm{O}(s,t) \mid \det A = 1, \det A_{22} > 0\}.$$

This also follows from the isomorphism in Lemma 6.1.8.

Remark 6.1.19 It can be shown that as a smooth manifold $\mathrm{SO}^+(s,t)$ is diffeomorphic to $\mathrm{SO}(s) \times \mathrm{SO}(t) \times \mathbb{R}^{st}$ (if $s,t \geq 1$), because $\mathrm{SO}(s) \times \mathrm{SO}(t)$ is the maximal compact subgroup of $\mathrm{SO}^+(s,t)$ (cf. [13, 83]). This also determines the fundamental group of $\mathrm{SO}^+(s,t)$.

Example 6.1.20 For applications concerning the Standard Model, the most important of these Lie groups is the proper orthochronous Lorentz group $\mathrm{SO}^+(1,3) \cong \mathrm{SO}^+(3,1)$ of 4-dimensional Minkowski spacetime. This is a connected, non-compact Lie group of dimension 6. As a smooth manifold it is diffeomorphic to $\mathrm{SO}(3) \times \mathbb{R}^3$ and has fundamental group $\mathbb{Z}_2$.

6.2 Clifford Algebras

We saw in Sect. 6.1 that a vector space (V, Q) with a (non-degenerate) symmetric bilinear form defines a canonical *Lie group* $\mathrm{O}(V, Q)$ and there is an associated *Lie algebra* $\mathfrak{o}(V, Q)$. It is less obvious that the vector space (V, Q) also defines another canonical algebraic object, an *associative algebra* $\mathrm{Cl}(V, Q)$, called the *Clifford algebra* of (V, Q).

Definition 6.2.1 Let $\mathbb{K} = \mathbb{R}, \mathbb{C}$. An **associative $\mathbb{K}$-algebra with unit element** 1 is a $\mathbb{K}$-vector space A of finite dimension together with a bilinear, associative product

$$\cdot : A \times A \longrightarrow A$$

and an element $1 \in A$ such that $1 \cdot a = a = a \cdot 1$ for all $a \in A$. In particular, the product on A is distributive and associative, but in general not commutative. The **direct sum** of associative algebras A, B with unit element is defined as the vector space $A \oplus B$ with the product

$$(a, b) \cdot (a', b') = (a \cdot a', b \cdot b') \quad \forall a, a' \in A, b, b' \in B.$$

The **tensor product** of associative algebras A, B with unit element is defined as the vector space $A \otimes_{\mathbb{K}} B$ with the product

$$(a \otimes b) \cdot (a' \otimes b') = (a \cdot a') \otimes (b \cdot b') \quad \forall a, a' \in A, b, b' \in B.$$

Definition 6.2.2 A **homomorphism** between $\mathbb{K}$-algebras A, B with unit elements is a $\mathbb{K}$-linear map $\phi: A \to B$ such that $\phi(1) = 1$ and

$$\phi(a \cdot a') = \phi(a) \cdot \phi(a') \quad \forall a, a' \in A.$$

An **isomorphism** is a bijective homomorphism. An **automorphism** is an isomorphism $\phi: A \to A$. A **representation** of A on a vector space V is a homomorphism $\phi: A \to \mathrm{End}(V)$ into the endomorphism algebra of V. A representation is called **faithful** if the homomorphism $\phi: A \to \mathrm{End}(V)$ is injective.

Definition 6.2.3 If C is an associative algebra and $a, b \in C$, we define the **commutator** and **anticommutator** by

$$[a, b] = a \cdot b - b \cdot a,$$

$$\{a, b\} = a \cdot b + b \cdot a.$$

Definition 6.2.4 Let V be a $\mathbb{K}$-vector space with a symmetric bilinear form Q. A **Clifford algebra** of (V, Q) is a pair $(\mathrm{Cl}(V, Q), \gamma)$, where

1. $\mathrm{Cl}(V, Q)$ is an associative $\mathbb{K}$-algebra with unit element 1.
2. **Clifford relation:** $\gamma: V \to \mathrm{Cl}(V, Q)$ is a linear map with

$$\{\gamma(v), \gamma(w)\} = -2Q(v, w) \cdot 1 \quad \forall v, w \in V.$$

3. **Universal property:** If A is some other associative $\mathbb{K}$-algebra with unit element 1 and $\delta: V \to A$ a $\mathbb{K}$-linear map with

$$\{\delta(v), \delta(w)\} = -2Q(v, w) \cdot 1 \quad \forall v, w \in V,$$

then there exists a unique algebra homomorphism $\phi: \mathrm{Cl}(V, Q) \to A$ such that the following diagram commutes:

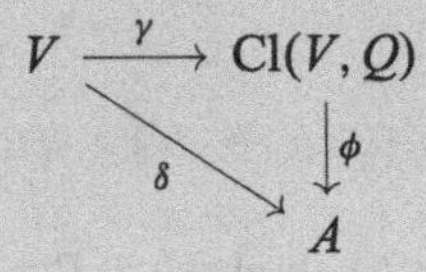

Remark 6.2.5 We can think of the linear map γ as a *linear square root* of the symmetric bilinear form $-Q$: in the definition of Clifford algebras, it suffices to demand that

$$\gamma(v)^2 = -Q(v, v) \cdot 1 \quad \forall v \in V,$$

because, considering this equation for vectors $v, w, v + w$, the equation

$$\{\gamma(v), \gamma(w)\} = -2Q(v, w) \cdot 1 \quad \forall v, w \in V$$

follows. The element $\gamma(v)$ in the Clifford algebra associated to a vector $v \in V$ is thus a "square root" of $-Q(v, v) \cdot 1$, depending linearly on v.

Remark 6.2.6 Clifford algebras arose in physics in the work of P.A.M. Dirac, who tried to find a "square root" of the Laplacian Δ, i.e. a differential operator D on a pseudo-Euclidean vector space $(\mathbb{R}^{s,t}, \eta)$ of dimension $n = s + t$ such that

$$D^2 = D \circ D = \Delta = -\sum_{i=1}^{n} \eta_i \frac{\partial^2}{\partial x_i^2},$$

where $\eta_i = \eta(e_i, e_i)$ in an orthonormal basis $e_1, \ldots, e_{s+t}$.

If we expand formally

$$D = \sum_{i=1}^{n} \gamma(e_i) \frac{\partial}{\partial x_i},$$

it follows that

$$\begin{aligned} D^2 &= \sum_{i,j=1}^{n} \gamma(e_i)\gamma(e_j) \frac{\partial^2}{\partial x_i \partial x_j} \\ &= \frac{1}{2} \sum_{i,j=1}^{n} \left(\gamma(e_i)\gamma(e_j) + \gamma(e_j)\gamma(e_i)\right) \frac{\partial^2}{\partial x_i \partial x_j}. \end{aligned}$$

This implies that the symbols $\gamma(e_i)$ have to satisfy

$$\gamma(e_i)\gamma(e_j) + \gamma(e_j)\gamma(e_i) = -2\eta(e_i, e_j),$$

i.e. the relation of the Clifford algebra. We conclude that if we have a representation of the Clifford algebra of $(\mathbb{R}^{s,t}, \eta)$ on a vector space Δ, then we can define a *Dirac operator* D, a square root of the Laplacian, for maps on $\mathbb{R}^{s,t}$ with values in Δ.

Remark 6.2.7 The Clifford relation has another important consequence, that we will discuss in detail in Sect. 6.5: if $v \in V$ is a vector with $Q(v, v) = \pm 1$, then for all $x \in V$

$$\pm\gamma(v) \cdot \gamma(x) \cdot \gamma(v) = \begin{cases} -\gamma(x) \text{ if } x \parallel v, \\ \gamma(x) \text{ if } x \perp v. \end{cases}$$

We will see in Corollary 6.2.18 that the linear map γ is always injective and we can identify V with its image $\gamma(V)$. This implies that the vectors $v \in V$ with $Q(v, v) = \pm 1$ act on V as reflections in the hyperplane $v^\perp$ and thus arbitrary products of such vectors act as *pseudo-orthogonal transformations*. Conversely, demanding that $\pm\gamma(v) \cdot \gamma(x) \cdot \gamma(v)$ is the reflection in $v^\perp$ almost inevitably leads to the Clifford relation.

Note that v and $-v$ define the same reflection in $v^\perp$. Since the expression $\pm\gamma(v) \cdot \gamma(x) \cdot \gamma(v)$ depends quadratically on v, we can think of $\gamma(v)$ as a "square root" of the reflection in $v^\perp$. This construction, together with the *Cartan–Dieudonné Theorem* on reflections, will enable us to define the *spin groups*, certain double coverings of the pseudo-orthogonal groups, essentially using square roots of pseudo-orthogonal transformations.

6.2.1 Existence and Uniqueness of Clifford Algebras

Our first aim is to prove the existence and uniqueness of Clifford algebras.

Theorem 6.2.8 (Existence of Clifford Algebras) *For every finite-dimensional $\mathbb{K}$-vector space V with a symmetric bilinear form Q there exists a Clifford algebra $(\mathrm{Cl}(V,Q),\gamma)$.*

Proof We denote by $T(V)$ the tensor algebra of V:

$$T(V) = \bigoplus_{n\geq 0} V^{\otimes n} = \mathbb{K} \oplus V \oplus (V\otimes V) \oplus (V \otimes V \otimes V) \oplus \dots$$

Let $I(Q)$ denote the two-sided ideal in $T(V)$ generated by the set

$$\{v\otimes v + Q(v,v)\cdot 1 \mid v \in V\}.$$

The tensor algebra $T(V)$ is determined by V alone (and infinite-dimensional), whereas the ideal $I(Q)$ depends also on the symmetric bilinear form Q.

We set

$$\mathrm{Cl}(V,Q) = T(V)/I(Q).$$

Then $\mathrm{Cl}(V,Q)$ is an associative algebra with product

$$[a]\cdot[b] = [a\otimes b] \quad \forall a,b \in T(V).$$

If $Q \equiv 0$, it follows that the Clifford algebra is the exterior algebra $\Lambda^* V$. In general, let

$$i\colon V \longrightarrow T(V)$$

and

$$\pi\colon T(V) \longrightarrow \mathrm{Cl}(V,Q)$$

denote the canonical embedding and projection. Then

$$\gamma = \pi \circ i\colon V \longrightarrow \mathrm{Cl}(V,Q)$$

is linear and satisfies the identity

$$\begin{aligned}\gamma(v)^2 &= [v\otimes v]\\ &= -Q(v,v)\cdot 1 \quad \forall v \in V.\end{aligned}$$

This implies by polarization

$$\{\gamma(v), \gamma(w)\} = -2Q(v, w) \cdot 1 \quad \forall v, w \in V.$$

Since the vector space V generates $T(V)$ by taking tensor products, it follows that the image $\gamma(V)$ generates $\mathrm{Cl}(V, Q)$ multiplicatively.

By definition of the tensor algebra, every linear map

$$\delta: V \longrightarrow A$$

to an associative algebra A extends to an algebra homomorphism

$$\Delta: T(V) \longrightarrow A.$$

If δ satisfies the identity

$$\{\delta(v), \delta(w)\} = -2Q(v, w) \cdot 1 \quad \forall v, w \in V,$$

then $I(Q)$ is a subset of $\ker \Delta$ and the map Δ descends to a homomorphism

$$\phi: \mathrm{Cl}(V, Q) \longrightarrow A$$

with $\phi \circ \gamma = \delta$. Given A and δ, the homomorphism ϕ with $\phi \circ \gamma = \delta$ is uniquely determined, because $\gamma(V)$ generates $\mathrm{Cl}(V, Q)$ multiplicatively and ϕ is fixed on the image $\gamma(V)$. □

Corollary 6.2.9 (Uniqueness of Clifford Algebras) *If the associative algebras* $(\mathrm{Cl}(V, Q), \gamma)$ *and* $\left(\mathrm{Cl}'(V, Q), \gamma'\right)$ *are both Clifford algebras for the same vector space* (V, Q), *then there exists a unique algebra isomorphism*

$$f: \mathrm{Cl}(V, Q) \longrightarrow \mathrm{Cl}'(V, Q)$$

so that $f \circ \gamma = \gamma'$.

Proof This follows from the universal property of Clifford algebras. □

From the proof of Theorem 6.2.8 we see:

Corollary 6.2.10 *The image of the vector space* V *under* γ *generates* $\mathrm{Cl}(V, Q)$ *multiplicatively.*

Corollary 6.2.11 *If* $Q \equiv 0$, *then there exists an algebra isomorphism*

$$(\mathrm{Cl}(V, 0), \cdot) \cong (\Lambda^* V, \wedge),$$

where γ *is given by the standard embedding of* V *into* $\Lambda^* V$.

Corollary 6.2.12 *Suppose that* $\dim_{\mathbb{K}} V = n$ *and* $e_1, \ldots, e_n$ *is an orthonormal basis for* (V, Q). *Then the set of elements of the form*

$$\gamma(e_{i_1}) \cdot \gamma(e_{i_2}) \cdots \gamma(e_{i_k}),$$

where $1 \leq i_1 < i_2 < \ldots < i_k \leq n$, $0 \leq k \leq n$, *and the empty product for* $k = 0$ *is equal to* 1, *span* $\mathrm{Cl}(V, Q)$ *as a vector space. This implies that*

$$\dim_{\mathbb{K}} \mathrm{Cl}(V, Q) \leq 2^n.$$

Proof The elements of $\gamma(V)$ generate $\mathrm{Cl}(V, Q)$ multiplicatively, hence the collection of all products of the basis vectors $\{\gamma(e_i)\}$ span $\mathrm{Cl}(V, Q)$ as a vector space. If in such a product the same vector $\gamma(e_i)$ appears twice, then we can cancel it (after possibly permuting the vectors) because of the Clifford relation. □

Example 6.2.13 The simplest non-trivial example of a Clifford algebra is the Clifford algebra over a 1-dimensional vector space $(\mathbb{K}, Q)$. Let e be a non-zero element of $\mathbb{K}$. Then $\mathrm{Cl}(\mathbb{K}, Q)$ is 2-dimensional and is spanned as a $\mathbb{K}$-vector space by $\{1, \gamma(e)\}$ with

$$\gamma(e) \cdot \gamma(e) = -Q(v, v) \cdot 1.$$

Definition 6.2.14 Let $T^0(V)$ and $T^1(V)$ denote the vector subspaces of elements of the tensor algebra of even and odd degree. We set

$$\mathrm{Cl}^0(V, Q) = T^0(V) / \left(T^0(V) \cap I(Q)\right),$$
$$\mathrm{Cl}^1(V, Q) = T^1(V) / \left(T^1(V) \cap I(Q)\right)$$

and call these vector subspaces the **even** and **odd part** of the Clifford algebra. Since

$$I(Q) = \left(T^0(V) \cap I(Q)\right) \oplus \left(T^1(V) \cap I(Q)\right),$$

it follows that

$$\mathrm{Cl}(V, Q) = \mathrm{Cl}^0(V, Q) \oplus \mathrm{Cl}^1(V, Q)$$

and $\mathrm{Cl}(V, Q)$ has the structure of a $\mathbb{Z}_2$**-graded associative algebra** or **superalgebra**:

$$\mathrm{Cl}^i(V, Q) \cdot \mathrm{Cl}^j(V, Q) \subset \mathrm{Cl}^{i+j \bmod 2}(V, Q).$$

In particular, $\mathrm{Cl}^0(V, Q)$ is a subalgebra of $\mathrm{Cl}(V, Q)$ and is spanned by products $v_1 \cdots v_{2k}$ of an even number of vectors $v_i \in V$.

6.2.2 Clifford Algebras and Exterior Algebras

We saw in Corollary 6.2.11 that for $Q \equiv 0$ the Clifford algebra $\mathrm{Cl}(V, Q)$ is isomorphic to the exterior algebra $\Lambda^* V$ with the wedge product. In this section we will show that for an arbitrary symmetric bilinear form Q the Clifford algebra $\mathrm{Cl}(V, Q)$ is still (canonically) isomorphic to the exterior algebra as a *vector space*. The multiplication in $\mathrm{Cl}(V, Q)$ can thus be thought of as a deformation (depending on Q) of the wedge product on $\Lambda^* V$ (we follow the exposition in [20]).

Let V be a finite-dimensional $\mathbb{K}$-vector space with a symmetric bilinear form Q.

Definition 6.2.15 For $v \in V$ and $\sigma \in \Lambda^k V$ there is a unique $(k-1)$-form $v \lrcorner \sigma \in \Lambda^{k-1} V$, called the **contraction** of v and σ, with the following properties:

1. if $\sigma \in V$, then $v \lrcorner \sigma = Q(v, \sigma)$;
2. for all $\sigma \in \Lambda^k V, \omega \in \Lambda^l V$ we have

$$v \lrcorner (\sigma \wedge \omega) = (v \lrcorner \sigma) \wedge \omega + (-1)^k \sigma \wedge (v \lrcorner \omega).$$

The proof is left as an exercise.

Theorem 6.2.16 (The Clifford Algebra Is Isomorphic to the Exterior Algebra as a Vector Space) *There exists a canonical isomorphism of vector spaces*

$$\Lambda^* V \longrightarrow \mathrm{Cl}(V, Q).$$

In any orthonormal basis $e_1, \ldots, e_n$ of (V, Q) this isomorphism is given by

$$e_{i_1} \wedge e_{i_2} \wedge \ldots \wedge e_{i_k} \longmapsto \gamma(e_{i_1}) \cdot \gamma(e_{i_2}) \cdots \gamma(e_{i_k}).$$

In particular, the dimension of the Clifford algebra is

$$\dim_{\mathbb{K}} \mathrm{Cl}(V, Q) = 2^n,$$

where $n = \dim_{\mathbb{K}} V$.

Proof We define a linear map

$$\delta \colon V \longrightarrow \mathrm{End}(\Lambda^* V)$$

by

$$\begin{aligned} \delta(v) \colon \Lambda^* V &\longrightarrow \Lambda^* V \\ \alpha &\longmapsto v \wedge \alpha - v \lrcorner \alpha. \end{aligned}$$

We have

$$\{\delta(v), \delta(w)\}\alpha = v \wedge (w \wedge \alpha - w \lrcorner \alpha) - v \lrcorner (w \wedge \alpha - w \lrcorner \alpha)$$
$$+ w \wedge (v \wedge \alpha - v \lrcorner \alpha) - w \lrcorner (v \wedge \alpha - v \lrcorner \alpha)$$
$$= -2Q(v, w)\alpha.$$

By the universal property of Clifford algebras, the linear map δ extends to an algebra homomorphism

$$\phi: \mathrm{Cl}(V, Q) \longrightarrow \mathrm{End}(\Lambda^* V).$$

Consider the linear map

$$f: \mathrm{Cl}(V, Q) \longrightarrow \Lambda^* V$$
$$x \longmapsto (\phi(x))(1_{\Lambda^* V}).$$

If $e_1, \ldots, e_n$ is an orthonormal basis for (V, Q), then

$$f(\gamma(e_{i_1}) \cdot \gamma(e_{i_2}) \cdots \gamma(e_{i_k})) = e_{i_1} \wedge e_{i_2} \wedge \ldots e_{i_k}.$$

In particular, f is surjective. Corollary 6.2.12 then implies that f is an isomorphism of vector spaces. □

Remark 6.2.17 The linear map

$$f: \mathrm{Cl}(V, Q) \longrightarrow \Lambda^* V$$

is called the **symbol map** in [15], its inverse f^{-1} the **quantization map**.

Corollary 6.2.18 (Linear Map γ Is an Embedding) *Let* $(\mathrm{Cl}(V, Q), \gamma)$ *be a Clifford algebra. Then the linear map*

$$\gamma: V \longrightarrow \mathrm{Cl}(V, Q)$$

is injective and we can therefore identify V with its image under γ.

Remark 6.2.19 Since we now know the dimension of Clifford algebras, the universal property can be used to find isomorphisms of the Clifford algebra $\mathrm{Cl}(V, Q)$ to other associative algebras A as follows:

- Find a linear map $\delta: V \to A$ that satisfies

$$\delta(v)^2 = -Q(v, v) \cdot 1 \quad \forall v \in V.$$

  It then induces an algebra homomorphism $\phi: \mathrm{Cl}(V, Q) \to A$.

- Let $e_1, \ldots, e_n$ be an orthonormal basis of V. Show that the products of the images $\delta(e_i)$ span A. Then ϕ is surjective.
- Suppose that the algebras $\mathrm{Cl}(V, Q)$ and A have the same dimensions. Then ϕ is an algebra isomorphism.

We will use this strategy several times in the following sections, for instance, in Lemma 6.3.2, Lemma 6.3.3 and Lemma 6.3.20.
Here is an application of Remark 6.2.19.

Lemma 6.2.20 *The linear map*

$$-\mathrm{Id}\colon V \longrightarrow V$$

induces an algebra automorphism

$$\alpha\colon \mathrm{Cl}(V, Q) \longrightarrow \mathrm{Cl}(V, Q)$$

with $\alpha^2 = 1$. *The subspace* $\mathrm{Cl}^j(V, Q)$ *is equal to the* $(-1)^j$*-eigenspace of* α. *In particular,*

$$\dim_{\mathbb{K}} \mathrm{Cl}^0(V, Q) = \dim_{\mathbb{K}} \mathrm{Cl}^1(V, Q) = \frac{1}{2} \dim_{\mathbb{K}} \mathrm{Cl}(V, Q).$$

6.3 The Clifford Algebras for the Standard Symmetric Bilinear Forms

Definition 6.3.1
1. For the standard vector space $(V, Q) = (\mathbb{R}^{s,t}, \eta)$ we denote the Clifford algebra by $\mathrm{Cl}(s, t)$. This is a real associative algebra. For $(s, t) = (n, 0)$ we also denote the Clifford algebra by $\mathrm{Cl}(n)$.
2. For the standard vector space $(V, Q) = \left(\mathbb{C}^d, q\right)$ we denote the Clifford algebra by $\mathbb{C}\mathrm{l}(d)$ (note that q is the standard non-degenerate complex bilinear form and not a Hermitian form). This is a complex associative algebra.

Lemma 6.3.2 (Complexified Clifford Algebra) *There exists an isomorphism of complex associative algebras*

$$\mathbb{C}\mathrm{l}(s + t) \cong \mathrm{Cl}(s, t) \otimes_{\mathbb{R}} \mathbb{C}.$$

Complex representations of $\mathrm{Cl}(s, t)$ *are equivalent to complex representations of* $\mathbb{C}\mathrm{l}(s + t)$.

Proof The complex linear map

$$\begin{aligned}\delta\colon \mathbb{C}^{s+t} \cong \mathbb{R}^{s,t} \otimes_{\mathbb{R}} \mathbb{C} &\longrightarrow \mathrm{Cl}(s,t) \otimes_{\mathbb{R}} \mathbb{C} \\ (v \otimes z) &\longmapsto \gamma(v) \otimes z\end{aligned}$$

satisfies

$$\begin{aligned}\delta(v \otimes z)^2 &= -\eta(v,v)z^2 \\ &= -q(v \otimes z, v \otimes z).\end{aligned}$$

The claim follows from Remark 6.2.19. □

Lemma 6.3.3 (Even Complex Clifford Algebra) *For any $n \geq 1$*

$$\mathbb{C}\mathrm{l}^0(n) \cong \mathbb{C}\mathrm{l}(n-1).$$

Proof Let e_n be the n-th standard basis vector of $\mathbb{C}^n$ and define

$$\begin{aligned}\delta\colon \mathbb{C}^{n-1} &\longrightarrow \mathbb{C}\mathrm{l}^0(n) \\ x &\longmapsto x \cdot e_n.\end{aligned}$$

Then δ satisfies

$$\begin{aligned}\delta(x)^2 &= x \cdot e_n \cdot x \cdot e_n \\ &= x \cdot x \\ &= -q(x,x).\end{aligned}$$

The claim follows from Remark 6.2.19 and the dimension formula in Lemma 6.2.20. □

6.3.1 Gamma Matrices

The Clifford algebra $\mathrm{Cl}(V,Q)$ is generated multiplicatively by the subspace $\gamma(V)$. It is therefore important to know the elements $\gamma(e_i)$ for a basis $e_1, \ldots, e_n$ of V.

Definition 6.3.4 Let $(V, Q) = (\mathbb{R}^{s,t}, \eta)$ with standard basis $e_1, \ldots, e_n$. Suppose

$$\rho: \mathrm{Cl}(V, Q) \longrightarrow \mathrm{End}(\Sigma)$$

is a representation of $\mathrm{Cl}(V, Q)$ on a $\mathbb{K}$-vector space $\Sigma = \mathbb{K}^N$. Then we define for $a = 1, \ldots, n$ the **mathematical gamma matrices** by

$$\gamma_a = \rho \circ \gamma(e_a)$$

and the **physical gamma matrices** by

$$\Gamma_a = (-i)\gamma_a.$$

The anticommutators are given by

$$\{\gamma_a, \gamma_b\} = -2\eta_{ab} I_N,$$
$$\{\Gamma_a, \Gamma_b\} = 2\eta_{ab} I_N,$$

where I_N denotes the $N \times N$-unit matrix. We also set

$$\gamma_{ab} = \frac{1}{2}[\gamma_a, \gamma_b] = \frac{1}{2}(\gamma_a\gamma_b - \gamma_b\gamma_a),$$
$$\Gamma_{ab} = \frac{1}{2}[\Gamma_a, \Gamma_b] = \frac{1}{2}(\Gamma_a\Gamma_b - \Gamma_b\Gamma_a).$$

6.3.2 *The Chirality Operator in Even Dimensions*

For an even-dimensional real vector space, a choice of orientation defines an important element in the Clifford algebra.

Suppose that $n = s+t = 2k$ is even. A **chirality element** for $\mathrm{Cl}(s, t)$ is a Clifford element of the form

$$\omega = \lambda e_1 \cdots e_n \in \mathrm{Cl}(s, t) \otimes \mathbb{C},$$

where $e_1, \ldots, e_n$ is an oriented orthonormal basis of $\mathbb{R}^{s,t}$ and λ is a complex constant, determined below.

Lemma 6.3.5 *A chirality element ω does not depend on the choice of oriented orthonormal basis $e_1, \ldots, e_n$.*

Proof Suppose that

$$e_i' = \sum_{j=1}^{n} A_{ij} e_j$$

with a matrix $A \in \mathrm{SO}(s,t)$. Under the vector space isomorphism

$$\psi\colon \Lambda^* \mathbb{R}^{s,t} \longrightarrow \mathrm{Cl}(s,t)$$

from Theorem 6.2.16 we have

$$\begin{aligned} e_1 \cdots e_n &= \psi(e_1 \wedge e_2 \ldots \wedge e_n), \\ e_1' \cdots e_n' &= \psi\left(e_1' \wedge e_2' \ldots \wedge e_n'\right). \end{aligned}$$

However

$$\begin{aligned} e_1' \wedge e_2' \wedge \ldots \wedge e_n' &= \det(A) e_1 \wedge e_2 \wedge \ldots \wedge e_n \\ &= e_1 \wedge e_2 \ldots \wedge e_n. \end{aligned}$$

Hence $e_1' \cdots e_n' = e_1 \cdots e_n$. □

Lemma 6.3.6 *Every chirality element ω satisfies*

$$\begin{aligned} \{\omega, e_a\} &= 0, \\ [\omega, e_a \cdot e_b] &= 0 \end{aligned}$$

for all $1 \leq a, b \leq n$.

Proof The first equation follows from

$$\begin{aligned} e_a \cdot \omega &= \lambda e_a \cdot e_1 \cdots e_n \\ &= (-1)^{a-1} \lambda e_1 \cdots e_a \cdot e_a \cdots e_n, \\ \omega \cdot e_a &= \lambda e_1 \cdots e_n \cdot e_a \\ &= (-1)^{n-a} \lambda e_1 \cdots e_a \cdot e_a \cdots e_n \\ &= -e_a \cdot \omega, \end{aligned}$$

since n is even. The second equation is a consequence of the first. □

Lemma 6.3.7 *If $\lambda^2 = (-1)^{k+t}$, then the chirality element satisfies*

$$\omega^2 = 1.$$

Proof

$$\begin{aligned}\omega^2 &= \lambda^2 e_1 \cdots e_n \cdot e_1 \cdots e_n \\ &= \lambda^2 (-1)^{n(n-1)/2} (e_1)^2 \cdots (e_n)^2 \\ &= \lambda^2 (-1)^{k+t} 1 \\ &= 1,\end{aligned}$$

where we used that

$$\frac{1}{2} n(n-1) = k(2k-1) \equiv k \bmod 2.$$

□

Different choices for λ with $\lambda^2 = (-1)^{k+t}$ are possible and several different choices appear in the literature. The simplest choice is probably

$$\lambda = i^{k+t}.$$

We will use in the following this choice up to a sign (because of certain conventions for Weyl spinors, described later) and set:

Definition 6.3.8 The **chirality element** in even dimension $n = s + t = 2k$ for $\mathrm{Cl}(s,t)$ is defined by

$$\omega = -i^{k+t} e_1 \cdots e_n \in \mathrm{Cl}(s,t) \otimes \mathbb{C}.$$

If $k + t$ is even, then ω is an element of the real Clifford algebra $\mathrm{Cl}(s,t)$.

If γ_a are mathematical gamma matrices in a complex representation of $\mathrm{Cl}(s,t)$, then the **mathematical chirality operator** is defined by

$$\gamma_{n+1} = -i^{k+t} \gamma_1 \cdots \gamma_n.$$

If Γ_a are physical gamma matrices, then the **physical chirality operator** is defined by

$$\Gamma_{n+1} = -i^{k+t} \Gamma_1 \cdots \Gamma_n.$$

We have $\gamma_{n+1} = (-1)^k \Gamma_{n+1}$. The chirality operators do not depend on the choice of oriented orthonormal basis for $\mathbb{R}^{s,t}$.

Remark 6.3.9 By analogy, we can define a chirality element for the complex Clifford algebra $\mathbb{C}\mathrm{l}(2k)$ by

$$\omega = -i^k e_1 \cdots e_{2k} \in \mathbb{C}\mathrm{l}(2k),$$

where $e_1, \ldots, e_{2k}$ is an orthonormal basis of $(\mathbb{C}^{2k}, q)$. Taking an orthonormal basis

$$e_1, \ldots, e_s, e_{s+1}, \ldots, e_{s+t}$$

of $\mathbb{R}^{s,t}$ and the corresponding orthonormal basis

$$e_1, \ldots, e_s, ie_{s+1}, \ldots, ie_{s+t}$$

of $\mathbb{C}^{2k}$, we see that the chirality elements for the real and complex Clifford algebra coincide under the isomorphism $\mathrm{Cl}(s,t) \otimes \mathbb{C} \cong \mathbb{C}\mathrm{l}(2k)$.

6.3.3 Raising Indices of Gamma Matrices

Let $\Gamma_1, \ldots, \Gamma_n$ be physical gamma matrices. We set

$$\begin{aligned}
\Gamma^a &= \eta^{ac}\Gamma_c, \\
\Gamma^{bc} &= \frac{1}{2}\left[\Gamma^b, \Gamma^c\right] \\
&= \frac{1}{2}\left(\Gamma^b\Gamma^c - \Gamma^c\Gamma^b\right), \\
\Gamma^{n+1} &= -i^{k+t}\Gamma^1 \cdots \Gamma^n
\end{aligned}$$

and similarly for the mathematical γ-matrices (in the first equation there is an implicit sum over c; this is an instance of the Einstein summation convention). These matrices satisfy by Lemma 6.3.6

$$\begin{aligned}
\left\{\Gamma^{n+1}, \Gamma^a\right\} &= 0, \\
\left[\Gamma^{n+1}, \Gamma^{bc}\right] &= 0, \\
\gamma^{bc} &= -\Gamma^{bc}.
\end{aligned}$$

6.3.4 Examples of Clifford Algebras in Low Dimensions

In the following examples we use the Pauli matrices

$$\sigma_1 = \begin{pmatrix} 0 & 1 \\ 1 & 0 \end{pmatrix}, \quad \sigma_2 = \begin{pmatrix} 0 & -i \\ i & 0 \end{pmatrix}, \quad \sigma_3 = \begin{pmatrix} 1 & 0 \\ 0 & -1 \end{pmatrix}.$$

It is easy to check that they satisfy the identities

$$\begin{aligned} \sigma_j^2 &= I_2 \quad j = 1,2,3, \\ \sigma_j\sigma_{j+1} &= -\sigma_{j+1}\sigma_j = i\sigma_{j+2} \quad j = 1,2,3, \end{aligned}$$

where in the second equation $j+1$ and $j+2$ are taken mod3.

Example 6.3.10 The Clifford algebra $\mathrm{Cl}(1,0)$ is spanned as a real vector space by elements $1, \gamma(e_1)$ with

$$\gamma(e_1)^2 = -1.$$

It follows that $\mathrm{Cl}(1,0)$ is isomorphic as a real algebra to the 2-dimensional algebra $\mathbb{C}$ with $\gamma(e_1) = i$.

Example 6.3.11 The Clifford algebra $\mathrm{Cl}(0,1)$ is spanned as a real vector space by elements $1, \gamma(e_1)$ with

$$\gamma(e_1)^2 = 1.$$

It follows that $\mathrm{Cl}(0,1)$ is isomorphic as a real algebra to the 2-dimensional algebra $\mathbb{R} \oplus \mathbb{R}$ with multiplication $(a,b)\cdot(a',b') = (aa',bb')$, unit element $1 = (1,1)$ and $\gamma(e_1) = (1,-1)$.

Example 6.3.12 The algebra $\mathbb{C} \oplus \mathbb{C}$ is spanned as a vector space by $1 = (1,1)$ and $\gamma(e_1) = (i,-i)$. It follows that

$$\mathbb{C}\mathrm{l}(1) \cong \mathbb{C} \oplus \mathbb{C}.$$

Example 6.3.13 The matrices

$$\begin{aligned} \gamma_1 &= i\sigma_1, \\ \gamma_2 &= i\sigma_2 \end{aligned}$$

satisfy

$$\begin{aligned} \gamma_1^2 = \gamma_2^2 &= -I_2, \\ \gamma_1\gamma_2 + \gamma_2\gamma_1 &= 0. \end{aligned}$$

Therefore γ_1, γ_2 are mathematical gamma matrices for Cl(2, 0) and generate a faithful representation of this Clifford algebra on $\mathbb{C}^2$. We can identify $\gamma(e_i)$ with γ_i. The Clifford algebra is 4-dimensional and is spanned as a real vector space by $\{1, \gamma_1, \gamma_2, \gamma_1\gamma_2\}$.

Example 6.3.14 The matrices

$$\begin{aligned} \gamma_1 &= i\sigma_1, \\ \gamma_2 &= \sigma_2 \end{aligned}$$

satisfy

$$\begin{aligned} \gamma_1^2 &= -I_2, \\ \gamma_2^2 &= I_2, \\ \gamma_1\gamma_2 + \gamma_2\gamma_1 &= 0. \end{aligned}$$

Therefore γ_1, γ_2 are mathematical gamma matrices for Cl(1, 1) and generate a faithful representation of this Clifford algebra on $\mathbb{C}^2$. Again, the Clifford algebra is spanned as a real vector space by $\{1, \gamma_1, \gamma_2, \gamma_1\gamma_2\}$.

Example 6.3.15 Note that the matrices

$$I_2, \quad \sigma_1, \quad \sigma_2, \quad \sigma_3 = -i\sigma_1\sigma_2$$

span $\text{Mat}(2 \times 2, \mathbb{C}) = \text{End}\left(\mathbb{C}^2\right)$ as a complex vector space. It follows that

$$\text{Cl}(2) \cong \text{End}\left(\mathbb{C}^2\right).$$

Example 6.3.16 The matrices

$$\begin{aligned} \gamma_0 &= \begin{pmatrix} 0 & iI_2 \\ iI_2 & 0 \end{pmatrix}, \\ \gamma_k &= \begin{pmatrix} 0 & \sigma_k \\ -\sigma_k & 0 \end{pmatrix} \quad \forall k = 1, 2, 3 \end{aligned}$$

satisfy

$$\begin{aligned} (\gamma_0)^2 &= -I_4, \\ (\gamma_k)^2 &= -I_4 \quad \forall k = 1, 2, 3, \\ \gamma_a\gamma_b + \gamma_b\gamma_a &= 0 \quad \forall a \neq b. \end{aligned}$$

They are mathematical gamma matrices for $\mathrm{Cl}(4,0)$, i.e. for the Clifford algebra of Euclidean space of dimension 4. Another choice for the same Clifford algebra is

$$\gamma_0 = \begin{pmatrix} 0 & I_2 \\ -I_2 & 0 \end{pmatrix},$$

$$\gamma_k = \begin{pmatrix} 0 & i\sigma_k \\ i\sigma_k & 0 \end{pmatrix} \quad \forall k = 1, 2, 3.$$

The γ^a generate a faithful representation of $\mathrm{Cl}(4,0)$ on $\mathbb{C}^4$. We identify $\gamma(e_i)$ with γ_i. As a real vector space $\mathrm{Cl}(4,0)$ has dimension 16 and is spanned by

$$\begin{aligned} &1 \quad (1 \text{ element}), \\ &\gamma_a \quad (4 \text{ elements}), \\ &\gamma_a\gamma_b \quad (a < b,\ 6 \text{ elements}), \\ &\gamma_a\gamma_b\gamma_c \quad (a < b < c,\ 4 \text{ elements}), \\ &\gamma_0\gamma_1\gamma_2\gamma_3 \quad (1 \text{ element}). \end{aligned}$$

Example 6.3.17 The matrices

$$\Gamma_0 = \begin{pmatrix} 0 & I_2 \\ I_2 & 0 \end{pmatrix},$$

$$\Gamma_k = \begin{pmatrix} 0 & \sigma_k \\ -\sigma_k & 0 \end{pmatrix} \quad \forall k = 1, 2, 3$$

satisfy

$$\begin{aligned} (\Gamma_0)^2 &= I_4, \\ (\Gamma_k)^2 &= -I_4 \quad \forall k = 1, 2, 3, \\ \Gamma_a\Gamma_b + \Gamma_b\Gamma_a &= 0 \quad \forall a \neq b. \end{aligned}$$

They are physical gamma matrices for $\mathrm{Cl}(1,3)$, i.e. for the Clifford algebra of Minkowski spacetime with signature $(+,-,-,-)$, in the so-called **Weyl representation** or **chiral representation**. The associated mathematical gamma matrices $\gamma_a = i\Gamma_a$ satisfy

$$\begin{aligned} (\gamma_0)^2 &= -I_4, \\ (\gamma_k)^2 &= I_4 \quad k = 1, 2, 3, \\ \gamma_a\gamma_b + \gamma_b\gamma_a &= 0 \quad \forall a \neq b. \end{aligned}$$

The γ_a generate a faithful representation of $\mathrm{Cl}(1,3)$ on $\mathbb{C}^4$. As a real vector space $\mathrm{Cl}(1,3)$ has dimension 16 and is spanned by the matrices corresponding to the ones above.

Example 6.3.18 Let Γ_a and $\gamma_a = i\Gamma_a$ be the physical and mathematical gamma matrices for $\mathrm{Cl}(1,3)$ considered in Example 6.3.17. If we set

$$\Gamma'_a = \gamma_a,$$
$$\gamma'_a = i\Gamma'_a = -\Gamma_a,$$

then these are physical and mathematical gamma matrices for the Clifford algebra $\mathrm{Cl}(3,1)$ of Minkowski spacetime with signature $(-,+,+,+)$.

Example 6.3.19 A similar argument to the one in Example 6.3.15 shows that

$$\mathbb{C}\mathrm{l}(4) \cong \mathrm{End}\left(\mathbb{C}^4\right).$$

6.3.5 The Structure of the Standard Clifford Algebras

Lemma 6.3.20 (Complex Clifford Algebras Are Periodic) *The complex Clifford algebras satisfy the periodicity*

$$\begin{aligned}\mathbb{C}\mathrm{l}(n+2) &\cong \mathbb{C}\mathrm{l}(n) \otimes_{\mathbb{C}} \mathbb{C}\mathrm{l}(2)\\ &\cong \mathbb{C}\mathrm{l}(n) \otimes_{\mathbb{C}} \mathrm{End}\left(\mathbb{C}^2\right).\end{aligned}$$

Here $\otimes_{\mathbb{C}}$ denotes the standard tensor product of associative algebras.

Proof We follow [20]. Write $\mathbb{C}^{n+2}$ as $\mathbb{C}^n \oplus \mathbb{C}^2$ and decompose an element of $\mathbb{C}^{n+2}$ accordingly as (x,y). Let e_1, e_2 be the standard basis of $\mathbb{C}^2$ and $\omega = -ie_1e_2$ the corresponding chirality element. Define the linear map

$$\begin{aligned}\delta\colon \mathbb{C}^{n+2} &\longrightarrow \mathbb{C}\mathrm{l}(n) \otimes_{\mathbb{C}} \mathbb{C}\mathrm{l}(2)\\ (x,y) &\longmapsto x \otimes \omega + 1 \otimes y.\end{aligned}$$

Then

$$\begin{aligned}\delta(x,y)^2 &= x^2 \otimes \omega^2 + x \otimes \omega y + x \otimes y\omega + 1 \otimes y^2\\ &= -(q(x,x) + q(y,y)) \cdot 1,\end{aligned}$$

because $\omega^2 = 1$ and $\{\omega, y\} = 0$. The first isomorphism then follows by Remark 6.2.19. The second isomorphism follows from Example 6.3.15. □

Together with Example 6.3.12, Example 6.3.15 and Lemma 6.3.3 this implies the following structure theorem.

Theorem 6.3.21 (Structure Theorem for Complex Clifford Algebras) *As complex algebras the complex Clifford algebra and its even part are given by Table 6.1.*

Example 6.3.22 In dimension $n = 4$ we have

$$\mathbb{C}\mathrm{l}(4) \cong \mathrm{End}\left(\mathbb{C}^4\right),$$
$$\mathbb{C}\mathrm{l}^0(4) \cong \mathrm{End}\left(\mathbb{C}^2\right) \oplus \mathrm{End}\left(\mathbb{C}^2\right).$$

Without proof we mention the following theorem (see [28, 40, 49]).

Theorem 6.3.23 (Structure Theorem of Real Clifford Algebras) *The structure of the real Clifford algebras* $\mathrm{Cl}(s,t)$ *and* $\mathrm{Cl}^0(s,t)$ *is given by Tables 6.2 and 6.3, where we set* $\rho = s-t$ *and* $n = s+t$ *and all endomorphism algebras are understood as real algebras.*

Example 6.3.24 For Minkowski spacetime in dimension 4 we have

Table 6.1 Complex Clifford algebras

n	$\mathbb{C}\mathrm{l}(n)$	$\mathbb{C}\mathrm{l}^0(n)$	N
Even	$\mathrm{End}\left(\mathbb{C}^N\right)$	$\mathrm{End}\left(\mathbb{C}^{N/2}\right) \oplus \mathrm{End}\left(\mathbb{C}^{N/2}\right)$	$2^{n/2}$
Odd	$\mathrm{End}\left(\mathbb{C}^N\right) \oplus \mathrm{End}\left(\mathbb{C}^N\right)$	$\mathrm{End}\left(\mathbb{C}^N\right)$	$2^{(n-1)/2}$

Table 6.2 Real Clifford algebras

ρ mod 8	$\mathrm{Cl}(s,t)$	N
0	$\mathrm{End}\left(\mathbb{R}^N\right)$	$2^{n/2}$
1	$\mathrm{End}\left(\mathbb{C}^N\right)$	$2^{(n-1)/2}$
2	$\mathrm{End}\left(\mathbb{H}^N\right)$	$2^{(n-2)/2}$
3	$\mathrm{End}\left(\mathbb{H}^N\right) \oplus \mathrm{End}\left(\mathbb{H}^N\right)$	$2^{(n-3)/2}$
4	$\mathrm{End}\left(\mathbb{H}^N\right)$	$2^{(n-2)/2}$
5	$\mathrm{End}\left(\mathbb{C}^N\right)$	$2^{(n-1)/2}$
6	$\mathrm{End}\left(\mathbb{R}^N\right)$	$2^{n/2}$
7	$\mathrm{End}\left(\mathbb{R}^N\right) \oplus \mathrm{End}\left(\mathbb{R}^N\right)$	$2^{(n-1)/2}$

Table 6.3 Even part of real Clifford algebras

ρ mod 8	$\mathrm{Cl}^0(s,t)$	N
0	$\mathrm{End}\left(\mathbb{R}^N\right) \oplus \mathrm{End}\left(\mathbb{R}^N\right)$	$2^{(n-2)/2}$
1	$\mathrm{End}\left(\mathbb{R}^N\right)$	$2^{(n-1)/2}$
2	$\mathrm{End}\left(\mathbb{C}^N\right)$	$2^{(n-2)/2}$
3	$\mathrm{End}\left(\mathbb{H}^N\right)$	$2^{(n-3)/2}$
4	$\mathrm{End}\left(\mathbb{H}^N\right) \oplus \mathrm{End}\left(\mathbb{H}^N\right)$	$2^{(n-4)/2}$
5	$\mathrm{End}\left(\mathbb{H}^N\right)$	$2^{(n-3)/2}$
6	$\mathrm{End}\left(\mathbb{C}^N\right)$	$2^{(n-2)/2}$
7	$\mathrm{End}\left(\mathbb{R}^N\right)$	$2^{(n-1)/2}$

$$\mathrm{Cl}(1,3) \cong \mathrm{End}\left(\mathbb{R}^4\right),$$
$$\mathrm{Cl}(3,1) \cong \mathrm{End}\left(\mathbb{H}^2\right),$$
$$\mathrm{Cl}^0(1,3) \cong \mathrm{Cl}^0(3,1) \cong \mathrm{End}\left(\mathbb{C}^2\right).$$

6.4 The Spinor Representation

Definition 6.4.1 The vector space of **(Dirac) spinors** is $\Delta_n = \mathbb{C}^N$, where N is given by the values in Table 6.1. The **(Dirac) spinor representation of the complex Clifford algebra**

$$\rho\colon \mathbb{C}\mathrm{l}(n) \longrightarrow \mathrm{End}\,(\Delta_n),$$

defined by the structure theorem of the complex Clifford algebras $\mathbb{C}\mathrm{l}(n)$, is given by Table 6.4. There are induced complex spinor representations of $\mathrm{Cl}(s,t)$.

Notice that the space Δ_n of Dirac spinors is a complex vector space, whose dimension is always even and grows exponentially with n.

Definition 6.4.2 The bilinear map

$$\begin{aligned} \mathbb{R}^{s,t} \times \Delta_n &\longrightarrow \Delta_n \\ (X,\psi) &\longmapsto X\cdot\psi = \rho(\gamma(X))\psi \end{aligned}$$

is called **mathematical Clifford multiplication** of a spinor with a vector. Similarly, **physical Clifford multiplication** is given by $(-i)$ times mathematical Clifford multiplication. More generally, via the isomorphism of vector spaces

$$\Lambda^*\mathbb{R}^{s,t} \longrightarrow \mathrm{Cl}(s,t)$$

from Theorem 6.2.16, followed by the complex spinor representation, we can define a (mathematical) **Clifford multiplication of spinors with forms**.

Example 6.4.3 Suppose that the restriction of the Dirac spinor representation to $\gamma\left(\mathbb{R}^{s,t}\right) \subset \mathrm{Cl}(s,t)$ is given by physical gamma matrices Γ_a and mathematical

Table 6.4 Complex spinor representation

n	Representation
Even	$\mathbb{C}\mathrm{l}(n) \xrightarrow{\cong} \mathrm{End}\,(\Delta_n)$
Odd	$\mathbb{C}\mathrm{l}(n) \xrightarrow{\cong} \mathrm{End}\,(\Delta_n) \oplus \mathrm{End}\,(\Delta_n) \xrightarrow{\mathrm{pr}_1} \mathrm{End}\,(\Delta_n)$

gamma matrices $\gamma_a = i\Gamma_a$ in Mat($N \times N, \mathbb{C}$). Then physical Clifford multiplication of a basis vector $e_a \in \mathbb{R}^{s,t}$ with a spinor $\psi \in \mathbb{C}^N$ is equal to

$$e_a \cdot \psi = \Gamma_a \psi,$$

whereas mathematical Clifford multiplication is equal to

$$e_a \cdot \psi = \gamma_a \psi = i\Gamma_a \psi.$$

For the following result, recall that according to Lemma 6.3.3 there is an isomorphism

$$\mathbb{C}l^0(n) \cong \mathbb{C}l(n-1).$$

Corollary 6.4.4 (Induced Spinor Representation on the Even Clifford Algebra) *Consider the restriction of the Dirac spinor representation to the even subalgebra* $\mathbb{C}l^0(n)$.

- *If n is odd, then the induced representation is irreducible:*

$$\mathbb{C}l^0(n) \xrightarrow{\cong} \mathrm{End}\,(\Delta_n)\,.$$

- *If n is even, then the induced representation splits into* **left-handed (positive)** *and* **right-handed (negative) Weyl spinors**:

$$\mathbb{C}l^0(n) \xrightarrow{\cong} \mathrm{End}\left(\Delta_n^+\right) \oplus \mathrm{End}\left(\Delta_n^-\right).$$

See Table 6.5.

If n is odd we identify here $\mathbb{C}l^0(n)$ with the first summand in

$$\mathbb{C}l(n) \cong \mathrm{End}\,(\Delta_n) \oplus \mathrm{End}\,(\Delta_n)\,,$$

so that the restriction of the spinor representation is non-trivial.

The result in the even-dimensional case can be clarified as follows.

Table 6.5 Weyl spinor representations

n	Induced representation	Spinor space	N
Even	$\mathbb{C}l^0(n) \xrightarrow{\cong} \mathrm{End}\left(\Delta_n^+\right) \oplus \mathrm{End}\left(\Delta_n^-\right)$	$\Delta_n^+ \cong \Delta_n^- \cong \mathbb{C}^{N/2}$	$2^{n/2}$
Odd	$\mathbb{C}l^0(n) \xrightarrow{\cong} \mathrm{End}\,(\Delta_n)$	$\Delta_n \cong \mathbb{C}^N$	$2^{(n-1)/2}$

Proposition 6.4.5 (Weyl Spinors and Chirality in Even Dimensions) *Let $n = 2k$ be even, Δ_n the Dirac spinor representation and Γ_{n+1} the chirality operator.*

1. *$\Delta_n^{\pm}$ can be identified with the (± 1)-eigenspaces of Γ_{n+1} on Δ_n.*
2. *The induced representation of $\mathbb{C}\text{l}^0(n)$ maps $\Delta_n^{\pm}$ to itself, while elements in $\mathbb{C}\text{l}^1(n)$ (such as vectors in $\mathbb{R}^{s,t}$) map $\Delta_n^{\pm}$ to $\Delta_n^{\mp}$. It follows that*

$$\mathbb{C}\text{l}^0(n) \cong \text{Hom}\left(\Delta_n^+, \Delta_n^+\right) \oplus \text{Hom}\left(\Delta_n^-, \Delta_n^-\right),$$
$$\mathbb{C}\text{l}^1(n) \cong \text{Hom}\left(\Delta_n^+, \Delta_n^-\right) \oplus \text{Hom}\left(\Delta_n^-, \Delta_n^+\right).$$

Proof We can split the space $\Delta_n = \mathbb{C}^N$ into the (± 1)-eigenspaces of Γ_{n+1}, because $\Gamma_{n+1}^2 = 1$. We call these spaces $\Delta_n^{\pm}$. We have

$$[\Gamma_{n+1}, \Gamma_a \Gamma_b] = 0 \quad \text{and} \quad \{\Gamma_{n+1}, \Gamma_a\} = 0 \quad \forall 1 \leq a, b \leq n$$

by Lemma 6.3.6. This shows that $\Delta_n^{\pm}$ are invariant under the representation of $\mathbb{C}\text{l}^0(n)$ and also implies the second claim in 2. The final claim then follows from $\mathbb{C}\text{l}(n) \cong \text{End}(\Delta_n)$. □

Remark 6.4.6 Note that the definition of positive and negative Weyl spinors depends on the sign of the chirality operator, which can be chosen arbitrarily. Moreover, in the literature sometimes positive Weyl spinors are called right-handed and negative Weyl spinors left-handed. We continue to use our conventions.

6.5 The Spin Groups

In this section we want to discuss the spin groups, which are certain subgroups embedded in the Clifford algebra. Recall how we defined linear Lie groups in Chap. 1: for $\mathbb{K} = \mathbb{R}, \mathbb{C}, \mathbb{H}$ we started with the associative endomorphism algebra

$$\text{End}\left(\mathbb{K}^n\right) = \mathbb{K}^{n \times n}$$

and considered the group of invertible elements

$$\text{GL}(n, \mathbb{K}) \subset \text{End}\left(\mathbb{K}^n\right),$$

which is an open subset of $\text{End}(\mathbb{K}^n)$. The linear Lie groups were then defined as closed subgroups of the Lie groups $\text{GL}(n, \mathbb{K})$.

To define spin groups we will follow a similar approach, where we replace the endomorphism algebra $\text{End}(\mathbb{K}^n)$ by the Clifford algebra $\text{Cl}(s, t)$. It turns out that spin groups are certain double coverings of (pseudo-)orthogonal groups. In particular, the Lie algebra given by the Clifford algebra with the canonical commutator of associative algebras

$$[a, b] = a \cdot b - b \cdot a \quad \forall a, b \in \mathrm{Cl}(s, t)$$

contains as a Lie subalgebra the Lie algebra of the pseudo-orthogonal group $\mathrm{SO}^+(s, t)$. We follow in this section [13] and [55].

6.5.1 The Pin and Spin Groups

Definition 6.5.1 The **group of invertible elements** in the Clifford algebra $\mathrm{Cl}(s, t)$ is defined by

$$\mathrm{Cl}^\times(s, t) = \{x \in \mathrm{Cl}(s, t) \mid \exists y \in \mathrm{Cl}(s, t) : xy = yx = 1\}.$$

There is an analogous definition for $\mathbb{C}\mathrm{l}(n)$.

Lemma 6.5.2 *The group* $\mathrm{Cl}^\times(s, t)$ *is an open subset of* $\mathrm{Cl}(s, t)$ *and therefore a Lie group.*

Proof Let $n = s + t$. According to Lemma 6.3.2

$$\mathbb{C}\mathrm{l}(n) \cong \mathrm{Cl}(s, t) \otimes_{\mathbb{R}} \mathbb{C}.$$

Decomposing $y \in \mathbb{C}\mathrm{l}(n)$ into $y = u + iv$ with $u, v \in \mathrm{Cl}(s, t)$ it follows that

$$\mathrm{Cl}^\times(s, t) = \mathbb{C}\mathrm{l}^\times(n) \cap \mathrm{Cl}(s, t).$$

However, Theorem 6.3.21 on the structure of complex Clifford algebras implies that $\mathbb{C}\mathrm{l}^\times(n)$ is an open subset of $\mathbb{C}\mathrm{l}(n)$ for all n, because the general linear group $\mathrm{GL}(N, \mathbb{C})$ is open in $\mathrm{End}\left(\mathbb{C}^N\right)$. This implies that the intersection $\mathbb{C}\mathrm{l}^\times(n) \cap \mathrm{Cl}(s, t)$ is open in $\mathrm{Cl}(s, t)$. □

Definition 6.5.3 We define the following subsets of $\mathbb{R}^{s,t}$:

$$\begin{aligned} S_+^{s,t} &= \{v \in \mathbb{R}^{s,t} \mid \eta(v, v) = +1\}, \\ S_-^{s,t} &= \{v \in \mathbb{R}^{s,t} \mid \eta(v, v) = -1\}, \\ S_\pm^{s,t} &= S_+^{s,t} \cup S_-^{s,t}. \end{aligned}$$

Here η is the standard symmetric bilinear form of signature (s, t).

Proposition 6.5.4 *The following subsets of* $\mathrm{Cl}(s,t)$ *form subgroups of the group* $\mathrm{Cl}^{\times}(s,t)$*:*

$$\begin{aligned} \mathrm{Pin}(s,t) &= \left\{ v_1 v_2 \cdots v_r \mid v_i \in S_{\pm}^{s,t}, r \geq 0 \right\}, \\ \mathrm{Spin}(s,t) &= \mathrm{Pin}(s,t) \cap \mathrm{Cl}^0(s,t) \\ &= \left\{ v_1 v_2 \cdots v_{2r} \mid v_i \in S_{\pm}^{s,t}, r \geq 0 \right\}, \\ \mathrm{Spin}^+(s,t) &= \left\{ v_1 \cdots v_{2p} w_1 \cdots w_{2q} \mid v_i \in S_{+}^{s,t}, w_j \in S_{-}^{s,t}, p, q \geq 0 \right\}. \end{aligned}$$

We endow these subsets with the subset topology from the vector space $\mathrm{Cl}(s,t)$*. We also set*

$$\begin{aligned} \mathrm{Pin}(n) &= \mathrm{Pin}(n,0), \\ \mathrm{Spin}(n) &= \mathrm{Spin}(n,0). \end{aligned}$$

We call the group $\mathrm{Pin}(s,t)$ *the* **pin group**, *the group* $\mathrm{Spin}(s,t)$ *the* **spin group** *and the group* $\mathrm{Spin}^+(s,t)$ *the* **orthochronous spin group**.

Proof The proof is not difficult and left as an exercise. Note that in the definition of $\mathrm{Spin}^+(s,t)$ both numbers $2p$ and $2q$ are even. □

Remark 6.5.5 In the literature the group $\mathrm{Spin}^+(s,t)$ is also often simply called the spin group. We added the adjective *orthochronous* to distinguish it from the spin group $\mathrm{Spin}(s,t)$.

Definition 6.5.6 For an element $u = v_1 v_2 \cdots v_r \in \mathrm{Pin}(s,t)$ we set

$$\deg(u) = \begin{cases} 0 & \text{if } u \in \mathrm{Cl}^0(s,t), \\ 1 & \text{if } u \in \mathrm{Cl}^1(s,t). \end{cases}$$

$$\deg^t(u) = \begin{cases} 0 & \text{if an even number of } v_i \in S_{+}^{s,t}, \\ 1 & \text{if an odd number of } v_i \in S_{+}^{s,t}. \end{cases}$$

Hence

$$\begin{aligned} u \in \mathrm{Spin}(s,t) &\Leftrightarrow \deg(u) = 0, \\ u \in \mathrm{Spin}^+(s,t) &\Leftrightarrow \deg(u) = 0 = \deg^t(u). \end{aligned}$$

Definition 6.5.7 We identify the vector space $\mathbb{R}^{s,t}$ in the canonical way via the embedding γ with a vector subspace of $\mathrm{Cl}(s,t)$ and consider the following map:

$$\begin{aligned} R\colon \mathrm{Pin}(s,t) \times \mathbb{R}^{s,t} &\longrightarrow \mathbb{R}^{s,t} \\ (u,x) &\longmapsto (-1)^{\deg(u)} u \cdot x \cdot u^{-1}. \end{aligned}$$

We prove in Lemma 6.5.10 below that this map is well-defined and yields a continuous homomorphism

$$\begin{aligned} \lambda\colon \mathrm{Pin}(s,t) &\longrightarrow \mathrm{O}(s,t) \\ u &\longmapsto R_u = R(u,\cdot). \end{aligned}$$

Remark 6.5.8 Notice that the definition of the map R is very similar to the definition of the maps

$$\begin{aligned} \mathrm{SU}(2) \times \mathbb{R}^3 &\longrightarrow \mathbb{R}^3 \\ (A,X) &\longmapsto AXA^{-1}, \end{aligned}$$

and

$$\begin{aligned} \mathrm{SU}(2) \times \mathrm{SU}(2) \times \mathbb{R}^4 &\longrightarrow \mathbb{R}^4 \\ (A_-, A_+, X) &\longmapsto A_- \cdot X \cdot A_+^{-1}, \end{aligned}$$

from Exercise 1.9.20 and Exercise 1.9.21.

Remark 6.5.9 It will follow from Theorem 6.5.13 that

$$\begin{aligned} u \in \mathrm{Spin}(s,t) &\Leftrightarrow \lambda(u) \in \mathrm{SO}(s,t), \\ u \in \mathrm{Spin}^+(s,t) &\Leftrightarrow \lambda(u) \in \mathrm{SO}^+(s,t). \end{aligned}$$

Hence the degrees deg and deg$'$ of an element $u \in \mathrm{Pin}(s,t)$ measure whether $\lambda(u)$ preserves orientation or time-orientation.

Lemma 6.5.10

1. *The map R is well-defined, i.e. it has image in the subspace $\mathbb{R}^{s,t}$ of the Clifford algebra.*
2. *For any vector $v \in S^{s,t}_{\pm}$ the map*

$$R_v = R(v,\cdot)\colon \mathbb{R}^{s,t} \longrightarrow \mathbb{R}^{s,t}$$

is the reflection in the hyperplane $v^{\perp} \subset \mathbb{R}^{s,t}$.

3. The map R yields a continuous homomorphism

$$\lambda \colon \mathrm{Pin}(s,t) \longrightarrow \mathrm{O}(s,t).$$

Proof Fix a vector $v \in \mathrm{R}^{s,t}$ with $\eta(v,v) = \pm 1$. Then $v^{-1} = \mp v$ and

$$\begin{aligned} R_v(x) &= -v \cdot x \cdot v^{-1} \\ &= \pm v \cdot x \cdot v. \end{aligned}$$

We get

$$R_v(x) = \begin{cases} -x \text{ if } x \parallel v, \\ \;\; x \text{ if } x \perp v, \end{cases}$$

hence $R_v(x) \in \mathbb{R}^{s,t}$ and R_v is reflection in $v^{\perp}$. Since

$$R_{v_1 v_2 \cdots v_r}(x) = R_{v_1} \circ R_{v_2} \circ \ldots \circ R_{v_r}(x),$$

we conclude that $R(u,x)$ is indeed an element of $\mathbb{R}^{s,t}$ for all $u \in \mathrm{Pin}(s,t)$ and $x \in \mathbb{R}^{s,t}$. Since reflections are elements of the orthogonal group, it follows that λ is a continuous homomorphism to the orthogonal group. □

We see that vectors $v \in \mathbb{R}^{s,t}$ with $\eta(v,v) = \pm 1$ act as reflections on $\mathbb{R}^{s,t}$ and a general element a of the pin group acts as a composition of such reflections, hence as an element of the pseudo-orthogonal group. Note that the vectors $\pm v$ define the same reflections, hence elements $\pm a$ of the pin group define the same element of the pseudo-orthogonal group. As we will see in Theorem 6.5.13, this implies that the pin group is a *double covering* of the pseudo-orthogonal group.

We need the following algebraic theorem.

Theorem 6.5.11 (Cartan–Dieudonné Theorem) *Every element of* $\mathrm{O}(s,t)$ *can be written as a composition of at most* $2(s+t)$ *reflections in hyperplanes* $v_i^{\perp}$ *with vectors* $v_i \in S_{\pm}^{s,t}$.

Proof The idea is to set for an element $A \in \mathrm{O}(s,t)$

$$a_i = A^{-1} e_i \quad \forall i = 1, \ldots, s+t$$

with the standard basis $e_1, \ldots, e_{s+t}$. We can then find a composition R of reflections which map the orthonormal basis $a_1, \ldots, a_{s+t}$ to $e_1, \ldots, e_{s+t}$, implying that $A = R$:

we first find a composition R_1 of reflections that maps

$$a_1 \longmapsto e_1$$

as follows:

1. If $a_1 = e_1$, then R_1 is the identity.
2. If the norm of $a_1 - e_1$ is non-zero, then R_1 is the reflection in the hyperplane $(a_1 - e_1)^{\perp}$.
3. If the norm of $a_1 - e_1$ is zero, then R_1 is the composition of the reflection in the hyperplane $(a_1 + e_1)^{\perp}$ followed by the reflection in the hyperplane $e_1^{\perp}$.
4. In the second and third case we normalize the vectors $a_1 - e_1$ and $a_1 + e_1$ to norm ± 1.

Since $R_1(a_1) = e_1$, it follows that R_1 maps

$$a_j \longmapsto e_1^{\perp} = \mathrm{span}(e_2, \ldots, e_{s+t}) \quad \forall j = 2, \ldots, s+t.$$

To prove the claim we then proceed by induction on $s + t$. □

Theorem 6.5.12 (Special Orthogonal Groups and Reflections) *Let $R \in \mathrm{O}(s,t)$ be a composition of reflections in hyperplanes $v_i^{\perp}$ with vectors $v_i \in S_{\pm}^{s,t}$.*

1. *R is an element of $\mathrm{SO}(s,t)$ if and only if the number of vectors v_i is even.*
2. *R is an element of $\mathrm{SO}^+(s,t)$ if and only if both the number of vectors $v_i \in S_+^{s,t}$ and the number of vectors $v_i \in S_-^{s,t}$ are even.*

Proof Notice that for any vector $v \in S_{\pm}^{s,t}$ we can split $\mathbb{R}^{s,t}$ into maximally positive and negative definite subspaces $W_+ \oplus W_-$ so that, with respect to a suitable basis,

$$R_v = \left(\begin{array}{cccc|cccc} -1 & & & & & & & \\ & 1 & & & & & & \\ & & \ddots & & & & & \\ & & & 1 & & & & \\ \hline & & & & 1 & & & \\ & & & & & 1 & & \\ & & & & & & \ddots & \\ & & & & & & & 1 \end{array}\right) \quad \text{if } \eta(v,v) = +1,$$

$$R_v = \left(\begin{array}{cccc|cccc} 1 & & & & & & & \\ & 1 & & & & & & \\ & & \ddots & & & & & \\ & & & 1 & & & & \\ \hline & & & & -1 & & & \\ & & & & & 1 & & \\ & & & & & & \ddots & \\ & & & & & & & 1 \end{array}\right) \quad \text{if } \eta(v, v) = -1.$$

This implies that an even number of reflections in hyperplanes $v^\perp$ with $\eta(v, v) = \pm 1$ is in $\mathrm{SO}(s, t)$, i.e. has determinant 1. If in addition the number of reflections in hyperplanes $v^\perp$ with $\eta(v, v) = +1$ is even, then the map is in $\mathrm{SO}^+(s, t)$. See [13, Theorem 1.5] for more details. □

Theorem 6.5.13 (Relation Between the Pin and Spin Groups and the Orthogonal Groups) *Consider the homomorphism*

$$\begin{aligned} \lambda \colon \mathrm{Pin}(s, t) &\longrightarrow \mathrm{O}(s, t) \\ u &\longmapsto R_u. \end{aligned}$$

1. *The homomorphism λ is open and surjective with kernel equal to $\{\pm 1\}$.*
2. *The preimages under λ of the subgroups* $\mathrm{SO}(s, t)$ *and* $\mathrm{SO}^+(s, t)$ *are equal to* $\mathrm{Spin}(s, t)$ *and* $\mathrm{Spin}^+(s, t)$*, which are therefore open subgroups of* $\mathrm{Pin}(s, t)$.
3. *The homomorphism λ restricts to surjective homomorphisms*

$$\begin{aligned} \lambda \colon \mathrm{Spin}(s, t) &\longrightarrow \mathrm{SO}(s, t), \\ \lambda \colon \mathrm{Spin}^+(s, t) &\longrightarrow \mathrm{SO}^+(s, t), \end{aligned}$$

with kernel equal to $\{\pm 1\}$.

4. *As a topological space the orthochronous spin group* $\mathrm{Spin}^+(s, t)$ *is connected if $s \geq 2$ or $t \geq 2$.*

Proof The statement that λ is surjective and open follows from Theorem 6.5.11 and its proof. We show that the kernel of λ is equal to $\{\pm 1\}$: Suppose that

$$\lambda(u) = I \in \mathrm{O}(s, t).$$

Then $\deg(u) = 0$, since R_u has to be composed of an even number of reflections. It follows that

$$u \cdot e_i \cdot u^{-1} = e_i$$

and therefore

$$e_i \cdot u \cdot e_i = -\eta(e_i, e_i)u \quad \forall i = 1, \ldots, s+t. \tag{6.2}$$

Expanding u in the standard basis for the Clifford algebra, suppose that

$$u = ae_{i_1} \cdots e_{i_{2k}}$$

with $k \geq 1$ and $a \in \mathbb{R}$. Applying Eq. (6.2) with $i = i_{2k}$, it follows that $a = 0$. This implies that $u \in \mathbb{R} \cdot 1$. Since $u \in \mathrm{Pin}(s,t)$, we conclude that $u = \pm 1$. Statements 2. and 3. then follow from Proposition 6.5.12.

We finally show that $\mathrm{Spin}^+(s,t)$ is connected if $s \geq 2$ or $t \geq 2$. Since $\mathrm{SO}^+(s,t)$ is connected and the kernel of λ is $\{\pm 1\}$, it suffices to show that every $u \in \mathrm{Spin}^+(s,t)$ can be connected to $-u$ by a continuous curve in $\mathrm{Spin}^+(s,t)$. Suppose that $s \geq 2$ and consider the curve

$$\gamma(\tau) = -u \cdot (e_1 \cos(\tau) + e_2 \sin(\tau)) \cdot (e_1 \cos(\tau) - e_2 \sin(\tau)),$$

where $\tau \in \mathbb{R}$. It is easy to check that this is indeed a curve in $\mathrm{Spin}^+(s,t)$ with

$$\gamma(0) = u,$$
$$\gamma\left(\frac{\pi}{2}\right) = -u.$$

There is a similar argument in the case $t \geq 2$. □

Since the homomorphism λ is continuous, open and surjective and has kernel equal to $\{\pm 1\}$, it follows that:

Corollary 6.5.14 *We can define a unique Lie group structure on the groups*

$$\mathrm{Pin}(s,t), \quad \mathrm{Spin}(s,t), \quad \mathrm{Spin}^+(s,t)$$

so that λ becomes a smooth double covering of Lie groups.

Remark 6.5.15 There is a more natural way to define a smooth structure on the pin and spin groups: Using a different definition of the pin group, which can be found in the classic paper [7] of Atiyah, Bott and Shapiro, it is possible to show that $\mathrm{Pin}(s,t)$ is a closed subset of $\mathrm{Cl}^\times(s,t)$ and therefore by Cartan's Theorem 1.1.44 an embedded Lie subgroup. Theorem 6.5.13 then implies that $\mathrm{Spin}(s,t)$ and $\mathrm{Spin}^+(s,t)$ are also embedded Lie subgroups of $\mathrm{Cl}^\times(s,t)$. The Lie group structure on these groups as closed subgroups of $\mathrm{Cl}^\times(s,t)$ coincides with the one from Corollary 6.5.14.

Corollary 6.5.16 *For all $n \geq 3$ the homomorphisms*

$$\lambda\colon \mathrm{Spin}(n) \longrightarrow \mathrm{SO}(n),$$
$$\lambda\colon \mathrm{Spin}^+(n,1) \longrightarrow \mathrm{SO}^+(n,1),$$
$$\lambda\colon \mathrm{Spin}^+(1,n) \longrightarrow \mathrm{SO}^+(1,n)$$

are the universal coverings.

Proof This follows, because according to Proposition 2.6.3 and Remark 6.1.19 for $n \geq 3$ the Lie groups $\mathrm{SO}(n)$, $\mathrm{SO}^+(n,1)$ and $\mathrm{SO}^+(1,n)$ have fundamental group $\mathbb{Z}_2$. □

Example 6.5.17 Exercises 1.9.20 and 1.9.21 imply that there exist isomorphisms

$$\mathrm{Spin}(3) \cong \mathrm{SU}(2),$$
$$\mathrm{Spin}(4) \cong \mathrm{SU}(2) \times \mathrm{SU}(2).$$

Similarly, it can be shown that

$$\mathrm{Spin}^+(1,3) \cong \mathrm{SL}(2,\mathbb{C}).$$

See Sect. 6.8.2 and Exercise 6.13.17.

6.5.2 The Spinor Representation of the Orthochronous Spin Group

Definition 6.5.18 We denote by

$$\kappa\colon \mathrm{Spin}^+(s,t) \longrightarrow \mathrm{GL}(\Delta_n)$$

the **spinor representation** induced by restriction of the spinor representation of the even Clifford algebra $\mathrm{Cl}^0(s,t)$.

Proposition 6.5.19 *The spinor representation is compatible with Clifford multiplication in the following way:*

$$\kappa(g)(x \cdot \psi) = (\lambda(g)x) \cdot (\kappa(g)\psi),$$

for all $g \in \mathrm{Spin}^+(s,t)$, $x \in \mathbb{R}^{s,t}$ and $\psi \in \Delta_n$.

Proof Let

$$\rho: \mathrm{Cl}(s,t) \longrightarrow \mathrm{End}(\Delta_n)$$

denote the spinor representation. Then

$$\begin{aligned}
\kappa(g)(x \cdot \psi) &= \rho(g)\rho(x)(\psi) \\
&= \rho\left(gxg^{-1}\right)\rho(g)(\psi) \\
&= \rho(\lambda(g)x)\rho(g)(\psi) \\
&= (\lambda(g)x) \cdot (\kappa(g)\psi).
\end{aligned}$$

□

6.5.3 The Lie Algebra of the Spin Group

In this subsection we determine the Lie algebra of $\mathrm{Spin}^+(s,t)$ and calculate the differential of the covering homomorphism λ from the orthochronous spin group to the proper orthochronous orthogonal group.

The Lie group $\mathrm{Cl}^\times(s,t)$ is an open subset of $\mathrm{Cl}(s,t)$, hence its Lie algebra $\mathfrak{cl}^\times(s,t)$ is canonically isomorphic as a vector space to $\mathrm{Cl}(s,t)$ with commutator

$$[x,y] = x \cdot y - y \cdot x, \quad \forall x, y \in \mathrm{Cl}(s,t).$$

Since $\mathrm{Pin}(s,t)$ and $\mathrm{Spin}^+(s,t)$ are Lie subgroups of $\mathrm{Cl}^\times(s,t)$, it follows that the Lie algebras $\mathfrak{pin}(s,t)$ and $\mathfrak{spin}^+(s,t)$ are Lie subalgebras of $\mathfrak{cl}^\times(s,t)$.

Definition 6.5.20 Let $e_1, \ldots, e_{s+t}$ be an orthonormal basis of $\mathbb{R}^{s,t}$. We define

$$M(s,t) = \mathrm{span}\{e_i e_j \in \mathrm{Cl}(s,t) \mid 1 \le i < j \le s+t\}.$$

Lemma 6.5.21 *The vector space $M(s,t)$ is a Lie subalgebra of $\mathfrak{cl}^\times(s,t)$ of dimension*

$$\dim M(s,t) = \frac{1}{2}(s+t)(s+t-1).$$

Proof This is Exercise 6.13.9. □

Proposition 6.5.22 (Description of the Lie Algebra of the Spin Group) *For all $s,t \ge 0$ the following identity holds:*

$$\mathfrak{spin}^+(s,t) = M(s,t).$$

Proof We follow an idea in [99]. To show that e_ie_j for $i < j$ is an element of the Lie algebra of the spin group, it suffices to find a curve in $\mathrm{Spin}^+(s,t)$ through the unit element 1 with velocity vector e_ie_j. We set $\eta_i = \eta(e_i, e_i) = ||e_i||^2_\eta$.

1. Suppose that $i \neq j$ and $\eta_i = \eta_j$. Then the curve

$$\gamma(\tau) = \cos(\tau)1 + \sin(\tau)e_ie_j, \quad \tau \in \mathbb{R}$$

through 1 is contained in $\mathrm{Spin}^+(s,t)$, because we can write

$$\cos(\tau)1 + \sin(\tau)e_ie_j = e_i(-\eta_i \cos(\tau)e_i + \sin(\tau)e_j)$$

and

$$\begin{aligned} || - \eta_i \cos(\tau)e_i + \sin(\tau)e_j||^2_\eta &= \eta_i \cos^2(\tau) + \eta_j \sin^2(\tau) \\ &= \eta_i \\ &= ||e_i||^2_\eta. \end{aligned}$$

2. In the case $i \neq j$ and $\eta_i = -\eta_j$ a similar argument shows that the curve

$$\gamma(\tau) = \cosh(\tau)1 + \sinh(\tau)e_ie_j, \quad \tau \in \mathbb{R}$$

through 1 is contained in $\mathrm{Spin}^+(s,t)$.

Taking the derivative in $\tau = 0$ of both curves, it follows that

$$e_ie_j \in \mathfrak{spin}^+(s,t) \quad \forall i \neq j.$$

This implies that

$$M(s,t) \subset \mathfrak{spin}^+(s,t).$$

However, the dimensions of both vector spaces agree and we conclude that $M(s,t) = \mathfrak{spin}^+(s,t)$. □

Let E_{ij} be the elementary $(s+t) \times (s+t)$-matrix with a 1 at the intersection of the i-th row and the j-th column and zeros elsewhere. We define the matrix

$$\epsilon_{ij} = \eta_i E_{ji} - \eta_j E_{ij}.$$

Corollary 6.5.23 (Differential of the Covering Homomorphism λ) *The differential of the homomorphism*

$$\lambda\colon \mathrm{Spin}^+(s,t) \longrightarrow \mathrm{SO}^+(s,t)$$

is given by

$$\lambda_*\colon \mathfrak{spin}^+(s,t) \longrightarrow \mathfrak{so}^+(s,t)$$

with

$$\begin{aligned}\lambda_*(z)x &= [z,x]\\ &= z\cdot x - x\cdot z\end{aligned}$$

for all $x \in \mathbb{R}^{s,t}$*. This implies the following explicit formula:*

$$\lambda_*(e_ie_j) = 2\epsilon_{ij} \quad \forall i<j.$$

We can recover any $z \in \mathfrak{spin}^+(s,t)$ *from its image* $\lambda_*(z)$ *via*

$$z = \frac{1}{2}\sum_{k<l} \eta(\lambda_*(z)e_k, e_l)\eta_k\eta_l e_k e_l.$$

Proof The formula $\lambda_*(z)x = [z,x]$ follows from

$$\lambda(u)x = u\cdot x\cdot u^{-1} \quad \forall u \in \mathrm{Spin}^+(s,t).$$

The remaining formulas follow by direct calculation. □

The isomorphism λ_* between $\mathfrak{spin}^+(s,t)$ and $\mathfrak{so}^+(s,t)$ will appear in the formula for the spin covariant derivative in Sect. 6.10.2 and also, in a rather unexpected way, in the discussion of the Grand Unified Theory with gauge group Spin(10) in Sect. 9.5.5.

6.6 *Majorana Spinors

In Sect. 6.6 and Sect. 6.7 we discuss two concepts – Majorana spinors and spin invariant scalar products – that are less well-known in the mathematical literature, but play quite an important role in physics. Majorana spinors are related to so-called *real* or *quaternionic structures* for the spinor representation. It turns out that in every dimension the spinor representation admits a real or quaternionic structure, which in even dimensions may or may not be compatible with the decomposition into Weyl spinors.

Definition 6.6.1 Let V be a complex vector space with a representation of a Lie group G.

1. A **real structure** on V is a complex antilinear G-equivariant map $\sigma\colon V \to V$ with $\sigma^2 = \mathrm{Id}$.
2. A **quaternionic structure** on V is a complex antilinear G-equivariant map $J\colon V \to V$ with $J^2 = -\mathrm{Id}$.

Proposition 6.6.2 *Let σ be a real structure on the complex G-representation space V and V^σ the real subspace*

$$V^\sigma = \{v \in V \mid \sigma(v) = v\}.$$

Then

$$V = V^\sigma \oplus iV^\sigma,$$

hence we can decompose elements in V into a "real" and "imaginary" part. The complex representation on V induces real representations of G on V^σ and iV^σ, which are both isomorphic (as real representations).

Proof We set

$$V^\sigma = \{v \in V \mid \sigma(v) = v\},$$
$$V'^\sigma = \{v \in V \mid \sigma(v) = -v\}.$$

It is clear that V^σ and V'^σ are real subspaces of V and

$$v = \frac{1}{2}(v + \sigma(v)) + \frac{1}{2}(v - \sigma(v))$$

is a decomposition of a vector $v \in V$ into vectors in V^σ and V'^σ. Hence we can decompose $V = V^\sigma \oplus V'^\sigma$. Since σ is complex antilinear, we have

$$v \in V^\sigma \Leftrightarrow iv \in V'^\sigma.$$

This proves $V'^\sigma = iV^\sigma$. If $g \in G$ and $v \in V^\sigma$, then

$$\sigma(g \cdot v) = g \cdot \sigma(v) = g \cdot v,$$

because σ is G-equivariant. This shows that $g \cdot v \in V^\sigma$, hence G preserves V^σ, and similarly V'^σ. It follows that the representation of G restricts to representations on these real vector spaces. The map

$$\begin{aligned} f\colon V^\sigma &\longrightarrow V'^\sigma \\ v &\longmapsto iv \end{aligned}$$

is G-equivariant, because the G-representation on V is complex, and thus defines a real isomorphism between the G-representations on V^σ and V'^σ. □

Proposition 6.6.3 *Let J be a quaternionic structure on the complex G-representation space V and*

$$I: V \longrightarrow V$$
$$v \longmapsto iv.$$

Then I, J and $K = IJ$ define the structure of a G-equivariant quaternionic vector space on V.

Proof It is clear that I is G-equivariant, because the G-representation on V is complex, and J is G-equivariant by definition. Hence K is also G-equivariant. Moreover, $I^2 = -1$ and $J^2 = -1$. Since J is complex antilinear, we have $JI = -IJ$. This implies that I, J, K satisfy the quaternionic identities. □

Definition 6.6.4 Let $\kappa: \mathrm{Spin}^+(r,s) \rightarrow \mathrm{GL}(\Delta)$ with $\Delta = \Delta_n$ be the complex spinor representation.

1. The spinor representation Δ is called **Majorana** if it admits a real structure σ. In this case there exists in Δ a real subspace

$$\Delta^\sigma = \{s \in \Delta \mid \sigma(s) = s\}$$

of half dimension

$$\dim_{\mathbb{R}} \Delta^\sigma = \frac{1}{2}\dim_{\mathbb{R}} \Delta = \dim_{\mathbb{C}} \Delta$$

and the complex spinor representation on Δ induces a real representation of $\mathrm{Spin}^+(s,t)$ on Δ^σ. Elements of Δ^σ are called Majorana spinors. We can decompose any $\psi \in \Delta$ uniquely as

$$\psi = \phi_1 + i\phi_2$$

with Majorana spinors ϕ_1, ϕ_2. Spinors $\alpha_1, \ldots, \alpha_k$ which form a real basis for Δ^σ satisfy

$$\sigma(\alpha_i) = \alpha_i \quad \forall i.$$

2. The spinor representation Δ is called **symplectic Majorana** if it admits a quaternionic structure J. In this case Δ has a $\mathrm{Spin}^+(r,s)$-equivariant structure $I, J, K = IJ$ of a quaternionic vector space. Elements of Δ are called symplectic Majorana spinors. We can find a complex basis $\chi_1, \lambda_1, \ldots, \chi_k, \lambda_k$ of Δ so that

the spinors $\chi_1, \ldots, \chi_k$ form a quaternionic basis of Δ and J is given by

$$J(\chi_i) = \lambda_i,$$
$$J(\lambda_i) = -\chi_i \quad \forall i,$$

hence they are related by the standard symplectic matrix

$$\begin{pmatrix} 0 & 1 \\ -1 & 0 \end{pmatrix}.$$

This explains the name *symplectic* Majorana spinors.

Remark 6.6.5 In the physics literature (see, for instance, [54]) one writes for a spinor $\psi \in \Delta$

$$\psi^C = \sigma(\psi)$$

if σ is a real structure and

$$\psi^C = J(\psi)$$

if J is a quaternionic structure for the spinor representation Δ. The spinor ψ^C is called the **charge conjugate**. Choosing a complex basis for the spinor space Δ, we can identify $\Delta = \mathbb{C}^N$. Real and quaternionic structures then correspond to a matrix B with

$$\psi^C = B^{-1}\psi^*, \tag{6.3}$$

where * denotes the complex conjugate, B satisfies $B^*B = I_N$ for a real and $B^*B = -I_N$ for a quaternionic structure (with the unit matrix I_N), and $\mathrm{Spin}^+(s,t)$-equivariance corresponds to

$$\kappa(g)^* = B\kappa(g)B^{-1} \quad \forall g \in \mathrm{Spin}^+(s,t),$$

where $\kappa\colon \mathrm{Spin}^+(s,t) \to \mathrm{GL}(\Delta)$ denotes the spinor representation.
Recall that in even dimensions the complex spinor representation Δ splits into the complex Weyl spinors $\Delta = \Delta_+ \oplus \Delta_-$. We denote by $\pi_+\colon \Delta \to \Delta_+$ the projection along Δ_-.

Definition 6.6.6 Let $\kappa\colon \mathrm{Spin}^+(r,s) \to \mathrm{GL}(\Delta)$ with $\Delta = \Delta_n$ be the complex spinor representation over an even-dimensional vector space.

1. The spinor representation Δ is called **Majorana–Weyl** if it admits a real structure σ that commutes with π_+. In this case σ induces on both Weyl spinor spaces $\Delta_\pm$ a real structure. The elements of

$$\Delta_\pm^\sigma = \{s \in \Delta_\pm \mid \sigma(s) = s\}$$

are called left-handed and right-handed Majorana–Weyl spinors.

Table 6.6 Majorana spinors

ρ mod 8	Type of spinors	Minimal real spinor representation
0	Majorana–Weyl	$\Delta^\sigma_\pm$ of dimension $2^{(n-2)/2}$
1	Majorana, not Weyl	Δ^σ of dimension $2^{(n-1)/2}$
2	Majorana and Weyl, but not Majorana–Weyl	Δ^σ and $\Delta_\pm$ of dimension $2^{n/2}$
3	Symplectic Majorana, not Weyl	Δ of dimension $2^{(n+1)/2}$
4	Symplectic Majorana–Weyl	$\Delta_\pm$ of dimension $2^{n/2}$
5	Symplectic Majorana, not Weyl	Δ of dimension $2^{(n+1)/2}$
6	Majorana and Weyl, but not Majorana–Weyl	Δ^σ and $\Delta_\pm$ of dimension $2^{n/2}$
7	Majorana, not Weyl	Δ^σ of dimension $2^{(n-1)/2}$

2. The spinor representation Δ is called **symplectic Majorana–Weyl** if it admits a quaternionic structure J that commutes with π_+. In this case $I, J, K = IJ$ induce on both Weyl spinor spaces $\Delta_\pm$ the structure of a quaternionic vector space. Elements of $\Delta_\pm$ are called left-handed and right-handed symplectic Majorana–Weyl spinors.

It is known precisely in which dimensions and signatures these types of spinors exist: For the standard scalar product η on $\mathbb{R}^{s,t}$ of signature (s,t) we set again $\rho = s - t$ and $n = s + t$. We then have, without proof, Table 6.6 (see [40] and compare with Theorem 6.3.23).

Example 6.6.7 For Minkowski spacetime $\mathbb{R}^{n-1,1}$ of dimension n we have $n = \rho + 2$. We see that in Minkowski spacetime of dimension 4 there exist both Majorana and Weyl spinors of real dimension 4, but not Majorana–Weyl spinors. Note that this implies that Majorana spinors in this dimension always have components both in Δ_+ and Δ_-.

In Minkowski spacetime of dimension 2 and 10 (relevant in superstring theory) there exist Majorana–Weyl spinors of real dimension 1 and 16, respectively, and in Minkowski spacetime of dimension 11 (relevant in M-theory) there exist Majorana spinors of real dimension 32.

6.7 *Spin Invariant Scalar Products

In this section we will study complex bilinear and Hermitian scalar products on the spinor space Δ, which are invariant under the action of the spin group. It turns out that the combination of both types of scalar products is related to the existence of real and quaternionic structures and thus to Majorana spinors. Hermitian scalar products are particularly important, because we need them in Chap. 7 to define Lorentz invariant Lagrangians involving spinors.

6.7.1 Majorana Forms

Definition 6.7.1 Let $\Delta = \Delta_n$ be the complex spinor representation of $\mathrm{Cl}(s,t)$. We fix constants $\mu, \nu = \pm 1$ and consider complex bilinear, non-degenerate forms

$$(\cdot,\cdot)\colon \Delta \times \Delta \longrightarrow \mathbb{C}$$

with the following properties:

1. $(X \cdot \psi, \phi) = \mu(\psi, X \cdot \phi)$ for all $X \in \mathbb{R}^{s,t}$ and all $\psi, \phi \in \Delta$.
2. $(\psi, \phi) = \nu(\phi, \psi)$ for all $\psi, \phi \in \Delta$.

We call such a form a **Majorana form**.

Lemma 6.7.2 *For a Majorana form let $\{\chi_\alpha\}$ be a complex basis of Δ and C the matrix with entries*

$$C_{\alpha\beta} = (\chi_\alpha, \chi_\beta).$$

If we expand

$$\psi = \sum_\alpha \psi_\alpha \chi_\alpha, \quad \phi = \sum_\alpha \phi_\alpha \chi_\alpha$$

and identify ψ, ϕ with the column vectors with entries ψ_α, ϕ_α, then

$$(\psi, \phi) = \psi^T C \phi.$$

Furthermore, property 1. and 2. in Definition 6.7.1 are equivalent to

1. $\gamma_a^T = \mu C \gamma_a C^{-1}$ *for all* $a = 1, \ldots, s+t$.
2. $C^T = \nu C$.

The first equation also holds with the physical Clifford matrices Γ_a instead of the mathematical gamma matrices γ_a.

Proof This is left as an exercise. □

Definition 6.7.3 The matrix C is called the **charge conjugation matrix** (this convention is a bit confusing, because the matrix C rather than B from Sect. 6.6 is called the charge conjugation matrix).

Lemma 6.7.4 *Every Majorana form is invariant under the action of* $\mathrm{Spin}^+(s,t)$.

Proof Property 1. in Definition 6.7.1 implies

$$\begin{aligned}(X \cdot \psi, X \cdot \phi) &= \mu(\psi, (X \cdot X) \cdot \phi) \\ &= -\mu\eta(X,X)(\psi, \phi) \quad \forall X \in \mathbb{R}^{s,t}.\end{aligned}$$

Since every element $g \in \mathrm{Spin}^+(s,t)$ is of the form

$$g = X_1 \cdots X_{2p} Y_1 \cdots Y_{2q}$$

with $\eta(X_i, X_i) = 1, \eta(Y_j, Y_j) = -1$ it follows that

$$\begin{aligned}(g \cdot \psi, g \cdot \phi) &= \mu^{2p+2q} \eta(X_1, X_1) \cdots \eta(X_{2p}, X_{2p}) \eta(Y_1, Y_1) \cdots \eta(Y_{2q}, Y_{2q})(\psi, \phi) \\ &= (\psi, \phi).\end{aligned}$$

□

Not all combinations of μ and ν are possible. Table 6.7 lists without proof the combinations that are allowed, depending on the dimension $n = s + t$ (see [40, 54, 140]). In even dimensions n there are always two possibilities.

Example 6.7.5 In dimension 4 a charge conjugation matrix is necessarily antisymmetric

$$C^T = -C$$

and can satisfy either

$$\gamma_a^T = C \gamma_a C^{-1}$$

or

$$\gamma_a^T = -C \gamma_a C^{-1}$$

for all $a = 1, 2, 3, 4$.

Table 6.7 Signs in Majorana forms

n mod 8	μ	ν
0	−1	+1
0	+1	+1
1	+1	+1
2	+1	+1
2	−1	−1
3	−1	−1
4	−1	−1
4	+1	−1
5	+1	−1
6	+1	−1
6	−1	+1
7	−1	+1

Definition 6.7.6 The **Majorana conjugate** $\tilde{\psi}$ of a spinor $\psi \in \Delta$ with respect to a Majorana form is defined by

$$\tilde{\psi} = (\psi, \cdot) \in \Delta^*.$$

In a basis $\{\chi_\alpha\}$ for Δ where $\psi = \sum_\alpha \psi_\alpha \chi_\alpha$ we have with respect to the dual basis of Δ^*

$$\tilde{\psi} = \psi^T C.$$

In the literature (for example, [54]) the Majorana conjugate is often denoted by a bar.

Majorana forms have very interesting applications in **neutrino physics**, because they can be used to define a **Majorana mass term** for neutrinos; see Sect. 7.8 and Sect. 9.2.4.

Remark 6.7.7 In our discussion so far we assumed that the components of spinors in expressions like

$$(\psi, \phi) = \psi^T C \phi$$

are commuting complex numbers. In quantum field theory, spinors become fields of operators on spacetime acting on a Hilbert space. In the classical limit $\hbar \to 0$ these operators are **anticommuting**. If we treat spinors as anticommuting, then we have to introduce another minus sign in property 2. in Definition 6.7.1:

$$(\psi, \phi) = -\nu(\phi, \psi) \quad \forall \psi, \phi \in \Delta$$

with ν still given by Table 6.7 (property 1. stays the same). This has the paradoxical consequence that symmetric (antisymmetric) Majorana forms become antisymmetric (symmetric). For instance, in the situation of Example 6.7.5, a Majorana form in dimension 4 with anticommuting spinors is symmetric. We will always use commuting spinors except where stated otherwise.

6.7.2 Dirac Forms

Definition 6.7.8 Let $\Delta = \Delta_n$ be the complex spinor representation of $\mathrm{Cl}(s, t)$. We fix a constant $\delta = \pm 1$ and consider non-degenerate $\mathbb{R}$-bilinear forms

$$\langle \cdot, \cdot \rangle : \Delta \times \Delta \longrightarrow \mathbb{C}$$

with the following properties, where * denotes complex conjugation:

1. $\langle X \cdot \psi, \phi \rangle = \delta \langle \psi, X \cdot \phi \rangle$ for all $X \in \mathbb{R}^{s,t}$ and all $\psi, \phi \in \Delta$.
2. $\langle \psi, \phi \rangle = \langle \phi, \psi \rangle^*$ for all $\psi, \phi \in \Delta$.
3. $\langle \psi, c\phi \rangle = c\langle \psi, \phi \rangle = \langle c^* \psi, \phi \rangle$ for all $\psi, \phi \in \Delta$ and all $c \in \mathbb{C}$.

We call such a form a **Dirac form** (we do *not* assume that the form is positive definite).

Lemma 6.7.9 *For a Dirac form let $\{\chi_\alpha\}$ be a complex basis of Δ and A the matrix with entries*

$$A_{\alpha\beta} = \langle \chi_\alpha, \chi_\beta \rangle.$$

If we expand

$$\psi = \sum_\alpha \psi_\alpha \chi_\alpha, \quad \phi = \sum_\alpha \phi_\alpha \chi_\alpha$$

and also denote by ψ, ϕ the column vectors with entries ψ_α, ϕ_α, then

$$\langle \psi, \phi \rangle = \psi^\dagger A \phi.$$

Furthermore, property 1. and 2. in Definition 6.7.8 are equivalent to

1. $\gamma_a^\dagger = \delta A \gamma_a A^{-1}$ for all $a = 1, \ldots, s + t$.
2. $A^\dagger = A$.

There is an equivalent equation to the first one with physical Clifford matrices Γ_a:

1. $\Gamma_a^\dagger = -\delta A \Gamma_a A^{-1}$ for all $a = 1, \ldots, s + t$.

Proof This is left as an exercise. □

Lemma 6.7.10 *Every Dirac form is invariant under the action of* $\text{Spin}^+(s, t)$.

Proof This is left as an exercise. □

Definition 6.7.11 We call a complex representation of the Clifford algebra $\text{Cl}(s, t)$ **basis unitary** if all gamma matrices γ_a are unitary, or equivalently if all Γ_a are unitary.

It can be shown that the spinor representations of Clifford algebras can be chosen basis unitary, see Exercise 6.13.5. We will choose in all examples that we discuss a Hermitian scalar product on the spinor space such that the spinor representation is basis unitary. Then the identity

$$\Gamma_a^{-1} = \eta(e_a, e_a)\Gamma_a$$

Table 6.8 Two options for matrix A defining Dirac forms

δ	A	ϵ
$(-1)^{t+1}$	$\epsilon\Gamma_{s+1}\cdots\Gamma_{s+t}$	$\epsilon^* = (-1)^{t(t+1)/2}\epsilon$
$(-1)^s$	$\epsilon\Gamma_1\cdots\Gamma_s$	$\epsilon^* = (-1)^{s(s-1)/2}\epsilon$

implies

$$\Gamma_a^\dagger = \Gamma_a \quad \forall a = 1, \ldots, s,$$
$$\Gamma_a^\dagger = -\Gamma_a \quad \forall a = s+1, \ldots, s+t.$$

Remark 6.7.12 If a representation of the Clifford algebra is basis unitary, then all matrices

$$\sum_{a=1}^n v_a\Gamma_a \quad \text{with} \quad v_a \in \mathbb{R}, \sum_{a=1}^n v_a^2 = 1$$

are unitary. Hence for $t = 0$ the induced representations of the Lie groups $\mathrm{Pin}(n)$ and $\mathrm{Spin}(n)$ are unitary. This is not the case if $t \geq 1$.
The following result can be found in [49].

Proposition 6.7.13 (Standard Expressions for Dirac Forms) *For a basis unitary spinor representation of* $\mathrm{Cl}(s,t)$ *consider one of the two options for the matrix A in Table 6.8, where* $\epsilon \in \mathbb{C}$ *(we always choose* $\epsilon = 1$ *or* $\epsilon = i$*). In each case A satisfies the properties of Lemma 6.7.9 and defines a Dirac form for the spinor representation. The matrix A is unitary for both options.*

Proof We only prove the first case, the second one follows similarly. We calculate for $a = 1, \ldots, s$

$$A\Gamma_a = (-1)^t\Gamma_a A = -\delta\Gamma_a^\dagger A$$

and for $a = s+1, \ldots, s+t$

$$\begin{aligned} A\Gamma_a &= (-1)^{a-s-t}\epsilon\Gamma_{s+1}\cdots\Gamma_a\Gamma_a\cdots\Gamma_{s+t} \\ &= (-1)^{t+1}\Gamma_a A \\ &= -\delta\Gamma_a^\dagger A. \end{aligned}$$

This verifies property 1. in Definition 6.7.8. To check that A is Hermitian we calculate

$$\begin{aligned} A^\dagger &= \epsilon^*\Gamma_{s+t}^\dagger\cdots\Gamma_{s+1}^\dagger \\ &= \epsilon^*(-1)^t\Gamma_{s+t}\cdots\Gamma_{s+1} \end{aligned}$$

$$= \epsilon^*(-1)^t(-1)^{t(t-1)/2}\Gamma_{s+1}\cdots\Gamma_{s+t}$$
$$= A.$$

It is also easy to check that A is unitary. □

Remark 6.7.14 If $t = 0$, the first choice yields $A = I$ with $\delta = -1, \epsilon = 1$. In this case all physical gamma matrices are Hermitian (and unitary by assumption), while the mathematical gamma matrices are skew-Hermitian.

Definition 6.7.15 The **Dirac conjugate** $\overline{\psi}$ of a spinor $\psi \in \Delta$ with respect to a Dirac form is defined by

$$\overline{\psi} = \langle\psi, \cdot\rangle \in \Delta^*.$$

In a basis $\{\chi_\alpha\}$ for Δ where $\psi = \sum_\alpha \psi_\alpha\chi_\alpha$ we have with respect to the dual basis of Δ^*

$$\overline{\psi} = \psi^\dagger A.$$

We write the Dirac scalar product of spinors ψ, ϕ as

$$\langle\psi, \phi\rangle = \overline{\psi}\phi.$$

Dirac forms are used in the Standard Model to define a **Dirac mass term** in the Lagrangian for all fermions (except possibly the neutrinos) and, together with the Dirac operator, the **kinetic term** and the **interaction term**; see Sect. 7.6.

Remark 6.7.16 If we treat spinors as anticommuting, then contrary to the situation in Remark 6.7.7, we still have

$$\langle\psi, \phi\rangle = \langle\phi, \psi\rangle^* \quad \forall\psi, \phi \in \Delta.$$

The explanation is that $\langle\psi, \psi\rangle$ has to be real for all $\psi \in \Delta$. More generally, complex anticommuting Grassmann numbers α, β satisfy $(\alpha\beta)^* = \beta^*\alpha^*$.

6.7.3 *Relation Between Invariant Forms and Majorana Spinors*

We follow in this subsection [140]. Let $\Delta = \Delta_n$ be the complex spinor representation of $\mathrm{Cl}(s, t)$. We fix a Majorana form $(\cdot\,, \cdot)$ and a Dirac form $\langle\cdot\,, \cdot\rangle$, described in a basis $\{\chi_\alpha\}$ for Δ by matrices C, A with

$$\Gamma_a^T = \mu C \Gamma_a C^{-1} \quad \forall a = 1, \ldots, s+t,$$
$$C^T = \nu C$$

and

$$\Gamma_a^\dagger = -\delta A \Gamma_a A^{-1} \quad \forall a = 1, \ldots, s+t,$$
$$A^\dagger = A.$$

Lemma 6.7.17 *There exists a unique complex antilinear map*

$$\tau : \Delta \longrightarrow \Delta$$

such that

$$(\psi, \phi) = \langle \tau(\psi), \phi \rangle.$$

In the basis $\{\chi_\alpha\}$ the map τ is given by

$$\begin{aligned} \tau : \Delta &\longrightarrow \Delta \\ \psi &\longmapsto B^{-1}\psi^* \end{aligned}$$

with matrix B equal to

$$B = (C^\dagger)^{-1} A.$$

Proof The existence of a unique map τ follows, because the Majorana form is non-degenerate. The map τ is complex antilinear, because the Dirac form is complex antilinear in the first entry. A simple calculation shows that B is the correct matrix. Note that

$$(B^{-1})^\dagger = CA^{-1}.$$

□

Remark 6.7.18 In the literature (see [140]) one sometimes finds the definition $B^T = CA^{-1}$. We continue to use our definition.

Lemma 6.7.19 *The map τ satisfies*

$$\tau(X \cdot \psi) = \mu\delta X \cdot \tau(\psi) \quad \forall \psi \in \Delta, X \in \mathbb{R}^{s,t}.$$

Equivalently, the matrix B satisfies

$$\Gamma_a^* = -\mu\delta B\Gamma_a B^{-1} \quad \forall a = 1, \ldots, s+t.$$

Proof We calculate:

$$\begin{aligned}\langle \tau(X \cdot \psi), \phi \rangle &= (X \cdot \psi, \phi) \\ &= \mu(\psi, X \cdot \phi) \\ &= \mu\langle \tau(\psi), X \cdot \phi \rangle \\ &= \mu\delta\langle X \cdot \tau(\psi), \phi \rangle.\end{aligned}$$

This implies the claim, because the Dirac form is non-degenerate. The formula for B follows from this or by direct calculation. □

Theorem 6.7.20 (Scalar Products and Real and Quaternionic Structures) *Suppose that the spinor representation Δ is basis unitary, the matrix C is unitary and (A, δ) is one of the two choices in Proposition 6.7.13. Then $B = CA$ is unitary.*

1. *In the first case, if $A = \epsilon\Gamma_{s+1} \cdots \Gamma_{s+t}$, then the matrix B satisfies*

$$B^*B = \nu\mu^t(-1)^{t(t-1)/2}I.$$

2. *In the second case, if $A = \epsilon\Gamma_1 \cdots \Gamma_s$, then the matrix B satisfies*

$$B^*B = \nu\mu^s(-1)^{s(s-1)/2}I.$$

3. *In both choices for A the matrix B^*B is $+I$ if and only if*

$$\rho \equiv 0, 1, 7 \bmod 8 \quad \text{and all } \mu, \nu \text{ for both choices for } A,$$

$$\rho \equiv 2 \bmod 8 \quad \text{and} \quad \begin{cases} \mu(-1)^{n/2} = -1 & \text{if } A = \epsilon\Gamma_{s+1} \cdots \Gamma_{s+t}, \\ \mu(-1)^{n/2} = +1 & \text{if } A = \epsilon\Gamma_1 \cdots \Gamma_s, \end{cases}$$

$$\rho \equiv 6 \bmod 8 \quad \text{and} \quad \begin{cases} \mu(-1)^{n/2} = +1 & \text{if } A = \epsilon\Gamma_{s+1} \cdots \Gamma_{s+t}, \\ \mu(-1)^{n/2} = -1 & \text{if } A = \epsilon\Gamma_1 \cdots \Gamma_s, \end{cases}$$

where $\rho = s - t$. In these situations B defines a real structure for the spinor representation of the orthochronous spin group.

4. *The matrix B^*B is $-I$ if and only if*

$$\rho \equiv 3,4,5 \bmod 8 \quad \textit{and all } \mu, \nu \textit{ for both choices for } A,$$

$$\rho \equiv 2 \bmod 8 \quad \textit{and} \quad \begin{cases} \mu(-1)^{n/2} = +1 & \textit{if } A = \epsilon\Gamma_{s+1}\cdots\Gamma_{s+t}, \\ \mu(-1)^{n/2} = -1 & \textit{if } A = \epsilon\Gamma_1\cdots\Gamma_s, \end{cases}$$

$$\rho \equiv 6 \bmod 8 \quad \textit{and} \quad \begin{cases} \mu(-1)^{n/2} = -1 & \textit{if } A = \epsilon\Gamma_{s+1}\cdots\Gamma_{s+t}, \\ \mu(-1)^{n/2} = +1 & \textit{if } A = \epsilon\Gamma_1\cdots\Gamma_s. \end{cases}$$

In these situations B defines a quaternionic structure for the spinor representation of the orthochronous spin group.

Proof

1. We calculate

$$\begin{aligned} B^*B &= C^*A^*CA \\ &= \nu C^{-1}A^*CA \\ &= \nu(-1)^t C^{-1}\Gamma_{s+1}^T\cdots\Gamma_{s+t}^T C\Gamma_{s+1}\cdots\Gamma_{s+t} \\ &= \nu(-1)^t\mu^t\Gamma_{s+1}\cdots\Gamma_{s+t}\Gamma_{s+1}\cdots\Gamma_{s+t} \\ &= \nu\mu^t(-1)^{t(t-1)/2}I. \end{aligned}$$

2. This follows similarly.
3. and 4. The remaining claims follow by considering Table 6.7. Note that the value of $(-1)^{t(t-1)/2}$ depends only on the value of t mod 4, and similarly for s.

□

Remark 6.7.21 This theorem explains why Majorana spinors exist if $\rho = 0,1,7 \bmod 8$ and for the right choice of μ also if $\rho = 2,6 \bmod 8$ (see Sect. 6.6).

Corollary 6.7.22 *For a spinor $\psi \in \Delta$ the Majorana conjugate is equal to the Dirac conjugate, $\tilde{\psi} = \overline{\psi}$, i.e.*

$$(\psi,\cdot) = \langle\psi,\cdot\rangle$$

if and only if $\tau(\psi) = \psi$ or, equivalently, $B\psi = \psi^*$. This happens for a non-zero spinor ψ if and only if B defines a real structure and ψ is a Majorana spinor.
This corollary finally explains the relation between Majorana spinors and invariant forms for the spinor representations.

6.8 Explicit Formulas for Minkowski Spacetime of Dimension 4

We collect some explicit formulas concerning Clifford algebras and spinors for the case of 4-dimensional Minkowski spacetime.

6.8.1 The Lorentz Clifford Algebra

In Minkowski spacetime of dimension 4 and signature $(+,-,-,-)$ (usually used in quantum field theory) there exist both Weyl and Majorana spinors, but not Majorana–Weyl spinors.

Recall from Example 6.3.17 that the matrices

$$\Gamma_0 = \begin{pmatrix} 0 & I_2 \\ I_2 & 0 \end{pmatrix}$$

$$\Gamma_k = \begin{pmatrix} 0 & \sigma_k \\ -\sigma_k & 0 \end{pmatrix} \quad \forall k = 1,2,3$$

are physical gamma matrices for $\mathrm{Cl}(1,3)$. This is the so-called Weyl representation of the Clifford algebra. It is a basis unitary representation. The physical chirality operator is given by

$$\Gamma_5 = -i\Gamma_0\Gamma_1\Gamma_2\Gamma_3 = i\Gamma^0\Gamma^1\Gamma^2\Gamma^3 = \begin{pmatrix} I_2 & 0 \\ 0 & -I_2 \end{pmatrix}.$$

Hence in our convention the first two components of a Dirac 4-spinor correspond to a left-handed Weyl spinor and the last two components to a right-handed Weyl spinor (this is the standard convention in quantum field theory and the Standard Model).

For the matrix A defining the Dirac form we use the second choice in Proposition 6.7.13. Then

$$A = \Gamma_0$$

with $\delta = -1, \epsilon = 1$. This implies

$$\Gamma_a^\dagger = \Gamma_0\Gamma_a\Gamma_0 \quad \forall a = 0,1,2,3$$

and the Dirac conjugate of a spinor ψ is given by

$$\overline{\psi} = \psi^\dagger\Gamma_0$$

with Hermitian scalar product

$$\langle \psi, \phi \rangle = \overline{\psi}\phi = \psi^\dagger \Gamma_0 \phi,$$

satisfying

$$\langle X \cdot \psi, \phi \rangle = -\langle \psi, X \cdot \phi \rangle \quad \forall X \in \mathbb{R}^{1,3}.$$

If we write a Dirac spinor as

$$\psi = \begin{pmatrix} \chi \\ \xi \end{pmatrix},$$

then

$$\overline{\psi}\psi = \chi^\dagger \xi + \xi^\dagger \chi.$$

Note that this Hermitian scalar product is not positive definite. In fact the subspaces of left-handed and right-handed Weyl spinors are each *null*.

For the matrix C defining the Majorana form we choose

$$C = i\Gamma_0\Gamma_2 = i\Gamma^2\Gamma^0 = \begin{pmatrix} -i\sigma_2 & 0 \\ 0 & i\sigma_2 \end{pmatrix} = \begin{pmatrix} 0 & -1 & & \\ 1 & 0 & & \\ & & 0 & 1 \\ & & -1 & 0 \end{pmatrix}.$$

Then

$$\Gamma_a^T = -C\Gamma_a C^{-1} \quad \forall a = 0, 1, 2, 3,$$
$$C^T = -C,$$

hence $\mu = \nu = -1$. The matrix C is unitary and

$$B = CA = -i\Gamma_2.$$

The matrix C corresponds to the matrix ϵ in Sect. 2.1.3. We have

$$B^{-1} = B = B^* = -i\Gamma_2 = i\Gamma^2$$

and the charge conjugate of a Dirac spinor is given by

$$\psi^C = B^{-1}\psi^* = i\Gamma^2\psi^*.$$

This implies (as expected from Theorem 6.7.20)

$$B^*B = I$$

and B defines a real structure on the spinor space Δ. Majorana spinors ψ are characterized by

$$\psi^* = B\psi.$$

If we write

$$\psi = \begin{pmatrix} \chi \\ \xi \end{pmatrix},$$

then Majorana spinors are precisely those of the form

$$\psi = \begin{pmatrix} \chi \\ i\sigma_2 \chi^* \end{pmatrix}.$$

6.8.2 *The Orthochronous Lorentz Spin Group*

Our aim in this subsection is to prove that the orthochronous spin group $\mathrm{Spin}^+(1,3)$ of 4-dimensional Minkowski spacetime is isomorphic to the 6-dimensional real Lie group $\mathrm{SL}(2,\mathbb{C})$. We identify $\mathbb{R}^{1,3}$ with the Hermitian 2×2-complex matrices via the real vector space isomorphism

$$\mathbb{R}^{1,3} \longrightarrow \mathrm{Herm}(2,\mathbb{C})$$

$$x = (x^0, x^1, x^2, x^3) \longmapsto x^\mu \sigma_\mu = X = \begin{pmatrix} x^0 + x^3 & x^1 - ix^2 \\ x^1 + ix^2 & x^0 - x^3 \end{pmatrix}.$$

Here $\sigma_1, \sigma_2, \sigma_3$ are the Pauli matrices and $\sigma_0 = I_2$. The following is easy to verify:

Lemma 6.8.1 *Under this identification,* $\eta(x,x) = \det X$ *for all* $x \in \mathbb{R}^{1,3}$.

We can then prove the following (see [146, Appendix A]).

Theorem 6.8.2 (Explicit Description of the Orthochronous Lorentz Spin Group) *Under the identification above, the following map is well-defined*

$$\mathrm{SL}(2,\mathbb{C}) \times \mathbb{R}^{1,3} \longrightarrow \mathbb{R}^{1,3}$$

$$(M, X) \longmapsto MXM^\dagger.$$

This maps yields a Lie group homomorphism

$$\psi\colon \mathrm{SL}(2,\mathbb{C}) \longrightarrow \mathrm{SO}^+(1,3)$$

which is surjective and has kernel $\{I, -I\}$. Since $\mathrm{SL}(2,\mathbb{C})$ *is simply connected, it follows that* $\mathrm{SL}(2,\mathbb{C}) \cong \mathrm{Spin}^+(1,3)$.

Proof This is Exercise 6.13.17. Compare with Exercises 1.9.20 and 1.9.21. □

6.9 Spin Structures and Spinor Bundles

In this section we discuss the notion of *spin structure* that is needed to define spinors globally on manifolds. If the tangent bundle TM of a manifold M is a trivial vector bundle, then spin structures always exist. If TM is non-trivial, then there may exist a topological obstruction to the existence of a spin structure, measured by the *second Stiefel–Whitney class* $w_2(M)$. If a spin structure for a pseudo-Riemannian manifold (M, g) exists, then we can define an associated complex vector bundle, called the *spinor bundle*. The sections of this bundle are called *spinor fields* or *spinors* on the manifold M.

6.9.1 Spin Structures

Definition 6.9.1 Let M be a smooth manifold. A **pseudo-Riemannian metric** g of signature (s, t), where

$$(\underbrace{+,\ldots,+}_{s},\underbrace{-,\ldots,-}_{t}),$$

is a section $g \in \Gamma(T^*M \otimes T^*M)$ that defines at each point $x \in M$ a non-degenerate, symmetric bilinear form

$$g_x \colon T_xM \times T_xM \longrightarrow \mathbb{R}$$

of signature (s, t).
Let M be a smooth manifold with a pseudo-Riemannian metric g of signature (s, t). The frame bundle then has the structure of a principal $\mathrm{O}(s, t)$-bundle. For the following definitions recall the notion of a G-reduction of a principal bundle from Definition 4.2.17.

Definition 6.9.2
1. The pseudo-Riemannian manifold (M, g) is called **orientable** if the frame bundle can be reduced to a principal $\mathrm{SO}(s, t)$-bundle under the embedding $\mathrm{SO}(s, t) \subset \mathrm{O}(s, t)$.
2. The pseudo-Riemannian manifold (M, g) is called **time-orientable** if the frame bundle can be reduced to a principal $\mathrm{O}^+(s, t)$-bundle under the embedding $\mathrm{O}^+(s, t) \subset \mathrm{O}(s, t)$.

3. The pseudo-Riemannian manifold (M, g) is called **orientable and time-orientable** if the frame bundle can be reduced to a principal $\mathrm{SO}^+(s,t)$-bundle under the embedding $\mathrm{SO}^+(s,t) \subset \mathrm{O}(s,t)$.
   If such reductions are chosen, we call the pseudo-Riemannian manifold (M, g) **oriented, time-oriented** or **oriented and time-oriented**.

An orientation of M is just an orientation of the tangent bundle TM. A time-orientation has the following interpretation: using some homotopy theory and Lemma 6.1.14 it is not difficult to see that the tangent bundle TM admits maximally g-positive definite vector subbundles $W \to M$ and that any two such subbundles are homotopic. A time-orientation of (M, g) then corresponds to a choice of orientation of such a maximally g-positive definite vector subbundle W. On an orientable pseudo-Riemannian manifold there are precisely two different orientations and similarly for time-orientations on time-orientable pseudo-Riemannian manifolds.

Suppose that (M, g) is oriented and time-oriented. We denote the $\mathrm{SO}^+(s,t)$-frame bundle by

$$\pi_{\mathrm{SO}} \colon \mathrm{SO}^+(M) \longrightarrow M.$$

Recall the double covering

$$\lambda \colon \mathrm{Spin}^+(s,t) \longrightarrow \mathrm{SO}^+(s,t)$$

from Sect. 6.5.1.

Definition 6.9.3 A **spin structure** on M is a $\mathrm{Spin}^+(s,t)$-principal bundle

$$\pi_{\mathrm{Spin}} \colon \mathrm{Spin}^+(M) \longrightarrow M$$

with a double covering

$$\Lambda \colon \mathrm{Spin}^+(M) \longrightarrow \mathrm{SO}^+(M)$$

such that the following diagram commutes:

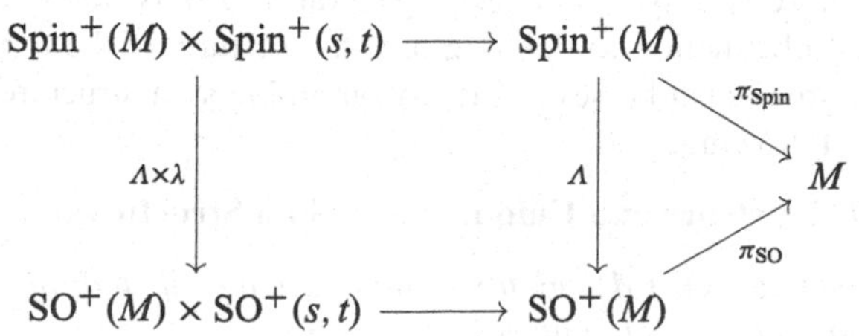

$$\begin{array}{ccccc}
\mathrm{Spin}^+(M) \times \mathrm{Spin}^+(s,t) & \longrightarrow & \mathrm{Spin}^+(M) & & \\
\Big\downarrow{\scriptstyle \Lambda\times\lambda} & & \Big\downarrow{\scriptstyle \Lambda} & \searrow^{\pi_{\mathrm{Spin}}} & \\
 & & & & M \\
 & & & \nearrow_{\pi_{\mathrm{SO}}} & \\
\mathrm{SO}^+(M) \times \mathrm{SO}^+(s,t) & \longrightarrow & \mathrm{SO}^+(M) & &
\end{array}$$

Here the horizontal arrow in the top and bottom line are the right actions of the structure groups on the principal bundles.

According to Definition 4.2.17, a spin structure is thus a **λ-equivariant bundle morphism** $\Lambda\colon \mathrm{Spin}^+(M) \to \mathrm{SO}^+(M)$, i.e. a **$\lambda$-reduction** of $\mathrm{SO}^+(M)$.

Definition 6.9.4 Two spin structures

$$\Lambda\colon \mathrm{Spin}^+(M) \longrightarrow \mathrm{SO}^+(M)$$
$$\Lambda'\colon \mathrm{Spin}^+(M)' \longrightarrow \mathrm{SO}^+(M)$$

are called **isomorphic** if there exists a $\mathrm{Spin}^+(s,t)$-equivariant bundle isomorphism

$$F\colon \mathrm{Spin}^+(M) \longrightarrow \mathrm{Spin}^+(M)'$$

such that the following diagram commutes:

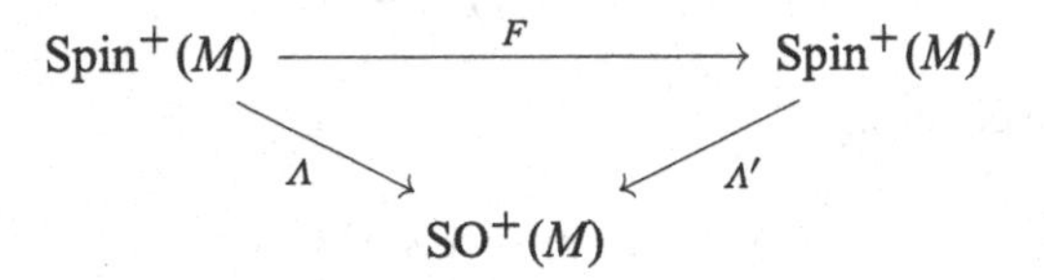

Remark 6.9.5 Note that a spin structure is more than just a double covering of the frame bundle which fibrewise looks like the covering λ. We demand in addition that the actions of the structure groups on both principal bundles are compatible. This additional structure is needed, for example, in the proof of Proposition 6.9.13 to define Clifford multiplication on the level of bundles and in Proposition 6.10.7 to define a connection 1-form on $\mathrm{Spin}^+(M)$ associated to the Levi-Civita connection on TM.

Remark 6.9.6 We can define in the same way spin structures for any principal $\mathrm{SO}^+(s,t)$-bundle, not only for the tangent bundle of a pseudo-Riemannian manifold. If the tangent bundle TM of the manifold M is trivial, then $\mathrm{SO}^+(M)$ is trivial and a spin structure is easy to define. In general, if TM is non-trivial, there may be a topological obstruction to defining a spin structure. If the topology of M is non-trivial there may also be several non-isomorphic spin structures. The precise statement is the following.

Theorem 6.9.7 (Existence and Uniqueness of Spin Structures)

1. *The frame bundle* $\mathrm{SO}^+(M)$ *admits a spin structure if and only if the second Stiefel–Whitney class of M vanishes,* $w_2(M) = 0$.
2. *If* $\mathrm{SO}^+(M)$ *admits a spin structure, then there is a bijection between the set of isomorphism classes of spin structures on M and the cohomology group* $H^1(M;\mathbb{Z}_2)$.

Proof The details of the proof would take us too far afield, because we have not discussed characteristic classes like the Stiefel–Whitney classes. A proof can be found in [55] and [88] in the Riemannian case and in [13] in the general pseudo-Riemannian case. Note that a manifold M is orientable if and only if $w_1(M) = 0$ and the number of different orientations is given by the number of elements of $H^0(M; \mathbb{Z}_2)$, hence the existence of a spin structure can be understood as the existence of a "higher orientation" for M. □

We call a manifold M **spin** if $w_2(M) = 0$. The following manifolds can be shown to be spin:

- All manifolds M with trivial tangent bundle TM, in particular, all Euclidean spaces $\mathbb{R}^n$ and tori T^n.
- All spheres S^n.
- All orientable 2-dimensional manifolds.
- Complex projective spaces $\mathbb{CP}^n$ are spin if and only if n is odd.
- If M and N are spin manifolds, then so is $M \times N$.

The cohomology group $H^1(M; \mathbb{Z}_2)$ vanishes, for example, if $\pi_1(M)$ is trivial. In particular we get:

Corollary 6.9.8 *The manifold* $\mathbb{R}^{s,t}$ *admits for all* $s, t \geq 0$ *a unique spin structure.*

Note that if the tangent bundle of M is trivial, then there always exists (after a choice of trivialization of TM) a canonical (trivial) spin structure, but there exist additional (non-trivial) ones if $H^1(M; \mathbb{Z}_2) \neq 0$. This happens, for instance, in the case of the torus T^n, where

$$H^1(T^n; \mathbb{Z}_2) \cong (\mathbb{Z}_2)^n$$

has 2^n elements.

Remark 6.9.9 If we think of the structure group of a principal bundle as a symmetry group, then the existence of a reduction of the structure group means that the bundle admits a more *fundamental* or *hidden* symmetry group. In particular, the frame bundle $\mathrm{SO}^+(M)$ of a pseudo-Riemannian manifold M admits a spin structure if and only if it has a more fundamental underlying symmetry, given by the orthochronous spin group $\mathrm{Spin}^+(s,t)$.

We can recover all tensor bundles and the tangent bundle itself from the principal bundle $\mathrm{Spin}^+(M)$ as associated vector bundles. However, we also get additional vector bundles that cannot be associated to the $\mathrm{SO}^+(M)$-frame bundle. In general, the spin structure $\Lambda\colon \mathrm{Spin}^+(M) \to \mathrm{SO}^+(M)$ is not unique: several non-isomorphic spin structures may induce the same frame bundle $\mathrm{SO}^+(M)$.

We briefly want to discuss how a section of the bundle $\mathrm{SO}^+(M)$ determines sections of $\mathrm{Spin}^+(M)$.

Definition 6.9.10 We call a local section $e = (e_1, \ldots, e_n)$ of $\mathrm{SO}^+(M)$ an ***n*-bein** or **vielbein** (and **tetrad** in dimension $n = 4$).

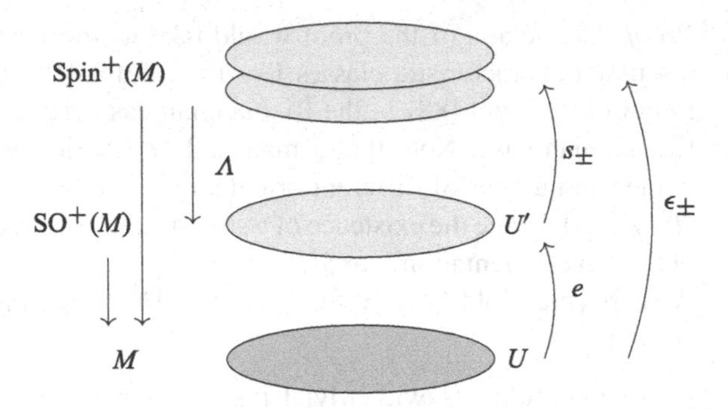

Fig. 6.1 Local frames for $\mathrm{Spin}^+(M)$

Lemma 6.9.11 *Suppose we have chosen a spin structure on M. Then for every vielbein e on a contractible open set $U \subset M$ there exist precisely two local sections $\epsilon_\pm$ of* $\mathrm{Spin}^+(M)$ *over U such that $\Lambda \circ \epsilon_\pm = e$.*

Proof The image of e is a contractible open subset U' of $\mathrm{SO}^+(M)$ diffeomorphic to U and thus

$$\Lambda|_{\Lambda^{-1}(U')}: \Lambda^{-1}(U') \longrightarrow U'$$

is a trivial two-sheeted covering, admitting precisely two sections

$$s_\pm: U' \longrightarrow \Lambda^{-1}(U') \subset \mathrm{Spin}^+(M).$$

Then $\epsilon_\pm = s_\pm \circ e$ have the claimed properties. See Fig. 6.1. □

6.9.2 Spinor Bundles

Definition 6.9.12 Let $\mathrm{Spin}^+(M) \to M$ be a spin structure on M and

$$\kappa: \mathrm{Spin}^+(s,t) \longrightarrow \mathrm{GL}(\Delta)$$

the spinor representation. Then the **(Dirac) spinor bundle** is the associated complex vector bundle

$$S = \mathrm{Spin}^+(M) \times_\kappa \Delta$$

over M. Sections of S are called **spinor fields** or **spinors**. Note that the spinor bundle may depend on the choice of spin structure.

Proposition 6.9.13 *Let $S \rightarrow M$ be the spinor bundle associated to a spin structure.*

1. *There exists a well-defined bilinear* **Clifford multiplication**

$$\begin{aligned} TM \times S &\longrightarrow S \\ (X, \Psi) &\longmapsto X \cdot \Psi \end{aligned}$$

on the level of bundles, restricting to a map $T_pM \times S_p \rightarrow S_p$ in every point $p \in M$. This map also induces a well-defined Clifford multiplication of forms with spinors.

2. *If the dimension n of M is even, then S splits as a direct sum of complex* **Weyl spinor bundles** $S = S_+ \oplus S_-$, *defined by*

$$S_\pm = \mathrm{Spin}^+(M) \times_\kappa \Delta^\pm.$$

In this case, Clifford multiplication with a vector maps $S_\pm$ to $S_\mp$.

Proof

1. Let ρ_{SO} be the standard representation of $\mathrm{SO}^+(s,t)$ on $\mathbb{R}^{s,t}$. Then the tangent bundle TM is isomorphic to the associated vector bundle

$$\mathrm{SO}^+(M) \times_{\rho_{\mathrm{SO}}} \mathbb{R}^{s,t}.$$

Suppose ϵ is an element of $\mathrm{Spin}^+(M)$. Then $\Lambda(\epsilon)$ is an element of $\mathrm{SO}^+(M)$ and we can define Clifford multiplication $TM \times S \rightarrow S$ as follows:

$$\begin{aligned} \left(\mathrm{SO}^+(M) \times_{\rho_{\mathrm{SO}}} \mathbb{R}^{s,t}\right) \times \left(\mathrm{Spin}^+(M) \times_\kappa \Delta\right) &\longrightarrow \left(\mathrm{Spin}^+(M) \times_\kappa \Delta\right) \\ ([\Lambda(\epsilon), x], [\epsilon, \psi]) &\longmapsto [\epsilon, x \cdot \psi]. \end{aligned}$$

Here $x \cdot \psi$ is the standard Clifford multiplication between elements $x \in \mathbb{R}^{s,t}$ and $\psi \in \Delta$. A direct argument using Definition 6.9.3 of spin structures, Definition 4.7.3 of associated vector bundles and Proposition 6.5.19 on the compatibility of the spinor representation with Clifford multiplication shows that the map $TM \times S \rightarrow S$ is well-defined.

2. This follows, because by Lemma 6.3.5 the chirality element

$$\omega = i^{k+t} e_1 \cdots e_n$$

is well-defined, independent of the choice of vielbein, and Clifford multiplication of the chirality element with a spinor is well-defined by part 1. of the proposition.

□

It is sometimes convenient to describe spinors locally: Let e be a local vielbein and $\epsilon_\pm$ the associated local sections of the spin structure principal bundle on a contractible, open subset U in M. If $\Psi: U \to S$ is a local section of the spinor bundle, then there exist two maps

$$\psi_\pm: U \longrightarrow \Delta$$

such that Ψ can be written as

$$\Psi = [\epsilon_\pm, \psi_\pm].$$

We have $\psi_\pm = -\psi_\mp$. We choose one of the two local sections ϵ and the associated map ψ, so that

$$\Psi = [\epsilon, \psi].$$

If we define a formula locally using ϵ and ψ, we always have to check that the result is independent of this choice. Generally speaking, this will always be the case if the expression is linear in ψ. For example, physical Clifford multiplication with a basis vector can be expressed as

$$e_a \cdot \Psi = [\epsilon, \Gamma_a \psi].$$

The right-hand side is indeed independent of the choice of ϵ, since $\Gamma_a \psi$ is linear in ψ.

6.9.3 Structures on Spinor Bundles

In Sect. 6.6 and Sect. 6.7 we considered several structures on the spinor representation space Δ:

1. Real structures σ and quaternionic structures J.
2. Majorana forms $(\cdot\,,\cdot)$ and Dirac forms $\langle\cdot\,,\cdot\rangle$.

It is clear that these structures extend fibrewise to globally well-defined, smooth structures on the spinor bundle $S \to M$: this follows, because by definition real and quaternionic structures commute with the spinor representation of the orthochronous spin group and Majorana and Dirac forms are invariant under the orthochronous spin group, cf. Lemma 6.7.4 and Lemma 6.7.10. We can therefore define the following structures:

1. Real bundle automorphisms σ_S and quaternionic bundle automorphisms J_S of the spinor bundle S (depending on the dimension and signature of the manifold)

together with (symplectic) Majorana or (symplectic) Majorana–Weyl sections of S.

2. Majorana bundle metrics $(\cdot\,,\cdot)_S$ and Dirac bundle metrics $\langle\cdot\,,\cdot\rangle_S$ on S.

6.10 The Spin Covariant Derivative

Given a spin structure on a pseudo-Riemannian manifold and the spinor bundle S, we would like to have a covariant derivative on S so that we can define field equations involving derivatives of spinors. It turns out that the standard Levi-Civita connection on the tangent bundle, determined by the pseudo-Riemannian metric, defines a unique compatible covariant derivative on S, called the *spin covariant derivative*.

Let (M, g) be an oriented and time-oriented pseudo-Riemannian manifold of signature (s, t). We assume that M is spin and a spin structure has been chosen. Let ∇ denote the Levi-Civita connection on TM associated to the metric g.

6.10.1 Spin Connection

Definition 6.10.1 In a local vielbein $e = (e_1, \ldots, e_n)$ on an open set $U \subset M$ we can write

$$\nabla e_a = \omega_{ab}\eta^{bc} \otimes e_c,$$

with certain uniquely determined real-valued 1-forms ω_{ab} on U (and the Einstein convention is understood here and in the following formulas). We also set

$$\omega_{cab} = \omega_{ab}(e_c).$$

Remark 6.10.2 In the physics literature the forms ω_{ab} are often defined with the opposite sign. We continue to use our definition.

Lemma 6.10.3 *The* 1*-forms* ω_{ab} *are antisymmetric in the indices* a, b*:*

$$\omega_{ab} = -\omega_{ba} \quad \forall a, b = 1, \ldots, n.$$

Proof This is Exercise 6.13.19. □

Definition 6.10.4 The **anholonomy coefficients** $\Omega_{ab}{}^{c}$ of a local vielbein e are defined by

$$[e_a, e_b] = \Omega_{ab}{}^{c} e_c.$$

With respect to local coordinates x^μ in a local chart on $U \subset M$ one also defines in the physics literature real functions E_a^μ on U by

$$e_a = E_a^\mu \partial_\mu.$$

Lemma 6.10.5 *The* 1*-forms* ω_{ab} *are determined by the anholonomy coefficients as follows:*

$$\omega_{cab} = \omega_{ab}(e_c) = \frac{1}{2}(\Omega_{cab} - \Omega_{abc} + \Omega_{bca}),$$

where $\Omega_{abc} = \Omega_{ab}{}^d \eta_{dc}$.

Proof This is Exercise 6.13.20. □

The Levi-Civita connection ∇ induces a connection 1-form

$$A_{\mathrm{SO}} \in \Omega^1\left(\mathrm{SO}^+(M), \mathfrak{so}^+(s,t)\right)$$

on the frame bundle $\mathrm{SO}^+(M)$. The Levi-Civita connection is the associated covariant derivative on the tangent bundle

$$TM = \mathrm{SO}^+(M) \times_{\rho_{\mathrm{SO}}} \mathbb{R}^{s,t}.$$

This means that if $Y = \sum_a y^a e_a$ is the local expansion of a vector field Y on M, then

$$\nabla_X Y = \left(L_X y^c + A^e_{\mathrm{SO}}(X)^c{}_a y^a\right) e_c,$$

where $A^e_{\mathrm{SO}} = e^* A_{\mathrm{SO}}$ is the local connection 1-form. This implies:

Lemma 6.10.6 *The local connection* 1*-form* A^e_{SO} *is given by*

$$A^e_{\mathrm{SO}}(X)^c{}_a = \omega_{ab}(X)\eta^{bc} \quad \forall X \in T_pM, p \in U.$$

Consider the Lie group homomorphism

$$\lambda\colon \mathrm{Spin}^+(s,t) \longrightarrow \mathrm{SO}^+(s,t).$$

Since this map is a covering, it induces an isomorphism

$$\lambda_*\colon \mathfrak{spin}^+(s,t) \xrightarrow{\cong} \mathfrak{so}^+(s,t).$$

Proposition 6.10.7 *Let*

$$\Lambda\colon \mathrm{Spin}^+(M) \longrightarrow \mathrm{SO}^+(M)$$

be the covering map given by the spin structure. Then

$$A_{\mathrm{Spin}} = (\lambda_*)^{-1} \circ \left(\Lambda^* A_{\mathrm{SO}}\right) \in \Omega^1\left(\mathrm{Spin}^+(M), \mathfrak{spin}^+(s,t)\right)$$

is a connection 1*-form on the principal bundle* $\mathrm{Spin}^+(M) \rightarrow M$*, called the* **spin connection.**

Proof We have to verify the properties of connection 1-forms.

1. Let $g \in \mathrm{Spin}^+(s,t)$ and $Y \in T\mathrm{Spin}^+(M)$. Then by the defining properties of a spin structure

$$\begin{aligned} r_g^* A_{\mathrm{Spin}}(Y) &= (\lambda_*)^{-1} \circ A_{\mathrm{SO}}(\Lambda_* r_{g*} Y) \\ &= (\lambda_*)^{-1} \circ A_{\mathrm{SO}}(r_{\lambda(g)*} \Lambda_* Y) \\ &= (\lambda_*)^{-1} \circ \mathrm{Ad}_{\lambda(g)^{-1}} \circ \lambda_* \circ A_{\mathrm{Spin}}(Y) \\ &= \mathrm{Ad}_{g^{-1}} \circ A_{\mathrm{Spin}}(Y). \end{aligned}$$

2. Let $X \in \mathfrak{spin}^+(s,t)$ and $\epsilon \in \mathrm{Spin}^+(M)$. Then

$$\begin{aligned} A_{\mathrm{Spin}}(\tilde{X}_\epsilon) &= (\lambda_*)^{-1} \circ A_{\mathrm{SO}}\left(\left.\frac{d}{dt}\right|_{t=0} \Lambda(\epsilon \cdot \exp(tX))\right) \\ &= (\lambda_*)^{-1} \circ A_{\mathrm{SO}}\left(\widetilde{\lambda_* X}_{\Lambda(\epsilon)}\right) \\ &= X. \end{aligned}$$

□

Note that the proof of this proposition crucially needs the compatibility of the structure group actions in Definition 6.9.3 of spin structures.

6.10.2 Spin Covariant Derivative

Definition 6.10.8 We call the associated covariant derivative on the spinor bundle S defined by the spin connection A_{Spin} the **spin covariant derivative**, again denoted by ∇.

The spin covariant derivative is completely determined by the Levi-Civita connection, once a spin structure has been chosen. We want to derive a local formula for the spin covariant derivative. Suppose that U is contractible and $\epsilon_\pm$ the associated local trivializations of the principal bundle $\mathrm{Spin}^+(M)$ on U. We choose one of those trivializations, called ϵ. Then we get associated local connection 1-forms

$$A_{\mathrm{Spin}}^\epsilon = \epsilon^* A_{\mathrm{Spin}} \in \Omega^1(U, \mathfrak{spin}^+(s,t)).$$

We write a section Ψ of the spinor bundle S as

$$\Psi = [\epsilon, \psi]$$

with $\psi: U \rightarrow \Delta$. Then the spin covariant derivative on the spinor bundle can be written locally as

$$\nabla_X \Psi = [\epsilon, \nabla_X \psi]$$

where

$$\nabla_X \psi = d\psi(X) + A^{\epsilon}_{\text{Spin}}(X) \cdot \psi$$

and $A^{\epsilon}_{\text{Spin}}(X)$ acts through the representation κ_* of $\mathfrak{spin}^+(s,t)$ on Δ induced from the spinor representation κ.

Proposition 6.10.9 (Explicit Formula for the Spin Covariant Derivative) *The spin covariant derivative can be calculated by the explicit formula*

$$\begin{aligned}\nabla_X \psi &= d\psi(X) + \frac{1}{4}\omega_{ab}(X)\gamma^{ab}\psi \\ &= d\psi(X) - \frac{1}{4}\omega_{ab}(X)\Gamma^{ab}\psi.\end{aligned}$$

We need the following lemma:

Lemma 6.10.10 *Write the components of a matrix $A \in \mathfrak{so}^+(s,t)$ as*

$$A^c{}_a = w_{ab}\eta^{bc}.$$

Then the map

$$\kappa_* \circ (\lambda_*)^{-1}: \mathfrak{so}^+(s,t) \xrightarrow{\cong} \mathfrak{spin}^+(s,t) \longrightarrow \text{End}(\Delta),$$

where κ is the spinor representation, is given by

$$\kappa_*(\lambda_*)^{-1}(A) = \frac{1}{4}w_{ab}\gamma^{ab}.$$

Proof This follows from the last formula in Corollary 6.5.23. □

We can now prove Proposition 6.10.9.

Proof We want to prove that

$$\kappa_*\left(A^{\epsilon}_{\text{Spin}}(X)\right) = \frac{1}{4}\omega_{ab}(X)\gamma^{ab},$$

where κ_* is the homomorphism induced by the spinor representation. We have

$$\begin{aligned} A^{\epsilon}_{\mathrm{Spin}} &= \epsilon^* A_{\mathrm{Spin}} \\ &= (\lambda_*)^{-1} \circ A^{e}_{\mathrm{SO}}. \end{aligned}$$

The claim then follows immediately from Lemma 6.10.6 and Lemma 6.10.10. □

Remark 6.10.11 As mentioned above, in physics the 1-forms ω_{ab} are sometimes defined with the opposite sign,

$$\nabla e_a = -\omega_{ab}\eta^{bc} \otimes e_c.$$

Then $\nabla_X \psi$ has to be defined by

$$\nabla_X \psi = d\psi(X) + \frac{1}{4}\omega_{bc}(X)\Gamma^{bc}\psi.$$

Lemma 6.10.12 *If the dimension n of the manifold M is even, then the spin covariant derivative preserves the splitting of the spinor bundle S into the Weyl spinor bundles S_+ and S_-. This means that*

$$\nabla_X \Psi \in \Gamma(S_\pm) \quad \forall \Psi \in \Gamma(S_\pm), X \in \mathfrak{X}(M).$$

Proof This follows from $\left[\Gamma^{n+1}, \Gamma^{ab}\right] = 0$. □

The spin covariant derivative has the following property:

Theorem 6.10.13 *The spin covariant derivative is compatible with the Levi-Civita connection in the following way: for all vector fields $X, Y \in \mathfrak{X}(M)$ and spinors $\Psi \in \Gamma(S)$ the identity*

$$\nabla_X(Y \cdot \Psi) = (\nabla_X Y) \cdot \Psi + Y \cdot \nabla_X \Psi$$

holds.

Proof This is Exercise 6.13.21. □

The following result is often useful:

Theorem 6.10.14 *Let $(\cdot\,,\cdot)_S$ be a Majorana bundle metric and $\langle \cdot\,,\cdot \rangle_S$ a Dirac bundle metric on the spinor bundle S, defined by Majorana and Dirac forms for the spinor representation. Then these metrics are compatible with the spin covariant derivative in the following way:*

$$\begin{aligned} L_X(\Psi, \Phi)_S &= (\nabla_X \Psi, \Phi)_S + (\Psi, \nabla_X \Phi)_S, \\ L_X\langle \Psi, \Phi \rangle_S &= \langle \nabla_X \Psi, \Phi \rangle_S + \langle \Psi, \nabla_X \Phi \rangle_S \end{aligned}$$

for all vector fields $X \in \mathfrak{X}(M)$ and spinors $\Psi, \Phi \in \Gamma(S)$.

Proof This follows from Proposition 5.9.7, because both types of forms on the spinor space Δ are invariant under $\mathrm{Spin}^+(s,t)$. □

6.10.3 Dirac Operator

We can now define the Dirac operator.

Definition 6.10.15 The **Dirac operator** $D\colon \Gamma(S) \to \Gamma(S)$ on the spinor bundle S is defined by

$$D\Psi = \eta^{ab} e_a \cdot \nabla_{e_b} \Psi$$

(with mathematical Clifford multiplication and the Einstein summation convention). In a given vielbein e where $\Psi = [\epsilon, \psi]$ we can write

$$D\Psi = [\epsilon, D\psi]$$

where

$$\begin{aligned} D\psi &= \gamma^a \nabla_{e_a} \psi \\ &= i\Gamma^a \nabla_{e_a} \psi \\ &= i\Gamma^a \left(d\psi(e_a) - \frac{1}{4}\omega_{abc}\Gamma^{bc}\psi \right). \end{aligned}$$

In the physics literature the Dirac operator is often denoted by $\not{D}$.
It is easy to check that the definition of the Dirac operator is independent of the local vielbein e: This can be checked directly or by noticing that the Dirac operator D is the composition of the maps

$$\Gamma(S) \xrightarrow{\nabla} \Gamma(T^*M \otimes S) \xrightarrow{\eta} \Gamma(TM \otimes S) \xrightarrow{\gamma} \Gamma(S),$$

where $\eta\colon T^*M \to TM$ is the isomorphism induced by the pseudo-Riemannian metric on M and γ is Clifford multiplication. The Dirac operator is obviously a first-order differential operator on the spinor bundle S.

In even dimensions n, ∇ preserves the splitting $S = S_+ \oplus S_-$, while Clifford multiplication with a vector interchanges S_+ and S_-. This implies:

Corollary 6.10.16 *If the dimension n of the manifold M is even, then the Dirac operator D maps*

$$D\colon \Gamma(S_\pm) \longrightarrow \Gamma(S_\mp),$$

i.e. it takes sections of S_+ to sections of S_- and vice versa.

In any dimension n, suppose that $\langle \cdot\,,\cdot \rangle_S$ a Dirac bundle metric with $\delta = -1$, i.e.

$$\langle X \cdot \Phi, \Psi \rangle_S = -\langle \Phi, X \cdot \Psi \rangle_S \quad \forall X \in TM, \forall \Phi, \Psi \in S.$$

On the complex vector space $\Gamma_0(S)$ of sections of S with compact support in M we define an **L^2-scalar product of spinors**

$$\langle \cdot\,,\cdot \rangle_{S,L^2} \colon \Gamma_0(S) \times \Gamma_0(S) \longrightarrow \mathscr{C}^\infty(M)$$

by

$$\langle \Phi, \Psi \rangle_{S,L^2} = \int_M \langle \Phi, \Psi \rangle_S \mathrm{dvol}_g.$$

Here dvol_g is the volume element associated to the pseudo-Riemannian metric g and the orientation of M (cf. Sect. 7.2.1). We need to restrict to sections with compact support, because otherwise the integral may not be finite.

We can then prove the following:

Theorem 6.10.17 (Dirac Operator Is Formally Self-Adjoint) *Let M be a manifold without boundary. If the Dirac form satisfies $\delta = -1$, then the Dirac operator $D\colon \Gamma_0(S) \to \Gamma_0(S)$ is formally self-adjoint, i.e.*

$$\langle D\Phi, \Psi \rangle_{S,L^2} = \langle \Phi, D\Psi \rangle_{S,L^2}$$

for all spinors $\Phi, \Psi \in \Gamma_0(S)$.

A proof (of a more general theorem) can be found in Exercise 7.9.12.

6.11 Twisted Spinor Bundles

Let $P \to M$ be a principal G-bundle and $\rho\colon G \to \mathrm{GL}(V)$ a complex representation with associated vector bundle $E = P \times_\rho V$. Let $S \to M$ be the spinor bundle associated to a spin structure on M.

Definition 6.11.1 We call the bundle $S \otimes E$ a **twisted spinor bundle** or **gauge multiplet spinor bundle**.

Let $s\colon U \to P$ be a local gauge. Then we can write a local section τ of the associated vector bundle E as

$$\tau = [s, v]$$

with a map $v\colon U \to V$. We identify V with $\mathbb{C}^r$ using a basis $v_1, \ldots, v_r$ for V. This defines a local frame $\tau_1, \ldots, \tau_r$ for E, given by $\tau_i = [s, v_i]$.

Let e be a local vielbein for TM and $\epsilon: U \rightarrow \text{Spin}^+(M)$ a corresponding local trivialization.

Lemma 6.11.2 *Any section Ψ of the twisted spinor bundle can be written locally as*

$$\Psi = \sum_{i=1}^{r} \Psi_i \otimes \tau_i,$$

with sections Ψ_i of S. Equivalently,

$$\Psi = [\epsilon \times s, \psi],$$

where ψ is a ***multiplet*** *of the form*

$$\psi = \begin{pmatrix} \psi_1 \\ \psi_2 \\ \vdots \\ \psi_r \end{pmatrix} : U \longrightarrow \Delta \otimes \mathbb{C}^r,$$

and ψ_i are maps from U to Δ. This decomposition is unique, once local sections of P and $\text{Spin}^+(M)$ *as well as a basis for V have been chosen.*

We want to consider covariant derivatives on twisted spinor bundles. Let A be a connection 1-form on the principal G-bundle P.

Definition 6.11.3 The **twisted spin covariant derivative** ∇^A on the twisted spinor bundle $S \otimes E$ is defined by

$$\nabla_X^A \Psi = \left[\epsilon \times s, \nabla_X^A \psi\right]$$

where

$$\nabla_X^A \psi = d\psi(X) - \frac{1}{4}\omega_{bc}(X)\Gamma^{bc}\psi + (\rho_* A_s(X))\psi.$$

Here ψ is a map on U to $\Delta \otimes V$ and the matrices Γ^{bc} act on the Δ-part of ψ, i.e. on each of the spinor components ψ_i separately, while $\rho_* A_s(X)$ acts on the V-part of ψ, i.e. mixes the components of the multiplet ψ.

It is easy to check that the definition of ∇^A does not depend on the choice of ϵ and s. We also get a Dirac operator on twisted spinor bundles.

Definition 6.11.4 The **Dirac operator**

$$D_A: \Gamma(S \otimes E) \longrightarrow \Gamma(S \otimes E)$$

on a twisted spinor bundle $S \otimes E$ is defined by

$$D_A \Psi = \eta^{ab} e_a \cdot \nabla^A_{e_b} \Psi.$$

Equivalently, D_A is the composition of the maps

$$\Gamma(S \otimes E) \xrightarrow{\nabla^A} \Gamma(T^*M \otimes S \otimes E) \xrightarrow{\eta} \Gamma(TM \otimes S \otimes E) \xrightarrow{\gamma} \Gamma(S \otimes E).$$

Locally we have

$$D_A \Psi = [\epsilon \times s, D_A \psi]$$

with

$$D_A \psi = i\Gamma^a \left(d\psi(e_a) - \frac{1}{4}\omega_{abc}\Gamma^{bc}\psi + (\rho_* A_a)\psi \right).$$

In the physics literature the Dirac operator is sometimes denoted by $\not{D}_A$.
Suppose again that $\langle \cdot , \cdot \rangle_S$ is a Dirac bundle metric with $\delta = -1$, $\langle \cdot , \cdot \rangle_E$ is a Hermitian bundle metric on E and $\langle \cdot , \cdot \rangle_{S \otimes E}$ the induced bundle metric on $S \otimes E$. On the complex vector space $\Gamma_0(S \otimes E)$ of sections of $S \otimes E$ with compact support in M we define an L^2**-scalar product of twisted spinors**

$$\langle \cdot , \cdot \rangle_{S\otimes E, L^2} \colon \Gamma_0(S \otimes E) \times \Gamma_0(S \otimes E) \longrightarrow \mathscr{C}^\infty(M)$$

by

$$\langle \Phi, \Psi \rangle_{S\otimes E, L^2} = \int_M \langle \Phi, \Psi \rangle_{S\otimes E} \mathrm{dvol}_g.$$

We then get the following analogue of Theorem 6.10.17.

Theorem 6.11.5 (Twisted Dirac Operator Is Formally Self-Adjoint) *Let M be a manifold without boundary. If the Dirac form satisfies $\delta = -1$, then the twisted Dirac operator $D\colon \Gamma_0(S \otimes E) \to \Gamma_0(S \otimes E)$ is formally self-adjoint, i.e.*

$$\langle D_A \Phi, \Psi \rangle_{S\otimes E, L^2} = \langle \Phi, D_A \Psi \rangle_{S\otimes E, L^2}.$$

This is proved in Exercise 7.9.12.

6.12 Twisted Chiral Spinors

Suppose the dimension n of the manifold M is even. Then the spinor bundle splits into Weyl spinor bundles $S = S_+ \oplus S_-$. Let $P \rightarrow M$ be a principal G-bundle and

$$\rho_\pm : G \longrightarrow \mathrm{GL}(V_\pm)$$

two (possibly distinct) representations of G on complex vector spaces $V_\pm$. Let $E_\pm = P \times_{\rho_\pm} V_\pm$ be the associated vector bundles.

Definition 6.12.1 We call

$$(S \otimes E)_+ = (S_+ \otimes E_+) \oplus (S_- \otimes E_-)$$

a **twisted chiral spinor bundle**. We also consider the twisted bundle

$$(S \otimes E)_- = (S_- \otimes E_+) \oplus (S_+ \otimes E_-).$$

We can write a section of the twisted chiral spinor bundle as

$$\psi = \psi_+ + \psi_-,$$

where

$$\psi_\pm : M \longrightarrow (S_\pm \otimes E_\pm).$$

Suppose A is a connection 1-form on P.

Definition 6.12.2 The **twisted chiral spin covariant derivative** ∇^A on the twisted chiral spinor bundle $(S \otimes E)_+$ is defined by

$$\nabla^A_X \Psi = \left[\epsilon \times s, \nabla^A_X \psi\right],$$

where

$$\nabla^A_X \psi = d\psi(X) - \frac{1}{4}\omega_{bc}(X)\Gamma^{bc}\psi + (\rho_{+*}A_s(X))\psi_+ + (\rho_{-*}A_s(X))\psi_-.$$

We can again define a Dirac operator.

Definition 6.12.3 The **Dirac operator**

$$D_A : \Gamma((S \otimes E)_+) \longrightarrow \Gamma((S \otimes E)_-)$$

is locally given by

$$D_A \psi = i\Gamma^a \left(d\psi(e_a) - \frac{1}{4}\omega_{abc}\Gamma^{bc}\psi + (\rho_{+*}A_a)\psi_+ + (\rho_{-*}A_a)\psi_- \right).$$

We can decompose the Dirac operator D_A into

$$D_{A\pm}\colon \Gamma(S_\pm \otimes E_\pm) \longrightarrow \Gamma(S_\mp \otimes E_\pm).$$

We again denote the Dirac operator also by $\not{D}_A$.

Remark 6.12.4 Similar to Remark 6.2.6 for the classical Dirac operator it is possible to calculate the square of (twisted) Dirac operators over general manifolds. This square is again given by the Laplacian plus certain correction terms that depend on the curvature of the vector bundles. The formula for the square of the Dirac operator is known as the **Lichnerowicz–Weitzenböck formula** (see [15, 88] and [115] for more details).

Remark 6.12.5 It can be proved that on a closed (compact without boundary) Riemannian manifold the twisted chiral Dirac operator

$$D_A\colon \Gamma((S \otimes E)_+) \longrightarrow \Gamma((S \otimes E)_-),$$

as a linear map between infinite-dimensional vector spaces, has finite-dimensional kernel and cokernel. This includes as a special case a twisted Dirac operator of the form

$$D_A\colon \Gamma(S_+ \otimes E) \longrightarrow \Gamma(S_- \otimes E).$$

The **index** of a twisted chiral Dirac operator is the integer

$$\mathrm{ind}(D_A) = \dim\ker(D_A) - \dim\mathrm{coker}(D_A).$$

The famous **Atiyah–Singer Index Theorem** gives a formula for this index in terms of characteristic classes of the vector bundles TM and $E_\pm$. See [15, 88] and [115] for detailed expositions of the index theorem ([15] considers specifically the case of a twisted chiral spinor bundle and calls the vector bundle $E = E_+ \oplus E_-$ a **superbundle**).

Remark 6.12.6 The notion of twisted chiral spinor bundles seems like an unnecessary complication from a physics point of view. However, as we shall see in Chap. 8, they are crucial when describing the weak interactions in the Standard Model. In fact, in the Standard Model where

$$G = \mathrm{SU}(3) \times \mathrm{SU}(2) \times \mathrm{U}(1)$$

(continued)

Remark 6.12.6 (continued)
the vector spaces V_+ and V_- even have different dimensions: V_+ has dimension 24 and V_- has dimension 21. As G-representations these spaces decompose into irreducible representations of dimensions

$$\begin{aligned}\dim V_+ &= 24 = (6+2) + (6+2) + (6+2),\\ \dim V_- &= 21 = (3+3+1) + (3+3+1) + (3+3+1).\end{aligned}$$

This is related to the fact that the weak interaction in the Standard Model is not invariant under parity inversion that exchanges left-handed with right-handed fermions. See Sect. 8.5 for more details.

Remark 6.12.7 It is sometimes useful to form the **fibre product** of the principal $\mathrm{Spin}^+(s,t)$-bundle

$$\pi_S \colon \mathrm{Spin}^+(M) \longrightarrow M$$

and the principal G-bundle

$$\pi_P \colon P \longrightarrow M,$$

defined by

$$\mathrm{Spin}^+(M) \times_M P = \{(s,p) \in \mathrm{Spin}^+(M) \times P \mid \pi_P(p) = \pi_S(s)\}.$$

This is a principal bundle for the group $\mathrm{Spin}^+(s,t) \times G$. Twisted spinor bundles $S \otimes E$ are then vector bundles associated to this principal bundle via a representation of the group $\mathrm{Spin}^+(s,t) \times G$ (the same is true for vector bundles of the form $T \otimes E$ where T is associated to any $\mathrm{Spin}^+(s,t)$-representation, for example, the scalar or vector representation). The group $\mathrm{Spin}^+(s,t) \times G$ can thus be considered as the *full symmetry group* of the gauge theory.

6.13 Exercises for Chap. 6

6.13.1 Find a canonical isomorphism between $\mathrm{O}(s,t)$ and $\mathrm{O}(t,s)$ for all s,t.

6.13.2 Show that if both $s,t \neq 0$, then $\mathrm{O}(s,t)$ is not compact.

6.13.3 Prove that the subgroup $\mathrm{SO}^+(s,t)$ is the connected component of the identity in $\mathrm{O}(s,t)$.

6.13.4

1. The general form of a matrix $A \in \mathrm{O}(1,1)$ depends on one real parameter $a \geq 1$ and three signs δ, ϵ, γ. Determine this general form of A.
2. Determine the general form of a matrix $A \in \mathrm{SO}^+(1,1)$.

6.13.5 Consider the Clifford algebra $\mathrm{Cl}(s,t)$ with $s+t=n$. The set of all products of the basis vectors $e_1, \ldots, e_{s+t} \in \mathbb{R}^{s,t}$ forms a subgroup of the multiplicative group of $\mathrm{Cl}(s,t)$ of order 2^n. Prove that every complex (real) representation of $\mathrm{Cl}(s,t)$ admits a Hermitian (Euclidean) scalar product such that all gamma matrices are unitary (orthogonal).

6.13.6 Use Table 6.2 to show that for all $n \in \mathbb{N}_0$

$$\mathrm{Cl}(n+8) \cong \mathrm{End}\left(\mathbb{R}^{16}\right) \otimes_{\mathbb{R}} \mathrm{Cl}(n).$$

This result is called **Bott periodicity**.

6.13.7 Consider the Clifford algebras $\mathrm{Cl}(n)$ for $n \in \mathbb{N}_0$.

1. For $m \in \mathbb{N}_0$ define an integer $\rho'(m)$ inductively by

$$\rho'(0) = 0, \quad \rho'(1) = 1, \quad \rho'(2) = 3, \quad \rho'(3) = 7$$

and

$$\rho'(4+m) = 8 + \rho'(m) \quad \forall m \in \mathbb{N}_0.$$

Use Table 6.2 and Bott periodicity to prove that $\mathbb{R}^{2^m}$ admits a representation of $\mathrm{Cl}(\rho'(m))$ for all $m \in \mathbb{N}_0$.

2. Write an integer $n \in \mathbb{N}_0$ as $n = (2a+1)2^m$ with $a, m \in \mathbb{N}_0$, i.e. 2^m is the largest power of 2 dividing n. Define an integer $\rho(n)$ by

$$\rho(n) = \rho'(m) + 1.$$

Prove that $\mathbb{R}^n$ admits a representation of $\mathrm{Cl}(\rho(n) - 1)$ for all $n \in \mathbb{N}_0$.

6.13.8 Let $n \geq 1$ be an integer and consider the integer $\rho(n)$ from Exercise 6.13.7. Prove that the unit sphere $S^{n-1} \subset \mathbb{R}^n$ admits a set of $\rho(n) - 1$ orthonormal tangent vector fields. Determine those n for which this construction gives an orthonormal trivialization of the tangent bundle TS^{n-1}.

Remark According to a theorem of Adams [3] the sphere S^{n-1} admits no more than $\rho(n) - 1$ linearly independent tangent vector fields.

6.13.9 Let $e_1, \ldots, e_{s+t}$ be an orthonormal basis of $\mathbb{R}^{s,t}$ and define

$$M(s,t) = \mathrm{span}\{e_i e_j \in \mathrm{Cl}(s,t) \mid 1 \leq i < j \leq s+t\}.$$

Prove that the vector space $M(s,t)$ is a Lie subalgebra of $\mathfrak{cl}^{\times}(s,t) \cong \mathrm{Cl}(s,t)$ with the commutator

$$[x,y] = x \cdot y - y \cdot x.$$

Show that $M(s,t)$ has dimension $\frac{1}{2}(s+t)(s+t-1)$.

6.13.10

1. Let $n \geq 3$. Show that to every connected and simply connected Lie subgroup $H \subset \mathrm{SO}(n)$ we can associate a canonical Lie subgroup in $\mathrm{Spin}(n)$ isomorphic to H.
2. Find embeddings of the Lie groups $\mathrm{SU}(m)$ (for $m \geq 2$), $\mathrm{Sp}(k)$ (for $k \geq 1$) and G_2 into suitable spin groups $\mathrm{Spin}(n)$.

Remark According to Exercise 3.12.6 these embeddings define simply connected homogeneous spaces.

6.13.11 Consider the spinor representation of $\mathrm{Cl}(1,1)$ on $\Delta = \mathbb{C}^2$, defined using the mathematical gamma matrices from Example 6.3.14. Find an explicit map

$$\sigma \colon \Delta \longrightarrow \Delta$$

which is complex antilinear, commutes with γ_k for $k = 1,2$, preserves both Weyl spinor spaces Δ_+ and Δ_- and satisfies $\sigma^2 = \mathrm{Id}_\Delta$. This proves that the spinor representation of $\mathrm{Cl}(1,1)$ is Majorana–Weyl. Determine the subspaces of left-handed and right-handed Majorana–Weyl spinors.

6.13.12 Consider the spinor representation of $\mathrm{Cl}(4,0)$ on $\Delta = \mathbb{C}^4$, defined using the mathematical gamma matrices

$$\gamma_0 = \begin{pmatrix} 0 & I_2 \\ -I_2 & 0 \end{pmatrix}$$

$$\gamma_k = \begin{pmatrix} 0 & i\sigma_k \\ i\sigma_k & 0 \end{pmatrix} \quad \forall k = 1,2,3$$

from Example 6.3.16.

1. Find an explicit map

$$J \colon \Delta \longrightarrow \Delta$$

which is complex antilinear, commutes with γ_k for all $k = 0,1,2,3$, preserves both Weyl spinor spaces Δ_+ and Δ_- and satisfies $J^2 = -\mathrm{Id}_\Delta$.
2. Show that $I = i$, J and $K = IJ$ turn the Weyl spinor spaces $\Delta_\pm$ into quaternionic vector spaces. The spinor representation of $\mathrm{Cl}(4,0)$ is therefore symplectic Majorana–Weyl.

6.13.13 Find mathematical gamma matrices for the spinor representation of $\mathrm{Cl}(3,0)$ and show by an explicit calculation that it is symplectic Majorana.

6.13.14 Consider the spinor representation of $\mathrm{Cl}(1,1)$ on $\mathbb{C}^2$, defined using the mathematical gamma matrices from Example 6.3.14.

1. Find explicit unitary charge conjugation matrices C for both cases in the table in Sect. 6.7.1.
2. Show that the representation is basis unitary and determine the matrices A for both choices in Proposition 6.7.13.
3. Calculate all four combinations for the matrix $B = CA$ and show that in each case $B^*B = I_2$, hence B defines a real structure. Determine the Majorana spinors for each choice of B.

6.13.15 Consider the spinor representation of $\mathrm{Cl}(4,0)$ on $\mathbb{C}^4$, defined using the mathematical gamma matrices

$$\gamma_0 = \begin{pmatrix} 0 & I_2 \\ -I_2 & 0 \end{pmatrix}$$

$$\gamma_k = \begin{pmatrix} 0 & i\sigma_k \\ i\sigma_k & 0 \end{pmatrix} \quad \forall k = 1,2,3$$

from Example 6.3.16.

1. Find explicit unitary charge conjugation matrices C for both cases in the table in Sect. 6.7.1.
2. Show that the representation is basis unitary and determine the two choices for the matrix A from Proposition 6.7.13.
3. Calculate all four combinations for the matrix $B = CA$ and show that in each case $B^*B = -I_2$, hence B defines a quaternionic structure.

6.13.16 Consider the spinor representation of $\mathrm{Cl}(1,3)$ as in Sect. 6.8.

1. Determine a charge conjugation matrix C' such that

$$\Gamma_a^T = C'\Gamma_a C'^{-1} \quad \forall a = 0,1,2,3$$

$$C'^T = -C'.$$

2. Calculate the matrix $B = C'A$ with $A = \Gamma_0$ and show that $B^*B = -I$, hence B defines a quaternionic structure.
3. Determine the matrix A' corresponding to the first choice in Proposition 6.7.13.
4. Calculate the matrices $B = CA'$ (with C as in Sect. 6.8) and $B = C'A'$ and determine whether they define a real or quaternionic structure.

6.13.17 Consider the identification $\mathbb{R}^{1,3} \cong \mathrm{Herm}(2,\mathbb{C})$ from Sect. 6.8.2 and set

$$\begin{aligned} \mathrm{SL}(2,\mathbb{C}) \times \mathbb{R}^{1,3} &\longrightarrow \mathbb{R}^{1,3} \\ (M,X) &\longmapsto MXM^{\dagger}. \end{aligned}$$

1. Prove that this map is well-defined and yields a homomorphism

$$\psi\colon \mathrm{SL}(2,\mathbb{C}) \longrightarrow \mathrm{SO}^{+}(1,3)$$

   of Lie groups.
2. Show that ψ is surjective and has kernel $\{I, -I\}$.

6.13.18 Do a similar construction to the one in Exercise 6.13.17 to show that there exists a surjective Lie group homomorphism

$$\phi\colon \mathrm{SL}(2,\mathbb{R}) \longrightarrow \mathrm{SO}^{+}(1,2)$$

with kernel $\{I, -I\}$.

Remark The 3-dimensional Lie group $\mathrm{SL}(2,\mathbb{R})$ is isomorphic to $\mathrm{Spin}^{+}(1,2)$ (note that $\mathrm{SO}^{+}(1,2)$ has fundamental group $\mathbb{Z}$).

6.13.19 Show that the 1-forms ω_{ab} defined by the Levi-Civita connection with respect to a vielbein are antisymmetric in the indices a, b:

$$\omega_{ab} = -\omega_{ba} \quad \forall a, b = 1, \ldots, n.$$

6.13.20 Prove that the 1-forms ω_{ab} are determined by the anholonomy coefficients ${\Omega_{ab}}^{c}$ as follows:

$$\omega_{cab} = \omega_{ab}(e_c) = \frac{1}{2}(\Omega_{cab} - \Omega_{abc} + \Omega_{bca}),$$

where $\Omega_{abc} = {\Omega_{ab}}^{d}\eta_{dc}$.

6.13.21 Show that the spin covariant derivative is compatible with the Levi-Civita connection in the following way: For all vector fields $X, Y \in \mathfrak{X}(M)$ and spinors $\Psi \in \Gamma(S)$ the identity

$$\nabla_X(Y \cdot \Psi) = (\nabla_X Y) \cdot \Psi + Y \cdot \nabla_X \Psi$$

holds.

Part II
The Standard Model of Elementary Particle Physics

Chapter 7
The Classical Lagrangians of Gauge Theories

If we consider, from an abstract point of view, a field theory involving several types of fields on spacetime (scalar fields, gauge fields, spinors, etc.), then the **Lagrangian** of the field theory is the formula that contains the dynamics and all interactions between these fields. In classical field theory, the equations of motion, i.e. the field equations, that govern the evolution of the fields over time, are derived from the Lagrangian. In quantum field theory, the Lagrangian (through the **action**, the integral of the Lagrangian over spacetime) enters the formula for path integrals that are used to calculate correlators and scattering amplitudes for elementary particles.

Given that the structure of the common Lagrangians is quite simple, it is truly remarkable that the enormous complexity and intricacy of quantum field theories are already contained in the Lagrangians. The Lagrangians can be considered the fundamental cornerstones of field theories.

Lagrangians can be categorized depending on which types of fields and interactions they involve: there are Lagrangians for free fields, Lagrangians for a single interacting field and Lagrangians for several interacting fields. As a general rule, Lagrangians which are **harmonic**, i.e. quadratic in the fields, correspond to free theories, while Lagrangians which contain **anharmonic** terms of order three or higher in the fields lead in the quantum field theory to the creation and annihilation of particles and thus to interactions. **Interactions between fields** (in particular, in the case of weakly interacting, perturbative quantum field theories) are depicted using **Feynman diagrams**. Interacting quantum field theories are usually very complicated and in many cases (including the Standard Model) not fully understood.

There are *a priori* countless Lagrangians that one could consider for a given set of fields. The Lagrangians that are important in physics are mainly restricted by three principles:

1. Existence of symmetries.
2. The quantum field theory should be renormalizable.
3. The quantum field theory should be free of gauge anomalies.

M.J.D. Hamilton, *Mathematical Gauge Theory*, Universitext,
https://doi.org/10.1007/978-3-319-68439-0_7

We will briefly discuss how these principles restrict the possible Lagrangians and then study the Lagrangians that appear in the Standard Model of elementary particles. These Lagrangians are called:

- the Yang–Mills Lagrangian
- the Klein–Gordon and Higgs Lagrangian
- the Dirac Lagrangian
- Yukawa coupling (itself not a complete dynamic Lagrangian)

The Lagrangians in the Standard Model are all Lorentz invariant and gauge invariant. Lorentz invariance here means invariance under local Lorentz transformations of the spacetime manifold, acting on each tangent space. This implies that for fixed values of the fields the Lagrangian is a scalar function on spacetime. Lorentz invariance for a field theory involving spinors always means invariance under the orthochronous Lorentz spin group.

There are numerous books and articles on field theory and the Standard Model. In the present and the following chapter on the Standard Model we mainly rely on the following references:

- The book [16] by David Bleecker is one of the best mathematical treatments of symmetry breaking. Our discussion of symmetry breaking and the Higgs mechanism in Chap. 8 draws heavily from it.
- The article [9] by John C. Baez and John Huerta is an excellent mathematical exposition of the representations of the Standard Model and Grand Unified Theories. Our notation for the representations of the Standard Model in Chap. 8 mainly follows this reference.
- The book [100] by Ulrich Mosel is a concise summary of the Standard Model with a very good exposition of the Lagrangians and symmetry breaking. The explicit Lagrangians for the Standard Model that we derive in Chap. 8 mainly follow the notation in Mosel's book.
- The book [137] by Mark Thomson is an excellent modern and readable treatment of particle physics with many interesting details and explanations concerning experimental and theoretical aspects.
- The topic of the book [22] by Gustavo Castelo Branco, Luís Lavoura and João Paulo Silva is CP violation, but it also has a very clear description of the Standard Model and its Lagrangians.
- The book [62] by Carlo Giunti and Chung W. Kim focuses mainly on neutrino physics, but also contains in the first chapters a concise and modern description of the Standard Model, including details about the Lagrangians and quark mixing.
- Our main references for results from quantum field theory are the book [125] by Matthew Schwartz and the books [143–145] by Steven Weinberg.
- The website [105] of the Particle Data Group contains many up-to-date experimental values for elementary particles as well as some succinct theoretical discussions.

- A great source for the history and development of the Standard Model is the book [79] by Lillian Hoddeson, Laurie Brown, Michael Riordan and Max Dresden (editors).[1] Historical remarks can also be found on the official website [117] for the Nobel Prize in Physics. In particular, [118–120] and [121] contain very readable background material for the Nobel Prizes in Physics 2004, 2008, 2013 and 2015. A short history of gauge theory in physics and mathematics can be found in the book review [92] for Bleecker's book.

Further references are the books [33, 71] and [113] (on the complete Standard Model), [112] (on the electroweak theory), [42] (on QCD), [124, 132] (on QFT in general) and [39, 41, 101, 102, 114, 122, 123] (on the mathematics of the Standard Model) as well as the lecture notes [141].

7.1 Restrictions on the Set of Lagrangians

The Lagrangians that occur in physics are restricted from the infinite set of possible Lagrangians by certain principles that we want to discuss in this section.

7.1.1 Existence of Symmetries

The Lagrangian (or the action) of a field theory should be invariant under certain transformations of the fields, i.e. under certain symmetry groups. Particular examples are:

- Lorentz symmetry
- gauge symmetry
- conformal symmetry
- supersymmetry

We have to distinguish two meanings of symmetries in field theories. Here we think of the primary meaning: the Lagrangian for the fields and thus the *laws of physics*, not the field configurations or their initial values themselves, are invariant under symmetry transformations.

The secondary meaning of symmetry (invariance of the *actual field configuration*) is also sometimes of significance in physics. For example, the action of general relativity for the spacetime metric is invariant under the full orientation preserving diffeomorphism group of the spacetime manifold. A specific metric, however, is invariant only under a much smaller symmetry group, the *isometry group* of this metric (which could just consist of a single element, the identity map).

---

[1] I thank Anthony Britto for pointing out this reference.

Similarly the actions of supergravity theories are invariant under all local supersymmetries, but a specific supersymmetric configuration is invariant under a much smaller group of supersymmetries (a generic field configuration is not supersymmetric at all).

A third example is spontaneously broken gauge theories, which we consider in detail in Chap. 8. In this case the Lagrangian is invariant under gauge transformations with values in a Lie group G, but due to the existence of the Higgs condensate, the vacuum configuration is invariant only under gauge transformations with values in a subgroup $H \subset G$.

The existence of **gauge symmetries** is particularly important: it can be shown that a quantum field theory involving massless spin 1 bosons can be consistent (i.e. unitary, see Sect. 7.1.3) only if it is gauge invariant [125, 143]. This is the reason why we demand Lagrangians involving vector fields (or 1-forms) to be invariant under gauge transformations.

7.1.2 The Quantum Field Theory Should Be Renormalizable

The quantum field theory associated to the Lagrangian should be renormalizable to yield in the end (after renormalization of the parameters, such as coupling constants and masses, cf. Sect. B.2.8) finite results that can be compared with experiments and used to adjust the free parameters of the theory. A simple calculation of the *mass dimension* of summands in the Lagrangian determines which terms have a chance to yield renormalizable theories.

For example, let

$$\mathscr{L} = \mathscr{L}(\phi_1, \ldots, \phi_n)$$

be a renormalizable Lagrangian, where $\phi_1, \ldots, \phi_n$ denote certain fields on spacetime (not necessarily scalars). Suppose that $\mathscr{L}$ is Lorentz invariant and, say, gauge invariant (for instance, $\mathscr{L}$ could be the Yang–Mills Lagrangian or the Klein–Gordon Lagrangian). Then for all natural numbers k the k-th power $\mathscr{L}^k$ will also be Lorentz invariant and gauge invariant. However, in almost all cases, for $k \geq 2$, the Lagrangian $\mathscr{L}^k$ will be non-renormalizable, because it has the wrong mass dimension.

Demanding that the quantum field theory is renormalizable thus greatly restricts the possible terms that can appear in Lagrangians. Calculating mass dimensions (power counting), it can be shown that in 4-dimensional spacetime the only renormalizable and gauge invariant Lagrangians are sums of the Lagrangians that we discuss in this chapter[2] (where in the case of the Higgs Lagrangian for a scalar

[2] An exception, that we do not discuss in this book, is the *topological theta term* $\langle F_M^A, *F_M^A \rangle_{\mathrm{Ad}(P)}$, that appears in some modifications of QCD and in supersymmetric gauge theories.

field ϕ the potential has to be a polynomial in ϕ of degree less than or equal to 4). See [143, Sect. 12.3] for details.

This is very satisfying, because it means, from the point of view of quantum field theory in 4-dimensional spacetime, that there will be no additional types of interactions. It also turns out that all of the allowed Lagrangians actually appear in the Standard Model (in a certain specific form, i.e. with a specific gauge group G, specific charged fermions, etc.). The restriction of renormalizability does not hold for *effective* Lagrangians, i.e. Lagrangians that are only used for calculations at low energies.

7.1.3 The Quantum Field Theory Should Be Free of Gauge Anomalies

Symmetries of the classical field theory, like gauge symmetries, do not necessarily hold in the quantum field theory. The reason is that the measure involved in the definition of path integrals may not be invariant under the symmetry. If this happens, the symmetry is called **anomalous**.

In quantum theory, we demand that the Hilbert space of the system does not contain both vectors of positive norm and negative norm (states of negative norm are called **ghost states**). This property is sometimes called **unitarity** (a vector space with a positive definite Hermitian scalar product is also known as a unitary vector space).[3] If unitarity does not hold, i.e. there exist both states of positive and negative norm in the Hilbert space, then the scalar product does not have a probability interpretation (see Exercise 7.9.1), violating a fundamental axiom of quantum theory.

It is possible to show that in 4-dimensional Minkowski spacetime, anomalies of gauge symmetries imply that the quantum theory violates unitarity (this is related to the fact that the Lorentz metric is indefinite and that the scalar product on the Hilbert space of the quantum field theory must be Poincaré invariant; see [125, Chap. 8] for details). It follows that the quantum theory has to be free of gauge anomalies. In practice, this restricts the possible representations and charges of the fermions: the contributions of the fermions in the theory to the gauge anomaly depend on both the gauge groups and fermion representations and have to cancel each other. The Standard Model is anomaly free, see Sect. 8.5.8. For more details on anomalies, see [125, Chap. 30].

---

[3] There is another concept of unitarity (unitarity of the S-matrix, i.e. of time evolution) that we do not consider here.

One therefore has to be careful: even if a gauge theory is well-defined on the classical level, this may not be true for the associated quantum theory. In particular, vanishing of gauge anomalies has to be checked for every theory beyond the Standard Model, like Grand Unified Theories or supersymmetric extensions.

7.1.4 *The Lagrangian of the Standard Model*

Our aim in this chapter is to understand each term in the following Lagrangian, which is essentially the Lagrangian of the Standard Model and could be called the **Yang–Mills–Dirac–Higgs–Yukawa Lagrangian**:

$$\begin{aligned}\mathscr{L} &= \mathscr{L}_D[\Psi, A] + \mathscr{L}_H[\Phi, A] + \mathscr{L}_Y[\Psi_L, \Phi, \Psi_R] + \mathscr{L}_{YM}[A] \\ &= \mathrm{Re}\left(\overline{\Psi} D_A \Psi\right) + \langle d_A\Phi, d_A\Phi\rangle_E - V(\Phi) - 2g_Y \mathrm{Re}\left(\overline{\Psi}_L \Phi \Psi_R\right) - \frac{1}{2}\langle F_M^A, F_M^A\rangle_{\mathrm{Ad}(P)}\,.\end{aligned}$$

7.2 The Hodge Star and the Codifferential

Throughout this chapter, (M, g) is an n-dimensional oriented pseudo-Riemannian manifold. In physics, M is spacetime and g usually has Lorentzian signature. In mathematics, M is an arbitrary manifold and g is often taken to be Riemannian.

We first want to understand the *Yang–Mills Lagrangian* $\mathscr{L}_{YM}[A]$ for a connection A on a principal bundle $P \to M$ and derive the associated equation of motion, called the *Yang–Mills equation*. This equation is most easily stated using the *codifferential*, whose definition involves the *Hodge star operator*. The metric g on the manifold M enters the Yang–Mills equation precisely through the Hodge star. In this section we discuss as a mathematical preparation the Hodge star operator, the codifferential and some related concepts. We follow the exposition in [14].

7.2.1 *Scalar Products on Forms and the Hodge Star Operator*

The metric g together with the orientation of the manifold M define a **canonical volume form** dvol_g on M: If $e_1, \ldots, e_n$ is an oriented, orthonormal basis of T_pM, then dvol_g is characterized by

$$\mathrm{dvol}_g(e_1, \ldots, e_n) = +1.$$

Lemma 7.2.1 *If (U, ϕ) is an oriented chart for M with local coordinates x^μ, then*

$$\mathrm{dvol}_g = \sqrt{|g|}dx^1 \wedge \ldots \wedge dx^n,$$

where

$$|g| = |\det(g_{\mu\nu})|$$

is the absolute value of the determinant of the matrix with entries

$$g_{\mu\nu} = g(\partial_\mu, \partial_\nu).$$

Proof This is Exercise 7.9.2. □

We denote by $g^{\mu\nu}$ the entries of the matrix inverse to the matrix with entries $g_{\mu\nu}$. We can raise indices of tensors in the standard way using $g^{\mu\nu}$. For example,

$$T^{\mu\nu} = g^{\mu\rho}g^{\nu\sigma}T_{\rho\sigma},$$

where the Einstein summation convention is understood.

The semi-Riemannian metric g on M defines bundle metrics on the vector bundles of k-forms $\Lambda^k T^*M$ for all k. This yields scalar products between sections of these bundles that we can write explicitly as follows:

Definition 7.2.2 For $\mathbb{K} = \mathbb{R}, \mathbb{C}$ we define the **scalar product of forms**

$$\langle \cdot, \cdot \rangle : \Omega^k(M, \mathbb{K}) \times \Omega^k(M, \mathbb{K}) \longrightarrow \mathscr{C}^\infty(M, \mathbb{K})$$

as follows: for real-valued k-forms $\omega, \eta \in \Omega^k(M, \mathbb{R})$ on M we set

$$\begin{aligned}
\langle \omega, \eta \rangle &= \sum_{\mu_1 < \ldots < \mu_k} \omega_{\mu_1 \ldots \mu_k} \eta^{\mu_1 \ldots \mu_k} \\
&= \frac{1}{k!} \sum_{\mu_1 \ldots \mu_k} \omega_{\mu_1 \ldots \mu_k} \eta^{\mu_1 \ldots \mu_k} \\
&= \frac{1}{k!} \omega_{\mu_1 \ldots \mu_k} \eta^{\mu_1 \ldots \mu_k},
\end{aligned}$$

where

$$\omega_{\mu_1 \ldots \mu_k} = \omega(\partial_{\mu_1}, \ldots, \partial_{\mu_k})$$

in a local chart (U, ϕ) of M and the second and third sum extend over all k-tuples $\mu_1 \ldots \mu_k$.

For complex-valued k-forms $\omega, \eta \in \Omega^k(M, \mathbb{C}) \cong \Omega^k(M, \mathbb{R}) \otimes \mathbb{C}$ on M we set

$$\langle \omega, \eta \rangle = \sum_{\mu_1 < \ldots < \mu_k} \bar{\omega}_{\mu_1 \ldots \mu_k} \eta^{\mu_1 \ldots \mu_k}.$$

These scalar products are well-defined, independent of the choice of local chart. The associated **norm** is given in both cases by

$$|\omega|^2 = \langle \omega, \omega \rangle.$$

Remark 7.2.3 On a pseudo-Riemannian manifold the norm is in general not positive definite. In particular, $|\omega|^2 = 0$ does not imply $\omega = 0$. For this reason we usually try to avoid the notation $|\omega|^2$.

Definition 7.2.4 The **Hodge star operator**

$$*: \Omega^k(M, \mathbb{K}) \longrightarrow \Omega^{n-k}(M, \mathbb{K})$$

is the linear map defined for real-valued forms by

$$\langle \omega, \eta \rangle \mathrm{dvol}_g = \omega \wedge *\eta \quad \forall \omega, \eta \in \Omega^k(M, \mathbb{R})$$

and for complex-valued forms by

$$\langle \omega, \eta \rangle \mathrm{dvol}_g = \bar{\omega} \wedge *\eta \quad \forall \omega, \eta \in \Omega^k(M, \mathbb{C}).$$

Choosing a local frame, it can be shown that this uniquely defines $*$.

Remark 7.2.5 This definition of the Hodge star operator for pseudo-Riemannian manifolds does not necessarily coincide with the definition sometimes found in the literature. Baum [14], for instance, uses the definition

$$*' = (-1)^t * .$$

We continue to use our definition.

Suppose $e_1, \ldots, e_n$ is an oriented, orthonormal basis of tangent vectors with

$$g(e_i, e_i) = g_{ii} = g^{ii} = \pm 1.$$

Let $\alpha^1, \ldots, \alpha^n$ be the dual basis of 1-forms with $\alpha^i(e_j) = \delta^i_j$. Then

$$\mathrm{dvol}_g = \alpha^1 \wedge \ldots \wedge \alpha^n$$

and we have:

Lemma 7.2.6 *The Hodge star operator is given by*

$$*(\alpha^{m_1} \wedge \ldots \wedge \alpha^{m_k}) = g^{m_1 m_1} \cdots g^{m_k m_k} \epsilon_{m_1 \ldots m_k m_{k+1} \ldots m_n} \alpha^{m_{k+1}} \wedge \ldots \wedge \alpha^{m_n}.$$

In this formula there is on the right-hand side **no** *summation over indices, $\{m_{k+1}, \ldots, m_n\}$ is a complementary set to $\{m_1, \ldots, m_k\}$ and ϵ is totally antisymmetric with*

$$\epsilon_{123\ldots n} = 1.$$

In particular,

$$*\mathrm{dvol}_g = (-1)^t \cdot 1,$$
$$*1 = \mathrm{dvol}_g.$$

Definition 7.2.7 Let $\Omega_0^k(M, \mathbb{K})$ denote the differential forms with compact support on M. Then we define the L^2**-scalar product of forms**

$$\langle \cdot, \cdot \rangle_{L^2} \colon \Omega_0^k(M, \mathbb{K}) \times \Omega_0^k(M, \mathbb{K}) \longrightarrow \mathbb{K}$$

by

$$\langle \omega, \eta \rangle_{L^2} = \int_M \langle \omega, \eta \rangle \mathrm{dvol}_g.$$

We have to restrict the L^2-scalar product to forms with compact support, because otherwise the integral may not be finite.

We can generalize these constructions to twisted differential forms. Suppose that $E \to M$ is a $\mathbb{K}$-vector bundle with bundle metric $\langle \cdot, \cdot \rangle_E$. Together with the semi-Riemannian metric g we then get induced bundle metrics on the vector bundle $\Lambda^k T^*M \otimes E$ of twisted k-forms for all k. More explicitly we can write:

Definition 7.2.8 We define the **scalar product of twisted forms**

$$\langle \cdot, \cdot \rangle_E \colon \Omega^k(M, E) \times \Omega^k(M, E) \longrightarrow \mathscr{C}^\infty(M)$$

as follows: choose a local frame $e_1, \ldots, e_r$ for E over $U \subset M$ and expand k-forms F, G twisted with E as

$$F = \sum_{i=1}^r F_i \otimes e_i,$$

$$G = \sum_{j=1}^r G_j \otimes e_j,$$

with $F_i, G_j \in \Omega^k(U, \mathbb{K})$. Then we set

$$\langle F, G \rangle_E = \sum_{i,j=1}^{r} \langle F_i, G_j \rangle \langle e_i, e_j \rangle_E.$$

This scalar product is independent of the choice of local frame $\{e_i\}$.

We can also define a **Hodge star operator on twisted forms**

$$*\colon \Omega^k(M, E) \longrightarrow \Omega^{n-k}(M, E)$$

by

$$*F = \sum_{i=1}^{r} (*F_i) \otimes e_i$$

and an L^2**-scalar product of twisted forms**

$$\langle \cdot, \cdot \rangle_{E,L^2}\colon \Omega_0^k(M, E) \times \Omega_0^k(M, E) \longrightarrow \mathbb{K}$$

by

$$\langle \omega, \eta \rangle_{E,L^2} = \int_M \langle \omega, \eta \rangle_E \mathrm{dvol}_g.$$

7.2.2 *The Codifferential*

Let (M, g) be an oriented semi-Riemannian manifold of dimension n and signature (s, t). We have the usual exterior differential

$$d\colon \Omega^k(M) \longrightarrow \Omega^{k+1}(M)$$

on forms.

Definition 7.2.9 We define the **codifferential**

$$d^*\colon \Omega^{k+1}(M) \longrightarrow \Omega^k(M)$$

by

$$d^* = (-1)^{t+nk+1} * d * .$$

The codifferential has the following interesting property:

Theorem 7.2.10 (Codifferential on Forms Is Formal Adjoint of Differential) *Let M be a manifold without boundary. Then the codifferential d^* is the formal adjoint of the differential d with respect to the L^2-scalar product on forms with compact support, i.e.*

$$\langle d\omega, \eta \rangle_{L^2} = \langle \omega, d^* \eta \rangle_{L^2}$$

for all $\omega \in \Omega_0^k(M), \eta \in \Omega_0^{k+1}(M)$.

Proof We calculate the difference

$$\langle d\omega, \eta \rangle - \langle \omega, d^* \eta \rangle$$

with respect to the (pointwise) scalar product of forms. According to Exercise 7.9.3

$$**: \Omega^{n-k}(M) \longrightarrow \Omega^{n-k}(M)$$

is given by

$$** = (-1)^{t+(n-k)k}.$$

We have

$$\begin{aligned}
\left(\langle d\omega, \eta \rangle - \langle \omega, d^* \eta \rangle\right) \mathrm{dvol}_g &= (d\omega) \wedge *\eta - \omega \wedge * \left(d^* \eta\right) \\
&= (d\omega) \wedge *\eta + (-1)^k \omega \wedge (d * \eta) \\
&= d(\omega \wedge *\eta).
\end{aligned}$$

This implies the claim by Stokes' Theorem A.2.24. □

We want to generalize the definition of the codifferential to twisted forms. Let $E \to M$ be a $\mathbb{K}$-vector bundle with a scalar product $\langle \cdot, \cdot \rangle_E$ and a compatible covariant derivative ∇. In Sect. 5.12 we defined the associated exterior covariant derivative (or covariant differential)

$$d_\nabla: \Omega^k(M, E) \longrightarrow \Omega^{k+1}(M, E).$$

Definition 7.2.11 We define the **covariant codifferential**

$$d_\nabla^*: \Omega^{k+1}(M, E) \longrightarrow \Omega^k(M, E)$$

by

$$d_\nabla^* = (-1)^{t+nk+1} * d_\nabla * .$$

We then get the following analogue of Theorem 7.2.10.

Theorem 7.2.12 (Covariant Codifferential on Twisted Forms Is Formal Adjoint of Covariant Differential) *Let M be a manifold without boundary. Then the covariant codifferential d_∇^* is the formal adjoint of the exterior covariant differential d_∇ with respect to the L^2-scalar product on forms with compact support, i.e.*

$$\langle d_\nabla \omega, \eta \rangle_{E,L^2} = \langle \omega, d_\nabla^* \eta \rangle_{E,L^2}$$

for all $\omega \in \Omega_0^k(M,E), \eta \in \Omega_0^{k+1}(M,E)$.

Proof We follow the proof in [14]. Since d_∇ and d_∇^* are linear, it suffices to prove the statement for forms ω, η of the form

$$\omega = \sigma \otimes e, \quad \sigma \in \Omega_0^k(M), \quad e \in \Gamma(E),$$
$$\eta = \mu \otimes f, \quad \mu \in \Omega_0^{k+1}(M), \quad f \in \Gamma(E).$$

Then

$$d_\nabla \omega = (d\sigma) \otimes e + (-1)^k \sigma \wedge \nabla e$$

and

$$\begin{aligned} d_\nabla^* \eta &= (-1)^{t+nk+1} * d_\nabla * (\mu \otimes f) \\ &= (-1)^{t+nk+1} * \big((d * \mu) \otimes f + (-1)^{n-k-1} (*\mu) \wedge \nabla f \big) \\ &= (d^* \mu) \otimes f + (-1)^{t+nk+n-k} * ((*\mu) \wedge \nabla f) . \end{aligned}$$

In particular, with Exercise 7.9.3,

$$*d_\nabla^* \eta = -(-1)^k (d * \mu) \otimes f - (-1)^{n-1} (*\mu) \wedge \nabla f .$$

We introduce a scalar product

$$\langle \cdot , \cdot \rangle_E : \Omega^1(M,E) \otimes \Gamma(E) \longrightarrow \Omega^1(M)$$

by setting

$$\langle \omega \otimes a, b \rangle_E = \omega \langle a, b \rangle_E \quad \forall \omega \in \Omega^1(M), a, b \in \Gamma(E)$$

and extending linearly.

For the difference of the pointwise scalar products we then get

$$\begin{aligned}
\left(\langle d_\nabla \omega, \eta \rangle_E - \langle \omega, d_\nabla^* \eta \rangle_E\right) \mathrm{dvol}_g &= \langle (d\sigma) \otimes e, \mu \otimes f \rangle_E + (-1)^k \langle \sigma \wedge \nabla e, \mu \otimes f \rangle_E \\
&\quad - \langle \sigma \otimes e, (d^* \mu) \otimes f \rangle_E \\
&\quad + (-1)^{n-1} (\sigma \wedge *\mu) \wedge \langle \nabla f, e \rangle_E \\
&= d(\sigma \wedge *\mu) \langle e, f \rangle_E \\
&\quad + (-1)^{n-1} (\sigma \wedge *\mu) \wedge (\langle \nabla e, f \rangle_E + \langle \nabla f, e \rangle_E) \\
&= d\left((\sigma \wedge *\mu) \langle e, f \rangle_E\right).
\end{aligned}$$

In the final step we used that ∇ is compatible with the scalar product on E. The claim now follows by Stokes' Theorem A.2.24. □

7.3 The Yang–Mills Lagrangian

In this section we define the Yang–Mills Lagrangian and derive the associated Yang–Mills equation. We fix the following data:

- an n-dimensional oriented pseudo-Riemannian manifold (M, g)
- a principal G-bundle $P \to M$ with compact structure group G of dimension r
- an Ad-invariant positive definite scalar product $\langle \cdot, \cdot \rangle_{\mathfrak{g}}$ on $\mathfrak{g}$, determined by certain coupling constants, as in Sect. 2.5
- a $\langle \cdot, \cdot \rangle_{\mathfrak{g}}$-orthonormal vector space basis $T_1, \ldots, T_r$ for $\mathfrak{g}$.

The Ad-invariant scalar product $\langle \cdot, \cdot \rangle_{\mathfrak{g}}$ on $\mathfrak{g}$ determines a bundle metric on the associated real vector bundle $\mathrm{Ad}(P) = P \times_{\mathrm{Ad}} \mathfrak{g}$ that we denote by $\langle \cdot, \cdot \rangle_{\mathrm{Ad}(P)}$.

7.3.1 *The Yang–Mills Lagrangian*

Let A be a connection 1-form on the principal bundle P with curvature 2-form $F^A \in \Omega^2(P, \mathfrak{g})$. According to Corollary 5.13.5 the curvature defines a twisted 2-form

$$F_M^A \in \Omega^2(M, \mathrm{Ad}(P)).$$

Definition 7.3.1 The **Yang–Mills Lagrangian** is defined by

$$\mathscr{L}_{YM}[A] = -\frac{1}{2}\langle F_M^A, F_M^A\rangle_{\mathrm{Ad}(P)}.$$

For a fixed connection A, the Yang–Mills Lagrangian is a global smooth function

$$\mathscr{L}_{YM}[A]: M \longrightarrow \mathbb{R}.$$

Theorem 7.3.2 *The Yang–Mills Lagrangian is gauge invariant, i.e.*

$$\mathscr{L}_{YM}[f^*A] = \mathscr{L}_{YM}[A]$$

for all bundle automorphisms $f \in \mathscr{G}(P)$ and all connections A on P.

Proof Theorem 5.4.4 implies that the curvature form $F^A \in \Omega^2(P, \mathfrak{g})$ transforms as

$$F^{f^*A} = \mathrm{Ad}_{\sigma_f^{-1}} \circ F^A.$$

Let $f\cdot$ denote the action of f on the adjoint bundle, given by Theorem 5.3.8. Then

$$F_M^{f^*A} = f^{-1} \cdot F_M^A.$$

Since the scalar product $\langle\cdot,\cdot\rangle_{\mathfrak{g}}$ is Ad-invariant, it follows that $\langle\cdot,\cdot\rangle_{\mathrm{Ad}(P)}$ is invariant under the action of f^{-1}. This implies the claim. □

We want to find a formula for the Yang–Mills Lagrangian in local coordinates and in a local gauge. Let $s: U \to P$ be a local gauge. Then the local field strength is given by

$$F_s^A = s^*F^A \in \Omega^2(U, \mathfrak{g}).$$

The scalar product on the Lie algebra $\mathfrak{g}$ defines a scalar product

$$\langle\cdot,\cdot\rangle_{\mathfrak{g}}: \Omega^2(U, \mathfrak{g}) \times \Omega^2(U, \mathfrak{g}) \longrightarrow \mathscr{C}^\infty(U, \mathfrak{g}).$$

As before we set in a chart with coordinates x^μ

$$F_{\mu\nu}^A = F_s^A(\partial_\mu, \partial_\nu).$$

We can expand

$$F_s^A = F_s^{Aa} \otimes T_a$$

and

$$F^A_{\mu\nu} = F^{Aa}_{\mu\nu} T_a,$$

where $F^{Aa}_s \in \Omega^2(U)$ are real-valued differential forms, $F^{Aa}_{\mu\nu} \in \mathscr{C}^\infty(U)$ are real-valued smooth functions on U and we sum over the indices a.

We can then write the Yang–Mills Lagrangian locally as

$$\begin{aligned}\mathscr{L}_{YM}[A] &= -\frac{1}{2}\langle F^A_s, F^A_s\rangle_{\mathfrak{g}}\\ &= -\frac{1}{4}\left\langle F^A_{\mu\nu}, F^{A\mu\nu}\right\rangle_{\mathfrak{g}}\\ &= -\frac{1}{4}F^{Aa}_{\mu\nu}F^{A\mu\nu}_a,\end{aligned}$$

where we sum over all μ, ν. The local field strength is given by

$$F^A_{\mu\nu} = \partial_\mu A_\nu - \partial_\nu A_\mu + [A_\mu, A_\nu]$$

and

$$F^{Aa}_{\mu\nu} = \partial_\mu A^a_\nu - \partial_\nu A^a_\mu + f_{bca}A^b_\mu A^c_\nu$$

with the **structure constants** defined by

$$[T_a, T_b] = \sum_{c=1}^{r} f_{abc}T_c.$$

Lemma 7.3.3 *The structure constants of the Lie algebra $\mathfrak{g}$ with respect to a $\langle\cdot,\cdot\rangle_{\mathfrak{g}}$-orthonormal basis $\{T_a\}$ satisfy*

$$f_{abc} + f_{bac} = 0$$

and

$$f_{bca} + f_{bac} = 0$$

for all indices a, b, c. In particular,

$$f_{bca} = f_{abc}.$$

Proof The first claim is clear, because the Lie bracket is antisymmetric. The second claim follows, because the T_a are an orthonormal basis and the scalar product $\langle\cdot,\cdot\rangle_{\mathfrak{g}}$

on $\mathfrak{g}$ is Ad-invariant: we have

$$\langle [T_b, T_c], T_a \rangle_{\mathfrak{g}} + \langle T_c, [T_b, T_a] \rangle_{\mathfrak{g}} = 0,$$

which implies the claim. □

We can therefore also write the structure equation for the curvature as

$$F^{Aa}_{\mu\nu} = \partial_\mu A^a_\nu - \partial_\nu A^a_\mu + f_{abc} A^b_\mu A^c_\nu.$$

This implies the following explicit formula for the Yang–Mills Lagrangian:

$$\begin{aligned}
\mathscr{L}_{YM}[A] &= -\frac{1}{4} F^{Aa}_{\mu\nu} F^{A\mu\nu}_a \\
&= -\frac{1}{4}(\partial_\mu A^a_\nu - \partial_\nu A^a_\mu)(\partial^\mu A^\nu_a - \partial^\nu A^\mu_a) \\
&\quad - \frac{1}{2} f_{abc}(\partial_\mu A^a_\nu - \partial_\nu A^a_\mu) A^{b\mu} A^{c\nu} \\
&\quad - \frac{1}{4} f_{abc} f_{ade} A^b_\mu A^c_\nu A^{d\mu} A^{e\nu}.
\end{aligned} \tag{7.1}$$

The term in the second line is quadratic in the gauge field. It describes free (non-interacting) gauge bosons and is the only term if the group G is abelian. The terms in the third and fourth line are cubic and quartic in the gauge field and describe a direct interaction between the gauge bosons in non-abelian gauge theories. In the case of QCD these terms are called 3-gluon vertex and 4-gluon vertex. Figure 7.1 shows the Feynman diagrams for these vertices.

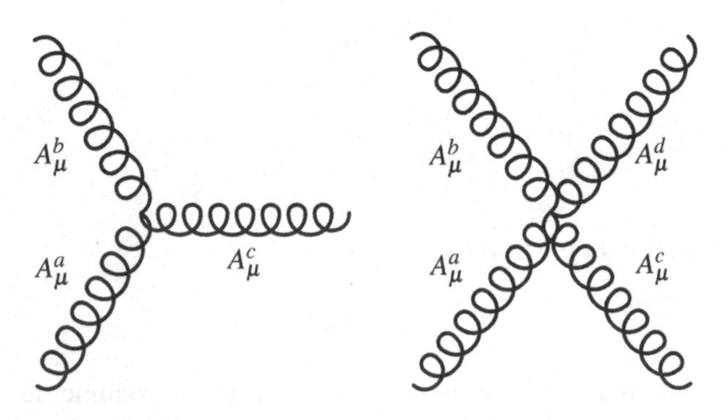

Fig. 7.1 Interaction vertices for non-abelian gauge bosons

Remark 7.3.4 In physics, the quantum field theory for a gauge field A determined by the Yang–Mills Lagrangian, without any additional matter fields, is known as **pure Yang–Mills theory** or **gluodynamics**. For non-abelian Lie groups, the quantum version of pure Yang–Mills theory predicts particles, known as **glueballs**, which only consist of gauge bosons (gluons in QCD). The Clay Millennium Prize Problem [37] on the mass gap is to prove that the masses of glueballs in a quantum pure Yang–Mills theory on $\mathbb{R}^4$ with compact simple gauge group G are bounded from below by a positive (non-zero) number.

Remark 7.3.5 The term "gauge invariance" was invented by Hermann Weyl in 1929 for the U(1) gauge theory of electromagnetism. Gauge theory for non-abelian structure groups G was first developed by Chen Ning Yang (Nobel Prize in Physics 1957) and Robert L. Mills (for $G = \mathrm{SU}(2)$) in the 1950s.

7.3.2 The Yang–Mills Equation

We assume now that

- the semi-Riemannian manifold (M, g) is closed, i.e. compact and without boundary.

Definition 7.3.6 Let $\mathscr{A}(P)$ denote the space of all connection 1-forms A on the principal bundle P. This is by the discussion in Sect. 5.13 a (usually infinite-dimensional) affine space over the vector space

$$\Omega^1_{\mathrm{hor}}(P, \mathfrak{g})^{\mathrm{Ad}} \xrightarrow{\cong} \Omega^1(M, \mathrm{Ad}(P)),$$

with isomorphism given by the map Λ. For $\alpha \in \Omega^1_{\mathrm{hor}}(P, \mathfrak{g})^{\mathrm{Ad}}$ we set

$$\alpha_M = \Lambda(\alpha) \in \Omega^1(M, \mathrm{Ad}(P)).$$

Definition 7.3.7 The **Yang–Mills action** for a principal G-bundle $P \to M$ is the smooth map

$$S_{YM}: \mathscr{A}(P) \longrightarrow \mathbb{R},$$

defined by

$$\begin{aligned} S_{YM}[A] &= -\frac{1}{2}\langle F_M^A, F_M^A \rangle_{\mathrm{Ad}(P),L^2} \\ &= -\frac{1}{2}\int_M \langle F_M^A, F_M^A \rangle_{\mathrm{Ad}(P)} \,\mathrm{dvol}_g. \end{aligned}$$

The integral is well-defined, because M is compact.

Definition 7.3.8 We call a connection A on the principal bundle P a **critical point** of the Yang–Mills action if

$$\left.\frac{d}{dt}\right|_{t=0} S_{YM}[A+t\alpha] = 0$$

for all variations

$$\alpha \in \Omega^1_{\mathrm{hor}}(P,\mathfrak{g})^{\mathrm{Ad}} \cong \Omega^1(M,\mathrm{Ad}(P)).$$

For a connection A on P we denote by d_A the associated covariant differential and by d_A^* the covariant codifferential. We want to prove:

Theorem 7.3.9 *A connection A on a principal bundle $P \to M$ is a critical point of the Yang–Mills action if and only if A satisfies the* ***Yang–Mills equation***

$$d_A^* F_M^A = 0,$$

i.e.

$$d_A * F_M^A = 0.$$

Proof We follow the proof in [14]. According to the structure equation in Theorem 5.5.4 we can calculate

$$\begin{aligned} F^{A+t\alpha} &= d(A+t\alpha) + \frac{1}{2}[A+t\alpha, A+t\alpha] \\ &= F^A + t(d\alpha + [A,\alpha]) + \frac{1}{2}t^2[\alpha,\alpha]. \end{aligned}$$

This implies

$$F_M^{A+t\alpha} = F_M^A + t\,(d_A\alpha_M) + \frac{1}{2}t^2[\alpha_M,\alpha_M].$$

We get with Theorem 7.2.12

$$\begin{aligned} \left.\frac{d}{dt}\right|_{t=0} \left\langle F_M^{A+t\alpha}, F_M^{A+t\alpha}\right\rangle_{\mathrm{Ad}(P),L^2} &= 2\left\langle d_A\alpha_M, F_M^A\right\rangle_{\mathrm{Ad}(P),L^2} \\ &= 2\left\langle \alpha_M, d_A^* F_M^A\right\rangle_{\mathrm{Ad}(P),L^2}. \end{aligned}$$

Since the scalar product on the Lie algebra $\mathfrak{g}$ is non-degenerate, the L^2-scalar product on $\Omega^1(M, \mathrm{Ad}(P))$ is non-degenerate. It follows that A is a critical point of the Yang–Mills Lagrangian if and only if $d_A^* F_M^A = 0$. □

In a local gauge $s: U \to P$ the Yang–Mills equation can be written as

$$d * F_s^A + \left[A_s, *F_s^A\right] = 0.$$

Remark 7.3.10 Recall that *any* connection A on the principal bundle P has to satisfy the **Bianchi identity**, which can be written according to Theorem 5.14.2 as

$$d_A F_M^A = 0.$$

Atiyah and Bott [6] have noted that the curvature F_M^A of a connection A that satisfies in addition to the Bianchi identity the Yang–Mills equation $d_A * F_M^A = 0$ can thus be considered as a **harmonic form** (in a non-linear sense if G is non-abelian) in $\Omega^2(M, \mathrm{Ad}(P))$ (compare with Exercise 7.9.5). The Yang–Mills equation is a second-order partial differential equation for the connection A.

Remark 7.3.11 Note that the Yang–Mills equation depends through the Hodge star operator on the pseudo-Riemannian metric g on M. If the equation holds for one metric, it does not necessarily hold for another metric.

Example 7.3.12 (Maxwell's Equations) In the case when $G = \mathrm{U}(1)$, the local curvature forms F_s are independent of the choice of local gauge s and define a global 2-form $F_M \in \Omega^2(M, \mathfrak{u}(1))$, see Corollary 5.6.4. The Bianchi identity and Yang–Mills equation are then given by

$$dF_M = 0,$$
$$d * F_M = 0.$$

These are **Maxwell's equations** for a source-free electromagnetic field (on a general n-dimensional oriented pseudo-Riemannian manifold). On Minkowski spacetime of dimension 4 we can use the construction in Sect. 5.7 to write Maxwell's equations in terms of the electric and magnetic field.

Maxwell's equations generalize to any abelian Lie group G. Note that in this case both the Bianchi identity and the Yang–Mills equation are linear. For a non-abelian structure group these equations are non-linear (and therefore much harder to solve).

We could study the Yang–Mills equation on any of the examples of principal bundles that we defined in Chap. 4, in particular, on the Hopf fibrations over projective spaces or on the canonical principal bundles over homogeneous spaces, once (pseudo-)Riemannian metrics on the base manifolds have been defined.

Definition 7.3.13 We call a connection A on a principal bundle a **Yang–Mills connection** if it satisfies the Yang–Mills equation.
Since the Yang–Mills equations do not depend on the choice of local gauge, the gauge group $\mathcal{G}(P)$ of the principal bundle $P \to M$ acts on the space of Yang–Mills connections. We can therefore set:

Definition 7.3.14 The **Yang–Mills moduli space** of a principal bundle $P \to M$ over a pseudo-Riemannian manifold (M, g) is the space of Yang–Mills connections A modulo the gauge group $\mathcal{G}(P)$.
The moduli space is usually the quotient of an infinite-dimensional space by the action of an infinite-dimensional group. It is therefore non-trivial to define, for example, a smooth structure on the moduli space.

Example 7.3.15 (Instantons) Let $P \to M$ be a principal G-bundle over an oriented *Riemannian* 4-*manifold* (M, g). In this case the Hodge star operator satisfies $** = 1$ on 2-forms on M. We consider connections A on P with curvature $F_M^A \in \Omega^2(M, \mathrm{Ad}(P))$ such that either

$$*F_M^A = F_M^A$$

or

$$*F_M^A = -F_M^A.$$

Connections that satisfy these identities are called **self-dual** and **anti-self-dual instantons**, respectively (see Exercise 7.9.3 for the notion of self-duality).

Since any connection A satisfies the Bianchi identity, instantons automatically satisfy the Yang–Mills equation. The instanton equations are examples of **BPS equations**, i.e. special first order equations (here for the gauge field A) whose solutions are (often) automatically solutions of the second order field equations (here the Yang–Mills equations). BPS equations appear in many other parts of physics, for example, in the theory of magnetic monopoles (Bogomolny equations) or in supergravity (Killing spinor equations).

The instanton equations are preserved under the action of the gauge group $\mathcal{G}(P)$ and we can define **instanton moduli spaces**. These moduli spaces, especially for structure groups $G = \mathrm{SU}(2)$ and $G = \mathrm{SO}(3)$, are the cornerstone of **Donaldson theory**, which revolutionized the understanding of smooth 4-manifolds in the 1980s.

7.3.3 *Massive Gauge Bosons*

The Yang–Mills Lagrangian

$$\begin{aligned}\mathscr{L}_{YM} &= -\frac{1}{2}\langle F_M^A, F_M^A\rangle_{\mathrm{Ad}(P)} \\ &= -\frac{1}{4}F_{\mu\nu}^{Aa}F_a^{A\mu\nu}\end{aligned}$$

describes *massless* gauge bosons. Arguments from physics show that gauge bosons of mass m are described by adding (in a local gauge) a term of the form

$$\frac{1}{2}m^2 A_a^\nu A_\nu^a \tag{7.2}$$

to the Yang–Mills Lagrangian. We could try to write this Lagrangian in an invariant form as above, such as

$$\frac{1}{2}m^2 \langle A_M, A_M\rangle_{\mathrm{Ad}(P)},$$

however the gauge field A does not define an element $A_M \in \Omega^1(M, \mathrm{Ad}(P))$ (only the difference of two gauge fields is such a twisted form). This indicates that the Lagrangian in Eq. (7.2) is not well-defined, independent of local gauge. It is also easy to see directly that local gauge transformations $g: U \rightarrow G$, which are not constant, in general do not leave the Lagrangian in Eq. (7.2) invariant.

Remark 7.3.16 One of the main features of the Higgs mechanism, discussed in Chap. 8, is that it allows us to introduce a non-zero mass for gauge bosons with a *gauge invariant* Lagrangian. Introducing a mass for gauge bosons is necessary to describe the weak interaction as a gauge theory, because experiments show that the W- and Z-gauge bosons of the weak interaction have a non-zero mass.

7.4 Mathematical and Physical Conventions for Gauge Theories

In mathematics and physics slightly different conventions are used for scalar products, coupling constants and covariant derivatives. We want to compare these conventions in this section. We fix the following data:

- a compact Lie group G which is either simple or $\mathrm{U}(1)$ (the conventions below can be generalized to any compact Lie group)
- an Ad_G-invariant positive definite scalar product $\langle\cdot,\cdot\rangle_{\mathfrak{g}}$ on the Lie algebra $\mathfrak{g}$ (if $\mathfrak{g}$ is simple we can take the negative of the Killing form and for $\mathfrak{u}(1) \cong \mathbb{R}$ we can take any positive definite scalar product)

- a $\langle \cdot, \cdot \rangle_{\mathfrak{g}}$-orthonormal basis $S_1, \ldots, S_r$ of the Lie algebra $\mathfrak{g}$
- a real coupling constant $g > 0$.

1. In **mathematics** we choose the scalar product

$$\langle \cdot, \cdot \rangle'_{\mathfrak{g}} = \frac{1}{g^2} \langle \cdot, \cdot \rangle_{\mathfrak{g}}$$

with orthonormal basis

$$T_a = gS_a, \quad a = 1, \ldots, r.$$

We expand the gauge field $A \in \Omega^1(P, \mathfrak{g})$ and curvature $F \in \Omega^2(P, \mathfrak{g})$ as

$$A = \sum_{a=1}^{r} A^a \otimes T_a,$$

$$F = \sum_{a=1}^{r} F^a \otimes T_a.$$

The covariant derivative (after a choice of local gauge) is

$$\nabla_\mu^A = \partial_\mu + A_\mu.$$

The local curvature is

$$F_{\mu\nu} = \partial_\mu A_\nu - \partial_\nu A_\mu + [A_\mu, A_\nu].$$

The Yang–Mills Lagrangian is

$$\mathscr{L}_{YM} = -\frac{1}{4} \left\langle F^{\mu\nu}, F_{\mu\nu} \right\rangle'_{\mathfrak{g}} = -\frac{1}{4} F^a_{\mu\nu} F^{\mu\nu}_a.$$

2. In **physics** we choose the Hermitian scalar product $\langle \cdot, \cdot \rangle_{i\mathfrak{g}}$ on $i\mathfrak{g}$ associated to $\langle \cdot, \cdot \rangle_{\mathfrak{g}}$ and the orthonormal basis

$$\frac{1}{i} S_a, \quad a = 1, \ldots, r$$

of $i\mathfrak{g}$. We expand the gauge field $B \in \Omega^1(P, i\mathfrak{g})$ and curvature $G \in \Omega^2(P, i\mathfrak{g})$ as

$$B = \frac{1}{i} \sum_{a=1}^{r} B^a \otimes S_a,$$

$$G = \frac{1}{i} \sum_{a=1}^{r} G^a \otimes S_a.$$

There are two different sign conventions for the covariant derivative:

$$\nabla_\mu^B = \partial_\mu \pm igB_\mu.$$

The local curvature is

$$G_{\mu\nu} = \partial_\mu B_\nu - \partial_\nu B_\mu \pm ig[B_\mu, B_\nu].$$

The Yang–Mills Lagrangian is

$$\mathscr{L}_{YM} = -\frac{1}{4}\langle G^{\mu\nu}, G_{\mu\nu}\rangle_{\mathfrak{g}} = -\frac{1}{4}G^a_{\mu\nu}G_a^{\mu\nu}.$$

3. The correspondence between the mathematical and physical conventions is given by setting

$$\begin{aligned} A &= \pm igB, \\ F &= \pm igG. \end{aligned}$$

If the representation of the Lie group G on a vector space V is unitary, then the field A_μ will act as a skew-Hermitian operator and B_μ will act as a Hermitian operator. We have $\nabla_\mu^A = \nabla_\mu^B$ and

$$\begin{aligned} A^a &= \pm B^a, \\ F^a &= \pm G^a \\ &= \mp G_a \\ &= F_a. \end{aligned}$$

Most of the time we shall use the mathematical convention and indicate when we use the physical convention.

Remark 7.4.1 Note one interesting point that can be seen most clearly in the physical convention: The coupling constant g appearing in the covariant derivative (describing the coupling of the gauge field to other fields, as we will see below) is the same as the coupling constant appearing in front of the term $[B_\mu, B_\nu]$ in the curvature $G_{\mu\nu}$, describing the coupling between the gauge bosons in non-abelian gauge theories.

7.5 The Klein–Gordon and Higgs Lagrangians

So far we have considered pure gauge theories that involve only a gauge field (connection) A. In physics, however, we are also interested in matter fields that couple to the gauge field. We first consider the case of scalar fields, like the Higgs field. We again fix an oriented pseudo-Riemannian manifold (M, g).

7.5.1 The Pure Scalar Field

Definition 7.5.1 A **complex scalar field** is a smooth map

$$\phi: M \longrightarrow \mathbb{C}.$$

A **multiplet of complex scalar fields** is a smooth map

$$\phi: M \longrightarrow \mathbb{C}^r$$

for some $r > 1$.
We consider the standard Hermitian scalar product

$$\langle v, w \rangle = v^\dagger w$$

on $\mathbb{C}^r$. If ϕ is a multiplet of scalar field with values in $\mathbb{C}^r$, then the differential $d\phi$ is an element

$$d\phi \in \Omega^1 (M, \mathbb{C}^r).$$

There is an induced Hermitian scalar product on the vector space-valued 1-forms $\Omega^1 (M, \mathbb{C}^r)$.

Definition 7.5.2 The **free Klein–Gordon Lagrangian** for a multiplet of complex scalar fields $\phi: M \to \mathbb{C}^r$ of mass m is defined by

$$\mathscr{L}_{KG}[\phi] = \langle d\phi, d\phi \rangle - m^2 \langle \phi, \phi \rangle.$$

For a given field ϕ the free Klein–Gordon Lagrangian defines a smooth map

$$\mathscr{L}_{KG}[\phi]: M \longrightarrow \mathbb{R}.$$

The expression $\langle d\phi, d\phi \rangle$ is called the **kinetic term** and the expression $-m^2 \langle \phi, \phi \rangle$ is called the **Klein–Gordon mass term**.

In local coordinates on M the kinetic term is given by

$$\langle d\phi, d\phi\rangle = \langle \partial^\mu \phi, \partial_\mu \phi\rangle.$$

It is also useful to consider a more general situation.

Definition 7.5.3 Let $V: \mathbb{R} \to \mathbb{R}$ be a smooth function, called a **potential**. Then the **Higgs Lagrangian** for a multiplet of complex scalar fields ϕ with potential V is defined by

$$\mathscr{L}_H[\phi] = \langle d\phi, d\phi\rangle - V(\phi),$$

where $V(\phi)$ denotes $V(\langle \phi, \phi\rangle)$ (of course it suffices to define the potential on $\mathbb{R}_{\geq 0}$). The Higgs field in the Standard Model, which we study in Chap. 8, is a multiplet of complex scalars described by a similar Lagrangian.
The potential V, if it contains terms of order higher than two in the field ϕ, describes a **direct interaction** between particles of the field ϕ. In the Standard Model, for instance, the potential V of the Higgs field is a quadratic polynomial in $\phi^\dagger\phi$, hence of order four in ϕ.

7.5.2 The Scalar Field Coupled to a Gauge Field

We now consider the case of a scalar field ϕ coupled to a gauge field A. We fix the following data:

- an n-dimensional oriented pseudo-Riemannian manifold (M, g)
- a principal G-bundle $P \to M$ with compact structure group G of dimension r
- a complex representation $\rho: G \to \mathrm{GL}(W)$ with associated complex vector bundle $E = P \times_\rho W \to M$
- a G-invariant Hermitian scalar product $\langle \cdot, \cdot\rangle_W$ on W with associated bundle metric $\langle \cdot, \cdot\rangle_E$ on the vector bundle E.

We then define:

Definition 7.5.4 If the dimension of W is one, then a smooth section of E is called a **complex scalar field** and if the dimension of W is greater than one, then a smooth section of E is called a **multiplet of complex scalar fields** (or simply a scalar field) and the vector space W is called a **multiplet space**.
With the exterior covariant derivative

$$d_A: \Gamma(E) \longrightarrow \Omega^1(M, E)$$

and the scalar product $\langle \cdot, \cdot \rangle_E$ on $\Omega^1(M, E)$ we set:

Definition 7.5.5 The **Klein–Gordon Lagrangian** for a multiplet of complex scalar fields $\Phi \in \Gamma(E)$ of mass m coupled to a gauge field A is defined by

$$\mathscr{L}_{KG}[\Phi, A] = \langle d_A\Phi, d_A\Phi \rangle_E - m^2 \langle \Phi, \Phi \rangle_E.$$

For given fields Φ and A the Klein–Gordon Lagrangian is a smooth function

$$\mathscr{L}_{KG}[\Phi, A]: M \longrightarrow \mathbb{R}.$$

The associated action $S_{KG}[\Phi, A]$ is the integral over the Klein–Gordon Lagrangian (on a closed manifold M).
In local coordinates on M we can write the kinetic term as

$$\langle d_A\Phi, d_A\Phi \rangle_E = \left\langle \nabla^{A\mu}\Phi, \nabla^A_\mu \Phi \right\rangle_E.$$

It is sometimes useful to have an even more explicit local formula for the Klein–Gordon Lagrangian: Choosing a local gauge $s: U \to P$, we can write

$$\Phi|_U = [s, \phi],$$

where $\phi: U \to W$ is a smooth function. The covariant derivative is given by

$$\nabla^A_\mu \Phi = [s, \nabla^A_\mu \phi], \quad \nabla^A_\mu \phi = \partial_\mu \phi + A_\mu \phi.$$

The term $A_\mu \phi$ is called the **minimal coupling** (we suppress in the notation the induced representation ρ_* of the Lie algebra $\mathfrak{g}$ on W). We identify W with $\mathbb{C}^r$ and the scalar product on W with the standard Hermitian product

$$\langle v, w \rangle = v^\dagger w$$

on $\mathbb{C}^r$. Since the representation of G on W is unitary and the gauge field A_μ has values in $\mathfrak{g}$, this implies that A_μ acts through skew-Hermitian matrices on $\mathbb{C}^r$:

$$A^\dagger_\mu = -A_\mu.$$

In a local gauge s for the principal bundle, the Klein–Gordon Lagrangian can then be written as

$$\begin{aligned} \mathscr{L}_{KG}[\Phi, A] = & (\partial^\mu \phi)^\dagger (\partial_\mu \phi) - m^2 \phi^\dagger \phi \\ & + (\partial^\mu \phi)^\dagger (A_\mu \phi) - (\phi^\dagger A_\mu)(\partial^\mu \phi) \\ & - \phi^\dagger A^\mu A_\mu \phi. \end{aligned} \tag{7.3}$$

The two terms in the first line, which are quadratic in the field ϕ with values in $W \cong \mathbb{C}^r$, are the Klein–Gordon Lagrangian for a free multiplet of complex scalar fields of mass m, consisting of the kinetic term and the mass term.

The terms in the second and third line are cubic and quartic in the fields ϕ and A_μ. These **interaction terms** describe an interaction (or coupling) between the gauge field and the multiplet of scalar fields and thus an indirect interaction between particles of the scalar field, mediated by the gauge bosons (see the Feynman diagrams after Remark 5.9.5 for a depiction of the interaction between a scalar field and a gauge field).

We see here (and later in the case of the Dirac Lagrangian for fermions) that in gauge theories where G does not act diagonally on the multiplet vector space $W = \mathbb{C}^s$, the action of the gauge group leads to two related kinds of mixing:

- The representation of the gauge group G on W, defining the associated bundle E, mixes different components of the multiplet, i.e. different components are gauge equivalent. In other words, the identification of a section of E with a map to V and the splitting into components depends on the choice of gauge.
- Via the induced representation of the Lie algebra $\mathfrak{g}$ on W, the gauge field A pairs different components of the multiplet in the interaction vertices.

This has important consequences for the Standard Model, where different particles like the up and down quark or the electron and electron neutrino form $\mathrm{SU}(2) \times \mathrm{U}(1)$-doublets.

Definition 7.5.6 Sections Φ of an associated vector bundle $E = P \times_\rho V$ with

$$\rho_* : \mathfrak{g} \longrightarrow \mathrm{End}(V)$$

non-trivial are called **charged scalars**. It follows that charged scalars have a non-trivial coupling to the gauge field A.

Theorem 7.5.7 *The Klein–Gordon Lagrangian of a multiplet of complex scalar fields, coupled to a gauge field, is gauge invariant:*

$$\mathscr{L}_{KG}\left[f^{-1}\Phi, f^*A\right] = \mathscr{L}_{KG}[\Phi, A]$$

for all bundle automorphisms $f \in \mathscr{G}(P)$.

We need the following lemma.

Lemma 7.5.8 *Let $f \in \mathscr{G}(P)$ be a bundle automorphism. Then*

$$d_{f^*A}\left(f^{-1}\Phi\right) = f^{-1}d_A\Phi.$$

Proof This follows from a calculation in local coordinates for $d_A\Phi(X) = \nabla^A_X\Phi$ with a vector field X. For a more invariant argument, note that by the definition of covariant derivatives using parallel transport in Sect. 5.9 it suffices to show that

$$D\left(f^{-1}\Phi, \gamma, x, f^*A\right) = D(\Phi, \gamma, x, A).$$

This follows from Exercise 5.15.9. □

We can now prove Theorem 7.5.7.

Proof The kinetic term $\langle d_A\Phi, d_A\Phi\rangle_E$ and the mass term $-m^2\langle\Phi, \Phi\rangle_E$ are both separately invariant under gauge transformations, because the scalar product $\langle\cdot\,,\cdot\rangle_W$ on the vector space W is G-invariant, hence $\langle\cdot\,,\cdot\rangle_E$ is invariant under the action of f^{-1}. □

In the Klein–Gordon Lagrangian for a scalar field the gauge field A is non-dynamic, i.e. does not appear with derivatives, and is just a fixed background field. The total Lagrangian that describes the dynamics of the scalar field, the gauge field and their interactions is the **Yang–Mills–Klein–Gordon Lagrangian**

$$\mathscr{L}_{KG}[\Phi, A] + \mathscr{L}_{YM}[A] = \langle d_A\Phi, d_A\Phi\rangle_E - m^2\langle\Phi, \Phi\rangle_E - \frac{1}{2}\left\langle F_M^A, F_M^A\right\rangle_{\mathrm{Ad}(P)}.$$

We can also consider the case of a scalar field with a potential coupled to a gauge field.

Definition 7.5.9 The **Higgs Lagrangian** for a multiplet of complex scalar fields coupled to a gauge field is defined by

$$\mathscr{L}_H[\Phi, A] = \langle d_A\Phi, d_A\Phi\rangle_E - V(\Phi),$$

where $V(\Phi)$ is a gauge invariant potential. We only consider the case where

$$V(\Phi) = V(\langle\Phi, \Phi\rangle_E),$$

with a function $V\colon \mathbb{R} \to \mathbb{R}$.

This Lagrangian describes an interaction between particles of the scalar field and particles of the gauge field and in addition a direct interaction between the particles of the scalar field (if the potential V contains terms of order three or higher in Φ).

A similar argument to the one in Theorem 7.5.7 shows:

Theorem 7.5.10 *The Higgs Lagrangian of a multiplet of complex scalar fields with potential V and coupled to a gauge field is gauge invariant:*

$$\mathscr{L}_H\left[f^{-1}\Phi, f^*A\right] = \mathscr{L}_H[\Phi, A]$$

for all bundle automorphisms $f \in \mathscr{G}(P)$.

The sum of the Higgs and Yang–Mills Lagrangians is called the **Yang–Mills–Higgs Lagrangian**

$$\mathscr{L}_H[\Phi, A] + \mathscr{L}_{YM}[A] = \langle d_A\Phi, d_A\Phi\rangle_E - V(\Phi) - \frac{1}{2}\left\langle F_M^A, F_M^A\right\rangle_{\mathrm{Ad}(P)}.$$

Remark 7.5.11 It is sometimes useful to consider **real scalar fields** Φ, which are sections in vector bundles E associated to real orthogonal representations of the Lie group G. The Klein–Gordon Lagrangian for a real scalar field of mass m coupled to a gauge field is

$$\mathscr{L}_{KG}[\Phi, A] = \frac{1}{2}\langle d_A\Phi, d_A\Phi\rangle_E - \frac{1}{2}m^2\langle \Phi, \Phi\rangle_E.$$

There is an analogous generalization to real scalar fields with a potential V.

7.6 The Dirac Lagrangian

Fermions are described classically by spinor fields on spacetime. In this section we define a Lagrangian for fermions. We fix the following data:

- an n-dimensional oriented and time-oriented pseudo-Riemannian spin manifold (M, g) of signature (s, t)
- a spin structure $\mathrm{Spin}^+(M)$ together with complex spinor bundle $S \to M$
- a Dirac form $\langle \cdot\,, \cdot\rangle$ (not necessarily positive definite) on the Dirac spinor space $\Delta = \Delta_n$ with associated Dirac bundle metric $\langle \cdot\,, \cdot\rangle_S$. We abbreviate $\langle \Psi, \Phi\rangle_S$ by $\overline{\Psi}\Phi$.

We can then define the Dirac Lagrangian for a free spinor field.

Definition 7.6.1 The **Dirac Lagrangian** for a free spinor field $\Psi \in \Gamma(S)$ of mass m is defined by

$$\begin{aligned}\mathscr{L}_D[\Psi] &= \mathrm{Re}\langle \Psi, D\Psi\rangle_S - m\langle \Psi, \Psi\rangle_S \\ &= \mathrm{Re}\left(\overline{\Psi} D\Psi\right) - m\overline{\Psi}\Psi,\end{aligned}$$

where $D\colon \Gamma(S) \to \Gamma(S)$ denotes the Dirac operator. The expression $\mathrm{Re}\left(\overline{\Psi} D\Psi\right)$ is called the **kinetic term** and $-m\overline{\Psi}\Psi$ is called the **Dirac mass term**.
Taking the real part in the kinetic term is necessary, because the Lagrangian has to be real. If the Dirac form $\langle \cdot, \cdot \rangle_S$ has $\delta = -1$, then the calculation in Exercise 7.9.12 implies that

$$(\langle \Psi, D\Psi\rangle_S - \langle D\Psi, \Psi\rangle_S)\mathrm{dvol}_g = d\alpha$$

for some $(n-1)$-form α on M depending on the spinor $\Psi \in \Gamma(S)$. As a consequence the kinetic term of the Dirac Lagrangian satisfies

$$\begin{aligned}\mathrm{Re}\left(\langle \Psi, D\Psi\rangle_S\right)\mathrm{dvol}_g &= \frac{1}{2}\left(\langle \Psi, D\Psi\rangle_S + \langle \Psi, D\Psi\rangle_S^*\right)\mathrm{dvol}_g \\ &= \frac{1}{2}\left(\langle \Psi, D\Psi\rangle_S + \langle D\Psi, \Psi\rangle_S\right)\mathrm{dvol}_g \\ &= \langle \Psi, D\Psi\rangle_S \mathrm{dvol}_g - \frac{1}{2}d\alpha.\end{aligned}$$

This implies by Stokes' Theorem A.2.24 that the action defined by $\langle \Psi, D\Psi\rangle_S$ and its real part are the same if the manifold M has no boundary and Ψ has compact support.

7.6.1 The Fermion Field Coupled to a Gauge Field

Similar to a scalar field, a spinor can be coupled to a gauge field. This construction is very important, because it defines the interaction between matter particles (fermions) and gauge bosons in gauge theories (for example, the interaction between electrons and photons in QED or the interaction between quarks and gluons in QCD). We fix in addition to the data above the following data:

- a principal G-bundle $P \to M$ with compact structure group G of dimension r
- a complex representation $\rho\colon G \to \mathrm{GL}(V)$ with associated complex vector bundle $E = P \times_\rho V \to M$

- a G-invariant Hermitian scalar product $\langle \cdot\,, \cdot \rangle_V$ on V with associated bundle metric $\langle \cdot\,, \cdot \rangle_E$ on the vector bundle E. Together with the Dirac form on the spinor bundle S we get a Hermitian scalar product $\langle \cdot\,, \cdot \rangle_{S\otimes E}$ on the twisted spinor bundle $S \otimes E$. We again abbreviate $\langle \Psi, \Phi \rangle_{S\otimes E}$ by $\overline{\Psi}\Phi$.

Choosing a local gauge $s: U \to P$ and an orthonormal basis $v_1, \ldots, v_s$ for V, the twisted spinors Ψ, Φ correspond to multiplets

$$\Psi = \begin{pmatrix} \Psi_1 \\ \vdots \\ \Psi_s \end{pmatrix}, \quad \Phi = \begin{pmatrix} \Psi_1 \\ \vdots \\ \Psi_s \end{pmatrix},$$

where Ψ_i and Φ_j are sections of the spinor bundle S over U. The scalar product on $S \otimes E$ can then be written as

$$\overline{\Psi}\Phi = \sum_{j=1}^{s} \overline{\Psi}_j \Phi_j.$$

Definition 7.6.2 The **Dirac Lagrangian** for a twisted spinor field $\Psi \in \Gamma(S \otimes E)$ of mass m coupled to a gauge field A on the principal bundle P is defined by

$$\begin{aligned} \mathscr{L}_D[\Psi, A] &= \mathrm{Re}\langle \Psi, D_A \Psi \rangle_{S\otimes E} - m\langle \Psi, \Psi \rangle_{S\otimes E} \\ &= \mathrm{Re}\left(\overline{\Psi} D_A \Psi\right) - m\overline{\Psi}\Psi, \end{aligned}$$

where $D_A: \Gamma(S \otimes E) \to \Gamma(S \otimes E)$ denotes the twisted Dirac operator. The associated action $S_D[\Psi, A]$ is the integral over the Dirac Lagrangian (on a closed manifold M).

Choosing in addition to the local gauge for P and the orthonormal basis for V a local vielbein e for the tangent bundle TM with associated local trivialization ϵ of $\mathrm{Spin}^+(M)$, we can write the Dirac Lagrangian as

$$\begin{aligned} \mathscr{L}_D[\Psi, A] =& \mathrm{Re} \sum_{j=1}^{s} i\overline{\psi}_j \Gamma^p \left(\partial_p - \frac{1}{4}\omega_{pqr}\Gamma^{qr} \right) \psi_j - \sum_{j=1}^{s} m\overline{\psi}_j \psi_j \\ &+ \mathrm{Re} \sum_{j=1}^{s} i\overline{\psi}_j \Gamma^p (A_p \psi)_j, \end{aligned} \tag{7.4}$$

(continued)

Definition 7.6.2 (continued)
where ψ is a map with values in $\Delta \otimes V$, ψ_j are maps with value in Δ, and Γ^p are physical gamma matrices. Here the two terms in the first line are the Dirac Lagrangian for a free multiplet of fermions, consisting of the **kinetic term**

$$\mathrm{Re} \sum_{j=1}^{s} i\overline{\psi}_j \Gamma^p \partial_p \psi_j,$$

a coupling between the spinor field and the metric g via ω_{pqr}, and the Dirac mass term. The term in the second line, which is cubic in the fields, is the **interaction term** that describes an interaction between the fermions and the gauge field and thus an indirect interaction between the fermions (see the Feynman diagram in Fig. 7.2 for the interaction between a fermion ψ and a gauge field A_p).

The gauge field A_p with values in the Lie algebra $\mathfrak{g}$ acts on the V part of ψ through the induced representation (suppressed in the notation). Since the gauge field A acts by skew-Hermitian matrices, the interaction term is automatically real and we can drop the symbol Re.

Definition 7.6.3 Sections Ψ of a twisted spinor bundle $S \otimes E$, where E is associated to a representation ρ of the gauge group G on a vector space V with

$$\rho_*: \mathfrak{g} \longrightarrow \mathrm{End}(V)$$

non-trivial, are called **charged fermions**. It follows that charged fermions have a non-trivial coupling to the gauge field A.

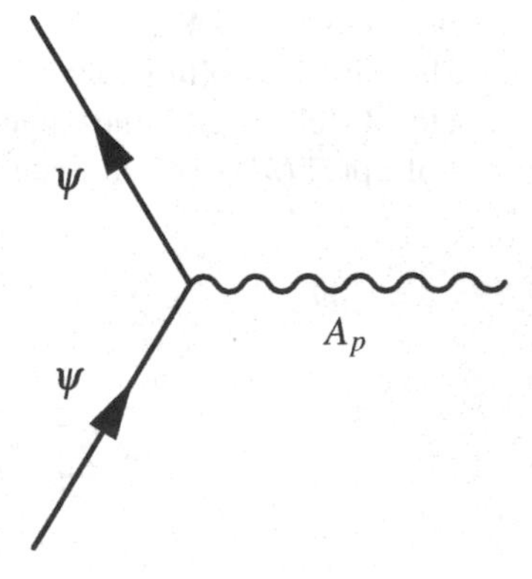

Fig. 7.2 Interaction vertex for fermion and gauge field

Theorem 7.6.4 *The Dirac Lagrangian for a twisted spinor field is gauge invariant:*

$$\mathscr{L}_D\left[f^{-1}\Psi, f^*A\right] = \mathscr{L}_D[\Psi, A]$$

for all bundle automorphisms $f \in \mathscr{G}(P)$.

Proof This is Exercise 7.9.11. □

Example 7.6.5 For the **strong interaction** (**QCD**) we have $G = \mathrm{SU}(3)$, $V \cong \mathbb{C}^3$ and there are six multiplets Ψ_f, called quarks, for the flavours $f = u, d, c, s, t, b$. The three components of every multiplet are called colours. The interaction term involving the gauge field A_μ with values in $\mathfrak{su}(3)$ (corresponding to the eight gluons) mixes different colours of a quark of a given flavour, but does not mix different flavours (different flavours of quarks are only mixed by the weak interaction). The Lagrangian for QCD can thus be written as

$$\mathscr{L}_D[\Psi, A] = \sum_f \left(\mathrm{Re}\left(\overline{\Psi}_f D_A \Psi_f\right) - m_f \overline{\Psi}_f \Psi_f\right),$$

where the sum runs over the six different flavours f.

Remark 7.6.6 Considering the mass term $-m\overline{\Psi}\Psi$ in the Dirac Lagrangian, it is clear that all components of the multiplet Ψ have the same mass m. We could try to generalize this and introduce different mass terms for different components of the multiplet. However, if the components with different masses are related by the action of a group element $g \in G$, then gauge invariance of the Lagrangian will be lost.

This could be a problem for the Standard Model, because we want to combine particles with very different masses, like the electron and electron neutrino, into $\mathrm{SU}(2) \times \mathrm{U}(1)$-doublets and at the same time keep the Lagrangian gauge invariant. It turns out that the situation in the Standard Model is even more difficult, because left-handed and right-handed fermions transform in different representations of $\mathrm{SU}(2) \times \mathrm{U}(1)$, so that a gauge invariant Dirac mass term is not defined, even if all components of the multiplet had the same mass. See Sect. 7.6.2 for more details. As we will discuss in Chap. 8, these problems can be solved by introducing a **Higgs field**.

We can again make both the spinor multiplet Ψ and the connection 1-form A dynamic by considering the **Yang–Mills–Dirac Lagrangian**

$$\mathscr{L}_D[\Psi, A] + \mathscr{L}_{YM}[A] = \mathrm{Re}\left(\overline{\Psi} D_A \Psi\right) - m\overline{\Psi}\Psi - \frac{1}{2}\langle F_M^A, F_M^A \rangle_{\mathrm{Ad}(P)}.$$

7.6.2 Lagrangians for Chiral Fermions

In this subsection we consider the case of an oriented and time-oriented Lorentzian spin manifold M of *even* dimension n with metric of signature $(1, n-1)$ or $(n-1, 1)$ (the most interesting case for the Standard Model is Minkowski spacetime of dimension $n = 4$). The Dirac bundle metric $\langle \cdot, \cdot \rangle_S$ on the spinor bundle $S \to M$ has a special property: both choices of the matrix A in Proposition 6.7.13 consist of a product of an odd number of gamma matrices. Hence if we decompose spinors $\Psi, \Phi \in \Gamma(S)$ into left-handed (positive) and right-handed (negative) components we get

$$\begin{aligned} \overline{\Psi}\Phi &= \langle \Psi, \Phi \rangle_S \\ &= \langle \Psi_L, \Phi_R \rangle_S + \langle \Psi_R, \Phi_L \rangle_S \\ &= \overline{\Psi}_L \Phi_R + \overline{\Psi}_R \Phi_L. \end{aligned} \tag{7.5}$$

In particular, we observe the following:

Proposition 7.6.7 (Dirac Bundle Metrics for Spinors on Lorentz Manifolds) *On even-dimensional oriented and time-oriented Lorentzian spin manifolds, for both choices of the matrix A in Proposition 6.7.13, the Dirac bundle metric is null on the subbundles S_L and S_R and pairs left-handed with right-handed spinors. In particular, the decomposition $S = S_L \oplus S_R$ of the spinor bundle into left-handed and right-handed Weyl spinors is not orthogonal with respect to the Dirac bundle metric.*

Remark 7.6.8 This is different from the situation on even-dimensional Riemannian spin manifolds, where we can take $A = I$ (the identity matrix) so that left-handed and right-handed Weyl spinors are orthogonal.

Formula (7.5) also holds if the spinors are sections of a twisted spinor bundle

$$S \otimes E = (S_L \otimes E) \oplus (S_R \otimes E).$$

We get:

Proposition 7.6.9 *On even-dimensional oriented and time-oriented Lorentzian manifolds, for both choices of the matrix A in Proposition 6.7.13, the (gauge invariant) Dirac Lagrangian for twisted spinors can be written as*

$$\begin{aligned} \mathscr{L}_D[\Psi, A] &= \mathrm{Re}\left(\overline{\Psi} D_A \Psi\right) - m\overline{\Psi}\Psi \\ &= \mathrm{Re}\left(\overline{\Psi}_L D_A \Psi_L + \overline{\Psi}_R D_A \Psi_R\right) - 2m\mathrm{Re}\left(\overline{\Psi}_L \Psi_R\right). \end{aligned}$$

In the second line all three Hermitian scalar products are taken in $S \otimes E$.

We want to generalize this discussion to the case of a twisted chiral spinor bundle. We consider a twisted chiral spinor bundle over a Lorentzian spin manifold of even dimension:

$$(S \otimes E)_+ = (S_L \otimes E_L) \oplus (S_R \otimes E_R).$$

Here E_L and E_R are complex vector bundles associated to representations

$$\rho_L: G \longrightarrow \mathrm{GL}(V_L),$$
$$\rho_R: G \longrightarrow \mathrm{GL}(V_R).$$

We fix G-invariant Hermitian scalar products on V_L and V_R which define Hermitian bundle metrics $\langle \cdot\, , \cdot \rangle_{E_L}$ and $\langle \cdot\, , \cdot \rangle_{E_R}$.

We can then define a *massless* Dirac Lagrangian as before:

$$\begin{aligned} \mathscr{L}_D[\Psi, A] &= \mathrm{Re}\left(\overline{\Psi} D_A \Psi\right) \\ &= \mathrm{Re}\left(\overline{\Psi}_L D_A \Psi_L + \overline{\Psi}_R D_A \Psi_R\right). \end{aligned}$$

In the second line the first scalar product is taken in $S \otimes E_L$ and the second scalar product in $S \otimes E_R$ (the Dirac operator only acts on the S-component and does not change the E-component). It is not difficult to check that this Lagrangian is gauge invariant.

However, if we now also want to define a Dirac mass term as before, we run into a problem that can ultimately be traced back to Proposition 7.6.7: the natural mass term

$$-m\overline{\Psi}\Psi = -2m\mathrm{Re}\left(\overline{\Psi}_L \Psi_R\right)$$

is so far *not* defined: it pairs a spinor with an E_L-component and a spinor with an E_R-component, but the Hermitian bundle metrics are only defined if both spinors have the same type of E-component.

We could try to introduce a Dirac mass term in this situation as follows:

Definition 7.6.10 Let V_R and V_L be unitary representations of a Lie group G. Then a **mass pairing** is a G-invariant form

$$\kappa: V_L \times V_R \longrightarrow \mathbb{C}$$

which is complex antilinear in the first argument and complex linear in the second. A mass pairing κ defines a form on the level of bundles

$$\kappa: E_L \times E_R \longrightarrow \mathbb{C}$$

that can then be used to define a gauge invariant Dirac mass term for chiral twisted spinors. However, the following theorem shows that in many cases a mass pairing vanishes identically:

Theorem 7.6.11 (Triviality of Mass Pairings) *Suppose that V_L and V_R are irreducible, unitary, non-isomorphic representations of G. Then every mass pairing κ is identically zero.*

Proof We can identify $\overline{V}_L^*$, the dual of the complex conjugate of V_L, with

$$\overline{V}_L^* = \{\alpha: V_L \to \mathbb{C} \mid \alpha \text{ is } \mathbb{C}\text{-antilinear}\}.$$

The induced G-representation on this space is defined by

$$(g \cdot \alpha)(v_L) = \alpha\left(g^{-1} \cdot v_L\right)$$

for $g \in G$, $v_L \in V_L$. The map

$$\begin{aligned} V_L &\longrightarrow \overline{V}_L^* \\ v_L &\longmapsto \langle \cdot\,, v_L \rangle_{V_L}, \end{aligned}$$

where $\langle \cdot\,, \cdot \rangle_{V_L}$ is the G-invariant Hermitian form on V_L, defines a *complex linear* G-equivariant isomorphism.

Suppose a mass pairing $\kappa \neq 0$ exists. Then

$$\begin{aligned} V_R &\longrightarrow \overline{V}_L^* \\ v_R &\longmapsto \kappa(\cdot, v_R) \end{aligned}$$

is a complex linear G-equivariant map. Combining both maps we get a complex linear G-equivariant map

$$V_R \longrightarrow V_L$$

which is non-zero, because $\kappa \neq 0$. By Schur's Lemma this map has to be an isomorphism of the representations V_R and V_L (because the kernel and image of the map are G-invariant), contradicting our assumption. □

It is known from experiments that a realistic theory of particle physics has to involve twisted chiral fermions with a non-zero mass, because the weak interaction is not invariant under parity inversion (see Sect. 8.5). Together with Remark 7.3.16 and Remark 7.6.6 it follows that there are three situations in which it is not clear how to define mass terms and at the same time keep the Lagrangian gauge invariant:

- non-zero masses for gauge bosons
- different masses for fermions in the same gauge multiplet
- non-zero masses for twisted chiral fermions.

We shall see in Chap. 8 that the introduction of the **Higgs field** allows a very elegant solution of these problems: using the Higgs field we can define a fully gauge invariant Lagrangian that contains certain interaction terms between the gauge bosons and the Higgs field and the fermions and the Higgs field. In a specific type of gauge, called a *unitary gauge*, these interaction terms take the form of mass terms for the gauge bosons and fermions.

7.7 Yukawa Couplings

In this section we discuss Yukawa couplings which are used in the Standard Model to define a mass for twisted chiral fermions. Yukawa couplings are certain trilinear forms involving two twisted chiral spinors and one scalar field. The idea is that the G-representation on the scalar field precisely cancels the difference between the representations on the twisted chiral spinors so that the whole trilinear expression is gauge invariant. We consider the case of an oriented and time-oriented Lorentzian spin manifold (M, g) of dimension n with signature $(1, n-1)$ or $(n-1, 1)$ together with a principal G-bundle $P \to M$.

Definition 7.7.1 Suppose that V_L, V_R, W are unitary representation spaces of the compact Lie group G. Then we define a **Yukawa form** as a map

$$\tau\colon V_L \times W \times V_R \longrightarrow \mathbb{C}$$

which is invariant under the action of G, complex antilinear in V_L, real linear in W and complex linear in V_R.
Suppose τ is a Yukawa form. We then define:

Definition 7.7.2 For a real constant g_Y the G-invariant scalar

$$\begin{aligned}(\Delta_L \otimes V_L) \times W \times (\Delta_R \otimes V_R) &\longrightarrow \mathbb{R}\\ (\lambda_L \otimes v_L, \phi, \lambda_R \otimes v_R) &\longmapsto -2g_Y \mathrm{Re}\left(\overline{\lambda}_L \lambda_R \tau(v_L, \phi, v_R)\right)\end{aligned}$$

is called a **Yukawa coupling** (the constant g_Y is also called a Yukawa coupling and sometimes already appears in the definition of the Yukawa form τ). It defines a gauge invariant Lagrangian for which we use the shorthand notation

$$\begin{aligned}\mathscr{L}_Y[\Psi_L, \Phi, \Psi_R] &= -2g_Y \mathrm{Re}\left(\overline{\Psi}_L \Phi \Psi_R\right) \\ &= -g_Y\left(\overline{\Psi}_L \Phi \Psi_R\right) - g_Y\left(\overline{\Psi}_L \Phi \Psi_R\right)^*,\end{aligned}$$

where the Yukawa form τ is implicit,

$$\begin{aligned}\Psi_L &\in \Gamma(S_L \otimes E_L), \\ \Phi &\in \Gamma(F), \\ \Psi_R &\in \Gamma(S_R \otimes E_R),\end{aligned}$$

and E_L, F, E_R are the complex vector bundles associated to the principal bundle P via the G-representations V_L, W, V_R.

We will discuss in Chap. 8 how Yukawa coupling between two twisted chiral fermions and the Higgs field leads to masses for the fermions. The Lagrangian of the **Standard Model** is then essentially the sum of all the Lagrangians that we discussed in this chapter, i.e. the following **Yang–Mills–Dirac–Higgs–Yukawa Lagrangian**:

$$\begin{aligned}\mathscr{L} &= \mathscr{L}_D[\Psi, A] + \mathscr{L}_H[\Phi, A] + \mathscr{L}_Y[\Psi_L, \Phi, \Psi_R] + \mathscr{L}_{YM}[A] \\ &= \mathrm{Re}\left(\overline{\Psi} D_A \Psi\right) + \langle d_A\Phi, d_A\Phi\rangle_E - V(\Phi) - 2g_Y \mathrm{Re}\left(\overline{\Psi}_L \Phi \Psi_R\right) - \frac{1}{2}\left\langle F_M^A, F_M^A\right\rangle_{\mathrm{Ad}(P)}.\end{aligned}$$

Remark 7.7.3 In the discussions in this chapter, the pseudo-Riemannian metric g on the manifold M has been considered as a fixed background. Classically we can add a Lagrangian for the metric (like the Einstein–Hilbert Lagrangian) to make g dynamic. However, this approach does not yield a well-defined quantum field theory. Since we are mainly interested in the Standard Model, which is defined on flat Minkowski spacetime of dimension 4, we will not discuss aspects of quantum gravity.

7.8 Dirac and Majorana Mass Terms

So far we have considered Dirac mass terms for spinor fields. For a spinor $\Psi \in \Gamma(S)$ such a mass term is given by

$$-m\langle\Psi, \Psi\rangle_S = -m\overline{\Psi}\Psi,$$

where $\langle \cdot, \cdot \rangle$ is a Dirac form on the spinor space. We want to discuss a second type of mass term that is important in neutrino physics.

Definition 7.8.1 Let $(\cdot, \cdot)$ denote a Majorana form on the spinor space Δ as in Sect. 6.7.1. Then

$$-m\mathrm{Re}(\Psi, \Psi)_S$$

is called a **Majorana mass term**.

It is clear that both Dirac and Majorana mass terms are invariant under the action of the spin group. We want to compare these forms in the case of Minkowski spacetime of dimension 4. Recall from Sect. 6.8 that the Dirac form is defined by the matrix

$$A = \Gamma_0 = \begin{pmatrix} 0 & I_2 \\ I_2 & 0 \end{pmatrix}$$

and the Majorana form is defined by the matrix

$$C = i\Gamma_0\Gamma_2 = \begin{pmatrix} -i\sigma_2 & 0 \\ 0 & i\sigma_2 \end{pmatrix}.$$

If we decompose a Dirac spinor Ψ into left-handed and right-handed Weyl spinors

$$\psi = \begin{pmatrix} \psi_L \\ \psi_R \end{pmatrix},$$

then the Dirac mass term is given by

$$\begin{aligned} -m\langle \psi, \psi \rangle &= -m\psi^\dagger A \psi \\ &= -m\left(\psi_L^\dagger \psi_R + \psi_R^\dagger \psi_L\right) \end{aligned}$$

and the Majorana mass term is given by

$$\begin{aligned} -m\mathrm{Re}(\psi, \psi) &= m\mathrm{Re}\left(\tilde{\psi}\psi\right) \\ &= -m\mathrm{Re}\left(\psi^T C \psi\right) \\ &= m\mathrm{Re}\left(i\psi_L^T \sigma_2 \psi_L - i\psi_R^T \sigma_2 \psi_R\right). \end{aligned}$$

Here we used the notation $\tilde{\psi} = \psi^T C$ for the Majorana conjugate from Definition 6.7.6. The important consequence is that the Dirac mass term is zero for spinors which have only one Weyl component ψ_L or ψ_R, while the Majorana mass term may be non-zero in this case.

We briefly want to discuss the extension of these Lorentz invariant mass terms to Lorentz and gauge invariant mass terms for charged fermions, i.e. sections of twisted spinor bundles $S \otimes E$. In the case of the Dirac mass term we saw in Sect. 7.6.2 that such an extension is always possible if both left-handed and right-handed Weyl spinor bundles are twisted with the same associated vector bundle E, using the Hermitian scalar product on $S \otimes E$, coming from the Dirac form on S and a Hermitian scalar product on E.

In the case of the Majorana mass term there is now a problem, because the *complex bilinear* Majorana form on S usually does not combine with the *Hermitian* scalar product on E. If E is the associated bundle $P \times_\rho V$, then we need a G-invariant complex bilinear form on the vector space V. However, even in simple situations such an invariant bilinear form does not exist:

Lemma 7.8.2 *Let*

$$\begin{aligned} \rho_k \colon \mathrm{U}(1) &\longrightarrow \mathrm{U}(1) \\ \alpha &\longmapsto \alpha^k \end{aligned}$$

be the complex representation of $\mathrm{U}(1)$ *on* $\mathbb{C}$ *of winding number k. Suppose that B is a* $\mathrm{U}(1)$*-invariant complex bilinear form on* $\mathbb{C}$. *Then* $B \equiv 0$.

Proof We have

$$\begin{aligned} B(z,z) &= B\left(\alpha^k z, \alpha^k z\right) \\ &= \alpha^{2k} B(z,z) \quad \forall \alpha \in \mathrm{U}(1), z \in \mathbb{C}. \end{aligned}$$

It follows that $B(z,z) = 0$ for all $z \in \mathbb{C}$, hence $B \equiv 0$. □

This indicates that there is no straightforward extension of the Majorana mass term to charged fermions.

7.9 Exercises for Chap. 7

7.9.1 (From [125]) Let H be a Hilbert space with a bilinear form

$$\langle \cdot , \cdot \rangle \colon H \times H \longrightarrow H,$$

satisfying

$$\langle \psi, \phi \rangle = \langle \phi, \psi \rangle^*,$$

where * denotes complex conjugation. We say that the bilinear form has a **probability interpretation** if the following holds: for all vectors $\phi, \psi \in H$ with

$$|\langle \phi, \phi \rangle|^2 = |\langle \psi, \psi \rangle|^2 = 1$$

the following inequality holds:

$$|\langle \phi, \psi \rangle|^2 \leq 1.$$

1. Suppose that the bilinear form is positive definite. Prove that the bilinear form has a probability interpretation.
2. Suppose that there exist vectors ϕ, ψ in H such that

$$\langle \phi, \phi \rangle > 0,$$
$$\langle \psi, \psi \rangle < 0.$$

Prove that the bilinear form does not have a probability interpretation.

7.9.2 Let (M, g) be an oriented pseudo-Riemannian manifold and (U, ϕ) an oriented chart for M with local coordinates x^μ. Prove that the volume form dvol_g is given by

$$\mathrm{dvol}_g = \sqrt{|g|}dx^1 \wedge \ldots \wedge dx^n,$$

where

$$|g| = |\det(g_{\mu\nu})|$$

is the absolute value of the determinant of the matrix with entries

$$g_{\mu\nu} = g(\partial_\mu, \partial_\nu).$$

7.9.3 Let (M, g) be an n-dimensional oriented pseudo-Riemannian manifold of signature (s, t) and $*$ the Hodge star operator.

1. Prove that

$$**: \Omega^k(M) \longrightarrow \Omega^k(M)$$

is given by

$$** = (-1)^{t+k(n-k)}.$$

2. Determine the even dimensions $n = 2k$ where $** = 1$ on $\Omega^k(M)$ if (M, g) is Riemannian or Lorentzian. In these dimensions we can define self-dual and anti-self-dual k-forms ω, satisfying $*\omega = \omega$ and $*\omega = -\omega$, respectively.

7.9.4 Let (M, g) be an n-dimensional oriented pseudo-Riemannian manifold of signature (s, t) and $*$ the Hodge star operator.

1. Let ∇ denote the Levi-Civita connection of g and suppose that $\alpha \in \Omega^1(M)$ is a 1-form. Prove that if α is parallel ($\nabla\alpha = 0$), then α is closed ($d\alpha = 0$).
2. Let $\eta \in \Omega^1(M)$ be a 1-form, $p \in M$ a point and $e_1, \ldots, e_n$ a local oriented g-orthonormal frame of the tangent bundle in an open neighbourhood of p with $(\nabla e_i)(p) = 0$ for all i. Let $\eta_i = \eta(e_i)$ and $\eta^i = g^{ii}\eta_i$ (no summation). Prove that at the point p

$$(*d * \eta)(p) = (-1)^t \left(\sum_{i=1}^{n} L_{e_i}\eta^i \right)(p).$$

7.9.5 Let (M, g) be a closed (compact without boundary) n-dimensional oriented pseudo-Riemannian manifold of signature (s, t). The **Laplace operator** on k-forms is defined by

$$\Delta = dd^* + d^*d \colon \Omega^k(M) \longrightarrow \Omega^k(M)$$

where d^* is the codifferential from Definition 7.2.9. A form ω is called **harmonic** if $\Delta\omega = 0$. Suppose that (M, g) is Riemannian.

1. Prove that

$$\omega \text{ is harmonic} \Leftrightarrow d\omega = 0 \text{ and } d^*\omega = 0.$$

2. Prove that

$$\omega \text{ is harmonic} \Leftrightarrow *\omega \text{ is harmonic}.$$

7.9.6 Let (M^4, g) be a pseudo-Riemannian 4-manifold with a principal bundle $P \to M$. Prove that the Yang–Mills action $S_{YM}[A]$ is invariant under a *conformal change* of the metric g:

$$g' = e^{2\lambda} g,$$

where $\lambda \in \mathscr{C}^\infty(M)$ is an arbitrary smooth function on M.

7.9.7

1. Prove that the connection A from Sect. 5.2.2 on the Hopf bundle $S^3 \to S^2$ with structure group U(1) satisfies the Yang–Mills equation (i.e. Maxwell's equations) if S^2 has the standard round Riemannian metric.
2. Prove that the Yang–Mills moduli space for the Hopf bundle $S^3 \to S^2$ over the round sphere S^2 consists of a single point.

7.9.8 Let $M = \mathbb{R}^{1,3}$ be Minkowski spacetime with the flat Minkowski metric η. Let $P \to M$ be a trivial principal G-bundle with a global gauge $s \colon M \to P$. For a connection A decompose the curvature $F = F^A$ as in Sect. 5.7 into generalized electric and magnetic fields E and B with values in the Lie algebra $\mathfrak{g}$.

1. Express the Bianchi identity and the Yang–Mills equation in terms of E, B and A.
2. Express the instanton equations $*F = F$ and $*F = -F$ in terms of E and B.

7.9.9

1. On the Hopf bundle $S^7 \to S^4$ with structure group SU(2) define in analogy to the construction in Sect. 5.2.2 an explicit connection 1-form $A \in \Omega^1\left(S^7, \mathfrak{su}(2)\right)$ using quaternions.
2. Prove that A is an anti-self-dual instanton for the standard round Riemannian metric on S^4.

7.9.10 Let (M, g) be a closed oriented pseudo-Riemannian manifold, $P \to M$ a principal G-bundle with compact structure group G and $E \to M$ an associated vector bundle with Hermitian bundle metric $\langle \cdot , \cdot \rangle_E$. We fix an Ad-invariant positive definite scalar product on the Lie algebra $\mathfrak{g}$ and consider the Yang–Mills–Higgs Lagrangian

$$\begin{aligned}\mathscr{L}_{YMH}[\Phi, A] &= \mathscr{L}_H[\Phi, A] + \mathscr{L}_{YM}[A] \\ &= \langle d_A\Phi, d_A\Phi\rangle_E - V(\Phi) - \frac{1}{2}\langle F_M^A, F_M^A\rangle_{\mathrm{Ad}(P)} .\end{aligned}$$

We are looking for critical points of the associated action S_{YMH} under variations of Φ and A.

1. Prove that variation of the field Φ leads to the field equation
$$d_A^* d_A \Phi = V'(\Phi)\Phi, \tag{7.6}$$
where V' is the derivative of $V\colon \mathbb{R} \to \mathbb{R}$ and $V'(\Phi) = V'(\langle \Phi, \Phi\rangle_E)$.
2. Show that elements $\alpha_M \in \Omega^1(M, \mathrm{Ad}(P))$, $\Phi \in \Gamma(E)$ define a canonical twisted 1-form $\alpha_M \cdot \Phi \in \Omega^1(M, E)$.
3. Prove that there exists a unique twisted 1-form
$$J_H(A, \Phi) \in \Omega^1(M, \mathrm{Ad}(P))$$
such that
$$\langle \alpha_M, J_H(A, \Phi)\rangle_{\mathrm{Ad}(P)} = 2\mathrm{Re}\left(\langle d_A\Phi, \alpha_M \cdot \Phi\rangle_E\right)$$
for all $\alpha_M \in \Omega^1(M, \mathrm{Ad}(P))$.
4. Show that variation of the connection A leads to the field equation
$$d_A^* F_M^A = J_H(A, \Phi). \tag{7.7}$$

Equations (7.6) and (7.7) are called **Yang–Mills–Higgs equations**.

7.9.11 Prove the statement in Theorem 7.6.4 concerning the gauge invariance of the Dirac Lagrangian.

7.9.12

1. Under the assumptions of Theorem 6.11.5, define a 1-form $\eta \in \Omega^1(M, \mathbb{C})$ by
$$\eta(X) = \langle X \cdot \Phi, \Psi \rangle_{S \otimes E} \quad \forall X \in \mathfrak{X}(M)$$
and prove that
$$\langle D_A \Phi, \Psi \rangle_{S \otimes E} - \langle \Phi, D_A \Psi \rangle_{S \otimes E} = (-1)^t * d * \eta$$
(Exercise 7.9.4 could be helpful).
2. Prove Theorem 6.11.5.
3. Discuss what can be said in the case $\delta = +1$ and the implications for the Dirac Lagrangian.

7.9.13 Let (M, g) be an n-dimensional closed oriented and time-oriented pseudo-Riemannian spin manifold, $S \to M$ a spinor bundle with Dirac bundle metric $\langle \cdot, \cdot \rangle_S$ with $\delta = -1$, $P \to M$ a principal G-bundle with compact structure group G and $E \to M$ an associated vector bundle with Hermitian bundle metric $\langle \cdot, \cdot \rangle_E$. We fix an Ad-invariant positive definite scalar product on the Lie algebra $\mathfrak{g}$ and consider the Yang–Mills–Dirac Lagrangian

$$\begin{aligned} \mathscr{L}_{YMD}[\Psi, A] &= \mathscr{L}_D[\Psi, A] + \mathscr{L}_{YM}[A] \\ &= \mathrm{Re}\left(\overline{\Psi} D_A \Psi\right) - m\overline{\Psi}\Psi - \frac{1}{2}\left\langle F_M^A, F_M^A \right\rangle_{\mathrm{Ad}(P)}. \end{aligned}$$

We are looking for critical points of the associated action S_{YMD} under variations of Ψ and A.

1. Prove that variation of the spinor Ψ leads to the **Dirac equation**
$$D_A \Psi = m\Psi.$$
2. Show that $\alpha_M \in \Omega^1(M, \mathrm{Ad}(P))$ and $\Psi \in \Gamma(S \otimes E)$ define via Clifford multiplication a canonical section $\alpha_M \cdot \Psi \in \Gamma(S \otimes E)$.
3. Prove that there exists a unique twisted 1-form
$$J_D(\Psi) \in \Omega^1(M, \mathrm{Ad}(P))$$
such that
$$\langle \alpha_M, J_D(\Psi) \rangle_{\mathrm{Ad}(P)} = \mathrm{Re}\left(\langle \Psi, \alpha_M \cdot \Psi \rangle_{S \otimes E}\right)$$
for all $\alpha_M \in \Omega^1(M, \mathrm{Ad}(P))$.
4. Show that variation of the connection A leads to the field equation
$$d_A^* F_M^A = J_D(\Psi).$$

Chapter 8
The Higgs Mechanism and the Standard Model

In this chapter we finally apply the formalism of mathematical gauge theory to physics. We first discuss gauge theories in which the gauge symmetry is *spontaneously broken*, leading to the existence of one or several Higgs bosons. We also study the Standard Model of elementary particles in some detail, including the particle content and the representations of the gauge group, the Higgs mechanism of mass generation and the explicit Lagrangians containing the interactions between all known elementary particles. For the Higgs mechanism we first consider the case of a general Lie group and Higgs vector space (with possibly several Higgs bosons) as well as the specific case of the Standard Model (with a single Higgs boson).

The predictions of the Standard Model, which was developed in the 1960s and 1970s, have been tested and verified with enormous accuracy, especially using different types of particle colliders. Typical colliders involve two accelerated and collimated beams of particles brought to collision, for example, of protons-protons, protons-antiprotons, electrons-positrons, protons-electrons or protons-positrons. The final particles in the Standard Model, which were postulated by the theory and then observed in experiments, were the top quark (1995 at Fermilab), the tau neutrino (2000 at Fermilab) and the Higgs boson (2012 at CERN).

It is known that the Standard Model is not a complete theory of particle physics. For example, the neutrinos in the Standard Model are massless, but experiments show that they have a small non-zero mass. Furthermore, observations of galaxies indicate that there is another form of matter in the universe, called *dark matter*, that only interacts with Standard Model particles through gravity and perhaps the weak force. This type of matter cannot be explained with the Standard Model. Finally, the gravitational interaction is not included in the Standard Model. The reason is that on the fundamental level the Standard Model is a quantum field theory and this kind of quantum theory most probably cannot be used to describe gravity. To find a quantum theory of gravity, and a unification with the other forces described by the Standard Model, is the primary aim of *Theories of Everything*, like string theory.

M.J.D. Hamilton, *Mathematical Gauge Theory*, Universitext,
https://doi.org/10.1007/978-3-319-68439-0_8

Even though the Standard Model is not complete, it is extremely well verified and will be a touchstone for any theory that tries to go beyond it. Beyond its significance for particle physics, the Standard Model has a remarkable mathematical structure that develops from basic principles (specific gauge groups, representations and Lagrangians) through a sequence of mathematical steps into a complex theory with rich and often unexpected properties. Our aim in this chapter is to understand to some degree the fundamental principles of the theory and how they generate the complexity of particle physics.

The references for this chapter are the same as those given at the beginning of Chap. 7.

8.1 The Higgs Field and Symmetry Breaking

8.1.1 The Yang–Mills–Higgs Lagrangian

We fix the following data:

- an n-dimensional oriented pseudo-Riemannian manifold (M, g)
- a principal G-bundle $P \to M$ with compact structure group G of dimension r
- a complex representation $\rho\colon G \to \mathrm{GL}(W)$ with associated complex vector bundle $E = P \times_\rho W \to M$ (sometimes the vector space W and the representation ρ are real)
- a G-invariant Hermitian scalar product $\langle\cdot\,,\cdot\rangle_W$ on W with associated bundle metric $\langle\cdot\,,\cdot\rangle_E$ on the vector bundle E. We denote by $\langle\langle\cdot\,,\cdot\rangle\rangle_W = \mathrm{Re}\langle\cdot\,,\cdot\rangle_W$ the associated positive definite Euclidean scalar product on the real vector space underlying W.

Definition 8.1.1 We call the vector space W the **Higgs vector space**, the associated vector bundle $E = P \times_\rho W$ the **Higgs bundle** and a section Φ of E the **Higgs field**. If the induced representation ρ_* of the Lie algebra $\mathfrak{g}$ is non-trivial, then the Higgs field is a charged scalar.

We can assume that $W = \mathbb{C}^n$ and the G-invariant scalar product on W is given by the standard Hermitian product

$$\langle v, w\rangle_W = v^\dagger w.$$

We consider a **potential**, also called the **Higgs potential**,

$$V\colon \mathbb{R} \longrightarrow \mathbb{R}.$$

The potential appears in the Higgs Lagrangian for the Higgs field Φ:

$$\mathscr{L}_H[A, \Phi] = \langle d_A\Phi, d_A\Phi\rangle_E - V(\Phi),$$

where

$$V(\Phi) = V(\langle \Phi, \Phi \rangle_E).$$

The combined Lagrangian for the Higgs field and the gauge field is then the Yang–Mills–Higgs Lagrangian

$$\mathscr{L}_H[\Phi, A] + \mathscr{L}_{YM}[A] = \langle d_A\Phi, d_A\Phi \rangle_E - V(\Phi) - \frac{1}{2}\langle F_M^A, F_M^A \rangle_{\mathrm{Ad}(P)}.$$

8.1.2 Spontaneously Broken Gauge Theories

Definition 8.1.2 A **vacuum configuration** or **vacuum** for the Yang–Mills–Higgs Lagrangian is a pair (Φ_0, A_0) comprising a Higgs field and a connection such that:

1. A_0 is a flat connection, $F^{A_0} \equiv 0$.
2. Φ_0 is covariantly constant, $d_{A_0}\Phi = \nabla^{A_0}\Phi_0 \equiv 0$.
3. The value of Φ_0 is at every point of M a minimum of the potential V.

Remark 8.1.3 It can be shown that vacuum configurations in this sense correspond to the minima of the **energy** determined by the Yang–Mills–Higgs Lagrangian and can thus be considered "stable". It follows from Exercise 7.9.10 that vacuum configurations are solutions of the classical equations of motion.
The structure of vacua of the Yang–Mills–Higgs Lagrangian for bundles E over general manifolds M can be quite complicated, due to the existence of non-trivial flat connections on the principal bundle P. For our purposes of understanding the Standard Model it suffices to restrict from now on to the following case:

- the manifold M is connected and simply connected.

Then Exercise 5.15.4 implies that a flat connection A_0 can only exist on trivial principal bundles P. Hence we also assume that

- the principal G-bundle P is trivial.

This also implies that the Higgs bundle E is trivial. We define:

Definition 8.1.4 A **vacuum vector** is an element $w_0 \in W$ which is a minimum of the real-valued function

$$V(w) = V(\langle w, w \rangle_W)$$

on W. The set of vacuum vectors in the Higgs vector space W is called the **space of vacua** or the **vacuum manifold** for V.

We then get:

Proposition 8.1.5 *Suppose that the manifold M is connected and simply connected and the principal bundle P is trivial. Let (Φ_0, A_0) be a vacuum configuration. Then there exists a global gauge $s_0: M \rightarrow P$, called the* **vacuum gauge**, *such that*

$$A_{0s_0} = s_0^* A_0 \equiv 0 \tag{8.1}$$

and

$$\Phi_0 = [s_0, w_0], \tag{8.2}$$

where $w_0 \in W$ is a constant vacuum vector. Conversely, for an arbitrary fixed global gauge s_0 of the principal bundle P, every vacuum vector w_0 determines a unique vacuum configuration (Φ_0, A_0) of the form in Eqs. (8.1) *and* (8.2).

Proof The first statement follows from Exercise 5.15.4. The second statement then follows, because with respect to the global gauge s, the covariant derivative on E is just the standard derivative on vector-valued functions from M to W. Hence Φ_0 is covariantly constant if and only if it is constant as a map to W. □

We collect certain assumptions that hold throughout Sect. 8.1 and Sect. 8.2. We fix from now on:

- a global vacuum gauge $s_0: M \rightarrow P$
- a vacuum vector $w_0 \in W$
- the associated vacuum configuration (Φ_0, A_0).

Recall that we have a unitary representation of the Lie group G on W.

Definition 8.1.6 The **unbroken subgroup** of the vacuum configuration is the isotropy group of the vacuum vector $w_0 \in W$:

$$H = G_{w_0} \subset G.$$

The group H is a closed Lie subgroup of G according to Proposition 3.2.9. Since G was assumed compact, H is compact as well. We call the gauge theory **spontaneously broken** if H is a proper subgroup of G, i.e. $H \subsetneq G$.

- We will assume from now on that the gauge theory is spontaneously broken.

Note that the unbroken subgroup H is only well-defined for a constant vacuum vector w_0 and that its explicit embedding in G depends on the choice of w_0.

- We will assume from now on that the Higgs potential $V(w)$ has a minimum, but not at $w = 0$, so that $w_0 \neq 0$. We call the spontaneous process (usually assumed to have happened moments after the Big Bang) where the Higgs field acquires from the unstable value $w = 0$ the value $w_0 \neq 0$ **(spontaneous) symmetry breaking**.

Mathematically the basic idea of symmetry breaking is that the isotropy group $H = G_{w_0}$ of the vector $w_0 \neq 0$ is smaller than the isotropy group $G = G_0$ of the vector 0, hence some of the original symmetry has *disappeared*, is *hidden* or *broken*.

Definition 8.1.7 We call the nowhere vanishing field Φ_0 the **Higgs condensate**. The Higgs condensate is a non-zero background field in which all other fields and the corresponding elementary particles propagate. The Higgs field is the only classical field in the Standard Model with a non-zero value in vacuum.

Remark 8.1.8 In quantum field theory, fields become fields of operators on the Hilbert space of the system and do not have classical values. In particular, the Higgs condensate is thought of as the **vacuum expectation value (vev)**

$$\langle \Phi_0 \rangle = \langle \Omega | \Phi_0 | \Omega \rangle$$

of an operator field Φ_0, where $|\Omega\rangle$ is the vacuum state. For our purposes it suffices to treat the Higgs condensate as a classical section Φ_0 of the vector bundle E.

In gauge theories we demand that the Lagrangian and hence the laws of physics are invariant under all gauge transformations for the structure group G. *This is still true in a spontaneously broken gauge theory.* However, in this case the Higgs condensate is only invariant under gauge transformations with values in the smaller subgroup H (compared to a gauge theory where the vacuum value is zero and hence invariant under all gauge transformations).

As we will see later, the non-zero value of the Higgs condensate after symmetry breaking and the coupling of the fields (gauge fields and matter fields) to the Higgs field are precisely the reasons why some elementary particles (gauge bosons and fermions) have a non-vanishing mass.

Example 8.1.9 We consider the case of the **electroweak interaction** in the Standard Model. The manifold M is 4-dimensional flat Minkowski spacetime with Minkowski metric η. We set

$$G = \mathrm{SU}(2) \times \mathrm{U}(1) = \mathrm{SU}(2)_L \times \mathrm{U}(1)_Y,$$

where the indices L and Y denote *weak* and *hypercharge*, and consider the Higgs vector space

$$W = \mathbb{C}^2$$

with the standard Hermitian scalar product and the unitary representation

$$\begin{aligned} (\mathrm{SU}(2) \times \mathrm{U}(1)) \times W &\longrightarrow W \\ (A, e^{i\alpha}) \cdot \begin{pmatrix} w_1 \\ w_2 \end{pmatrix} &= A \begin{pmatrix} e^{in_Y\alpha} & 0 \\ 0 & e^{in_Y\alpha} \end{pmatrix} \begin{pmatrix} w_1 \\ w_2 \end{pmatrix}. \end{aligned}$$

Here n_Y is a certain non-zero natural number that will be fixed to $n_Y = 3$ in Remark 8.3.1.

The potential of the Higgs field is

$$\begin{aligned} V(w) &= -\mu w^\dagger w + \lambda \left(w^\dagger w\right)^2 \\ &= -\mu||w||^2 + \lambda||w||^4 \end{aligned} \tag{8.3}$$

with certain constants $\mu, \lambda > 0$. It is clear that the Higgs potential is invariant under the action of G. The choice of Higgs potential is restricted by the conditions that

- it is G-invariant
- it is a polynomial of order at most four in w, so that the interaction defined by it is renormalizable
- it has a minimum, but not in $w = 0$.

Exercise 8.11.1 shows that the Higgs potential then must have the form $V(w)$ in Eq. (8.3).

A vector $w_0 \in W$ is a vacuum vector for the Higgs potential V if and only if

$$||w_0|| = \sqrt{\frac{\mu}{2\lambda}}.$$

We sometimes set

$$v = \sqrt{2}||w_0|| = \sqrt{\frac{\mu}{\lambda}}.$$

It follows that the vacuum manifold is a 3-sphere in $\mathbb{C}^2$ around the origin of radius $||w_0||$. All vacua are gauge equivalent under constant gauge transformations, i.e. the group G acts transitively on the vacuum manifold .

The unbroken subgroup H is isomorphic to U(1), but its embedding into G depends on the precise choice of w_0. We choose

$$w_0 = \begin{pmatrix} 0 \\ \sqrt{\frac{\mu}{2\lambda}} \end{pmatrix}.$$

Then H is the diagonal subgroup

$$H = \mathrm{U}(1)_Q = \left\{ \left(\begin{pmatrix} e^{i\delta/2} & 0 \\ 0 & e^{-i\delta/2} \end{pmatrix}, e^{i\delta/(2n_Y)} \right) \middle| \, \delta \in \mathbb{R} \right\} \subset G. \tag{8.4}$$

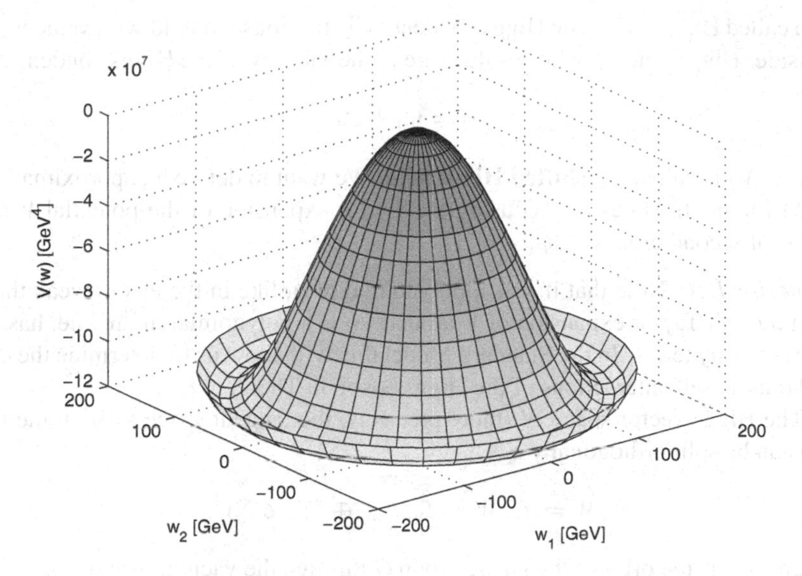

Fig. 8.1 The Higgs potential $V(w) = -\mu|w|^2 + \lambda|w|^4$ on $\mathbb{C}^2$ reduced to $\mathbb{C}$

Note that H is *not* the second factor of G. The index Q stands for *electromagnetic* (we will explain this later in Sect. 8.3).

If we reduce $\mathbb{C}^2$ to $\mathbb{C}$ and S^3 to S^1, then the Higgs potential V has the form of a *Mexican hat*; see Fig. 8.1. The values of the parameters μ and λ realized in nature have to be determined from experiments. It follows from the experimental values collected in Sect. 8.3.4 that

$$\mu \approx 7.82 \cdot 10^3\,\mathrm{GeV}^2,$$
$$\lambda \approx 0.129.$$

8.1.3 The Hessian of the Higgs Potential

Let w_0 be a vacuum vector and s_0 a global vacuum gauge of the principal bundle P. We write the Higgs field as

$$\Phi = [s_0, \phi]$$

with a globally defined map

$$\phi: M \longrightarrow W,$$

also called Higgs field. The Higgs condensate is the constant field with value w_0. We consider Higgs fields ϕ whose values are in the vicinity of the Higgs condensate:

$$\phi = w_0 + \Delta\phi,$$

where $\Delta\phi$ is called the **shifted Higgs field**. We want to derive an approximation to $V(\phi)$ for small values of $\Delta\phi$ using the Taylor expansion of the potential V up to terms of second order in $\Delta\phi$.

Remark 8.1.10 Note that if V is a polynomial in ϕ, like in the electroweak theory, then the full Taylor expansion of V around w_0 is a polynomial in $\Delta\phi$, i.e. has only finitely many terms. In the Standard Model this will allow us to determine the cubic and quartic self-interactions of the Higgs boson in Theorem 8.7.2.

The Higgs vector space W (more precisely, the tangent space to W at the point w_0) can be split orthogonally as

$$W = T_{w_0}W = T_{w_0}\mathcal{O}_{w_0} \oplus (T_{w_0}\mathcal{O}_{w_0})^{\perp},$$

where $\mathcal{O}_{w_0}$ is the orbit of the gauge group G through the vacuum vector w_0,

$$\mathcal{O}_{w_0} = G \cdot w_0,$$

and $(T_{w_0}\mathcal{O}_{w_0})^{\perp}$ is the orthogonal complement with respect to the positive definite scalar product

$$\langle\langle \cdot\,, \cdot \rangle\rangle_W = \mathrm{Re}\langle \cdot\,, \cdot \rangle_W.$$

It follows from Corollary 3.8.10, since the Lie group G is compact, that the orbit $\mathcal{O}_{w_0}$ is an embedded submanifold of W, diffeomorphic to the quotient space G/H, where H is the unbroken subgroup.

Definition 8.1.11 We set d for the dimension of $\mathcal{O}_{w_0} = G \cdot w_0$. This is equal to the codimension of the Lie subgroup H in G.
Let $\mathrm{Hess}(V)$ denote the **Hessian** of the potential V. The Hessian is a symmetric linear map

$$\mathrm{Hess}(V)_w \colon T_wW \longrightarrow T_wW$$

on the tangent space to W at a point w and can be defined on any Riemannian manifold as

$$\mathrm{Hess}(V)(X) = \nabla_X \mathrm{grad}\, V,$$

where in our situation X is a vector tangent to W and ∇ denotes the (flat) Levi-Civita connection induced by the positive definite inner product $\langle\langle \cdot\,, \cdot \rangle\rangle_W$ on W. This linear map is symmetric in the sense that

$$\langle\langle \mathrm{Hess}(V)(X), Y \rangle\rangle_W = \langle\langle X, \mathrm{Hess}(V)Y \rangle\rangle_W.$$

As a matrix the Hessian is given by the symmetric matrix of second derivatives

$$\mathrm{Hess}(V)_w = \left(\frac{\partial^2 V(w)}{\partial x_i \partial x_j}\right)$$

in standard coordinates $x_1, \ldots, x_{2n}$ on the real vector space underlying W.

We want to diagonalize this symmetric linear map in the vacuum point w_0. Since V has a minimum along the whole orbit of G through w_0 (because the potential is invariant under the unitary action of G), it follows that $\mathrm{grad}\, V$ vanishes along the orbit and hence

$$\mathrm{Hess}(V)_{w_0}(X) = 0$$

for vectors $X \in T_{w_0}\mathcal{O}_{w_0}$ tangent to the orbit. This implies:

Lemma 8.1.12 *The Hessian* $\mathrm{Hess}(V)$ *preserves the orthogonal splitting of* $T_{w_0}W$ *into the vector space tangent to the orbit and its orthogonal complement.*

Proof We have just argued that

$$\mathrm{Hess}(V)|_{T_{w_0}\mathcal{O}_{w_0}} \equiv 0.$$

The symmetry of the Hessian then implies for every $X \in (T_{w_0}\mathcal{O}_{w_0})^\perp$ and $Y \in T_{w_0}\mathcal{O}_{w_0}$ that

$$\langle\langle \mathrm{Hess}(V)(X), Y\rangle\rangle_W = \langle\langle X, \mathrm{Hess}(V)(Y)\rangle\rangle_W = 0$$

and thus

$$\mathrm{Hess}(V)\,(T_{w_0}\mathcal{O}_{w_0})^\perp \subset (T_{w_0}\mathcal{O}_{w_0})^\perp.$$

□

We can therefore find a diagonalization of the Hessian adapted to the splitting of the tangent space $T_{w_0}W$ into the vector space tangent to the orbit and its orthogonal complement.

Proposition 8.1.13 (Orthonormal Eigenbasis for Hessian of the Higgs Potential) *There exist real orthonormal bases*

- $e_1, \ldots, e_d$ *of* $T_{w_0}\mathcal{O}_{w_0}$
- $f_1, \ldots, f_{2n-d}$ *of* $(T_{w_0}\mathcal{O}_{w_0})^\perp$

consisting of eigenvectors of the Hessian $\mathrm{Hess}(V)$ *in* w_0*, where*

- *the* e_i *have eigenvalue* 0 *and*
- *the* f_j *have non-negative eigenvalues (because* w_0 *is a local minimum). We set* $2m_{f_j}^2$ *for the eigenvalue of* f_j *(with* $m_{f_j} \geq 0$*).*

8.1.4 The Nambu–Goldstone and Higgs Bosons

Definition 8.1.14 We expand the shifted Higgs field $\Delta\phi \in T_{w_0}W \cong W$ in the orthonormal eigenbasis of the Hessian from Proposition 8.1.13:

$$\Delta\phi = \frac{1}{\sqrt{2}}\sum_{i=1}^{d} \xi_i e_i + \frac{1}{\sqrt{2}}\sum_{j=1}^{2n-d} \eta_j f_j,$$

where ξ_i and η_j are real scalar fields on the spacetime manifold M. The ξ_i are called **Nambu–Goldstone bosons**, the η_j are called **Higgs bosons.**
It follows that

- the number d of Nambu–Goldstone bosons is equal to the dimension of G/H
- the number $2n - d$ of Higgs bosons is equal to the real dimension of the Higgs vector space W minus the dimension of G/H.

It is important to distinguish between the *Higgs field* and the *Higgs bosons*: the Nambu–Goldstone bosons correspond to perturbations of the Higgs field ϕ *along* the orbit of G through the vacuum vector w_0, while the Higgs bosons correspond to perturbations *orthogonal* to the orbit. As particles in the associated quantum field theory the Nambu–Goldstone and Higgs bosons are thus minimal excitations of the Higgs condensate.

Theorem 8.1.15 (Taylor Expansion of Higgs Potential) *Up to second order in the shifted Higgs field, we have by the Taylor expansion around the vacuum vector* w_0

$$V(\phi) \approx V(w_0) + \frac{1}{2}\sum_{j=1}^{2n-d} m_{f_j}^2 \eta_j^2.$$

Proof The Taylor formula up to second order in $\Delta\phi$ is

$$V(\phi) = V(w_0) + \langle\langle \mathrm{grad}\, V(w_0), \Delta\phi\rangle\rangle_W + \frac{1}{2}\langle\langle \Delta\phi, \mathrm{Hess}(V)_{w_0}\Delta\phi\rangle\rangle_W.$$

The claim now follows, because w_0 is a minimum of V, hence grad $V(w_0) = 0$, and e_i and f_j are an orthonormal basis of eigenvectors of the Hessian with eigenvalues 0 and $2m_{f_j}^2$. □

We conclude that the potential V is up to second order (and up to the irrelevant constant $V(w_0)$) the sum over the standard Klein–Gordon mass terms for real scalar fields η_j of mass $m_{f_j}^2$. We shall see in Sect. 8.2.3 that the Lagrangian $\mathscr{L}_H$ for the Higgs field ϕ expressed in terms of the shifted Higgs field contains Klein–Gordon summands of the form

$$\frac{1}{2}(\partial^\mu \eta_j)(\partial_\mu \eta_j) - \frac{1}{2} m_{f_j}^2 \eta_j^2$$

and

$$\frac{1}{2}(\partial^\mu \xi_i)(\partial_\mu \xi_i),$$

plus other terms, some of which we shall determine later. This discussion implies:

Corollary 8.1.16 *The Nambu–Goldstone bosons are real scalar fields of mass zero and the Higgs bosons are real scalar fields of mass* $m_{f_j} \geq 0$.
The terms of higher than quadratic order in the Taylor expansion of the potential V around w_0 can be interpreted as **interactions between Higgs bosons** (see Theorem 8.7.2 for the case of the Standard Model).

Example 8.1.17 We continue with Example 8.1.9. In the case of the electroweak theory we have $G = \mathrm{SU}(2)_L \times \mathrm{U}(1)_Y$ and $H = \mathrm{U}(1)_Q$, embedded in G as a diagonal subgroup, not as the second factor. Therefore we have **three Nambu–Goldstone bosons** ξ_1, ξ_2, ξ_3 and **one Higgs boson** η (also denoted by H, not to be confused with the unbroken gauge group), because W has complex dimension 2.

The Higgs potential is of the form

$$V(w) = -\mu w^\dagger w + \lambda \left(w^\dagger w\right)^2.$$

We can check that the **mass of the Higgs boson** is $m_H = \sqrt{2\mu}$: We choose as before the vacuum vector

$$w_0 = \begin{pmatrix} 0 \\ \sqrt{\frac{\mu}{2\lambda}} \end{pmatrix} = \frac{1}{\sqrt{2}} \begin{pmatrix} 0 \\ v \end{pmatrix}.$$

Then $T_{w_0}\mathcal{O}_{w_0}$ is the real span of the vectors

$$\begin{pmatrix} 1 \\ 0 \end{pmatrix}, \begin{pmatrix} i \\ 0 \end{pmatrix}, \begin{pmatrix} 0 \\ i \end{pmatrix}$$

and $(T_{w_0}\mathcal{O}_{w_0})^\perp$ is the real span of the vector

$$\begin{pmatrix} 0 \\ 1 \end{pmatrix}.$$

The Higgs field can be decomposed as

$$\phi = \begin{pmatrix} \phi_1 \\ \phi_2 \end{pmatrix} = \frac{1}{\sqrt{2}} \begin{pmatrix} \xi_1 + i\xi_2 \\ i\xi_3 \end{pmatrix} + \frac{1}{\sqrt{2}} \begin{pmatrix} 0 \\ v + H \end{pmatrix}$$

with real scalar fields ξ_j and H. In the standard coordinates $x_1 + ix_2, x_3 + ix_4$ for $\mathbb{C}^2$ we have

$$V(x) = -\mu||x||^2 + \lambda||x||^4.$$

Then

$$\frac{\partial V}{\partial x_i} = 2\left(-\mu x_i + 2\lambda||x||^2 x_i\right)$$

and

$$\frac{\partial^2 V}{\partial x_i \partial x_j} = 2\left(-\mu\delta_{ij} + 2\lambda(2x_i x_j + ||x||^2\delta_{ij})\right).$$

In the basis for $T_{w_0}W$

$$e_1 = \begin{pmatrix} 1 \\ 0 \end{pmatrix}, \quad e_2 = \begin{pmatrix} i \\ 0 \end{pmatrix}, \quad e_3 = \begin{pmatrix} 0 \\ i \end{pmatrix}, \quad f = \begin{pmatrix} 0 \\ 1 \end{pmatrix}$$

we get for the Hessian

$$\mathrm{Hess}(V)_{w_0} = \begin{pmatrix} 0 & 0 & 0 & 0 \\ 0 & 0 & 0 & 0 \\ 0 & 0 & 0 & 0 \\ 0 & 0 & 0 & 4\mu \end{pmatrix}.$$

This implies $2m_H^2 = 4\mu$, hence the mass of the Higgs boson is $m_H = \sqrt{2\mu}$. Note that the Hessian in the given basis is diagonal and the eigenvalues vanish in the direction $T_{w_0}\mathcal{O}_{w_0}$ along the orbit, while the eigenvalue along the direction of the vector f orthogonal to the orbit is positive, as expected.

The mass $m_H = \sqrt{2\mu}$ of the Higgs boson is thus determined by the quadratic self-coupling of the Higgs field or, if we write

$$\sqrt{2\mu} = v\sqrt{2\lambda},$$

by the absolute value of the Higgs condensate and the quartic self-coupling.

8.1.5 Unitary Gauge and the Nambu–Goldstone Bosons

Recall that the principal bundle $P \to M$ is trivial and we fixed a global gauge $s_0: M \to P$, called the vacuum gauge. The Higgs field Φ is given by

$$\Phi = [s_0, \phi]$$

with a smooth map $\phi: M \to W$.

We now consider gauge transformations on P. With respect to the global gauge s_0 they are given by physical gauge transformations

$$\tau: M \longrightarrow G$$

that act on ϕ by

$$\phi(x) \longmapsto \tau(x) \cdot \phi(x) \quad \forall x \in M,$$

where the G-representation on W is implicit. Note that, if we think of the manifold M as spacetime, physical gauge transformations τ in general are time-dependent.

Definition 8.1.18 For a given Higgs field ϕ, we call a smooth, physical gauge transformation $\tau: M \to G$ a **unitary gauge** with respect to a vacuum vector w_0 if all Nambu–Goldstone bosons of the transformed field $\tau \cdot \phi$ with respect to the vacuum vector w_0 vanish identically on M. We then say that the transformed Higgs field $\phi' = \tau \cdot \phi$ is in unitary gauge (with respect to the vacuum vector w_0).

An equivalent condition is:

Lemma 8.1.19 *A Higgs field ϕ is in unitary gauge with respect to the vacuum vector w_0 if the shifted Higgs field*

$$\Delta\phi(x) = \phi(x) - w_0 \in T_{w_0} W$$

is at every point $x \in M$ orthogonal to the tangent space $T_{w_0}\mathcal{O}_{w_0}$ to the orbit $\mathcal{O}_{w_0} = G \cdot w_0$.

The physical intuition behind unitary gauges is that Nambu–Goldstone bosons are not physical particles, but can be gauged away, in contrast to the Higgs bosons. Some details about the general existence of unitary gauges can be found in [16]. We only need the following statement.

Theorem 8.1.20 (Existence of Unitary Gauges in the Electroweak Theory) *Consider the electroweak gauge theory as in Example 8.1.9 with gauge group $G = \mathrm{SU}(2)_L \times \mathrm{U}(1)_Y$ and Higgs field of the form*

$$\phi = \begin{pmatrix} \phi_1 \\ \phi_2 \end{pmatrix},$$

where $\phi_1, \phi_2: M \to \mathbb{C}$. Assume that $\phi_2(x) \neq 0$ for all $x \in M = \mathbb{R}^4$. Then there exists a physical gauge transformation $\tau: M \to G$ such that

$$\tau \cdot \phi = \begin{pmatrix} 0 \\ \psi \end{pmatrix},$$

where $\psi: M \to \mathbb{R}$ is a real-valued function. The transformed Higgs field $\tau \cdot \phi$ is then in unitary gauge with respect to the vacuum vector

$$w_0 = \begin{pmatrix} 0 \\ \sqrt{\frac{\mu}{2\lambda}} \end{pmatrix}.$$

Proof We just sketch the proof and leave the details to Exercise 8.11.2. We write

$$\phi(x) = \begin{pmatrix} r(x)e^{i\theta(x)} \\ s(x)e^{i\mu(x)} \end{pmatrix},$$

where all functions r, θ, s, μ are real-valued and $s(x) \neq 0$ for all $x \in M$ by assumption. Since $H_1(M; \mathbb{Z}) = 0$, the functions θ and μ exist as globally well-defined single-valued functions (otherwise we would have to work directly with S^1-valued functions instead of $e^{i\theta}$ and $e^{i\mu}$).

We can first find an explicit smooth SU(2) gauge transformation to get ϕ into the form

$$\begin{pmatrix} 0 \\ s'(x)e^{i\mu'(x)} \end{pmatrix}.$$

A suitable U(1) smooth gauge transformation then yields ϕ in the form

$$\begin{pmatrix} 0 \\ s'(x) \end{pmatrix}.$$

Setting $\psi = s'$, the claim follows. □

8.2 Mass Generation for Gauge Bosons

The purpose of this section is to describe how masses of gauge bosons are generated through the Higgs field.

8.2.1 Broken and Unbroken Gauge Bosons

We fix in addition to the data above the following data:

- an Ad-invariant positive definite scalar product $\langle\cdot,\cdot\rangle_{\mathfrak{g}}$ on the Lie algebra $\mathfrak{g}$, determined by certain coupling constants with respect to an arbitrary fixed scalar product.

Let $H \subset G$ denote the stabilizer subgroup of the vacuum vector $w_0 \in W$. We denote by $\mathfrak{h} \subset \mathfrak{g}$ the Lie algebra of H and by $\mathfrak{h}^\perp$ its orthogonal complement with respect to the scalar product $\langle\cdot,\cdot\rangle_{\mathfrak{g}}$. We denote the dimension of $\mathfrak{h}^\perp \cong \mathfrak{g}/\mathfrak{h}$ as before by d.

Definition 8.2.1 We define the positive semi-definite, bilinear symmetric **mass form** m on $\mathfrak{g}$ by

$$\begin{aligned} m\colon \mathfrak{g} \times \mathfrak{g} &\longrightarrow \mathbb{R} \\ (A, B) &\longmapsto \langle\langle A \cdot w_0, B \cdot w_0 \rangle\rangle_W. \end{aligned}$$

Here $A \cdot w_0$ denotes the induced representation ρ_* of the Lie algebra $\mathfrak{g}$ on the Higgs vector space W.
According to Proposition 3.2.10 the kernel of the map

$$\begin{aligned} \mathfrak{g} &\longrightarrow T_{w_0} W \\ A &\longmapsto A \cdot w_0 \end{aligned}$$

is equal to the Lie algebra $\mathfrak{h}$ of the isotropy group. Hence $A \cdot w_0 = 0$ for all $A \in \mathfrak{h}$, while the map $A \mapsto A \cdot w_0$ is injective on the orthogonal complement $\mathfrak{h}^\perp$. This implies that $m(A, \cdot) \equiv 0$ if $A \in \mathfrak{h}$ and the restriction of m onto the orthogonal complement $\mathfrak{h}^\perp$ is positive definite. Since m is a symmetric form, we can diagonalize it and get:

Proposition 8.2.2 *We can find $\langle\cdot,\cdot\rangle_{\mathfrak{g}}$-orthonormal bases*

- $\alpha_1, \ldots, \alpha_d$ *of the subspace $\mathfrak{h}^\perp$, called* ***broken generators****, and*
- $\alpha_{d+1}, \ldots, \alpha_r$ *of the subspace $\mathfrak{h}$, called* ***unbroken generators***

such that the symmetric mass form m is diagonal in this basis. We can write

$$m(\alpha_a, \alpha_a) = \frac{1}{2} M_a^2 \geq 0,$$

with $M_a > 0$ for the broken generators and $M_a = 0$ for the unbroken generators.
If $A_\mu\colon M \to \mathfrak{g}$ is a gauge field, then we can decompose

$$A_\mu = \sum_{a=1}^{r} A_\mu^a \alpha_a,$$

where A^a_μ are the **broken and unbroken gauge bosons**, respectively. The numbers M_a are called the **masses of the gauge bosons**.

Remark 8.2.3 Note two interesting facts: the masses M_a are proportional to the norm $||w_0||$ of the value of the Higgs condensate and they also depend on the choice of the scalar product $\langle \cdot, \cdot \rangle_{\mathfrak{g}}$ on $\mathfrak{g}$ (and thus on the coupling constants) via the choice of orthonormal basis $\{\alpha_a\}$.

8.2.2 The Combined Lagrangian

For a gauge field A_μ we consider as before (in the mathematical convention) the covariant derivative

$$\nabla^A_\mu = \partial_\mu + A_\mu$$

and the curvature (or field strength)

$$F_{\mu\nu} = \partial_\mu A_\nu - \partial_\nu A_\mu + [A_\mu, A_\nu].$$

We can also express the field strength in our chosen orthonormal basis of broken and unbroken generators for the Lie algebra $\mathfrak{g}$:

$$F_{\mu\nu} = \sum_{a=1}^{r} F^a_{\mu\nu} \alpha_a.$$

The total Lagrangian for the Higgs field ϕ and the gauge field A_μ is:

$$\mathscr{L}_{YMH} = \left(\nabla^{A\mu}\phi\right)^\dagger \left(\nabla^A_\mu \phi\right) - V(\phi) - \frac{1}{4} F^{\mu\nu}_a F^a_{\mu\nu}.$$

There is a summation over Lie algebra indices a in the last term.

We write $\phi = w_0 + \Delta\phi$ with the shifted Higgs field $\Delta\phi$ as before. We would like to determine the terms up to second order in the fields $\Delta\phi$ and A_μ in the Lagrangian $\mathscr{L}_{YMH}$, i.e. the terms corresponding to "free" fields. Higher-order terms contain interactions between these fields.

Proposition 8.2.4 *Up to terms of second order the Lagrangian* $\mathscr{L}_{YMH}$ *is given by*

$$\begin{aligned}\mathscr{L}_{YMH} \approx {} & (\partial^\mu \Delta\phi)^\dagger (\partial_\mu \Delta\phi) + 2\mathrm{Re}\, (\partial^\mu \Delta\phi)^\dagger (A_\mu \cdot w_0) + (A^\mu \cdot w_0)^\dagger (A_\mu \cdot w_0) \\ & - V(\phi) - \frac{1}{4} \left(\partial_\mu A^a_\nu - \partial_\nu A^a_\mu\right) \left(\partial^\mu A^\nu_a - \partial^\nu A^\mu_a\right).\end{aligned}$$

Proof We have

$$\nabla^{A\mu}\phi = \partial^{\mu}\Delta\phi + A^{\mu}\cdot\Delta\phi + A^{\mu}\cdot w_0.$$

However, $A^{\mu}\cdot\Delta\phi$ is already quadratic in the fields and can be ignored. This implies the terms in the first line of the equation. The second line is clear (where $V(\phi)$ should be taken up to second order in $\Delta\phi$). We have also ignored terms of order 3 and 4 in the gauge field A_{μ}, appearing in the Yang–Mills Lagrangian for non-abelian gauge theories. □

8.2.3 Simplifying the Lagrangian

We want to simplify each of the summands in the Lagrangian in Proposition 8.2.4. Recall that according to Proposition 4.7.6 we can identify sections of associated vector bundles with vector space-valued maps once we have chosen a gauge for the principal bundle. So far we have not specified this choice of gauge in the formula in Proposition 8.2.4. We now make the following assumption:

- The Higgs field ϕ is in unitary gauge with respect to the vacuum vector w_0.

By this we mean that all Nambu–Goldstone bosons of ϕ vanish on all of the manifold M identically.

Remark 8.2.5 We have not discussed the existence of unitary gauges in general. However, according to Theorem 8.1.20 such gauges exist in the electroweak theory if the second component ϕ_2 of the Higgs field ϕ is everywhere on spacetime $M=\mathbb{R}^4$ non-zero. This is the case, in particular, if the value $\phi(x)$ of the Higgs field is everywhere on spacetime in an appropriate neighbourhood of the vacuum vector in the vector space W. It follows that after symmetry breaking we can assume in the electroweak theory that the Higgs field is in unitary gauge, at least if its fluctuations around the vacuum value are not too large.

Lemma 8.2.6 *If ϕ is in unitary gauge, then*

$$\mathrm{Re}\,(\partial^{\mu}\Delta\phi)^{\dagger}\left(A_{\mu}\cdot w_0\right) = 0.$$

Proof Note that $A_{\mu}\cdot w_0$ is tangential to the orbit of w_0 under G, while the shifted Higgs field $\Delta\phi$ and therefore its derivative is everywhere on M orthogonal to the orbit, by the assumption of unitary gauge. This implies the claim. □

Lemma 8.2.7 *If ϕ is in unitary gauge, then*

$$(\partial^{\mu}\Delta\phi)^{\dagger}\left(\partial_{\mu}\Delta\phi\right) = \frac{1}{2}\sum_{j=1}^{2n-d}\left(\partial^{\mu}\eta_j\right)\left(\partial_{\mu}\eta_j\right),$$

where η_j are the Higgs bosons.

Proof Follows immediately, because under our assumption that the Nambu–Goldstone bosons vanish we have

$$\Delta\phi = \frac{1}{\sqrt{2}}\sum_{j=1}^{2n-d}\eta_j f_j.$$

□

Lemma 8.2.8 *Up to second order in the Higgs bosons we have*

$$V(\phi) \approx V(w_0) + \frac{1}{2}\sum_{j=1}^{2n-d} m_{f_j}^2 \eta_j^2,$$

where m_{f_j} is the mass of the j-th Higgs boson.

Proof This formula can be found in Theorem 8.1.15. □

Lemma 8.2.9 *We have*

$$\left(A^\mu \cdot w_0\right)^\dagger \left(A_\mu \cdot w_0\right) = \frac{1}{2}\sum_{a=1}^{d} M_a^2 A_a^\mu A_\mu^a,$$

where M_a are the masses of the broken gauge bosons defined above.

Proof We have

$$\left(A^\mu \cdot w_0\right)^\dagger \left(A_\mu \cdot w_0\right) = m(A^\mu, A_\mu)$$

for the bilinear mass form m on $\mathfrak{g}$. This implies the claim by our choice of basis α_a. □

We now collect all terms and get:

Theorem 8.2.10 (Mass Generation for Gauge Bosons) *If the Higgs field ϕ is in unitary gauge after symmetry breaking, then the Lagrangian $\mathscr{L}_{YMH}$, up to terms of second order in the shifted Higgs field and the gauge field, is given by*

$$\begin{aligned}\mathscr{L}_{YMH} \approx\ & \frac{1}{2}\sum_{j=1}^{2n-d}\left(\partial^\mu \eta_j\right)\left(\partial_\mu \eta_j\right) - \frac{1}{2}\sum_{j=1}^{2n-d} m_{f_j}^2 \eta_j^2 \\ & - \frac{1}{4}\sum_{a=1}^{d}\left(\partial_\mu A_\nu^a - \partial_\nu A_\mu^a\right)\left(\partial^\mu A_a^\nu - \partial^\nu A_a^\mu\right) + \frac{1}{2}\sum_{a=1}^{d} M_a^2 A_a^\mu A_\mu^a \\ & - \frac{1}{4}\sum_{b=d+1}^{r}\left(\partial_\mu A_\nu^b - \partial_\nu A_\mu^b\right)\left(\partial^\mu A_b^\nu - \partial^\nu A_b^\mu\right).\end{aligned}$$

Here we removed the irrelevant constant $V(w_0)$. This Lagrangian has the following interpretation:

- The two terms in the first line are the Klein–Gordon Lagrangian for $2n - d$ real scalar Higgs bosons η_j of mass m_{f_j}.
- The two terms in the second line are the Lagrangian for d broken, massive gauge bosons $A^1_\mu, \ldots, A^d_\mu$ of mass M_a.
- The term in the third line is the Lagrangian for $r - d$ unbroken, massless gauge bosons $A^{d+1}_\mu, \ldots, A^r_\mu$.

This is the celebrated **Brout–Englert–Higgs mechanism** of creating in a gauge invariant way masses for gauge bosons. The exact Lagrangian $\mathscr{L}_{YMH}$ contains terms of order higher than two that describe interactions between Higgs bosons, between gauge bosons (in the non-abelian case) and between Higgs bosons and gauge bosons. For the Standard Model these terms can be found in Theorem 8.7.2, Corollary 8.7.5 and Theorem 8.7.6.

Remark 8.2.11 Notice why we get the mass term

$$\frac{1}{2}\sum_{a=1}^{d} M_a^2 A_a^\mu A_\mu^a$$

for the broken gauge bosons: in the Yang–Mills–Higgs Lagrangian we have the term

$$\left(\nabla^{A\mu}\phi\right)^\dagger \left(\nabla^A_\mu\phi\right)$$

involving the covariant derivative of the Higgs field, which describes the coupling between the Higgs field and the gauge field. We have then written

$$\phi = w_0 + \Delta\phi$$

which results in the term

$$\left(A^\mu \cdot w_0\right)^\dagger \left(A_\mu \cdot w_0\right).$$

This term is non-zero in general, because the value w_0 of the Higgs condensate is non-zero, and leads to the mass term for the broken gauge bosons. *We see that the ultimate reason for the masses of the broken gauge bosons is the coupling of the gauge field to the Higgs field via the covariant derivative (the Higgs field is a charged scalar) and the non-zero value w_0 of the Higgs condensate after symmetry breaking.* This also explains why the masses of the gauge bosons depend on both the coupling constants and the vacuum value $||w_0||$.

Remark 8.2.12 The mechanism of spontaneous symmetry breaking was developed by Philip W. Anderson (Nobel Prize in Physics 1977), Yoichiro Nambu (Nobel Prize in Physics 2008) and Jeffrey Goldstone in the early 1960s. It was extended to Yang–Mills theory independently by Robert Brout and François Englert [48], Peter W. Higgs [76], and Gerald S. Guralnik, Carl R. Hagen and Thomas W.B. Kibble [67] in 1964 (Nobel Prize in Physics 2013 for Englert and Higgs). The three papers on symmetry breaking from 1964 all appeared in vol. 13 of the *Physical Review Letters*. The theory was confirmed in July 2012 by the discovery of the Higgs boson at the Large Hadron Collider (LHC) at CERN in Geneva.

8.3 Massive Gauge Bosons in the SU(2) × U(1)-Theory of the Electroweak Interaction

The **weak interaction**, which describes particle decays such as the decay of a muon into an electron, an electron antineutrino and a muon neutrino

$$\mu \longrightarrow e^- + \nu_e^C + \nu_\mu,$$

was originally described by the so-called **4-Fermi interaction** with Lagrangian

$$\mathscr{L}_F = G_F \overline{\Psi}_\mu \psi_{\nu_\mu} \overline{\Psi}_e \psi_{\nu_e}$$

(G_F is a coupling constant) and associated Feynman diagram Fig. 8.2. This Lagrangian with a direct interaction between four fermions did not appear in Chap. 7 and, in fact, it is *non-renormalizable*.

In the Standard Model, the weak interaction together with the electromagnetic interaction are described by a gauge theory with gauge group $\mathrm{SU}(2)_L \times \mathrm{U}(1)_Y$. This theory is called the **electroweak theory**. The Feynman diagram of the 4-Fermi

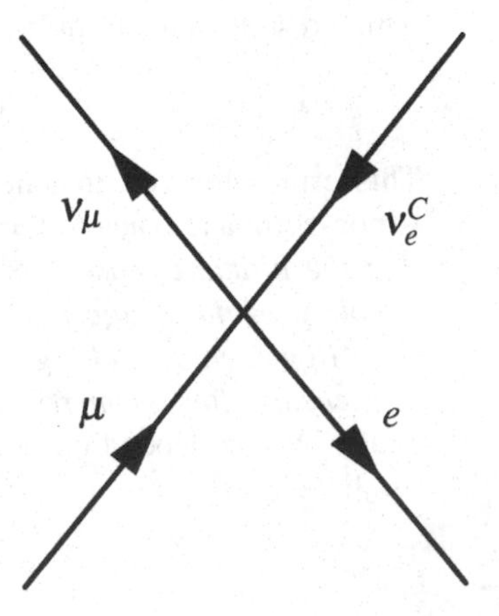

Fig. 8.2 Muon decay with 4-Fermi interaction

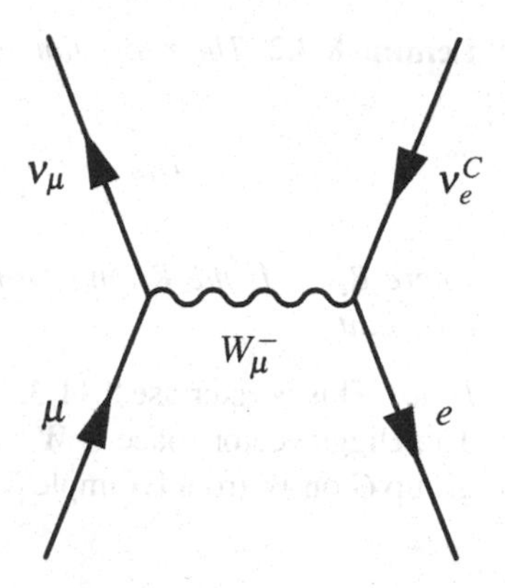

Fig. 8.3 Muon decay in electroweak theory

interaction is replaced by the Feynman diagram in Fig. 8.3, involving a virtual W^--gauge boson. The interactions in this diagram *are* renormalizable.

In this section we want to study the Higgs mechanism of mass generation in the special case of the electroweak interaction and discuss the associated massive gauge bosons $W^\pm$ and Z^0.

8.3.1 *The Lie Algebra* $\mathfrak{su}(2)_L \oplus \mathfrak{u}(1)_Y$

We continue with Example 8.1.9 and Example 8.1.17. Our gauge group is $G = \mathrm{SU}(2)_L \times \mathrm{U}(1)_Y$ and the manifold M is 4-dimensional flat Minkowski spacetime. We choose the Ad-invariant scalar product $\langle \cdot , \cdot \rangle_{\mathfrak{g}}$ on the Lie algebra $\mathfrak{g} = \mathfrak{su}(2)_L \oplus \mathfrak{u}(1)_Y$ in such a way that the following vectors form an orthonormal basis:

$$\beta_l = g_w \frac{i\sigma_l}{2} \in \mathfrak{su}(2)_L \quad (l = 1, 2, 3),$$

$$\beta_4 = g' \frac{i}{2n_Y} \in \mathfrak{u}(1)_Y,$$

where σ_l are the Pauli matrices

$$\sigma_1 = \begin{pmatrix} 0 & 1 \\ 1 & 0 \end{pmatrix}, \quad \sigma_2 = \begin{pmatrix} 0 & -i \\ i & 0 \end{pmatrix}, \quad \sigma_3 = \begin{pmatrix} 1 & 0 \\ 0 & -1 \end{pmatrix}$$

and the positive real numbers g_w and g' are the coupling constants corresponding to $\mathrm{SU}(2)_L$ and $\mathrm{U}(1)_Y$. The non-zero natural number n_Y is a normalization constant.

Remark 8.3.1 We will fix from now on $n_Y = 3$ (this is the convention used, for example, by [9] and [137]). Other references (such as [125]) use the convention $n_Y = 6$. We continue to use the traditional convention $n_Y = 3$.

Lemma 8.3.2 *The restriction of the scalar product $\langle\cdot,\cdot\rangle_{\mathfrak{g}}$ to $\mathfrak{su}(2)_L$ is given by*

$$\langle\cdot,\cdot\rangle_{\mathfrak{su}(2)_L} = -\frac{1}{2g_w^2}B_{\mathfrak{su}(2)}(X,Y) = -\frac{2}{g_w^2}\mathrm{tr}(X\cdot Y)$$

where $B_{\mathfrak{su}(2)}$ is the Killing form of $\mathfrak{su}(2)$. In particular, the scalar product is Ad-*invariant.*

Proof This is Exercise 8.11.3. Compare with Exercises 2.7.13 and 2.7.16. □

The Higgs vector space is $W = \mathbb{C}^2$. Recall the unitary representation of the gauge group G on W from Example 8.1.9. The basis elements $\beta_a \in \mathfrak{g}$ act on a vector

$$w = \begin{pmatrix} w_1 \\ w_2 \end{pmatrix} \in \mathbb{C}^2$$

as

$$\beta_l \cdot w = g_w \frac{i\sigma_l}{2}\begin{pmatrix} w_1 \\ w_2 \end{pmatrix} \quad (l = 1,2,3),$$

$$\beta_4 \cdot w = g'\frac{i}{2}\begin{pmatrix} w_1 \\ w_2 \end{pmatrix}.$$

The vacuum vector is given by

$$w_0 = \begin{pmatrix} 0 \\ \sqrt{\frac{\mu}{2\lambda}} \end{pmatrix} = \begin{pmatrix} 0 \\ ||w_0|| \end{pmatrix} \in \mathbb{C}^2,$$

where μ and λ are the parameters of the Higgs field potential that can be found in Example 8.1.9.

8.3.2 The Gauge Bosons

A direct calculation shows that the mass form

$$m\colon \mathfrak{g} \times \mathfrak{g} \longrightarrow \mathbb{R}$$

is given in the basis β_a by

$$m(\beta_a,\beta_b) = \frac{||w_0||^2}{4}\begin{pmatrix} g_w^2 & 0 & 0 & 0 \\ 0 & g_w^2 & 0 & 0 \\ 0 & 0 & g_w^2 & -g_w g' \\ 0 & 0 & -g_w g' & g'^2 \end{pmatrix}.$$

If we define a new orthonormal basis

$$\alpha_1 = \beta_1,$$
$$\alpha_2 = \beta_2,$$
$$\alpha_3 = \frac{1}{\sqrt{g_w^2 + g'^2}}(g_w\beta_3 - g'\beta_4),$$
$$\alpha_4 = \frac{1}{\sqrt{g_w^2 + g'^2}}(g'\beta_3 + g_w\beta_4),$$

then the bilinear form m becomes diagonal:

$$m(\alpha_a, \alpha_b) = \frac{||w_0||^2}{4}\begin{pmatrix} g_w^2 & 0 & 0 & 0 \\ 0 & g_w^2 & 0 & 0 \\ 0 & 0 & g_w^2 + g'^2 & 0 \\ 0 & 0 & 0 & 0 \end{pmatrix}.$$

We see that the subalgebra of the **stabilizer group** is given by

$$\mathfrak{h} = \mathrm{span}(\alpha_4).$$

Indeed, α_4 acts on the vacuum vector as

$$\begin{aligned}\alpha_4 \cdot w_0 &= \frac{1}{\sqrt{g_w^2 + g'^2}}\left(g'g_w\frac{i\sigma_3}{2} + g_w g'\frac{i}{2}\right) w_0 \\ &= \frac{g_w g'}{\sqrt{g_w^2 + g'^2}}\begin{pmatrix} i & 0 \\ 0 & 0 \end{pmatrix}\begin{pmatrix} 0 \\ ||w_0|| \end{pmatrix} \\ &= 0.\end{aligned}$$

Note that with respect to the standard unnormalized basis

$$\beta_3' = \frac{i\sigma_3}{2} \in \mathrm{su}(2)_L,$$
$$\beta_4' = \frac{i}{6} \in \mathrm{u}(1)_Y$$

the vector α_4 is just given by

$$\alpha_4 = \frac{g_w g'}{\sqrt{g_w^2 + g'^2}}(\beta_3' + \beta_4'),$$

hence proportional to $\beta_3' + \beta_4'$ as we expect from Example 8.1.9. We shall see in Sect. 8.3.5 that the factor

$$\frac{g_w g'}{\sqrt{g_w^2 + g'^2}}$$

can be identified with the elementary electric charge.

The subspace of **broken generators** is given by

$$\mathfrak{h}^\perp = \mathrm{span}(\alpha_1, \alpha_2, \alpha_3).$$

From the diagonal mass form we can read off the masses of the gauge bosons:

- There are **three massive gauge bosons**: two gauge bosons, corresponding to α_1, α_2, of mass

$$\frac{1}{\sqrt{2}}||w_0||g_w$$

  and one gauge boson, corresponding to α_3, of mass

$$\frac{1}{\sqrt{2}}||w_0||\sqrt{g_w^2 + g'^2}.$$

  As mentioned in Remark 8.2.3 we see precisely how the masses of the gauge bosons depend on the vacuum value $||w_0||$ of the Higgs field and the coupling constants.
- We also have **one massless gauge boson**, corresponding to α_4.

8.3.3 The Physics Notation

In physics the following notation is used: We set

$$\tan\theta_W = \frac{g'}{g_w},$$

where $\theta_W \in \left[0, \frac{\pi}{2}\right)$ is the **Weinberg angle** or **weak mixing angle**. Then

$$\alpha_3 = \cos\theta_W \beta_3 - \sin\theta_W \beta_4,$$
$$\alpha_4 = \sin\theta_W \beta_3 + \cos\theta_W \beta_4.$$

Hence the basis (α_3, α_4) is rotated by an angle θ_W with respect to (β_3, β_4) (clockwise in our situation). The Weinberg angle describes the direction of the unbroken generator α_4 of $\mathrm{U}(1)_Q$ with respect to the generators β_3 and β_4 of $\mathrm{SU}(2)_L \times \mathrm{U}(1)_Y$.

We can then decompose our gauge field

$$A_\mu = \sum_{a=1}^{4} A_\mu^a \beta_a$$

as

$$A_\mu = W_\mu^+ \frac{1}{\sqrt{2}}(\alpha_1 + i\alpha_2) + W_\mu^- \frac{1}{\sqrt{2}}(\alpha_1 - i\alpha_2) + Z_\mu^0 \alpha_3 + \gamma_\mu \alpha_4,$$

where

- $W_\mu^\pm = \frac{1}{\sqrt{2}}(A_\mu^1 \mp iA_\mu^2)$ are the **W-bosons** of mass

$$m_W = \frac{1}{\sqrt{2}}||w_0||g_w$$

- $Z_\mu^0 = \cos\theta_W A_\mu^3 - \sin\theta_W A_\mu^4$ is the **Z-boson** of mass

$$m_Z = \frac{1}{\sqrt{2}}||w_0||\sqrt{g_w^2 + g'^2}$$

- $\gamma_\mu = \sin\theta_W A_\mu^3 + \cos\theta_W A_\mu^4$ is the massless **photon** (not to be confused with a mathematical gamma matrix).

The masses of the W- and Z-bosons are related to the Weinberg angle via

$$\cos\theta_W = \frac{m_W}{m_Z}.$$

We shall see in Sect. 8.5.5 that $W^\pm$ have electric charge ± 1 whereas Z^0 and γ have electric charge 0. In general, gauge bosons are also called **vector bosons**, in particular, the W- and Z-bosons. The gauge field A_μ^3 is sometimes denoted by W_μ^0 and the gauge field A_μ^4 by B_μ.

8.3.4 Experimental Values

We discuss some experimental values (our reference is [106, 108]). The **Fermi constant**

$$G_F^0 = \frac{G_F}{(\hbar c)^3} = \frac{1}{4\sqrt{2}} \frac{g_w^2}{m_W^2},$$

which describes the effective strength of the weak interaction, can be determined from muon decay. Its value is

$$\frac{G_F}{(\hbar c)^3} = 1.1663787 \pm 0.0000006 \times 10^{-5}\,\mathrm{GeV}^{-2}.$$

This gives for

$$v = \sqrt{2}||w_0|| = 2\frac{m_W}{g_w} = \left(\sqrt{2}G_F^0\right)^{-1/2}$$

the value

$$v \approx 246.2197\,\mathrm{GeV}.$$

The weak mixing angle and the masses of the $W^{\pm}$ and Z bosons are

$$\begin{aligned} \sin^2\theta_W &= 0.23129 \pm 0.00005, \\ m_W &= 80.385 \pm 0.015\,\mathrm{GeV}, \\ m_Z &= 91.1876 \pm 0.0021\,\mathrm{GeV}. \end{aligned}$$

This implies

$$\theta_W \approx 28.75^\circ$$

and with the Fermi constant

$$g_w \approx 0.6530.$$

We will see in Sect. 8.3.5 that the electric coupling constant is given by

$$e = g_w \sin\theta_W = \frac{g_w g'}{\sqrt{g_w^2 + g'^2}}.$$

In particular,

$$e \approx 0.48 g_w,$$

so the weakness of the weak interaction compared to electromagnetism comes mainly from the large mass of the W- and Z-bosons. We write $e = \sqrt{4\pi\alpha}$ with fine-structure constant α given by

$$\begin{aligned} \alpha &= 7.2973525664 \pm 0.0000000017 \times 10^{-3} \\ &= 1/\left(137.035999139 \pm 0.000000031\right). \end{aligned}$$

The final value that was determined experimentally is the mass $m_H = \sqrt{2\mu}$ of the Higgs boson:

$$m_H = 125.09 \pm 0.21 \pm 0.11 \,\text{GeV}.$$

The various constants are not all independent, i.e. if we know some of the values, then we can predict others with the theory. Note that the values are not fully consistent with the formulas above, because there are higher-order corrections to these formulas coming from the associated quantum field theory. To be precise we have to indicate, for instance, the energy scale at which the coupling constants have been measured; see Sect. 9.4.

Remark 8.3.3 The theory of the electroweak interaction, the unification of electromagnetism and the weak nuclear force as an SU(2) × U(1) gauge theory together with spontaneous symmetry breaking, was developed during the 1960s by Sheldon Glashow, Abdus Salam and Steven Weinberg (Nobel Prize in Physics 1979). The discovery of the W- and Z-bosons at the Super Proton Synchrotron (SPS) at CERN was announced in 1983 (Nobel Prize in Physics 1984 for Carlo Rubbia and Simon Van der Meer). The quantum field theory of the electroweak interaction was shown to be renormalizable by Gerardus 't Hooft and Martinus J.G. Veltman (Nobel Prize in Physics 1999).

8.3.5 Charges

We continue to consider the case of the electroweak interaction with gauge group $G = \mathrm{SU}(2)_L \times \mathrm{U}(1)_Y$. Suppose V is a complex vector space of dimension m with a unitary representation of G. The generators β_l of $\mathfrak{su}(2)_L$ act as

$$\beta_l \to i g_w \mathbb{T}_l \quad (l = 1, 2, 3)$$

and the generator β_4 of $\mathfrak{u}(1)_Y$ acts as

$$\beta_4 \to i g' \frac{\mathbb{Y}}{2},$$

where $\mathbb{T}_l$ and $\mathbb{Y}$ are certain *Hermitian* operators on V.

Definition 8.3.4 The eigenvalues of $\mathbb{T}_3$ are called **weak isospin** and of $\mathbb{Y}$ **weak hypercharge**.

Since $\mathbb{T}_3$ and $\mathbb{Y}$ commute in every representation of G, we can find an orthonormal basis of V of common eigenvectors for both operators. If we identify V with $\mathbb{C}^m$ via this basis, then both $\mathbb{T}_3$ and $\mathbb{Y}$ act as diagonal matrices, whose entries are the charges of the multiplet component fields corresponding to the basis vectors for V.

Remark 8.3.5 The eigenvalues of the weak isospin operator $\mathbb{T}_3$ and thus the weak isospin charges are determined by the **weights** of the representation. The weights are elements in the dual space of the **Cartan subalgebra** in $\mathfrak{su}(2) \otimes \mathbb{C}$, which is spanned by $i\sigma_3$. See, for example, [153] for more details.

If we set

$$\mathbb{T}_+ = (\mathbb{T}_1 + i\mathbb{T}_2),$$
$$\mathbb{T}_- = (\mathbb{T}_1 - i\mathbb{T}_2),$$
$$\mathbb{Q} = \mathbb{T}_3 + \frac{\mathbb{Y}}{2},$$

(the third equation is known as the **Gell-Mann–Nishijima formula**), then a general gauge field $A_\mu: M \to \mathfrak{g}$ acts on the multiplets with values in V as

$$\begin{aligned}\mathbb{A}_\mu = &\frac{ig_w}{\sqrt{2}}(W_\mu^+\mathbb{T}_+ + W_\mu^-\mathbb{T}_-)\\ &+ Z_\mu^0 \frac{1}{\sqrt{g_w^2 + g'^2}}\left(ig_w^2\mathbb{T}_3 - ig'^2\frac{\mathbb{Y}}{2}\right)\\ &+ \gamma_\mu \frac{ig_w g'}{\sqrt{g_w^2 + g'^2}}\mathbb{Q}.\end{aligned} \tag{8.5}$$

It follows that the **elementary electric charge** is given by

$$e = \frac{g_w g'}{\sqrt{g_w^2 + g'^2}} = g_w \sin\theta_W.$$

For example, on the Higgs field

$$\phi = \begin{pmatrix}\phi_1\\ \phi_2\end{pmatrix}$$

with values in $W = \mathbb{C}^2$ the charge operators act as

$$\mathbb{T}_3 = \begin{pmatrix}\frac{1}{2} & 0\\ 0 & -\frac{1}{2}\end{pmatrix},$$
$$\mathbb{Y} = \begin{pmatrix}1 & 0\\ 0 & 1\end{pmatrix},$$
$$\mathbb{Q} = \begin{pmatrix}1 & 0\\ 0 & 0\end{pmatrix}.$$

We can then directly read off the charges of the components of ϕ. In order to indicate the electric charge of the Higgs field, it is often written as

$$\phi = \frac{1}{\sqrt{2}} \begin{pmatrix} \phi^+ \\ \phi^0 \end{pmatrix}$$

(the factor $\frac{1}{\sqrt{2}}$ is a convention).

Remark 8.3.6 If the convention $n_Y = 6$ instead of our $n_Y = 3$ is used, then the Gell-Mann–Nishijima formula becomes

$$\mathbb{Q} = \mathbb{T}_3 + \mathbb{Y}.$$

8.4 The SU(3)-Theory of the Strong Interaction (QCD)

Quantum chromodynamics (QCD), the theory of the strong interaction, is a gauge theory with gauge group $\mathrm{SU}(3)_C$, where C stands for *colour* (*chroma* is the Ancient Greek word for *colour*). There are eight gauge bosons for this gauge group, called **gluons**.

The matter particles (fermions) in QCD are called **quarks**. We will see later that there are six different types of quarks, called **quark flavours**. In nature, quarks are only observed in **bound states** called **hadrons**: **baryons** (like nucleons, i.e. the proton and neutron), that consist of three quarks (or antiquarks) and have half-integer spin, and **mesons**, that consist of a quark-antiquark pair and have integer spin. The lifetime of the heaviest quark, the *top quark*, is too short to form hadrons. All other quarks can form hadrons. There is evidence that exotic hadrons such as tetraquarks, consisting of two quarks and two antiquarks, and pentaquarks, consisting of four quarks (antiquarks) and one antiquark (quark), exist. New hadrons and hadron resonances (excited states) are still being discovered.

The gluons virtually "glue" the quarks together into hadrons. The exact reason why quarks appear only in colour neutral hadrons and not as isolated particles, known as **colour confinement**, is not fully understood. Basic properties of these bound states, like their masses, so far cannot be derived theoretically, but only through experiments or numerical simulations (lattice QCD). Hadrons like the proton and neutron are therefore surprisingly complex objects.

Even though protons and neutrons are colour neutral, there is a residual strong interaction between them (and other hadrons). For the nucleons this interaction is known as the strong nuclear force and leads to the formation of atomic nuclei. As a residual interaction it is comparable to the electromagnetic-chemical interaction between neutral atoms (atoms can be thought of as electromagnetic bound states), leading to the formation of molecules.

The quarks (**valence quarks**) that constitute a hadron, together with virtual gluons and **sea quarks**, which are virtual quark-antiquark pairs produced by virtual gluons, are collectively known as **partons**.

The decomposition of a hadron into partons (more precisely, the fraction of the hadron momentum carried by each type of parton) is complicated and is described by the **parton distribution function (PDF)** (or parton density function) of the hadron: the PDF at a given energy scale Q for a given type a of parton (quark flavour or gluon) is a probability density function $f_a(x, Q)$. The probability for finding a parton of type a carrying a fraction $x_a \in [x, x + dx]$ of the longitudinal momentum of the hadron is $f_a(x, Q)dx$ (see [42]). The PDF of the proton is very important for the correct interpretation of scattering experiments at colliders like the LHC. At present PDFs have to be determined experimentally and cannot be calculated from first principles, because they involve non-perturbative QCD [137].

The full gauge group of the Standard Model is therefore

$$G = \mathrm{SU}(3)_C \times \mathrm{SU}(2)_L \times \mathrm{U}(1)_Y.$$

The group $\mathrm{SU}(3)_C$ acts trivially on the Higgs vector space $W = \mathbb{C}^2$ and thus leaves the vacuum vector w_0 invariant. It follows that the gluons are *unbroken, massless* gauge bosons. The full unbroken gauge group of the Standard Model, the isotropy group of w_0, is

$$H = \mathrm{SU}(3)_C \times \mathrm{U}(1)_Q.$$

The full symmetry group of the Standard Model in the sense of Remark 6.12.7 is therefore the group

$$\mathrm{Spin}^+(1,3) \times \mathrm{SU}(3)_C \times \mathrm{SU}(2)_L \times \mathrm{U}(1)_Y$$

or, equivalently,

$$\mathrm{Spin}^+(3,1) \times \mathrm{SU}(3)_C \times \mathrm{SU}(2)_L \times \mathrm{U}(1)_Y,$$

depending on the signature chosen for Minkowski spacetime.

A very detailed discussion of QCD including theoretical and experimental aspects, especially concerning perturbative QCD, can be found in the book [42].

8.4.1 Basis for $\mathfrak{su}(3)_C$

Recall from Example 2.1.49 that $\mathfrak{su}(3)$ can be described as the vector space of skew-Hermitian tracefree matrices X, with adjoint representation given by

$$\mathrm{Ad}_Q X = Q \cdot X \cdot Q^{-1}, \quad Q \in \mathrm{SU}(3), X \in \mathfrak{su}(3).$$

We can identify complex 3×3-matrices X with elements of

$$\mathrm{End}\left(\mathbb{C}^3\right) = \mathbb{C}^3 \otimes \mathbb{C}^{3*}.$$

The standard Hermitian scalar product on $\mathbb{C}^3$ is invariant under the fundamental representation of SU(3). This implies that

$$\mathbb{C}^{3*} \cong \bar{\mathbb{C}}^3$$

as SU(3)-representations, hence

$$\mathrm{End}\left(\mathbb{C}^3\right) \cong \mathbb{C}^3 \otimes \bar{\mathbb{C}}^3.$$

For the fundamental representation $\mathbb{C}^3$ of $\mathrm{SU}(3)_C$ we choose the basis

$$r = \begin{pmatrix} 1 \\ 0 \\ 0 \end{pmatrix} \quad \text{“red”}, \qquad g = \begin{pmatrix} 0 \\ 1 \\ 0 \end{pmatrix} \quad \text{“green”}, \qquad b = \begin{pmatrix} 0 \\ 0 \\ 1 \end{pmatrix} \quad \text{“blue”}, \tag{8.6}$$

and write a general element of $\mathbb{C}^3$ as

$$\begin{pmatrix} z^r \\ z^g \\ z^b \end{pmatrix}.$$

We call $\mathbb{C}^3$ with this basis **colour space**. A basis of $\bar{\mathbb{C}}^3$ is given by the corresponding vectors

$$\bar{r} = \begin{pmatrix} 1 \\ 0 \\ 0 \end{pmatrix} \quad \text{“antired”}, \qquad \bar{g} = \begin{pmatrix} 0 \\ 1 \\ 0 \end{pmatrix} \quad \text{“antigreen”}, \qquad \bar{b} = \begin{pmatrix} 0 \\ 0 \\ 1 \end{pmatrix} \quad \text{“antiblue”}. \tag{8.7}$$

Hence under the canonical antilinear isomorphism $\mathbb{C}^3 \rightarrow \bar{\mathbb{C}}^3$, every colour gets mapped to its anticolour.

It follows that a complex basis of $\text{End}\left(\mathbb{C}^3\right)$ is given by the following nine matrices

$$r\bar{r}, \quad r\bar{g}, \quad r\bar{b}, \quad g\bar{r}, \quad g\bar{g}, \quad g\bar{b}, \quad b\bar{r}, \quad b\bar{g}, \quad b\bar{b},$$

where

$$r\bar{r} = r \otimes \bar{r} = \begin{pmatrix} 1 & 0 & 0 \\ 0 & 0 & 0 \\ 0 & 0 & 0 \end{pmatrix},$$

$$g\bar{b} = g \otimes \bar{b} = \begin{pmatrix} 0 & 0 & 0 \\ 0 & 0 & 1 \\ 0 & 0 & 0 \end{pmatrix},$$

$$\vdots$$

Since $\mathfrak{su}(3)$ is equal to the vector space of skew-Hermitian, tracefree elements of $\text{End}\left(\mathbb{C}^3\right)$, we get with the Gell-Mann matrices λ_a from Example 1.5.33:

Proposition 8.4.1 *A real basis of* $\mathfrak{su}(3)$ *is given by the matrices* $\nu_a = \frac{i}{2}\lambda_a$, *which can be written as*

$$\begin{aligned}
\nu_1 &= \frac{i}{2}(r\bar{g} + g\bar{r}), \\
\nu_2 &= \frac{1}{2}(r\bar{g} - g\bar{r}), \\
\nu_3 &= \frac{i}{2}(r\bar{r} - g\bar{g}), \\
\nu_4 &= \frac{i}{2}(r\bar{b} + b\bar{r}), \\
\nu_5 &= \frac{1}{2}(r\bar{b} - b\bar{r}), \\
\nu_6 &= \frac{i}{2}(g\bar{b} + b\bar{g}), \\
\nu_7 &= \frac{1}{2}(g\bar{b} - b\bar{g}), \\
\nu_8 &= \frac{i}{2\sqrt{3}}(r\bar{r} + g\bar{g} - 2b\bar{b}).
\end{aligned}$$

Lemma 8.4.2 *The basis vectors ν_a are orthonormal with respect to the positive definite scalar product*

$$\langle \cdot, \cdot \rangle_{\mathfrak{su}(3)} = -\frac{1}{3} B_{\mathfrak{su}(3)}(X, Y) = -2\mathrm{tr}(X \cdot Y),$$

where $B_{\mathfrak{su}(3)}$ is the Killing form of $\mathfrak{su}(3)$.

Proof This is Exercise 8.4.2. Compare with Exercises 2.7.16 and 2.7.17. □

It is sometimes convenient to consider the complexification of the Lie algebra $\mathfrak{su}(3)$:

Proposition 8.4.3 *A complex basis of $\mathfrak{su}(3) \otimes \mathbb{C}$ is given by the tracefree matrices*

$$\begin{aligned}
\mu_1 &= \frac{1}{\sqrt{2}} r\bar{g} = \frac{1}{\sqrt{2}}(\nu_2 - i\nu_1), \\
\mu_2 &= \frac{1}{\sqrt{2}} r\bar{b} = \frac{1}{\sqrt{2}}(\nu_5 - i\nu_4), \\
\mu_3 &= \frac{1}{\sqrt{2}} g\bar{r} = \frac{1}{\sqrt{2}}(-\nu_2 - i\nu_1), \\
\mu_4 &= \frac{1}{\sqrt{2}} g\bar{b} = \frac{1}{\sqrt{2}}(\nu_7 - i\nu_6), \\
\mu_5 &= \frac{1}{\sqrt{2}} b\bar{r} = \frac{1}{\sqrt{2}}(-\nu_5 - i\nu_4), \\
\mu_6 &= \frac{1}{\sqrt{2}} b\bar{g} = \frac{1}{\sqrt{2}}(-\nu_7 - i\nu_6), \\
\mu_7 &= \frac{1}{2}(r\bar{r} - g\bar{g}) = -i\nu_3, \\
\mu_8 &= \frac{1}{2\sqrt{3}}(r\bar{r} + g\bar{g} - 2b\bar{b}) = -i\nu_8.
\end{aligned}$$

These matrices are orthonormal with respect to the Hermitian *scalar product defined by the complexification of $-\frac{1}{3}B_{\mathfrak{su}(3)}$.*

Definition 8.4.4 The scalar product on $\mathfrak{su}(3)_C$ is defined by

$$\langle \cdot, \cdot \rangle_{\mathfrak{su}(3)_C} = -\frac{1}{3g_s^2} B_{\mathfrak{su}(3)}(X, Y) = -\frac{2}{g_s^2}\mathrm{tr}(X \cdot Y),$$

where g_s is the **strong coupling constant**. In particular, the scalar product is Ad-invariant. Orthonormal bases for $\mathfrak{su}(3)$ and $\mathfrak{su}(3) \otimes \mathbb{C}$ (with the complexification of the scalar product) are given by the vectors

$$g_s \nu_a$$

and

$$g_s \mu_a.$$

The choices of basis and scalar product are for QCD less standardized than for the electroweak interaction. We can expand the gluon gauge field G_μ with values in $\mathfrak{su}(3)_C$ in either of the bases $\{\nu_a\}, \{\mu_a\}$:

$$\begin{aligned} G_\mu &= \sum_{a=1}^{8} G_\mu^a g_s \nu_a = \sum_{a=1}^{8} G_\mu^a \frac{i}{2} g_s \lambda_a \\ &= \sum_{a=1}^{8} G_\mu'^a g_s \mu_a \\ &= \frac{1}{\sqrt{2}} g_s \Big(G_\mu'^{r\bar{g}} r\bar{g} + G_\mu'^{r\bar{b}} r\bar{b} + G_\mu'^{g\bar{r}} g\bar{r} + G_\mu'^{g\bar{b}} g\bar{b} + G_\mu'^{b\bar{r}} b\bar{r} + G_\mu'^{b\bar{g}} b\bar{g} \\ &\quad + G_\mu'^{r\bar{r}-g\bar{g}} \frac{1}{\sqrt{2}} (r\bar{r} - g\bar{g}) + G_\mu'^{r\bar{r}+g\bar{g}-2b\bar{b}} \frac{1}{\sqrt{3}} (r\bar{r} + g\bar{g} - 2b\bar{b}) \Big). \end{aligned} \tag{8.8}$$

8.5 The Particle Content of the Standard Model

8.5.1 *Fermions*

We want to add fermions, i.e. matter particles, to the Standard Model. In general, charged fermions, which couple to the gauge fields, are described by twisted chiral spinors, i.e. sections of twisted chiral spinor bundles

$$(S \otimes E)_+ = (S_L \otimes F_L) \oplus (S_R \otimes F_R), \tag{8.9}$$

where S_L is the left-handed and S_R the right-handed Weyl spinor bundle over 4-dimensional flat Minkowski spacetime M. The bundles F_L and F_R are associated vector bundles defined by complex unitary representations V_L and V_R of the gauge group G. We now want to describe the representations of G in the Standard Model, where

$$G = \mathrm{SU}(3)_C \times \mathrm{SU}(2)_L \times \mathrm{U}(1)_Y.$$

The complex vector spaces V_L and V_R have dimensions

$$\dim V_L = 24,$$
$$\dim V_R = 21.$$

As representations of G they decompose into orthogonal sums

$$V_L = V_L^1 \oplus V_L^2 \oplus V_L^3,$$
$$V_L = V_R^1 \oplus V_R^2 \oplus V_R^3$$

of G-subrepresentations V_L^i, V_R^i, $i = 1, 2, 3$, called the three **generations** or **families**. The generations have dimensions

$$\dim V_L^i = 8,$$
$$\dim V_R^i = 7.$$

The left-handed generations V_L^i for $i = 1, 2, 3$ are all isomorphic as G-representations and the same is true for the right-handed generations V_R^i. Each generation again decomposes into orthogonal sums

$$V_L^i = Q_L^i \oplus L_L^i,$$
$$V_R^i = Q_R^i \oplus L_R^i$$

of G-subrepresentations Q_L^i, Q_R^i and L_L^i, L_R^i, called **quark sectors** and **lepton sectors**. They have dimensions

$$\dim Q_L^i = 6,$$
$$\dim Q_R^i = 6,$$
$$\dim L_L^i = 2,$$
$$\dim L_R^i = 1.$$

Again the left-handed (right-handed) quark sectors are all isomorphic and the same is true for the left-handed (right-handed) lepton sectors across generations.

We denote by $\mathbb{C}^3$ the fundamental representation of $\mathrm{SU}(3)_C$ and by $\mathbb{C}^2$ the fundamental representation of $\mathrm{SU}(2)_L$, both with the standard invariant Hermitian scalar product. For both Lie groups we denote by $\mathbb{C}$ the trivial 1-dimensional representation. We also denote by $\mathbb{C}_y$ the representation of $\mathrm{U}(1)_Y$ where the generator $\beta_4 \in \mathfrak{u}(1)_Y$ acts as

$$\begin{aligned} \beta_4 \colon \mathbb{C} &\longrightarrow \mathbb{C} \\ z &\longmapsto ig'\frac{y}{2}z. \end{aligned}$$

Table 8.1 Fermion sectors

Sector	Representation	Physics notation	Complex dimension
Q_L^i	$\mathbb{C}^3 \otimes \mathbb{C}^2 \otimes \mathbb{C}_{1/3}$	$(\mathbf{3}, \mathbf{2})_{1/3}$	6
Q_R^i	$(\mathbb{C}^3 \otimes \mathbb{C} \otimes \mathbb{C}_{4/3}) \oplus (\mathbb{C}^3 \otimes \mathbb{C} \otimes \mathbb{C}_{-2/3})$	$(\mathbf{3}, \mathbf{1})_{4/3} \oplus (\mathbf{3}, \mathbf{1})_{-2/3}$	6
L_L^i	$\mathbb{C} \otimes \mathbb{C}^2 \otimes \mathbb{C}_{-1}$	$(\mathbf{1}, \mathbf{2})_{-1}$	2
L_R^i	$\mathbb{C} \otimes \mathbb{C} \otimes \mathbb{C}_{-2}$	$(\mathbf{1}, \mathbf{1})_{-2}$	1

The bold integers denote representations of certain dimensions. The usage of bold face seems to be standard in the physics literature; see also Definition 2.1.18

We note the following:

Lemma 8.5.1 *An element* $\alpha \in \mathrm{U}(1)_Y$ *acts on* $\mathbb{C}_y$ *by*

$$\begin{aligned} \mathbb{C}_y &\longrightarrow \mathbb{C}_y \\ z &\longmapsto \alpha^{3y} z. \end{aligned}$$

This representation is well-defined for all weak hypercharges y which are integer multiples of $\frac{1}{3}$. *The representation* $\mathbb{C}_y$ *has winding number* $3y$.

For instance, the representation $\mathbb{C}_{4/3}$ has winding number 4. Table 8.1 describes the G-representations of the quark and lepton sectors. As G-representations the vector spaces V_L and V_R thus decompose into irreducible representations of dimensions

$$\begin{aligned} \dim V_L &= 24 = (6+2) + (6+2) + (6+2), \\ \dim V_R &= 21 = (3+3+1) + (3+3+1) + (3+3+1). \end{aligned}$$

Interestingly, we see that the four simplest representations of the Lie group $\mathrm{SU}(3) \times \mathrm{SU}(2)$

$$(\mathbf{3}, \mathbf{2}), \quad (\mathbf{3}, \mathbf{1}), \quad (\mathbf{1}, \mathbf{2}) \quad \text{and} \quad (\mathbf{1}, \mathbf{1})$$

all appear in the Standard Model.

We now define bases for these representations, see Table 8.2. The basis for the quark sectors Q_L^i and Q_R^i are obtained as the tensor product of the basis vectors r, g, b from Eq. (8.6) and the basis vectors for the $\mathrm{SU}(2)_L \times \mathrm{U}(1)_Y$-representations. For both quarks and leptons the bases for the $\mathrm{SU}(2)_L \times \mathrm{U}(1)_Y$-representations are defined so that they consist of simultaneous eigenvectors for both charge operators $\mathbb{T}_3$ and $\mathbb{Y}$. We list these basis elements for the first generation together with their weak isospin, weak hypercharge and electric charge (recall that $\mathbb{Q} = \mathbb{T}_3 + \frac{\mathbb{Y}}{2}$).

Table 8.2 Fermion representations

Sector	$SU(2)_L \times U(1)_Y$ representation	Basis vectors	Particle	Charges		
				$\mathbb{T}_3$	$\mathbb{Y}$	$\mathbb{Q}$
Q_L^1	$\mathbb{C}^2 \otimes \mathbb{C}_{1/3}$	$\begin{pmatrix} 1 \\ 0 \end{pmatrix}$	u_L	$\frac{1}{2}$	$\frac{1}{3}$	$\frac{2}{3}$
		$\begin{pmatrix} 0 \\ 1 \end{pmatrix}$	d'_L	$-\frac{1}{2}$	$\frac{1}{3}$	$-\frac{1}{3}$
Q_R^1	$\mathbb{C} \otimes \mathbb{C}_{4/3}$	1	u_R	0	$\frac{4}{3}$	$\frac{2}{3}$
	$\mathbb{C} \otimes \mathbb{C}_{-2/3}$	1	d'_R	0	$-\frac{2}{3}$	$-\frac{1}{3}$
L_L^1	$\mathbb{C}^2 \otimes \mathbb{C}_{-1}$	$\begin{pmatrix} 1 \\ 0 \end{pmatrix}$	ν_{eL}	$\frac{1}{2}$	-1	0
		$\begin{pmatrix} 0 \\ 1 \end{pmatrix}$	e_L	$-\frac{1}{2}$	-1	-1
L_R^1	$\mathbb{C} \otimes \mathbb{C}_{-2}$	1	e_R	0	-2	-1

Remark 8.5.2 In references that use the convention $n_Y = 6$ instead of our $n_Y = 3$, the value of the hypercharge is one half of the value of our hypercharge Y.

It follows that the left-handed quarks and leptons of each generation form **isospin doublets (isodoublets)**, while the right-handed quarks and leptons are **isospin singlets (isosinglets)**. The quarks are at the same time **colour triplets**, while the leptons are **colour singlets**.

In the fourth column in Table 8.2 we state the corresponding names for the particles of specific weak isospin. We write

$$u_L, d'_L = \psi_L \otimes \begin{pmatrix} z^r \\ z^g \\ z^b \end{pmatrix} : M \longrightarrow \Delta_L \otimes \mathbb{C}^3,$$

$$u_R, d'_R = \psi_R \otimes \begin{pmatrix} z^r \\ z^g \\ z^b \end{pmatrix} : M \longrightarrow \Delta_R \otimes \mathbb{C}^3,$$

where ψ_L, ψ_R are arbitrary maps to the Weyl spinor spaces Δ_L, Δ_R and

$$\begin{pmatrix} z^r \\ z^g \\ z^b \end{pmatrix} \in \mathbb{C}^3$$

is a general element of the colour space. Here

$$d'_L = \psi_L \otimes r$$

is read as a "red left-handed down quark" and

$$u_R = \psi_R \otimes g$$

is read as a "green right-handed up quark". Similarly

$$\begin{aligned} e_L, \nu_{eL} &= \psi_L : M \longrightarrow \Delta_L \otimes \mathbb{C}, \\ e_R &= \psi_R : M \longrightarrow \Delta_R \otimes \mathbb{C} \end{aligned}$$

are the leptons. A general map to

$$\Delta_L \otimes V_L^1 = \left(\Delta_L \otimes Q_L^1\right) \oplus \left(\Delta_L \otimes L_L^1\right)$$

can then be written as

$$\begin{pmatrix} u_L \\ d'_L \end{pmatrix} \oplus \begin{pmatrix} \nu_{eL} \\ e_L \end{pmatrix}$$

and a general map to

$$\Delta_R \otimes V_R^1 = \left(\Delta_R \otimes Q_R^1\right) \oplus \left(\Delta_R \otimes L_R^1\right)$$

as

$$\left(u_R \oplus d'_R\right) \oplus e_R.$$

For the second generation we make the following replacements:

$$\begin{aligned} u &\to c \\ d' &\to s' \\ \nu_e &\to \nu_\mu \\ e &\to \mu. \end{aligned} \tag{8.10}$$

For the third generation we make the following replacements:

$$\begin{aligned} u &\to t \\ d' &\to b' \\ \nu_e &\to \nu_\tau \\ e &\to \tau. \end{aligned} \tag{8.11}$$

Table 8.3 Names of the fermions

Type	First generation	Second generation	Third generation
Quark	u Up	c Charm	t Top (Truth)
	d' Down	s' Strange	b' Bottom (Beauty)
Lepton	e Electron	μ Muon	τ Tau
	ν_e Electron neutrino	ν_μ Muon neutrino	ν_τ Tau neutrino

The representations and charges stay the same. Table 8.3 lists the names of these particles. The different types u, d, c, s, t, b of quarks are called **quark flavours** and the different types $e, \nu_e, \mu, \nu_\mu, \tau, \nu_\tau$ are called **lepton flavours**. The prime $'$ on the down-type quarks will be explained in Sect. 8.8.2. The electron, muon and tau are collectively known as the **charged leptons**. Note that in the Standard Model there are **no right-handed neutrinos**.

The remarkable exact repetition (except for the masses) of the first generation in two more generations cannot be explained in the Standard Model. In some theories beyond the Standard Model a right-handed neutrino singlet is added to each generation, making the leptons very similar to the quarks concerning the structure of SU(2) representations (the weak hypercharges and thus the electric charges are different); see Sect. 9.2.1.

Remark 8.5.3 The quark model was developed by Murray Gell-Mann (Nobel Prize in Physics 1969) and independently by George Zweig in 1964, originally containing only the up, down and strange flavour.

The Yang–Mills SU(3) gauge theory of the strong interaction, containing coloured quarks and a colour octet of gluons, was proposed in 1973 by Harald Fritzsch, Murray Gell-Mann and Heinrich Leutwyler. The bottom and top quark were postulated by Makoto Kobayashi and Toshihide Maskawa in 1972 (Nobel Prize in Physics 2008). The last quark in the three generations, the top quark, was experimentally observed for the first time in 1995 at the Collider Detector at Fermilab (CDF).

Remark 8.5.4 The fact that the $\mathrm{SU}(2)_L \times \mathrm{U}(1)_Y$-representations for left-handed and right-handed fermions are different implies that they interact differently with the W- and Z-bosons and thus the weak interaction is not invariant under inversion of parity (handedness). This was first predicted theoretically in 1956 by Tsung-Dao Lee and Chen Ning Yang (Nobel Prize in Physics 1957) and verified experimentally by Chien-Shiung Wu in 1957.

8.5.2 Antiparticles

Every fermion has an **antiparticle**. Antiparticles are sections of the complex conjugate bundle

$$\overline{(S \otimes E)_+} = \overline{(S_L \otimes F_L)} \oplus \overline{(S_R \otimes F_R)}.$$

We note the following useful fact:

Lemma 8.5.5 *There are complex linear isomorphisms of Lorentz spin representations*

$$\Delta_L \cong \Delta_L^* \cong \overline{\Delta}_R,$$
$$\Delta_R \cong \Delta_R^* \cong \overline{\Delta}_L,$$

where the first isomorphism is given by the Majorana form $(\cdot, \cdot)$ *and the second isomorphism by the Dirac form* $\langle \cdot, \cdot \rangle$.

In fact, these isomorphisms are given by the map τ from Lemma 6.7.17

$$\tau \colon \Delta \longrightarrow \overline{\Delta}$$
$$\psi \longmapsto \psi^C = B^{-1}\psi^*$$

where

$$B = CA$$

for the unitary matrices C and A from Sect. 6.8 and we used the notation for the charge conjugate from Eq. (6.3).

We can compare this with the constructions in Sect. 2.1.3, where we essentially defined Δ_R as $\overline{\Delta}_L^*$. We now understand that this isomorphism comes from the Dirac form $\langle \cdot, \cdot \rangle$. The second isomorphisms in Lemma 8.5.5 are given by the matrix ϵ, corresponding to the matrix C defining the Majorana form $(\cdot, \cdot)$.

If we set

$$V_L^C = \overline{V_R},$$
$$V_R^C = \overline{V_L}$$

and extend this notation to the representations Q_L, Q_R, L_L, L_R and the bundles F_L, F_R, we get from Lemma 8.5.5 that

$$\overline{(S \otimes E)_+} \cong (S_L \otimes F_L^C) \oplus (S_R \otimes F_R^C). \tag{8.12}$$

Table 8.4 Antifermion sectors

Sector	Representation	Physics notation	Complex dimension
Q_R^{iC}	$\bar{\mathbb{C}}^3 \otimes \bar{\mathbb{C}}^2 \otimes \mathbb{C}_{-1/3}$	$(\bar{\mathbf{3}}, \bar{\mathbf{2}})_{-1/3}$	6
Q_L^{iC}	$(\bar{\mathbb{C}}^3 \otimes \mathbb{C} \otimes \mathbb{C}_{-4/3}) \oplus (\bar{\mathbb{C}}^3 \otimes \mathbb{C} \otimes \mathbb{C}_{2/3})$	$(\bar{\mathbf{3}}, \mathbf{1})_{-4/3} \oplus (\bar{\mathbf{3}}, \mathbf{1})_{2/3}$	6
L_R^{iC}	$\mathbb{C} \otimes \bar{\mathbb{C}}^2 \otimes \mathbb{C}_1$	$(\mathbf{1}, \bar{\mathbf{2}})_1$	2
L_L^{iC}	$\mathbb{C} \otimes \mathbb{C} \otimes \mathbb{C}_2$	$(\mathbf{1}, \mathbf{1})_2$	1

The bold integers denote representations of certain dimensions. The usage of bold face seems to be standard in the physics literature; see also Definition 2.1.18

Then each generation of antiparticles is described by the representations in Table 8.4. Under the complex antilinear isomorphisms

$$\Delta_L \otimes V_L \longrightarrow \overline{\Delta_L \otimes V_L} \cong \Delta_R \otimes V_R^C,$$

$$\Delta_R \otimes V_R \longrightarrow \overline{\Delta_R \otimes V_R} \cong \Delta_L \otimes V_L^C$$

we map

$$u_L \longmapsto u_R^C$$

$$u_R \longmapsto u_L^C$$

$$d'_L \longmapsto d'^C_R$$

$$d'_R \longmapsto d'^C_L$$

$$\nu_{eL} \longmapsto \nu_{eR}^C$$

$$e_L \longmapsto e_R^C$$

$$e_R \longmapsto e_L^C,$$

and similarly for the second and third generation. It is clear that charge conjugation $\psi \mapsto \psi^C = B^{-1}\psi^*$ is an involution,

$$(\psi^C)^C = \psi,$$

since the matrix B defines a real structure on the spinor space.

We then get the $\mathrm{SU}(2)_L \times \mathrm{U}(1)_Y$-representations in Table 8.5. There are corresponding representations for the second and third generation. The antiquark

$$t_L^C = \psi_L \otimes \bar{b},$$

for example, is read as "antiblue left-handed top antiquark". The antiparticle of the electron is called a positron. All other antiparticles are named with the prefix "anti".

Table 8.5 Antifermion representations

Sector	$SU(2)_L \times U(1)_Y$ representation	Basis vectors	Particle	Charges		
				$\mathbb{T}_3$	$\mathbb{Y}$	$\mathbb{Q}$
Q_R^{1C}	$\bar{\mathbb{C}}^2 \otimes \mathbb{C}_{-1/3}$	$\begin{pmatrix} 1 \\ 0 \end{pmatrix}$	u_R^C	$-\frac{1}{2}$	$-\frac{1}{3}$	$-\frac{2}{3}$
		$\begin{pmatrix} 0 \\ 1 \end{pmatrix}$	$d_R'^C$	$\frac{1}{2}$	$-\frac{1}{3}$	$\frac{1}{3}$
Q_L^{1C}	$\mathbb{C} \otimes \mathbb{C}_{-4/3}$	1	u_L^C	0	$-\frac{4}{3}$	$-\frac{2}{3}$
	$\mathbb{C} \otimes \mathbb{C}_{2/3}$	1	$d_L'^C$	0	$\frac{2}{3}$	$\frac{1}{3}$
L_R^{1C}	$\bar{\mathbb{C}}^2 \otimes \mathbb{C}_1$	$\begin{pmatrix} 1 \\ 0 \end{pmatrix}$	ν_{eR}^C	$-\frac{1}{2}$	1	0
		$\begin{pmatrix} 0 \\ 1 \end{pmatrix}$	e_R^C	$\frac{1}{2}$	1	1
L_L^{1C}	$\mathbb{C} \otimes \mathbb{C}_2$	1	e_L^C	0	2	1

Table 8.6 Left-handed and right-handed particles and antiparticles

Left-handed fermions and antifermions	$\begin{pmatrix} u_L \\ d_L' \end{pmatrix}$	u_L^C	$d_L'^C$	$\begin{pmatrix} \nu_{eL} \\ e_L \end{pmatrix}$	e_L^C
Right-handed fermions and antifermions	$\begin{pmatrix} u_R^C \\ d_R'^C \end{pmatrix}$	u_R	d_R'	$\begin{pmatrix} \nu_{eR}^C \\ e_R^C \end{pmatrix}$	e_R

8.5.3 *Chirality of the Standard Model*

It is sometimes useful to separate fermions not into *particles* and *antiparticles*, but into *left-handed* and *right-handed* particles and antiparticles as in Table 8.6. Each generation of left-handed particles and antiparticles is described by the representation

$$\begin{aligned} V_L^i \oplus V_L^{iC} &= Q_L^i \oplus L_L^i \oplus Q_L^{iC} \oplus L_L^{iC} \\ &= \left(\mathbb{C}^3 \otimes \mathbb{C}^2 \otimes \mathbb{C}_{1/3}\right) \oplus \left(\mathbb{C} \otimes \mathbb{C}^2 \otimes \mathbb{C}_{-1}\right) \\ &\quad \oplus \left(\bar{\mathbb{C}}^3 \otimes \mathbb{C} \otimes \mathbb{C}_{-4/3}\right) \oplus \left(\bar{\mathbb{C}}^3 \otimes \mathbb{C} \otimes \mathbb{C}_{2/3}\right) \oplus \left(\mathbb{C} \otimes \mathbb{C} \otimes \mathbb{C}_2\right) \end{aligned}$$

and each generation of right-handed particles and antiparticles is described by the representation

$$\begin{aligned} V_R^i \oplus V_R^{iC} &= Q_R^i \oplus L_R^i \oplus Q_R^{iC} \oplus L_R^{iC} \\ &= \left(\mathbb{C}^3 \otimes \mathbb{C} \otimes \mathbb{C}_{4/3}\right) \oplus \left(\mathbb{C}^3 \otimes \mathbb{C} \otimes \mathbb{C}_{-2/3}\right) \oplus \left(\mathbb{C} \otimes \mathbb{C} \otimes \mathbb{C}_{-2}\right) \\ &\quad \oplus \left(\bar{\mathbb{C}}^3 \otimes \bar{\mathbb{C}}^2 \otimes \mathbb{C}_{-1/3}\right) \oplus \left(\mathbb{C} \otimes \bar{\mathbb{C}}^2 \otimes \mathbb{C}_1\right). \end{aligned}$$

These representations of $G = \mathrm{SU}(3)_C \times \mathrm{SU}(2)_L \times \mathrm{U}(1)_Y$ both have dimension 15, but are not isomorphic as complex representations. This follows from

$$\begin{aligned} \bar{\mathbb{C}}^3 &\ncong \mathbb{C}^3, \\ \mathbb{C}_{-y} &\ncong \mathbb{C}_y, \end{aligned}$$

even though

$$\bar{\mathbb{C}}^2 \cong \mathbb{C}^2$$

(see Exercise 2.7.3 and the remark following it). Hence we can say that the Standard Model is a **chiral gauge theory** in the following sense:

Definition 8.5.6 A gauge theory with fermions and gauge group G is called **chiral** if the G-representation for the right-handed particles and antiparticles is not complex linearly isomorphic to the G-representation for the left-handed particles and antiparticles.

Note that in any case

$$V_R^i \oplus V_R^{iC} \cong \overline{V_L^i \oplus V_L^{iC}}.$$

This implies that a gauge theory is chiral if the representation $V_L^i \oplus V_L^{iC}$ for the left-handed particles and antiparticles is not isomorphic to its complex conjugate representation. It follows that every gauge theory that aims at describing realistic physics has to have a gauge group that admits a complex representation not isomorphic to its complex conjugate. Such representations are sometimes called *complex* (in a different sense!). This is an important restriction on the possible gauge groups of Grand Unified Theories (GUTs).

The complete left-handed representations for fermions and antifermions of one generation is (in the physics notation)

$$V_L^i \oplus V_L^{iC} = (\mathbf{3}, \mathbf{2})_{1/3} \oplus (\bar{\mathbf{3}}, \mathbf{1})_{-4/3} \oplus (\bar{\mathbf{3}}, \mathbf{1})_{2/3} \oplus (\mathbf{1}, \mathbf{2})_{-1} \oplus (\mathbf{1}, \mathbf{1})_2. \quad (8.13)$$

Every realistic theory beyond the Standard Model should recover this representation. The complete right-handed representation is the complex conjugate of the left-handed representation.

Table 8.7 Higgs sector

Higgs vector space	Representation	Physics notation	Complex dimension
W	$\mathbb{C} \otimes \mathbb{C}^2 \otimes \mathbb{C}_1$	$(\mathbf{1}, \mathbf{2})_1$	2

The bold integers denote representations of certain dimensions. The usage of bold face seems to be standard in the physics literature; see also Definition 2.1.18

Table 8.8 Higgs field representation

Sector	$SU(2)_L \times U(1)_Y$ representation	Basis vectors	Particle	Charges		
				$\mathbb{T}_3$	$\mathbb{Y}$	$\mathbb{Q}$
W	$\mathbb{C}^2 \otimes \mathbb{C}_1$	$\begin{pmatrix} 1 \\ 0 \end{pmatrix}$	ϕ^+	$\frac{1}{2}$	1	1
		$\begin{pmatrix} 0 \\ 1 \end{pmatrix}$	ϕ^0	$-\frac{1}{2}$	1	0

8.5.4 Higgs Field

We saw above that the Higgs bundle is the vector bundle

$$E = \underline{\mathbb{C}} \otimes E \tag{8.14}$$

associated to the principal bundle P via a unitary representation on W. Here $\underline{\mathbb{C}}$ denotes the trivial complex line bundle associated to the trivial (scalar) representation of the Lorentz spin group. The representation W is given by Table 8.7. A basis with corresponding charges can be found in Table 8.8. The vector f corresponding to the Higgs boson is an element of $T_{w_0}W$ given by

$$f = \begin{pmatrix} 0 \\ 1 \end{pmatrix}.$$

The Higgs boson therefore also has

$$\mathbb{T}_3 = -\frac{1}{2}, \quad \mathbb{Y} = 1, \quad \mathbb{Q} = 0.$$

8.5.5 Gauge Fields

We finally want to summarize the representations of the gauge fields in the Standard Model. The gauge theory of the Standard Model is defined by a (trivial) principal G-bundle P over 4-dimensional Minkowski spacetime for the Lie group

$$G = SU(3)_C \times SU(2)_L \times U(1)_Y.$$

A connection A on the principal bundle P is a 1-form with values in the Lie algebra

$$\mathfrak{g} = \mathfrak{su}(3)_C \oplus \mathfrak{su}(2)_L \oplus \mathfrak{u}(1)_Y.$$

We decompose A accordingly into gauge fields

$$A = (G, W, B) = G + W + B,$$

which we call the **gluon gauge field**, the **weak gauge field** and the **hypercharge gauge field** (G is the standard notation for the gluon gauge field, not to be confused with the Lie group G).

We know from Sect. 5.13 that the difference ΔA of an arbitrary gauge field A minus a fixed reference gauge field A_0 can be thought of as a 1-form on spacetime M with values in the vector bundle $\mathrm{Ad}_\mathfrak{g}(P)$ associated to P via the adjoint representation of G on the Lie algebra $\mathfrak{g}$. Hence ΔA is a section of the twisted vector bundle

$$T^*M \otimes \mathrm{Ad}_\mathfrak{g}(P).$$

We write

$$\Delta A = \Delta G + \Delta W + \Delta B.$$

The adjoint representation of the Standard Model gauge group G on the Lie algebra $\mathfrak{g}$ splits into three orthogonal subrepresentations

$$\mathfrak{g} = \mathfrak{su}(3)_C \oplus \mathfrak{su}(2)_L \oplus \mathfrak{u}(1)_Y,$$

called the gluon sector, the weak sector and the hypercharge sector. Hence the bundle $T^*M \otimes \mathrm{Ad}_\mathfrak{g}(P)$ decomposes into a direct sum of twisted bundles

$$\left(T^*M \otimes \mathrm{Ad}_{\mathfrak{su}(3)_C}(P)\right) \oplus \left(T^*M \otimes \mathrm{Ad}_{\mathfrak{su}(2)_L}(P)\right) \oplus \left(T^*M \otimes \mathrm{Ad}_{\mathfrak{u}(1)_Y}(P)\right). \quad (8.15)$$

Let $\mathbb{R}^8$ denote the adjoint representation of $\mathrm{SU}(3)_C$, $\mathbb{R}^3$ the adjoint representation of $\mathrm{SU}(2)_L$ and $\mathbb{R}$ the trivial representation. Then the corresponding G-representations defining the adjoint bundles are given by Table 8.9. In the electroweak sector $\mathfrak{su}(2)_L \oplus \mathfrak{u}(1)_Y$, it can be shown that the orthonormal basis

$$\frac{1}{\sqrt{2}}(\alpha_1 \pm i\alpha_2), \alpha_3, \alpha_4$$

that we have chosen in Sect. 8.3.3 consists of simultaneous eigenvectors for the charge operators $\mathbb{T}_3$ and $\mathbb{Y}$ (here we use the *complexified* adjoint representation). The charges are summarized in Table 8.10 (see Exercise 8.11.6). Note that the W-

Table 8.9 Gauge sectors

Gauge sector	Representation	Physics notation	Real dimension
$\mathfrak{su}(3)_C$	$\mathbb{R}^8 \otimes \mathbb{R} \otimes \mathbb{R}$	$(\mathbf{8},\mathbf{1})_\mathbf{0}$	8
$\mathfrak{su}(2)_L$	$\mathbb{R} \otimes \mathbb{R}^3 \otimes \mathbb{R}$	$(\mathbf{1},\mathbf{3})_\mathbf{0}$	3
$\mathfrak{u}(1)_Y$	$\mathbb{R} \otimes \mathbb{R} \otimes \mathbb{R}$	$(\mathbf{1},\mathbf{1})_\mathbf{0}$	1

The bold integers denote representations of certain dimensions. The usage of bold face seems to be standard in the physics literature; see also Definition 2.1.18

Table 8.10 Representation of electroweak gauge bosons

Gauge sector	$\mathrm{SU}(2)_L \times \mathrm{U}(1)_Y$ representation	Basis vectors	Boson	Charges		
				$\mathbb{T}_3$	$\mathbb{Y}$	$\mathbb{Q}$
$\mathfrak{su}(2)_L \oplus \mathfrak{u}(1)_Y$	$(\mathbb{C}^3 \otimes \mathbb{C}) \oplus (\mathbb{C} \otimes \mathbb{C})$	$\frac{(\alpha_1 + i\alpha_2)}{\sqrt{2}}$	W^+	1	0	1
		$\frac{(\alpha_1 - i\alpha_2)}{\sqrt{2}}$	W^-	-1	0	-1
		α_3	Z^0	0	0	0
		α_4	γ	0	0	0

bosons have both a non-zero weak isospin and electric charge. The basis vectors

$$\frac{1}{\sqrt{2}}(\alpha_1 \pm i\alpha_2), \alpha_3, \alpha_4$$

are orthonormal with respect to the Hermitian scalar product on

$$(\mathfrak{su}(2)_L \oplus \mathfrak{u}(1)_Y) \otimes \mathbb{C}$$

defined by the complexification of the positive definite real scalar product on $\mathfrak{su}(2)_L \oplus \mathfrak{u}(1)_Y$, for which $\alpha_1, \alpha_2, \alpha_3, \alpha_4$ form an orthonormal basis.

8.5.6 *The Total Particle Content of the Standard Model*

If we like, we could now define the total particle content of the Standard Model as sections of the direct sum of the bundles in Eqs. (8.9), (8.12), (8.14) and (8.15) (with complexified adjoint representation).

8.5.7 *Hypercharges: Constraints from Group Theory*

The specific assignments of the values of the weak hypercharge Y, which determine the representations of the group $\mathrm{U}(1)_Y$ in the Standard Model, are not arbitrary, but have a certain pattern that we explain in this subsection and the following. These assignments have important consequences. For example, it is well-known that the sum of the electric charge of the proton and the electron is zero: the proton consists of two up valence quarks and one down valence quark and has electric charge

$$\frac{2}{3}+\frac{2}{3}-\frac{1}{3}=+1,$$

while the electron has electric charge -1. This equality, fundamental to the existence of neutral atoms, holds even though in the Standard Model the electric charges for the quarks are *a priori* independent of the electric charges for the leptons.

We begin by describing purely group theoretic constraints on the hypercharges in this section and quantum constraints in the following. Remarkably, it turns out that the Standard Model can only define a consistent quantum theory if, in particular, the sum of the electric charge of the proton and the electron is zero.

A $\mathbb{Z}_6$-Subgroup of the Standard Model Group

We first note the following (see [9]):

Theorem 8.5.7 ($\mathbb{Z}_6$-Subgroup of the Standard Model Group) *The subgroup*

$$K \cong \mathbb{Z}_6 \subset G = \mathrm{SU}(3)_C \times \mathrm{SU}(2)_L \times \mathrm{U}(1)_Y$$

of elements of the form

$$\left(\alpha^2 I_3, \alpha^{-3} I_2, \alpha\right), \quad \text{with } \alpha \in \mathrm{U}(1), \alpha^6 = 1,$$

acts trivially on the representations V_L, V_R and V_L^C, V_R^C. Here I_2 and I_3 denote the unit matrices.

Conversely, suppose the subgroup $K \cong \mathbb{Z}_6$ of the Standard Model group is given and each of the representations

$$(\mathbf{3},\mathbf{2})_\mathbf{y}, \quad (\mathbf{3},\mathbf{1})_\mathbf{y}, \quad (\mathbf{1},\mathbf{2})_\mathbf{y} \quad \text{and} \quad (\mathbf{1},\mathbf{1})_\mathbf{y}$$

is invariant under K. Then in each case the hypercharge y is related to the hypercharge y_{SM} in the Standard Model by

$$3y \equiv 3y_{\mathrm{SM}} \mod 6.$$

Proof This is Exercise 8.11.7. □

The first part of this theorem implies:

Corollary 8.5.8 *The representations V_L, V_R and V_L^C, V_R^C of the Standard Model group G descend to representations of*

$$\left(\mathrm{SU}(3)_C \times \mathrm{SU}(2)_L \times \mathrm{U}(1)_Y\right)/\mathbb{Z}_6.$$

As we saw in Exercise 1.9.12, there is a natural embedding

$$(\mathrm{SU}(3)_C \times \mathrm{SU}(2)_L \times \mathrm{U}(1)_Y)/\mathbb{Z}_6 \subset \mathrm{SU}(5).$$

Corollary 8.5.8 is one of the reasons why a SU(5) theory of Grand Unification (GUT) is possible (for more details, see Sect. 9.5).

Charge Quantization

The Lie algebra of the Standard Model is

$$\mathfrak{g} = \mathfrak{su}(3)_C \oplus \mathfrak{su}(2)_L \oplus \mathfrak{u}(1)_Y.$$

Concerning charges in general, we need to distinguish between semisimple Lie algebras and abelian Lie algebras. For example, using representation theory it is possible to show that in any representation of the Lie algebra $\mathfrak{su}(2)_L$ the weak isospin must have values which are integer multiples of $\frac{1}{2}$. The possible charges are thus **quantized**. To make this plausible, note that the commutation relation

$$[\tau_a, \tau_b] = \epsilon_{abc}\tau_c$$

implies that a non-trivial representation

$$\phi\colon \mathfrak{su}(2)_L \longrightarrow \mathrm{End}(V)$$

does not yield a representation after a rescaling $\lambda \cdot \phi$, with $\lambda \neq 0, 1 \in \mathbb{C}$. Similarly the charges for any semisimple Lie algebra are quantized (the charges are related to the discrete *weight lattice*).

On the other hand, if we consider the abelian Lie algebra $\mathfrak{u}(1)_Y$, then representations

$$\phi\colon \mathfrak{u}(1)_Y \longrightarrow \mathrm{End}(V)$$

yield representations after arbitrary rescalings $\lambda \cdot \phi$ with $\lambda \in \mathbb{C}$. Hence all values of weak hypercharge (even irrational ones) are possible and the charges are not quantized. They are only quantized if the representations of $\mathfrak{u}(1)_Y$ come from representations of the compact circle $\mathrm{U}(1)_Y$ (for a suitable, fixed circumference).

This shows that on the level of Lie algebras, there is no reason why the values of the weak hypercharge and electric charge for all particles in the Standard Model should be quantized and, in particular, be multiples of $\frac{1}{3}$. The quantization, however, would follow naturally for algebraic reasons if the Lie algebra of the Standard Model is a Lie subalgebra of some larger (Grand Unified) compact simple (or semisimple) Lie algebra.

8.5.8 *Hypercharges: Constraints from Vanishing of Anomalies*

If we take into account the quantum field theory defined by the Standard Model, there are additional constraints that restrict the assignments of weak hypercharges. These constraints result from demanding that the Standard Model is free of **gauge anomalies**, i.e. that all gauge symmetries of the classical theory still persist in the quantum theory. This means that the derivative (4-divergence) of certain Green's functions (correlators) has to vanish.

It can be shown that in 4-dimensional spacetime the only possible non-zero contribution to the 4-divergence of these Green's functions comes from triangle Feynman diagrams of the form in Fig. 8.4, with three external gauge bosons and one fermion loop (Feynman diagrams involving fermions and gauge bosons will be explained in more detail in Sect. 8.6. The appearance of loop diagrams indicates that the anomalies are indeed a quantum effect.) We denote the various hypercharges in the Standard Model (for one generation) as follows:

Y_Q: hypercharge of left-handed quark isodoublet

Y_u: hypercharge of right-handed up-type quark isosinglet

Y_d: hypercharge of right-handed down-type quark isosinglet

Y_L: hypercharge of left-handed lepton isodoublet

Y_e: hypercharge of right-handed electron-type isosinglet

Y_ν: hypercharge of (hypothetical) right-handed neutrino isosinglet.

For the Standard Model with gauge group

$$G = \mathrm{SU}(3)_C \times \mathrm{SU}(2)_L \times \mathrm{U}(1)_Y$$

the anomalies depend on which of the factors of G the three gauge bosons in the triangle diagram belong to. We denote the anomalies accordingly by $\mathrm{U}(1)_Y^3$, $\mathrm{SU}(3)_C^2\mathrm{U}(1)_Y$, etc.

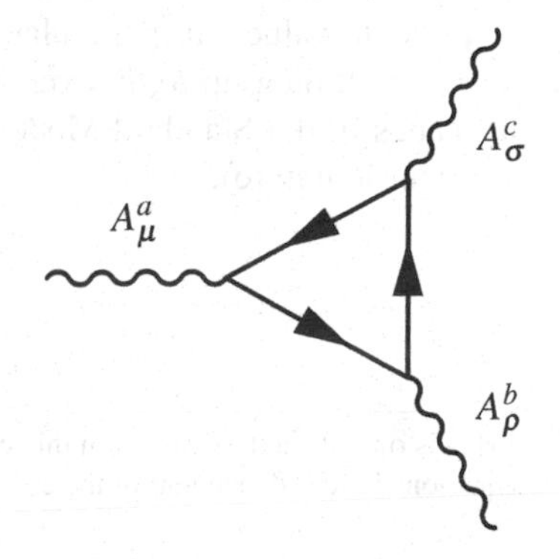

Fig. 8.4 Chiral anomaly

Table 8.11 Constraints on hypercharges from anomaly cancellations

Anomaly	Constraint
$U(1)_Y^3$	$(2Y_L^3 - Y_e^3 - Y_\nu^3) + 3(2Y_Q^3 - Y_u^3 - Y_d^3) = 0$
$SU(3)_C^2 U(1)_Y$	$2Y_Q - Y_u - Y_d = 0$
$SU(2)_L^2 U(1)_Y$	$Y_L + 3Y_Q = 0$
$\text{grav}^2 U(1)_Y$	$(2Y_L - Y_e - Y_\nu) + 3(2Y_Q - Y_u - Y_d) = 0$

Setting the gauge anomaly, also called the **chiral anomaly**, calculated from each of the triangle diagrams to zero, leads to constraints on the hypercharges that are summarized in Table 8.11 (the table is from [125, Sect. 30.4]; grav denotes a graviton).[1] All other anomalies, like the ones associated to $SU(3)_C^3$ or $SU(3)_C U(1)_Y^2$, vanish automatically.

Without trying to explain the calculation of these constraints in detail, the following is apparent from the table:

1. The assignments of hypercharges in the Standard Model

$$
\begin{aligned}
Y_Q &= \frac{1}{3} \\
Y_u &= \frac{4}{3} \\
Y_d &= -\frac{2}{3} \\
Y_L &= -1 \\
Y_e &= -2 \\
Y_\nu &= 0
\end{aligned}
$$

satisfy all constraints. Hence the Standard Model is **free of gauge anomalies**.

2. The first and third constraint are only satisfied if contributions from both quarks and leptons are taken together, i.e. these contributions to the triangle diagrams have to cancel each other.
3. Up to an overall factor the equations on the hypercharges strongly constrain their possible values. In particular, according to an argument in [125, Sect. 30.4], if $Y_\nu = 0$, then up to an overall factor the hypercharges must have precisely the values in the Standard Model (or satisfy $Y_Q = Y_L = Y_e = 0$, which does not occur in nature).

[1] This is one of the few places in this book where we cite a result from quantum field theory. The equations hold independent of the choice of normalization of hypercharge.

It follows that in a realistic situation, the assignments of hypercharges in the Standard Model are completely fixed (up to an overall factor) by the vanishing of gauge anomalies. In particular, vanishing of gauge anomalies implies that the hypercharge (and thus the electric charge) are quantized and that the proton charge plus the electron charge is zero. As mentioned above, in Grand Unified Theories, electric charge can be quantized automatically for purely group theoretic reasons, without invoking vanishing of gauge anomalies.

8.6 Interactions Between Fermions and Gauge Bosons

The interaction between fermions and gauge bosons comes from the following (massless) Dirac Lagrangian[2]

$$\mathscr{L}_D = \mathrm{Re}\left(\overline{\psi} D_A \psi\right) = \mathrm{Re}\left(\overline{\psi_L} D_A \psi_L + \overline{\psi_R} D_A \psi_R\right),$$

with Dirac operator given by

$$D_A = i\Gamma^\mu \left(\partial_\mu + A_\mu\right)$$

in a global Lorentz frame for flat 4-dimensional Minkowski spacetime M. The cubic term responsible for the interaction is the **interaction vertex**

$$\begin{aligned}\mathscr{L}_{int} &= i\overline{\psi}_L \Gamma^\mu A_\mu \psi_L + i\overline{\psi}_R \Gamma^\mu A_\mu \psi_R \\ &= i\langle \psi_L, \Gamma^\mu A_\mu \psi_L\rangle + i\langle \psi_R, \Gamma^\mu A_\mu \psi_R\rangle,\end{aligned}$$

where we have chosen a global trivialization ϵ of $\mathrm{Spin}^+(M)$ corresponding to the Lorentz frame. The scalar products are taken in $\Delta \otimes V_L$ and $\Delta \otimes V_R$. Recall that the interaction vertex is automatically real.

As in Sect. 8.5 we can decompose the gauge field

$$A = G + W + B$$

corresponding to the summands and the orthonormal bases for the Lie algebra

$$\mathfrak{g} = \mathfrak{su}(3)_C \oplus \mathfrak{su}(2)_L \oplus \mathfrak{u}(1)_Y$$

[2] Masses for fermions will be introduced in Sect. 8.8.

and we can decompose the maps

$$\psi_L : M \longrightarrow \Delta_L \otimes V_L,$$
$$\psi_R : M \longrightarrow \Delta_R \otimes V_R$$

in the bases for the quark and lepton sectors.

The interaction vertex decomposes into the electroweak and strong interaction vertex:

$$\mathscr{L}_{int} = \mathscr{L}_{int,ew} + \mathscr{L}_{int,s},$$

where $\mathscr{L}_{int,ew}$ involves only the gauge field $W + B$ and $\mathscr{L}_{int,s}$ only the gauge field G.

8.6.1 The Electroweak Interaction Vertex

We first discuss the electroweak weak interaction vertex.

Lemma 8.6.1 *In the representation $\mathbb{C}^2 \otimes \mathbb{C}_y$ of $\mathrm{SU}(2)_L \times \mathrm{U}(1)_Y$ the gauge field $W + B$ acts as*

$$\begin{aligned}\mathbb{W}_\mu + \mathbb{B}_\mu &= \frac{ig_w}{\sqrt{2}} \begin{pmatrix} 0 & W_\mu^+ \\ W_\mu^- & 0 \end{pmatrix} \\ &+ \frac{ig_w}{2} \cos\theta_W \begin{pmatrix} 1 - y\tan^2\theta_W & 0 \\ 0 & -1 - y\tan^2\theta_W \end{pmatrix} Z_\mu^0 \\ &+ \frac{ie}{2} \begin{pmatrix} 1+y & 0 \\ 0 & -1+y \end{pmatrix} \gamma_\mu.\end{aligned}$$

In the representation $\mathbb{C} \otimes \mathbb{C}_y$ of $\mathrm{SU}(2)_L \times \mathrm{U}(1)_Y$ the gauge field $W + B$ acts as

$$\mathbb{W}_\mu + \mathbb{B}_\mu = -\frac{ig_w y}{2} \cos\theta_W \tan^2\theta_W Z_\mu^0 + \frac{iey}{2} \gamma_\mu.$$

Here we have set $e = g_w \sin\theta_W$ for the elementary electric charge.

Proof This is Exercise 8.11.8. □

Using this lemma we get the following explicit formula (with Dirac conjugate $\overline{\psi} = \psi^\dagger \Gamma_0$ according to Sect. 6.8):

Theorem 8.6.2 (Electroweak Interaction Vertex) *The electroweak interaction vertex for the leptons and quarks is given by*

$$\mathscr{L}_{int,ew} = \mathscr{L}_{int,ew,l} + \mathscr{L}_{int,ew,q}, \tag{8.16}$$

where for the leptons we have

$$\begin{aligned}\mathcal{L}_{int,ew,l} = &-\frac{g_w}{\sqrt{2}}\left(\overline{\nu}_{eL}\Gamma^\mu W_\mu^+ e_L + \overline{e}_L\Gamma^\mu W_\mu^- \nu_{eL}\right)\\ &-\frac{g_w}{2\cos\theta_W}\overline{\nu}_{eL}\Gamma^\mu Z_\mu^0\nu_{eL} - \frac{g_w\left(-1+2\sin^2\theta_W\right)}{2\cos\theta_W}\overline{e}_L\Gamma^\mu Z_\mu^0 e_L\\ &+e\overline{e}_L\Gamma^\mu\gamma_\mu e_L\\ &-g_w\frac{\sin^2\theta_W}{\cos\theta_W}\overline{e}_R\Gamma^\mu Z_\mu^0 e_R\\ &+e\overline{e}_R\Gamma^\mu\gamma_\mu e_R\\ &+\textit{same terms for second generation}\\ &+\textit{same terms for third generation}\end{aligned} \tag{8.17}$$

and for the quarks we have (with the standard Hermitian scalar product over the components in colour space $\mathbb{C}^3$ *implicit)*

$$\begin{aligned}\mathcal{L}_{int,ew,q} = &-\frac{g_w}{\sqrt{2}}\left(\overline{u}_L\Gamma^\mu W_\mu^+ d_L' + \overline{d}_L'\Gamma^\mu W_\mu^- u_L\right)\\ &-\frac{g_w}{2\cos\theta_W}\left(1-\frac{4}{3}\sin^2\theta_W\right)\overline{u}_L\Gamma^\mu Z_\mu^0 u_L\\ &-\frac{g_w}{2\cos\theta_W}\left(-1+\frac{2}{3}\sin^2\theta_W\right)\overline{d}_L'\Gamma^\mu Z_\mu^0 d_L'\\ &-\frac{2e}{3}\overline{u}_L\Gamma^\mu\gamma_\mu u_L + \frac{e}{3}\overline{d}_L'\Gamma^\mu\gamma_\mu d_L'\\ &+\frac{2g_w}{3}\frac{\sin^2\theta_W}{\cos\theta_W}\overline{u}_R\Gamma^\mu Z_\mu^0 u_R - \frac{g_w}{3}\frac{\sin^2\theta_W}{\cos\theta_W}\overline{d}_R'\Gamma^\mu Z_\mu^0 d_R'\\ &-\frac{2e}{3}\overline{u}_R\Gamma^\mu\gamma_\mu u_R + \frac{e}{3}\overline{d}_R'\Gamma^\mu\gamma_\mu d_R'\\ &+\textit{same terms for second generation}\\ &+\textit{same terms for third generation.}\end{aligned} \tag{8.18}$$

For the second and third generation we make the replacements in Eqs. (8.10) *and* (8.11). *We have set* $e = g_W \sin\theta_W$ *for the elementary electric charge.*

Proof This is Exercise 8.11.9. □

These Lagrangians are sometimes written as (e.g. [62])

$$\mathscr{L}_{int,ew,l} = -\frac{g_w}{\sqrt{2}} j^{\mu}_{W,l} W^{+}_{\mu} - \frac{g_w}{\sqrt{2}} j^{\mu\dagger}_{W,l} W^{-}_{\mu} - \frac{g_w}{2\cos\theta_W} j^{\mu}_{Z,l} Z^0_{\mu} - e j^{\mu}_{\gamma,l} \gamma_{\mu},$$

$$\mathscr{L}_{int,ew,q} = -\frac{g_w}{\sqrt{2}} j^{\mu}_{W,q} W^{+}_{\mu} - \frac{g_w}{\sqrt{2}} j^{\mu\dagger}_{W,q} W^{-}_{\mu} - \frac{g_w}{2\cos\theta_W} j^{\mu}_{Z,q} Z^0_{\mu} - e j^{\mu}_{\gamma,q} \gamma_{\mu}$$

with the **currents** (for the first generation)

$$\begin{aligned} j^{\mu}_{W,l} &= \overline{\nu}_{eL} \Gamma^{\mu} e_L, \\ j^{\mu}_{Z,l} &= \overline{\nu}_{eL} \Gamma^{\mu} \nu_{eL} + \left(-1 + 2\sin^2\theta_W\right) \overline{e}_L \Gamma^{\mu} e_L + 2\sin^2\theta_W \overline{e}_R \Gamma^{\mu} e_R, \\ j^{\mu}_{\gamma,l} &= -\overline{e}_L \Gamma^{\mu} e_L - \overline{e}_R \Gamma^{\mu} e_R \end{aligned}$$

and

$$\begin{aligned} j^{\mu}_{W,q} &= \overline{u}_L \Gamma^{\mu} d'_L, \\ j^{\mu}_{Z,q} &= \left(1 - \frac{4}{3}\sin^2\theta_W\right) \overline{u}_L \Gamma^{\mu} u_L + \left(-1 + \frac{2}{3}\sin^2\theta_W\right) \overline{d}'_L \Gamma^{\mu} d'_L \\ &\quad - \frac{4}{3}\sin^2\theta_W \overline{u}_R \Gamma^{\mu} u_R + \frac{2}{3}\sin^2\theta_W \overline{d}'_R \Gamma^{\mu} d'_R, \\ j^{\mu}_{\gamma,q} &= \frac{2}{3}\overline{u}_L \Gamma^{\mu} u_L - \frac{1}{3}\overline{d}'_L \Gamma^{\mu} d'_L + \frac{2}{3}\overline{u}_R \Gamma^{\mu} u_R - \frac{1}{3}\overline{d}'_R \Gamma^{\mu} d'_R. \end{aligned}$$

The expressions for $j^{\mu}_{W,l}$ and $j^{\mu}_{W,q}$ hold for both commuting and anticommuting spinors. For anticommuting spinors we calculate

$$\begin{aligned} j^{\mu\dagger}_{W,l} &= -\left(e_L^T \Gamma^{\mu T} \Gamma^0 \nu^*_{eL}\right)^* \\ &= e_L^{\dagger} \Gamma^{\mu\dagger} \Gamma^0 \nu_{eL} \\ &= \overline{e}_L \Gamma^{\mu} \nu_{eL}, \end{aligned}$$

where we used $\Gamma^{\mu\dagger} = \Gamma^0 \Gamma^{\mu} \Gamma^0$. Similarly for $j^{\mu}_{W,q}$.

Definition 8.6.3 The interactions between fermions involving W-bosons are called **charged current interactions** and interactions involving Z-bosons **neutral current interactions**.

Remark 8.6.4 Note that the W-bosons pair different flavours of particles (neutrinos with electrons, up quarks with down quarks) with different electric charges and different weak isospin, because they act off-diagonally on $\mathbb{C}^2$. The sum of charges (weak isospin, weak hypercharge, electric charge) at each vertex is conserved. *Interactions involving W-bosons are the only vertices in the Standard Model that change flavour.*

The Z-boson and the photon γ on the other hand act diagonally and only pair particles of the same flavour (hence with the same charges). The charged W-bosons also couple only to left-handed fermions, while the Z-boson and the photon γ couple to both left-handed and right-handed fermions. Contrary to the Z-boson, the photon γ pairs both left-handed and right-handed fermions in exactly the same way, i.e. electromagnetism is invariant under parity inversion. The photon does not couple to the neutrino, because its electric charge is zero.

We summarize the electroweak interaction vertices in the corresponding Feynman diagrams in Figs. 8.5, 8.6, 8.7, 8.8 (note that, for instance, in the left diagram in Fig. 8.6, ν_{eL} only pairs with ν_{eL}, e_L with e_L, and e_R with e_R, etc.). In the associated quantum field theory these diagrams also describe the interactions involving antifermions. Each diagram can be interpreted as the following interactions between (possibly virtual) particles:

- A fermion or antifermion radiates off (or emits) a gauge boson (possibly changing flavour).
- A fermion or antifermion absorbs a gauge boson (possibly changing flavour).
- A fermion and antifermion annihilate in a gauge boson.
- A gauge boson decays into a fermion and antifermion.

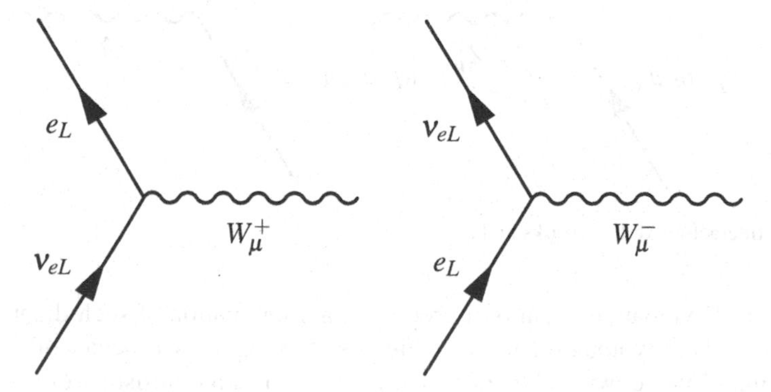

Fig. 8.5 Interaction vertex: leptons and $W^{\pm}$

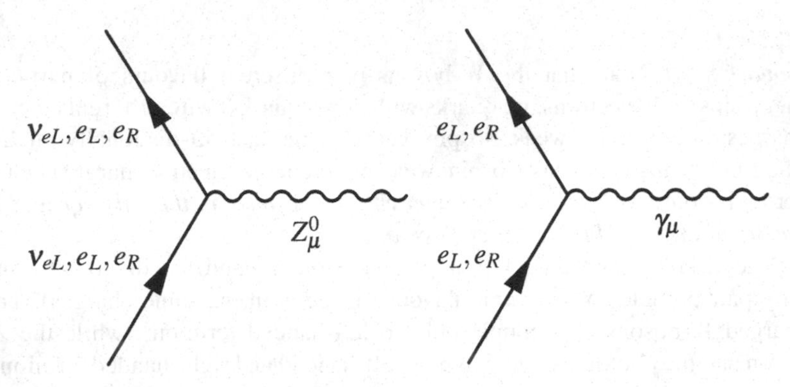

Fig. 8.6 Interaction vertex: leptons and Z^0, γ

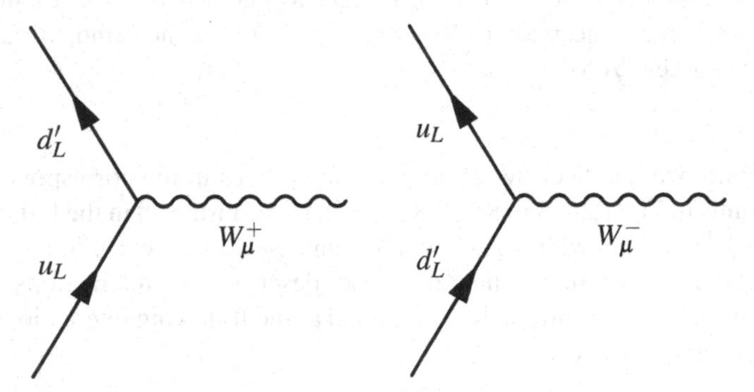

Fig. 8.7 Interaction vertex: quarks and $W^\pm$

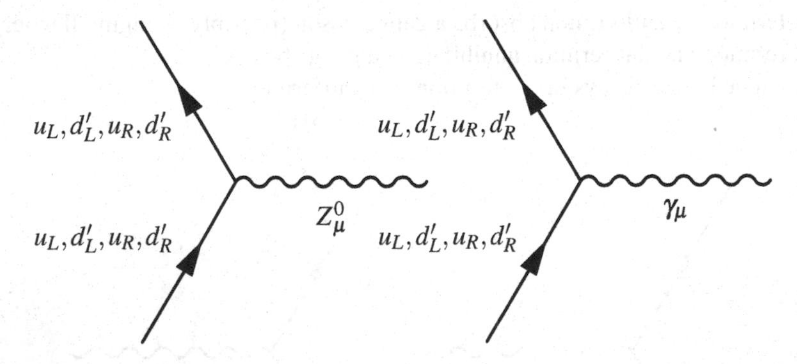

Fig. 8.8 Interaction vertex: quarks and Z^0, γ

The full Feynman diagram of a process is a combination of such diagrams. An example is the Feynman diagram in Fig. 8.9 that explains β-decay of a neutron (consisting of two down and one up quark) into a proton (consisting of two up and one down quark), an electron and an electron antineutrino, via a virtual W^--boson:

$$n \longrightarrow p + e + \nu^C_{eL}.$$

Fig. 8.9 β-decay of neutron

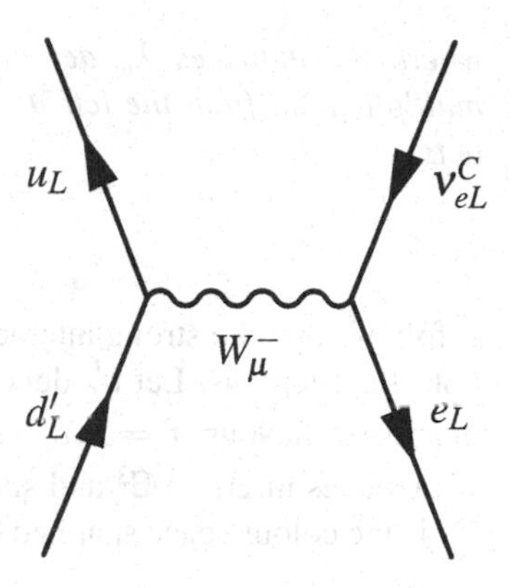

Neutrinos are produced by similar reactions in enormous amounts in the process of nucleosynthesis inside stars like the sun (**solar neutrinos**).

Remark 8.6.5 The interaction vertices in Theorem 8.6.2 are exactly the same for all three generations of leptons and quarks. For example, on a fundamental level the only difference in the Standard Model between the electron, muon and tau are their different masses (which lead to different lifetimes, etc.). This is known as **lepton flavour universality**. For quarks, the corresponding statement is true for the weak eigenstates considered above, but not for the so-called mass eigenstates because of quark mixing, to be discussed in Sect. 8.8.2.

8.6.2 *The Strong Interaction Vertex*

We briefly want to state a formula for the strong interaction vertex $\mathcal{L}_{int,s}$. Fermions appear in two representations of $SU(3)_C$: the trivial representation $\mathbb{C}$ and the fundamental representation $\mathbb{C}^3$.

If we expand the gluon field as in Eq. (8.8),

$$G_\mu = \sum_{a=1}^{8} G_\mu^a g_s \nu_a = \frac{ig_s}{2}\sum_{a=1}^{8} G_\mu^a \lambda_a,$$

we get:

Lemma 8.6.6 *In the representation $\mathbb{C}^3$ of $SU(3)_C$ the gluon field G acts as*

$$\mathbb{G}_\mu = \frac{ig_s}{2}\sum_{a=1}^{8} G_\mu^a \lambda_a,$$

where the matrices λ_a act in the standard way as endomorphisms of $\mathbb{C}^3$ by multiplication from the left. In the representation $\mathbb{C}$ of $\mathrm{SU}(3)_C$ *the gluon field G acts as*

$$\mathbb{G}_\mu = 0.$$

It follows that the strong interaction is restricted to quarks and antiquarks and does not affect leptons. Let q_L^f denote the left-handed quarks and q_R^f the right-handed quarks for flavours $f = u, d', c, s', t, b'$. We can think of q_L^f as a map on spacetime with values in $\Delta_L \otimes \mathbb{C}^3$ and similarly q_R^f as a map with values in $\Delta_R \otimes \mathbb{C}^3$, where $\mathbb{C}^3$ is the colour space spanned by the colour vectors r, g, b. We write

$$q_L^f = \begin{pmatrix} q_L^{fr} \\ q_L^{fg} \\ q_L^{fb} \end{pmatrix},$$

where $q_L^{fr}, q_L^{fg}, q_L^{fb}$ are ordinary left-handed Weyl spinors, i.e. maps with values in Δ_L, corresponding to the three different colours (analogously for the right-handed quarks). We then have:

Theorem 8.6.7 (Strong Interaction Vertex) *The strong interaction vertex for the quarks is given by*

$$\begin{aligned}
\mathscr{L}_{int,s} &= -\frac{g_s}{2}\sum_{a=1}^{8}\sum_f \left(\overline{q}_L^f \Gamma^\mu G_\mu^a \lambda_a q_L^f + \overline{q}_R^f \Gamma^\mu G_\mu^a \lambda_a q_R^f\right) \\
&= -\frac{g_s}{2}\sum_{a=1}^{8}(\overline{u}_L \Gamma^\mu G_\mu^a \lambda_a u_L + \overline{u}_R \Gamma^\mu G_\mu^a \lambda_a u_R \\
&\quad + \overline{d}'_L \Gamma^\mu G_\mu^a \lambda_a d'_L + \overline{d}'_R \Gamma^\mu G_\mu^a \lambda_a d'_R) \\
&\quad + \textit{same terms for second generation} \\
&\quad + \textit{same terms for third generation.}
\end{aligned} \tag{8.19}$$

There are implicit standard Hermitian scalar products in the colour space $\mathbb{C}^3$.

Note that the strong interaction does not pair quarks of different flavours like the weak interaction. If we expand the gluon field alternatively in the basis $\{\mu_a\}$,

$$G_\mu = g_s \sum_{a=1}^{8} G'^a_\mu \mu_a,$$

we see that the gluons

$$G'^{r\bar{g}}_\mu, \quad G'^{r\bar{b}}_\mu, \quad G'^{g\bar{r}}_\mu, \quad G'^{g\bar{b}}_\mu, \quad G'^{b\bar{r}}_\mu, \quad G'^{b\bar{g}}_\mu$$

pair quarks of different colours, because they act off-diagonally on the colour space, while

$$G'^{r\bar{r}-g\bar{g}}_\mu, \quad G'^{r\bar{r}+g\bar{g}-2b\bar{b}}_\mu$$

act diagonally and pair quarks of the same colour. The first type of gluons can thus be roughly compared to the W-bosons in the electroweak theory, while the second type of gluons corresponds to the Z-boson and photon γ.

See the Feynman diagrams in Figs. 8.10 and 8.11 for a generic strong interaction vertex and two examples of more specific ones. These diagrams can be interpreted as in Remark 8.6.4 (radiation/emission, absorption, annihilation, decay). The sum of colour charges at each vertex is conserved if the colour charge of gluons is defined suitably. Gluons can mediate interactions between quarks of different flavours like in the Feynman diagram in Fig. 8.12.

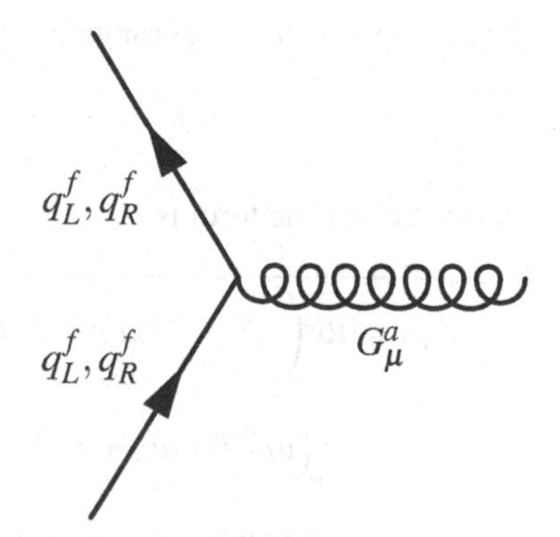

Fig. 8.10 Generic quark-gluon interaction vertex

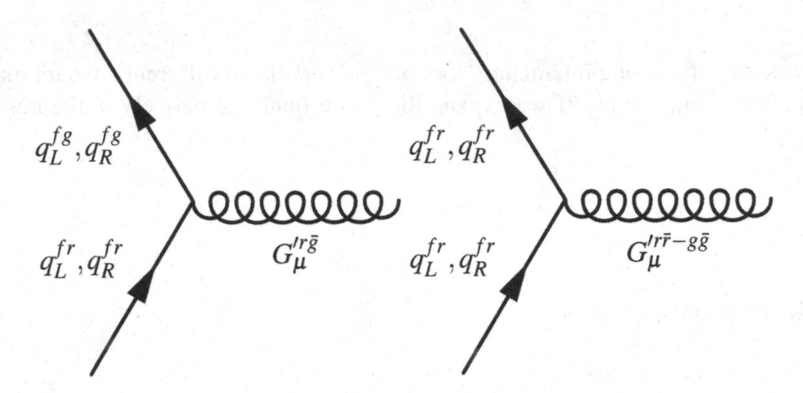

Fig. 8.11 Specific quark-gluon interaction vertices

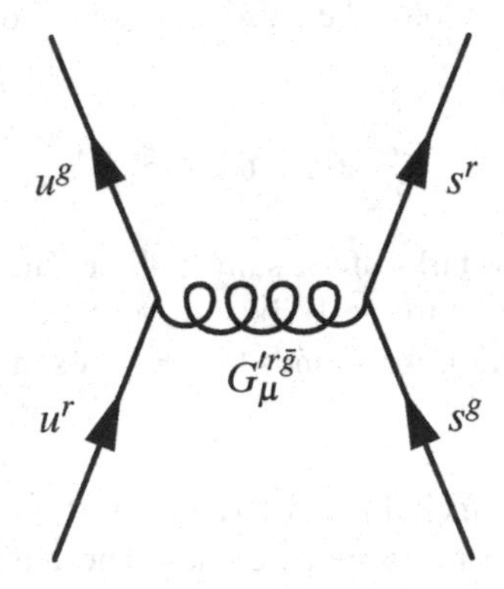

Fig. 8.12 Strong interaction between quarks of different flavours

8.6.3 The Dirac Lagrangian for Fermions

The complete Dirac Lagrangian for the fermions can now be written as

$$\mathscr{L}_D = \mathscr{L}_{D,\partial} + \mathscr{L}_{D,int} \tag{8.20}$$

where the kinetic term is

$$\boxed{\begin{aligned}\mathscr{L}_{D,\partial} = \mathrm{Re}\Big(& i\left(\overline{e}_L\Gamma^\mu\partial_\mu e_L + \overline{e}_R\Gamma^\mu\partial_\mu e_R + \overline{\nu}_{eL}\Gamma^\mu\partial_\mu\nu_{eL}\right) \\ & +i\left(\overline{u}_L\Gamma^\mu\partial_\mu u_L + \overline{u}_R\Gamma^\mu\partial_\mu u_R + \overline{d}'_L\Gamma^\mu\partial_\mu d'_L + \overline{d}'_R\Gamma^\mu\partial_\mu d'_R\right)\Big) \\ & + \text{same terms for second generation} \\ & + \text{same terms for third generation}\end{aligned}} \tag{8.21}$$

and

$$\mathscr{L}_{D,int} = \mathscr{L}_{int,ew} + \mathscr{L}_{int,s}$$

with the electroweak interaction vertex $\mathscr{L}_{int,ew}$ from Eq. (8.16) and the strong interaction vertex $\mathscr{L}_{int,s}$ from Eq. (8.19).

8.7 Interactions Between Higgs Bosons and Gauge Bosons

We want to determine the Yang–Mills–Higgs Lagrangian

$$\mathscr{L}_H[\phi, A] + \mathscr{L}_{YM}[A] = \left\langle \nabla^{A\mu}\phi, \nabla^A_\mu \phi \right\rangle - V(\phi) - \frac{1}{2}\left\langle F^A_M, F^A_M \right\rangle_{\mathrm{Ad}(P)}$$

for the electroweak interaction.

8.7.1 The Higgs Lagrangian

We first calculate the Higgs Lagrangian (the first two summands in the Yang–Mills–Higgs Lagrangian). We assume that we have chosen a unitary gauge so that the Higgs field is of the form

$$\phi = \frac{1}{\sqrt{2}} \begin{pmatrix} 0 \\ v + H \end{pmatrix},$$

where $H: M = \mathbb{R}^4 \to \mathbb{R}$ is the Higgs boson and $v = \sqrt{\frac{\mu}{\lambda}}$. The Higgs potential is given by

$$V(\phi) = -\mu\phi^\dagger\phi + \lambda\left(\phi^\dagger\phi\right)^2$$

and the mass of the Higgs boson is $m_H = \sqrt{2\mu}$. Let $A = W + B$ be the electroweak gauge field. The mass of the W-bosons is equal to

$$m_W = \frac{1}{2} g_w v$$

and the mass of the Z-boson is equal to

$$m_Z = \frac{1}{2\cos\theta_W} g_w v.$$

Lemma 8.7.1 *The covariant derivative $\nabla_\mu^A \phi$ is given by*

$$\nabla_\mu^A \phi = \frac{1}{\sqrt{2}} \begin{pmatrix} 0 \\ \partial_\mu H \end{pmatrix} + i m_W W_\mu^+ \begin{pmatrix} 1 + \frac{H}{v} \\ 0 \end{pmatrix} - \frac{i m_Z}{\sqrt{2}} Z_\mu^0 \begin{pmatrix} 0 \\ 1 + \frac{H}{v} \end{pmatrix}.$$

The potential $V(\phi)$ is (up to an irrelevant constant)

$$V(\phi) = \frac{1}{2} m_H^2 H^2 \left(1 + \frac{1}{v} H + \frac{1}{4v^2} H^2\right).$$

Proof This is Exercise 8.11.11. □

We get:

Theorem 8.7.2 (Electroweak Higgs Lagrangian) *After symmetry breaking the Higgs Lagrangian is given in unitary gauge by*

$$\boxed{\begin{aligned} \mathscr{L}_H \left[H, W^\pm, Z^0\right] = & \frac{1}{2} \left(\partial^\mu H\right) \left(\partial_\mu H\right) - \frac{1}{2} m_H^2 H^2 \\ & - \frac{1}{2} m_H^2 H^2 \left(\frac{1}{v} H + \frac{1}{4v^2} H^2\right) \\ & + m_W^2 W_\mu^- W^{+\mu} \\ & + m_W^2 W_\mu^- W^{+\mu} \left(\frac{2}{v} H + \frac{1}{v^2} H^2\right) \\ & + \frac{1}{2} m_Z^2 Z_\mu^0 Z^{0\mu} \\ & + \frac{1}{2} m_Z^2 Z_\mu^0 Z^{0\mu} \left(\frac{2}{v} H + \frac{1}{v^2} H^2\right). \end{aligned}} \tag{8.22}$$

Proof This is Exercise 8.11.12. □

These terms have the following interpretation:

- The term in the first line is the Lagrangian for a free real scalar Higgs boson of mass m_H.
- The term in the second line describes the interaction between Higgs bosons.
- The term in the third line is the mass term for the $W^\pm$-bosons.
- The term in the fourth line describes the residual interaction between the $W^\pm$-bosons and the Higgs boson.
- The term in the fifth line is the mass term for the Z^0-boson.
- The term in the sixth line describes the residual interaction between the Z^0-boson and the Higgs boson.

The free terms in this Lagrangian have already been determined in Theorem 8.2.10. We can also write

$$W_\mu^- W^{+\mu} = \left(W_\mu^+\right)^* W^{+\mu} = \frac{1}{2}\left(A_\mu^1 A^{1\mu} + A_\mu^2 A^{2\mu}\right)$$

using the definition

$$W_\mu^\pm = \frac{1}{\sqrt{2}}\left(A_\mu^1 \mp iA_\mu^2\right)$$

from Sect. 8.3.3. The couplings of the W- and Z-bosons to the Higgs boson are proportional to their masses squared, hence quite strong. Note that the photon does not couple to the Higgs boson. Figures 8.13 and 8.14 show Feynman diagrams for the interactions in Theorem 8.7.2. The diagram on the left in Fig. 8.14, for example, can be interpreted in one of the following ways:

- **Higgs–Strahlung** (vector bosons radiate off or emit a Higgs boson): $W^\pm \rightarrow W^\pm H$ and $Z^0 \rightarrow Z^0 H$
- **Absorption of a Higgs boson**: $W^\pm H \rightarrow W^\pm$ and $Z^0 H \rightarrow Z^0$

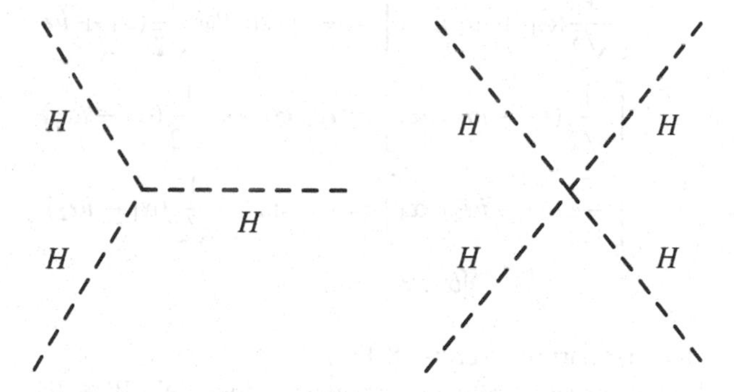

Fig. 8.13 Interaction vertices: Higgs boson

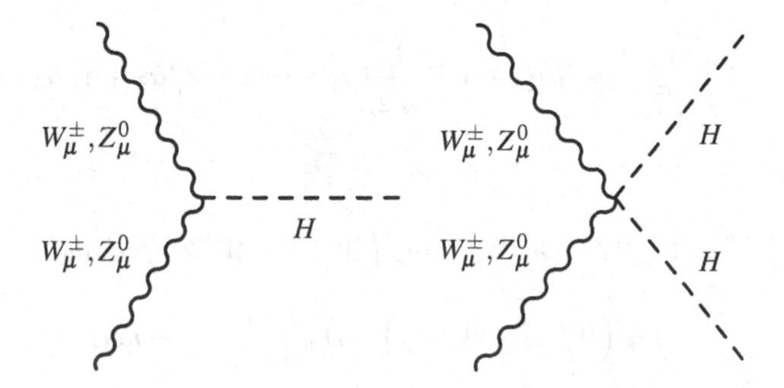

Fig. 8.14 Interaction vertices: electroweak gauge bosons and Higgs boson

- **Vector boson fusion**: $W^\pm W^\mp \to H$ and $Z^0 Z^0 \to H$
- **Vector boson decay**: $H \to W^\pm W^\mp$ and $H \to Z^0 Z^0$.

The electric charge at each vertex is conserved.

8.7.2 *The Yang–Mills Lagrangian*

We now calculate the Yang–Mills Lagrangian. We first determine the commutators of the basis vectors $\alpha_1, \alpha_2, \alpha_3, \alpha_4$ for $\mathfrak{su}(2)_L \oplus \mathfrak{u}(1)_Y$ in Sect. 8.3.2.

Lemma 8.7.3 *In the complexification of the Lie algebra $\mathfrak{su}(2)_L \oplus \mathfrak{u}(1)_Y$ we have*

$$[\beta_a, \beta_b] = -\epsilon_{abc} g_w \beta_c \quad \forall a, b, c \in \{1, 2, 3\}$$

$$\left[\frac{1}{\sqrt{2}}(\alpha_1 + i\alpha_2), \frac{1}{\sqrt{2}}(\alpha_1 - i\alpha_2)\right] = ig_w(\cos\theta_W \alpha_3 + \sin\theta_W \alpha_4)$$

$$\left[\frac{1}{\sqrt{2}}(\alpha_1 + i\alpha_2), \alpha_3\right] = -ig_w \cos\theta_W \frac{1}{\sqrt{2}}(\alpha_1 + i\alpha_2)$$

$$\left[\frac{1}{\sqrt{2}}(\alpha_1 + i\alpha_2), \alpha_4\right] = -ig_w \sin\theta_W \frac{1}{\sqrt{2}}(\alpha_1 + i\alpha_2)$$

$$\left[\frac{1}{\sqrt{2}}(\alpha_1 - i\alpha_2), \alpha_3\right] = ig_w \cos\theta_W \frac{1}{\sqrt{2}}(\alpha_1 - i\alpha_2)$$

$$\left[\frac{1}{\sqrt{2}}(\alpha_1 - i\alpha_2), \alpha_4\right] = ig_w \sin\theta_W \frac{1}{\sqrt{2}}(\alpha_1 - i\alpha_2)$$

$$[\alpha_3, \alpha_4] = 0.$$

Proof This is the first part of Exercise 8.11.13. □

This implies for the curvature of the electroweak gauge field $W + B$:

Theorem 8.7.4 *The curvature of the electroweak gauge field*

$$W_\mu^+ \frac{1}{\sqrt{2}}(\alpha_1 + i\alpha_2) + W_\mu^- \frac{1}{\sqrt{2}}(\alpha_1 - i\alpha_2) + Z_\mu^0 \alpha_3 + \gamma_\mu \alpha_4$$

is given by

$$F_{\mu\nu}^{W+B} = \Big[\partial_\mu W_\nu^+ - \partial_\nu W_\mu^+ - ig_w \left(W_\mu^+ Z_\nu^0 - W_\nu^+ Z_\mu^0\right) \cos\theta_W$$
$$-ig_w \left(W_\mu^+ \gamma_\nu - W_\nu^+ \gamma_\mu\right) \sin\theta_W\Big] \frac{1}{\sqrt{2}}(\alpha_1 + i\alpha_2)$$

$$+\left[\partial_\mu W_\nu^- - \partial_\nu W_\mu^- + ig_w\left(W_\mu^- Z_\nu^0 - W_\nu^- Z_\mu^0\right)\cos\theta_W \right.$$
$$\left. +ig_w\left(W_\mu^-\gamma_\nu - W_\nu^-\gamma_\mu\right)\sin\theta_W\right]\frac{1}{\sqrt{2}}(\alpha_1 - i\alpha_2)$$

$$+\left[\partial_\mu Z_\nu^0 - \partial_\nu Z_\mu^0 + ig_w\left(W_\mu^+ W_\nu^- - W_\nu^+ W_\mu^-\right)\cos\theta_W\right]\alpha_3$$

$$+\left[\partial_\mu \gamma_\nu - \partial_\nu \gamma_\mu + ig_w\left(W_\mu^+ W_\nu^- - W_\nu^+ W_\mu^-\right)\sin\theta_W\right]\alpha_4.$$

Proof This is the second part of Exercise 8.11.13. □

Let H be a 2-form on Minkowski spacetime M with values in the complex numbers. We write

$$||H_{\mu\nu}||^2 = \frac{1}{2}\overline{H}_{\mu\nu}H^{\mu\nu}.$$

Then we get:

Corollary 8.7.5 (Electroweak Yang–Mills Lagrangian) *The Yang–Mills Lagrangian for the electroweak gauge field $W + B$ is given by*

$$\begin{aligned}
\mathscr{L}_{YM}^{W+B} &= -\frac{1}{4}\left\langle F_{\mu\nu}^{W+B}, F_{W+B}^{\mu\nu}\right\rangle_{\mathfrak{su}(2)_L\oplus\mathfrak{u}(1)_Y} \\
&= -\frac{1}{2}\left|\left|\partial_\mu W_\nu^+ - \partial_\nu W_\mu^+ - ig_w\left(W_\mu^+ Z_\nu^0 - W_\nu^+ Z_\mu^0\right)\cos\theta_W \right.\right. \\
&\quad \left.\left. -ig_w\left(W_\mu^+\gamma_\nu - W_\nu^+\gamma_\mu\right)\sin\theta_W\right|\right|^2 \\
&\quad -\frac{1}{2}\left|\left|\partial_\mu W_\nu^- - \partial_\nu W_\mu^- + ig_w\left(W_\mu^- Z_\nu^0 - W_\nu^- Z_\mu^0\right)\cos\theta_W \right.\right. \\
&\quad \left.\left. +ig_w\left(W_\mu^-\gamma_\nu - W_\nu^-\gamma_\mu\right)\sin\theta_W\right|\right|^2 \\
&\quad -\frac{1}{2}\left|\left|\partial_\mu Z_\nu^0 - \partial_\nu Z_\mu^0 + ig_w\left(W_\mu^+ W_\nu^- - W_\nu^+ W_\mu^-\right)\cos\theta_W\right|\right|^2 \\
&\quad -\frac{1}{2}\left|\left|\partial_\mu \gamma_\nu - \partial_\nu \gamma_\mu + ig_w\left(W_\mu^+ W_\nu^- - W_\nu^+ W_\mu^-\right)\sin\theta_W\right|\right|^2.
\end{aligned} \tag{8.23}$$

More explicitly we can write

$$\mathscr{L}_{YM}^{W+B} = \mathscr{L}_{YM,2}^{W+B} + \mathscr{L}_{YM,3}^{W+B} + \mathscr{L}_{YM,4}^{W+B},$$

where (compare with [112])

$$
\begin{aligned}
\mathscr{L}_{YM,2}^{W+B} &= -\frac{1}{2}\left(\partial_\mu W_\nu^- - \partial_\nu W_\mu^-\right)\left(\partial^\mu W^{+\nu} - \partial^\nu W^{+\mu}\right) \\
&\quad - \frac{1}{4}\left(\partial_\mu Z_\nu^0 - \partial_\nu Z_\mu^0\right)\left(\partial^\mu Z^{0\nu} - \partial^\nu Z^{0\mu}\right) \\
&\quad - \frac{1}{4}\left(\partial_\mu \gamma_\nu - \partial_\nu \gamma_\mu\right)\left(\partial^\mu \gamma^\nu - \partial^\nu \gamma^\mu\right),
\end{aligned}
$$

$$
\begin{aligned}
\mathscr{L}_{YM,3}^{W+B} &= ig_w \cos\theta_W \Big[\left(\partial_\mu W_\nu^- - \partial_\nu W_\mu^-\right) W^{+\mu} Z^{0\nu} - \left(\partial_\mu W_\nu^+ - \partial_\nu W_\mu^+\right) W^{-\mu} Z^{0\nu} \\
&\quad - \left(\partial_\mu Z_\nu^0 - \partial_\nu Z_\mu^0\right) W^{+\mu} W^{-\nu}\Big] \\
&\quad + ig_w \sin\theta_W \Big[\left(\partial_\mu W_\nu^- - \partial_\nu W_\mu^-\right) W^{+\mu} \gamma^\nu - \left(\partial_\mu W_\nu^+ - \partial_\nu W_\mu^+\right) W^{-\mu} \gamma^\nu \\
&\quad - \left(\partial_\mu \gamma_\nu - \partial_\nu \gamma_\mu\right) W^{+\mu} W^{-\nu}\Big],
\end{aligned}
$$

$$
\begin{aligned}
\mathscr{L}_{YM,4}^{W+B} &= -\frac{1}{2} g_w^2 \left[\left(W_\mu^+ W^{-\mu}\right)^2 - W_\mu^+ W^{+\mu} W_\nu^- W^{-\nu}\right] \\
&\quad - g_w^2 \cos^2\theta_W \left[W_\mu^+ W^{-\mu} Z_\nu^0 Z^{0\nu} - W_\mu^+ Z^{0\mu} W_\nu^- Z^{0\nu}\right] \\
&\quad - g_w^2 \sin^2\theta_W \left[W_\mu^+ W^{-\mu} \gamma_\nu \gamma^\nu - W_\mu^+ \gamma^\mu W_\nu^- \gamma^\nu\right] \\
&\quad - g_w^2 \sin\theta_W \cos\theta_W \left[2 W_\mu^+ W^{-\mu} \gamma_\nu Z^{0\nu} - W_\mu^+ Z^{0\mu} W_\nu^- \gamma^\nu - W_\mu^+ \gamma^\mu W_\nu^- Z^{0\nu}\right].
\end{aligned}
$$

The Yang–Mills Lagrangian contains in addition to the quadratic kinetic terms various cubic and quartic couplings between the gauge bosons. The corresponding Feynman diagrams are depicted in Figs. 8.15 and 8.16 (electric charge is conserved at each vertex). We can do a similar calculation for the gluon gauge field G. We expand

$$
G_\mu = \sum_{a=1}^{8} G_\mu^a \frac{ig_s \lambda_a}{2}
$$

and define structure constants for the Lie algebra $\mathfrak{su}(3)$ in the Gell-Mann basis by

$$
\left[\frac{i\lambda_a}{2}, \frac{i\lambda_b}{2}\right] = f_{abc} \frac{i\lambda_c}{2}.
$$

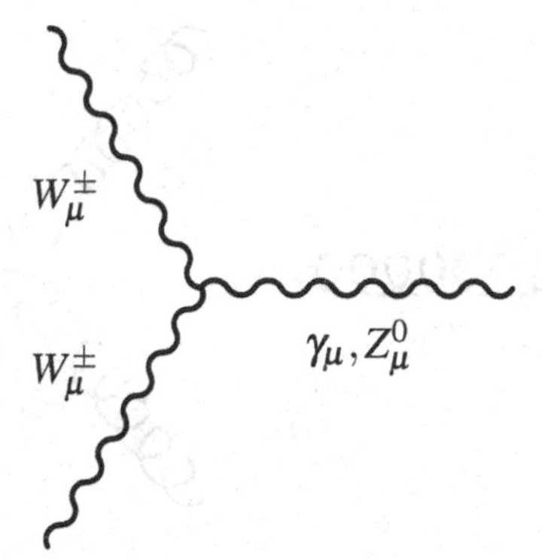

Fig. 8.15 Electroweak gauge bosons: 3-boson interaction vertex

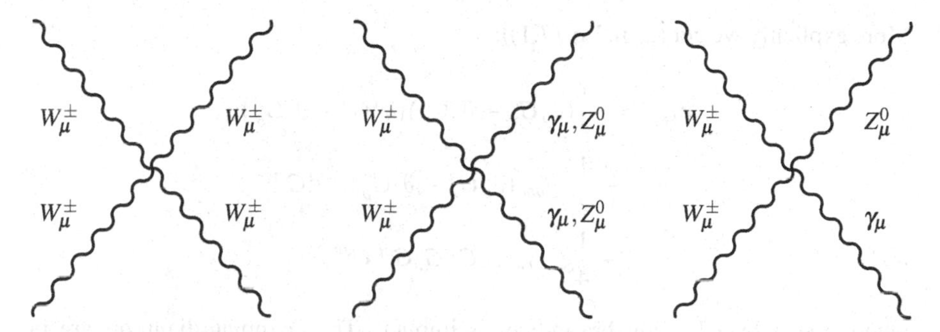

Fig. 8.16 Electroweak gauge bosons: 4-boson interaction vertices

Then the curvature of the gauge field G is

$$F^G_{\mu\nu} = \sum_{a=1}^{8} \left(\partial_\mu G^a_\nu - \partial_\nu G^a_\mu + g_s \sum_{b,c=1}^{8} G^b_\mu G^c_\nu f_{bca} \right) \frac{i g_s \lambda_a}{2}.$$

Theorem 8.7.6 (Gluon Yang–Mills Lagrangian) *The Yang–Mills Lagrangian for the gluon gauge field G is given by*

$$\boxed{\begin{aligned} \mathscr{L}^G_{YM} &= -\frac{1}{4} \left\langle F^G_{\mu\nu}, F^{\mu\nu}_G \right\rangle_{\mathfrak{su}(3)_C} \\ &= -\frac{1}{2} \left\| \sum_{a=1}^{8} \left(\partial_\mu G^a_\nu - \partial_\nu G^a_\mu + g_s \sum_{b,c=1}^{8} G^b_\mu G^c_\nu f_{bca} \right) \right\|^2 . \end{aligned}} \quad (8.24)$$

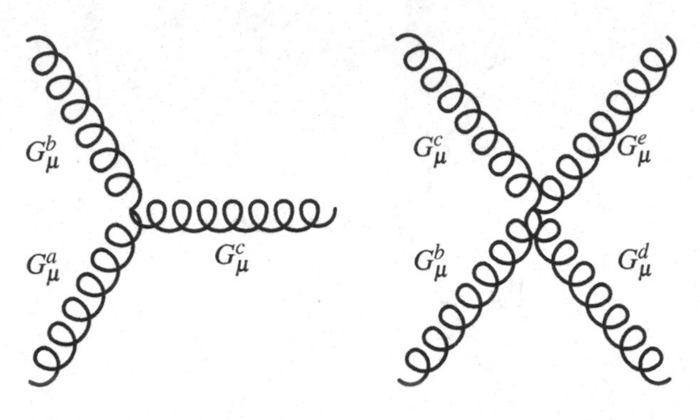

Fig. 8.17 3- and 4-gluon interaction vertices

More explicitly we get (as in Eq. (7.1)):

$$\begin{aligned}\mathscr{L}^G_{YM} &= -\frac{1}{4}(\partial_\mu G^a_\nu - \partial_\nu G^a_\mu)(\partial^\mu G^\nu_a - \partial^\nu G^\mu_a) \\ &\quad - \frac{1}{2} g_s f_{abc}(\partial_\mu G^a_\nu - \partial_\nu G^a_\mu) G^{b\mu} G^{c\nu} \\ &\quad - \frac{1}{4} g_s^2 f_{abc} f_{ade} G^b_\mu G^c_\nu G^{d\mu} G^{e\nu},\end{aligned}$$

where a sum over Lie algebra indices is implicit. The Feynman diagrams are in Fig. 8.17. The total Yang–Mills Lagrangian for both the electroweak and the gluon field is then

$$\mathscr{L}_{YM} = \mathscr{L}^{W+B}_{YM} + \mathscr{L}^G_{YM}. \tag{8.25}$$

8.8 Mass Generation for Fermions in the Standard Model

So far the fermions (leptons and quarks) in the Standard Model were assumed massless. In this section we define Yukawa couplings between the fermions and the Higgs field which lead to Dirac mass terms for the fermions (except for the neutrinos) after symmetry breaking. The masses of the fermions are proportional to the Yukawa coupling constants and the absolute value v of the Higgs condensate.

The Yukawa couplings also lead to a residual interaction vertex between two fermions of the same flavour and the Higgs boson. This means that fermions can interact by emitting and absorbing virtual Higgs bosons, hence the Higgs boson can be interpreted as the mediating particle of a new type of interaction, different from gauge interactions (in Sect. 8.7.1 we saw that the Higgs boson can also mediate an interaction between the weak gauge bosons).

8.8.1 Yukawa Couplings for Leptons

We begin with the leptons. Recall that as complex vector spaces the lepton and Higgs sector are equal to

$$\begin{aligned} L_L^i &\cong \mathbb{C}^2, \\ L_R^i &\cong \mathbb{C}, \\ W &\cong \mathbb{C}^2 \end{aligned}$$

for each generation. We set $i = e, \mu, \tau$ to denote the lepton generations and fix Yukawa couplings g_i.

Lemma 8.8.1 *For each generation $i = e, \mu, \tau$ the map*

$$\begin{aligned} \tau_l^i : L_L^i \times W \times L_R^i &\longrightarrow \mathbb{C} \\ (l_L, \phi, l_R) &\longmapsto g_i l_L^\dagger \phi l_R \end{aligned}$$

is an $\mathrm{SU}(2)_L \times \mathrm{U}(1)_Y$*-invariant Yukawa form.*

Proof This is the first part of Exercise 8.11.14. □

In physics the following notation is used: We write for the first generation

$$\begin{aligned} L_{1L} &= \begin{pmatrix} \nu_{eL} \\ e_L \end{pmatrix}, \\ \overline{L}_{1L} &= (\overline{\nu}_{eL}, \overline{e}_L), \\ e_{1R} &= e_R \end{aligned}$$

and similarly for generation 2 and 3 with (ν_e, e) replaced by (ν_μ, μ) and (ν_τ, τ), respectively. Then

$$\tau_L^i(v_L, \phi, v_R) = g_i \overline{L}_{iL} \phi e_{iR}.$$

Suppose that the Higgs field ϕ is in unitary gauge,

$$\phi = \frac{1}{\sqrt{2}} \begin{pmatrix} 0 \\ v + H \end{pmatrix}.$$

We define the lepton masses

$$m_i = \frac{1}{\sqrt{2}} g_i v$$

for $i = e, \mu, \tau$. Then we get:

Theorem 8.8.2 (Yukawa Coupling for Leptons) *After symmetry breaking the Yukawa Lagrangian for the three lepton generations associated to the Yukawa form in Lemma 8.8.1 is given in unitary gauge by*

$$\begin{aligned}\mathscr{L}_{Y,l} &= \mathscr{L}^{e}_{Y,l} + \mathscr{L}^{\mu}_{Y,l} + \mathscr{L}^{\tau}_{Y,l} \\ &= -2m_e \mathrm{Re}(\overline{e}_L e_R) - 2m_\mu \mathrm{Re}(\overline{\mu}_L \mu_R) - 2m_\tau \mathrm{Re}(\overline{\tau}_L \tau_R) \\ &\quad - \frac{2}{v} m_e \mathrm{Re}(\overline{e}_L e_R) H - \frac{2}{v} m_\mu \mathrm{Re}(\overline{\mu}_L \mu_R) H - \frac{2}{v} m_\tau \mathrm{Re}(\overline{\tau}_L \tau_R) H.\end{aligned} \tag{8.26}$$

Proof This is the second part of Exercise 8.11.14. □

The three terms in the second line are the Dirac mass terms for the electron, muon and tau. The terms in the third line are residual interactions between these leptons and the Higgs boson (see the Feynman diagram in Fig. 8.18). The coupling of the leptons to the Higgs boson is proportional to their mass. Note that the neutrinos do not appear in this Lagrangian. In particular, their mass is zero in the Standard Model.

8.8.2 *Yukawa Couplings for Quarks and Quark Mixing Across Generations*

We consider the following $\mathrm{SU}(2)_L \times \mathrm{U}(1)_Y$ representation spaces:

$$\bigoplus_{i=1}^{3} \left(\mathbb{C}^2 \otimes \mathbb{C}_{1/3}\right) \quad \text{(left-handed)} \tag{8.27}$$

and

$$\bigoplus_{i=1}^{3} \left(\left(\mathbb{C} \otimes \mathbb{C}_{4/3}\right) \oplus \left(\mathbb{C} \otimes \mathbb{C}_{-2/3}\right)\right) \quad \text{(right-handed).} \tag{8.28}$$

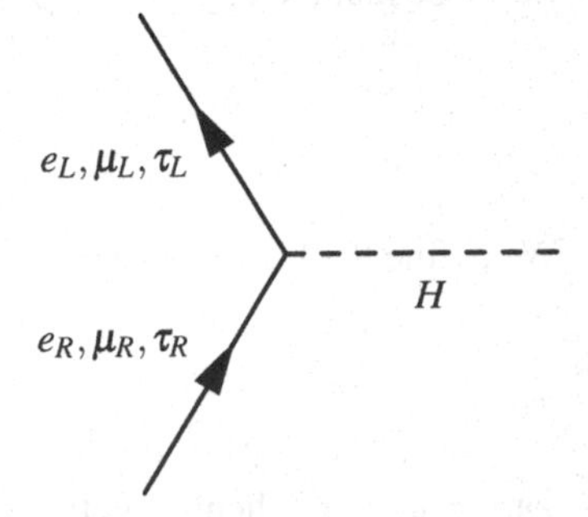

Fig. 8.18 Lepton-Higgs interaction vertex

The quark representation sectors Q_L and Q_R are obtained from these representations by tensoring with the fundamental representation $\mathbb{C}^3$ of $SU(3)_C$.

It turns out that the Yukawa couplings for the quarks are only diagonal in flavour space in another basis than the one we used so far. We write

$$u_1^I = u^I, u_2^I = c^I, u_3^I = t^I,$$
$$d_1^I = d^I, d_2^I = s^I, d_3^I = b^I$$

for the quarks that correspond to the standard basis elements for the irreducible summands in Eqs. (8.27) and (8.28) and indicate the left-handed basis by an index L and the right-handed basis by an index R. We also write for the left-handed isodoublets

$$Q_{iL}^I = \begin{pmatrix} u_{iL}^I \\ d_{iL}^I \end{pmatrix},$$
$$\overline{Q}_{iL}^I = (\overline{u}_{iL}^I, \overline{d}_{iL}^I).$$

Let

$$\phi = \frac{1}{\sqrt{2}} \begin{pmatrix} \phi^+ \\ \phi^0 \end{pmatrix}$$

be the Higgs field. We set

$$\phi_c = i\sigma_2 \phi^* = \frac{1}{\sqrt{2}} \begin{pmatrix} \phi^{0*} \\ -\phi^{+*} \end{pmatrix}.$$

In the physics literature, ϕ_c is often denoted by $\tilde{\phi}$.

Lemma 8.8.3 *The field ϕ_c satisfies*

$$(A\phi)_c = A\phi_c \quad \forall A \in SU(2), \phi \in \mathbb{C}^2.$$

Hence ϕ_c has the same weak isospin as ϕ and weak hypercharge $Y = -1$.

Proof This is Exercise 8.11.15. □

Lemma 8.8.4 *For arbitrary complex matrices Y^u and Y^d the expression*

$$\begin{aligned} \tau_Q &= Y_{ij}^d \overline{Q}_{iL}^I \phi d_{jR}^I + Y_{ij}^u \overline{Q}_{iL}^I \phi_c u_{jR}^I \\ &= \overline{Q}_L^I Y^d \phi d_R^I + \overline{Q}_L^I Y^u \phi_c u_R^I \end{aligned}$$

is an $\mathrm{SU}(3)_C \times \mathrm{SU}(2)_L \times \mathrm{U}(1)_Y$*-invariant Yukawa form (the second line is an abbreviation). The Hermitian scalar product over the components in colour space* $\mathbb{C}^3$ *in the terms on the right is implicit.*

Proof This is the first part of Exercise 8.11.16. □

We can find pairs of unitary matrices V_L^u, V_R^u and V_L^d, V_R^d that diagonalize the matrices Y^u and Y^d:

$$V_L^u Y^u V_R^{u\dagger} = D^u = \begin{pmatrix} g_u & & \\ & g_c & \\ & & g_t \end{pmatrix},$$

$$V_L^d Y^d V_R^{d\dagger} = D^d = \begin{pmatrix} g_d & & \\ & g_s & \\ & & g_b \end{pmatrix},$$

where all entries of the diagonal matrices are real and positive. We define new quarks

$$\begin{pmatrix} u_{L,R} \\ c_{L,R} \\ t_{L,R} \end{pmatrix} = V_{L,R}^u \begin{pmatrix} u_{L,R}^I \\ c_{L,R}^I \\ t_{L,R}^I \end{pmatrix},$$

$$\begin{pmatrix} d'_{L,R} \\ s'_{L,R} \\ b'_{L,R} \end{pmatrix} = V_{L,R}^u \begin{pmatrix} d_{L,R}^I \\ s_{L,R}^I \\ b_{L,R}^I \end{pmatrix}.$$

Since we transformed up-type and down-type quarks with the same matrices V_L^u and V_R^u, these quarks define a new splitting of the representation spaces into direct summands as in Eqs. (8.27) and (8.28). The quarks u, d', c, s', t, b' can be identified with the quarks we considered before in Sect. 8.5.

We also define new quarks

$$\begin{pmatrix} d_{L,R} \\ s_{L,R} \\ b_{L,R} \end{pmatrix} = V_{L,R}^d \begin{pmatrix} d_{L,R}^I \\ s_{L,R}^I \\ b_{L,R}^I \end{pmatrix}$$

$$= V_{L,R}^d V_{L,R}^{u\dagger} \begin{pmatrix} d'_{L,R} \\ s'_{L,R} \\ b'_{L,R} \end{pmatrix}.$$

Definition 8.8.5 We call the basis vectors corresponding to the quarks u, d, c, s, t, b **mass eigenstates** and the basis corresponding to u, d', c, s', t, b' **weak eigenstates** or **current eigenstates** (for reasons that become apparent below). We also define the quark masses

$$m_i = \frac{1}{\sqrt{2}} g_i v$$

for each flavour $i = u, d, c, s, t, b$.
We then get:

Theorem 8.8.6 (Yukawa Coupling for Quarks) *After symmetry breaking the Yukawa Lagrangian for the three quark generations associated to the Yukawa form in Lemma 8.8.4 is given in unitary gauge and in the mass eigenstate basis (u, d, c, s, t, b) by*

$$\boxed{\begin{aligned}\mathscr{L}_{Y,q} = &-2m_u\mathrm{Re}(\overline{u}_L u_R) - 2m_c\mathrm{Re}(\overline{c}_L c_R) - 2m_t\mathrm{Re}(\overline{t}_L t_R)\\ &- 2m_d\mathrm{Re}(\overline{d}_L d_R) - 2m_s\mathrm{Re}(\overline{s}_L s_R) - 2m_b\mathrm{Re}(\overline{b}_L b_R)\\ &- \frac{2}{v}m_u\mathrm{Re}(\overline{u}_L u_R)H - \frac{2}{v}m_c\mathrm{Re}(\overline{c}_L c_R)H - \frac{2}{v}m_t\mathrm{Re}(\overline{t}_L t_R)H\\ &- \frac{2}{v}m_d\mathrm{Re}(\overline{d}_L d_R)H - \frac{2}{v}m_s\mathrm{Re}(\overline{s}_L s_R)H - \frac{2}{v}m_b\mathrm{Re}(\overline{b}_L b_R)H.\end{aligned}} \tag{8.29}$$

Here the Hermitian scalar products over the colour space $\mathbb{C}^3$ are implicit.

Proof This is the second part of Exercise 8.11.16. □

The six terms in the first and second line are the Dirac mass terms for the quarks. The terms in the third and fourth line are residual interactions between the quarks and the Higgs boson (see the Feynman diagram in Fig. 8.19). The couplings of the quarks to the Higgs boson is again proportional to their mass. The top quark is the heaviest fermion and thus has the strongest coupling to the Higgs boson.

Remark 8.8.7 The process that leads to the strongest production of Higgs bosons at the hadron collider LHC is the so-called **gluon fusion process** [60, 147] with a

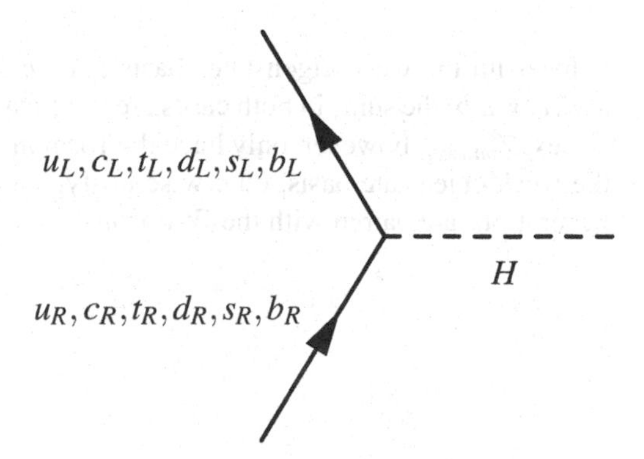

Fig. 8.19 Quark-Higgs interaction vertex

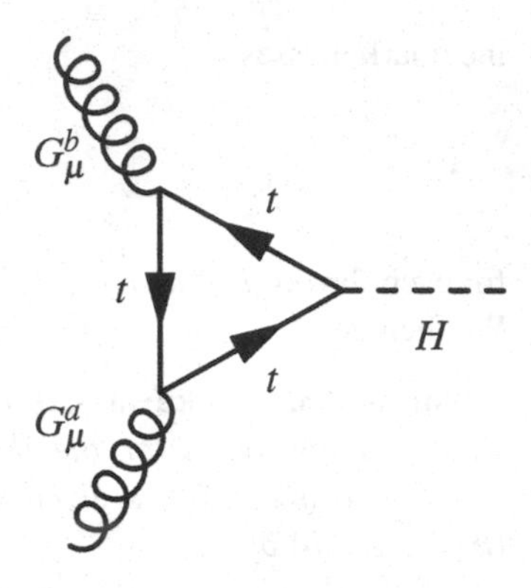

Fig. 8.20 Gluon-fusion production of Higgs boson

virtual top quark loop depicted in Fig. 8.20. There are corresponding processes with other quark flavours, which are, however, much weaker, because the Higgs boson couples most strongly to the top quark.

Definition 8.8.8 The matrix

$$V_{\text{CKM}} = V^u_L V^{d\dagger}_L \in \text{U}(3)$$

is called the **Cabibbo–Kobayashi–Maskawa (CKM)** matrix. The CKM matrix describes the physical effects of **left-handed quark mixing across generations** from the mass eigenstate basis to the weak eigenstate basis.

Since the matrices $V^d_{L,R} V^{u\dagger}_{L,R}$ are unitary, we can write the Dirac Lagrangian for the strong interaction of quarks

$$\text{Re} \sum_f i \left(\overline{q}^f_L \Gamma^\mu \partial_\mu q^f_L + \overline{q}^f_R \Gamma^\mu \partial_\mu q^f_R + \frac{ig_s}{2} \sum_{a=1}^{8} \left(\overline{q}^f_L \Gamma^\mu G^a_\mu \lambda_a q^f_L + \overline{q}^f_R \Gamma^\mu G^a_\mu \lambda_a q^f_R \right) \right)$$

either with the weak eigenstates basis u, d', c, s', t, b' or the mass eigenstate basis u, d, c, s, t, b (the sums in both cases are identical). The terms in the weak interaction vertex $\mathscr{L}_{int,ew,q}$, however, only have the form in Theorem 8.6.2 if they are written in the weak eigenstate basis, otherwise up-type and down-type quarks from different generations are paired with the W-bosons.

To see this explicitly consider the following charged current term in Eq. (8.18):

$$-\frac{g_w}{\sqrt{2}}\sum_{\alpha=1}^{3}\left(j^{\mu}_{W,q_\alpha}W^{+}_{\mu}+j^{\mu\dagger}_{W,q_\alpha}W^{-}_{\mu}\right) \tag{8.30}$$

with the quark current for the α-th generation

$$j^{\mu}_{W,q_\alpha}=\overline{u}_{\alpha L}\Gamma^{\mu}d'_{\alpha L}.$$

Using the mass eigenstate basis we can write the current equivalently as

$$j^{\mu}_{W,q_\alpha}=\sum_{k=1}^{3}\overline{u}_{\alpha L}\Gamma^{\mu}\left(V_{\text{CKM}}\right)_{\alpha k}d_{kL}. \tag{8.31}$$

It follows that the interactions with the W-bosons can connect quarks from different generations if the CKM matrix is not diagonal.

The Feynman diagram in Fig. 8.21 depicts a typical experimentally observed process where W-bosons pair quarks of different generations: the decay of a neutral kaon (consisting of a down quark and a strange antiquark) into a muon and antimuon:

$$K^0\left(ds^C\right)\longrightarrow\mu+\mu^C.$$

The W^+-boson pairs the up quark from the first generation and the strange quark from the second generation. This process was observed before the existence of a

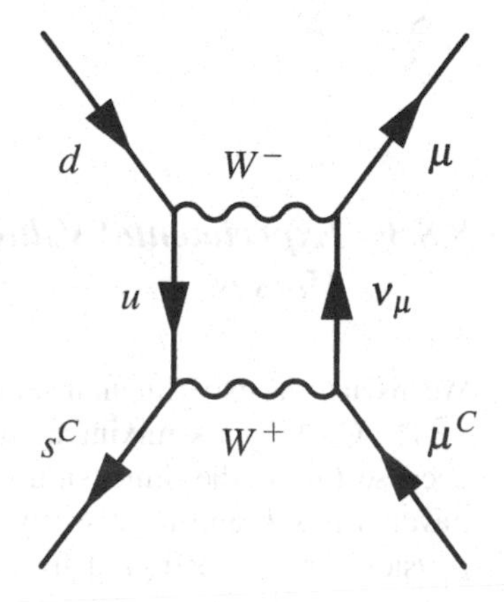

Fig. 8.21 Kaon decay via up quark

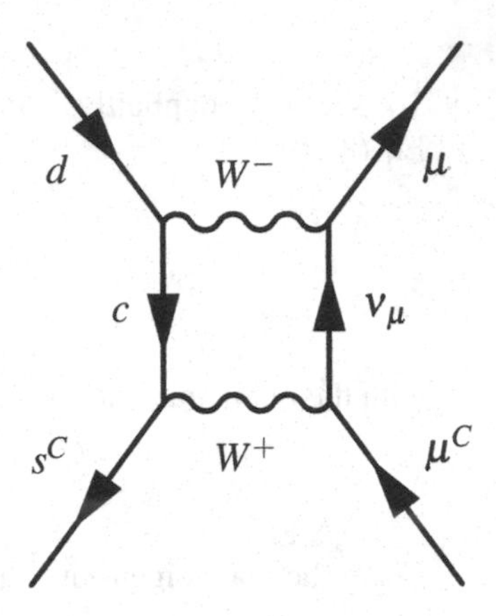

Fig. 8.22 Kaon decay via charm quark

fourth quark (the charm quark) was known. The decay rate of K^0 into μ and μ^C could not be explained with this process alone. This led S.L. Glashow, J. Iliopoulos and L. Maiani in 1970 to the postulation [64] of the charm quark with the process in Fig. 8.22: here the W^--boson pairs the down quark from the first generation with the charm quark from the second generation. These diagrams and much more details can be found in Thomson's book [137].

Summarizing we see that the charged current vertices with W-bosons are the only vertices in the Standard Model that

- connect different flavours of quarks and leptons *(flavour changing vertex)*
- can connect different generations of quarks *(generation changing vertex)*.

Remark 8.8.9 Similarly in general gauge theories the most interesting gauge bosons are those that act off-diagonally in the fermion representations, because they can connect different types of particles (in Grand Unified Theories this leads, for example, to the prediction of proton decay; see Sect. 9.5).

8.8.3 Experimental Values for the CKM Matrix and Fermion Masses

We discuss the experimental values for the CKM matrix (following [100] and [110]).

The CKM quark-mixing matrix $V_{\mathrm{CKM}} \in \mathrm{U}(3)$ has *a priori* nine real parameters, because this is the dimension of the Lie group U(3). We can multiply all quark flavours d, s, b and d', s', b' by an arbitrary phase in U(1) without changing the physics. Any collection of five out of these six real parameters change the matrix

V_{CKM} by multiplying rows and columns with a phase, so that 4 real parameters remain. These parameters are three **mixing angles** $\theta_{12}, \theta_{13}, \theta_{23} \in \left[0, \frac{\pi}{2}\right]$ and the **KM phase** $\delta \in [0, 2\pi)$. The KM phase is also called the **CP violating phase** for reasons that will become clear in Sect. 9.3.

It is possible to show that up to rephasings we can always write

$$V_{\text{CKM}} = \begin{pmatrix} 1 & 0 & 0 \\ 0 & c_{23} & s_{23} \\ 0 & -s_{23} & c_{23} \end{pmatrix} \begin{pmatrix} c_{13} & 0 & s_{13}e^{-i\delta} \\ 0 & 1 & 0 \\ -s_{13}e^{i\delta} & 0 & c_{13} \end{pmatrix} \begin{pmatrix} c_{12} & s_{12} & 0 \\ -s_{12} & c_{12} & 0 \\ 0 & 0 & 1 \end{pmatrix},$$

where

$$s_{ij} = \sin\theta_{ij}, \quad c_{ij} = \cos\theta_{ij}.$$

The matrix V_{CKM} is also often written in the following form:

$$V_{\text{CKM}} = \begin{pmatrix} V_{ud} & V_{us} & V_{ub} \\ V_{cd} & V_{cs} & V_{cb} \\ V_{td} & V_{ts} & V_{tb} \end{pmatrix}.$$

Here the entry V_{ud}, for instance, connects in Eq. (8.31) the down quark to the up quark. The absolute values of the entries of this matrix determined by experiments are approximately [110]

$$\begin{pmatrix} |V_{ud}| & |V_{us}| & |V_{ub}| \\ |V_{cd}| & |V_{cs}| & |V_{cb}| \\ |V_{td}| & |V_{ts}| & |V_{tb}| \end{pmatrix} \approx \begin{pmatrix} 0.97434 & 0.22506 & 0.00357 \\ 0.22492 & 0.97351 & 0.0411 \\ 0.00875 & 0.0403 & 0.99915 \end{pmatrix}. \tag{8.32}$$

The current experimental values for the masses of the quarks and leptons (excluding neutrinos) in absolute value and relative to the mass of the lightest fermion (the electron) can be found in Table 8.12 (see [106]).

Table 8.12 Fermion masses

First generation	Second generation	Third generation
$m_u = 2.15 \pm 0.15$ MeV $\approx 4.2 m_e$	$m_c = 1.28 \pm 0.025$ GeV $\approx 2500 m_e$	$m_t \approx 173$ GeV $\approx 340000 m_e$
$m_d = 4.70 \pm 0.20$ MeV $\approx 9.2 m_e$	$m_s = 93.5 \pm 2$ MeV $\approx 180 m_e$	$m_b = 4.18 \pm 0.03$ GeV $\approx 8200 m_e$
$m_e = 0.5109989461 \pm 0.0000000031$ MeV	$m_\mu = 105.6583745 \pm 0.0000024$ MeV $\approx 210 m_e$	$m_\tau = 1.77686 \pm 0.00012$ GeV $\approx 3500 m_e$

8.8.4 The Yukawa Lagrangian for Fermions

The complete Yukawa Lagrangian for the fermions is the sum

$$\mathscr{L}_Y = \mathscr{L}_{Y,l} + \mathscr{L}_{Y,q} \tag{8.33}$$

with the Yukawa Lagrangian $\mathscr{L}_{Y,l}$ for the leptons from Eq. (8.26) and $\mathscr{L}_{Y,q}$ for the quarks from Eq. (8.29).

8.9 The Complete Lagrangian of the Standard Model

After symmetry breaking the complete Lagrangian of the Standard Model in unitary gauge is the sum of all the boxed Lagrangians above, i.e. the sum of the Lagrangians in Eqs. (8.20), (8.22), (8.33) and (8.25):

$$\mathscr{L} = \mathscr{L}_D + \mathscr{L}_H + \mathscr{L}_Y + \mathscr{L}_{YM}.$$

The Standard Model discussed so far has the following **18 parameters** that have to be determined by experiments:

- The coupling constants g_s, g_w, g' for the gauge group $SU(3)_C \times SU(2)_L \times U(1)_Y$ (equivalently, the strong and electric coupling constants g_s and e, and the Weinberg angle θ_W).
- The parameters μ, λ of the Higgs potential (equivalently, the absolute value v of the Higgs condensate and the mass m_H of the Higgs boson).
- Three Yukawa couplings for the leptons and six Yukawa couplings for the quarks (equivalently, the masses of the leptons and the quarks).
- Three quark mixing angles $\theta_{12}, \theta_{13}, \theta_{23}$ and the KM phase δ that determine the CKM matrix.

8.10 Lepton and Baryon Numbers

We briefly discuss lepton numbers and baryon numbers.

Definition 8.10.1 We make the following definitions:

- the **electron lepton number** L_e is the number of electrons and electron neutrinos minus the number of their antiparticles.

- the **muon lepton number** L_μ and the **tau lepton number** L_τ are defined similarly.
- the **total lepton number** is $L = L_e + L_\mu + L_\tau$.

Inspecting all interaction vertices in the Feynman diagrams above it follows that:

Corollary 8.10.2 (Lepton Number Conservation) *The interactions in the Standard Model conserve all lepton numbers L_e, L_μ, L_τ, L separately.*
For example, an electron can change into an electron neutrino by emitting a W^--boson or an electron and an electron antineutrino can annihilate in a W^--boson. In each process the electron lepton number is conserved.

In contrast to the Standard Model, it turns out that the neutrinos in nature have small non-zero masses. This experimental observation is related to the phenomenon of neutrino oscillations (see Sect. 9.2.2), which implies that the lepton numbers L_e, L_μ, L_τ are in fact not conserved separately, but only the total lepton number L is conserved. This leads to the question of whether **charged lepton flavour violation (CLFV)** might also occur in nature, i.e. processes such as

$$\mu \longrightarrow e + \gamma$$

or

$$\tau \longrightarrow e + \mu^C + \mu,$$

which are not possible in the Standard Model and where some or all of the lepton numbers L_e, L_μ, L_τ are not conserved for the charged leptons e, μ, τ. CLFV is predicted by certain theories beyond the Standard Model. In addition, there may be processes involving lepton number violation (LNV) where the total lepton number L is not conserved. For more details, see [75].

Definition 8.10.3 We define the **baryon number** B to be one third of the number of quarks minus the number of antiquarks:

$$B = \frac{1}{3}\left(q - q^C\right).$$

Again, inspecting all interaction vertices in the Feynman diagrams above it follows that:

Corollary 8.10.4 (Baryon Number Conservation) *The interactions in the Standard Model conserve the baryon number B.*
Because of quark mixing we do not split the baryon number into different generations. Baryons consisting of three quarks, like the proton and neutron, have baryon number $+1$. Mesons consisting of quark-antiquark pairs have baryon number 0.

Some theories beyond the Standard Model, for instance, GUTs with **proton decay** (see Sect. 9.5.8), predict that the baryon number B is not conserved (baryon number violation).

Remark 8.10.5 Most matter particles observed in nature, leptons and hadrons, are unstable and **decay**, usually with very short lifetimes. Decays are due to the weak, electromagnetic and strong interaction, with decays via the weak interaction having the longest lifetime and decays via the strong interaction having the shortest lifetime. Decays are related to the creation and annihilation of particles and can be described by the Feynman diagrams discussed above.

In the Standard Model the only stable leptons are

- the electron e and
- the neutrinos ν_e, ν_μ, ν_τ (as mentioned above there exist, however, oscillations between neutrinos)

and the only stable hadron is

- the proton p,

together with the corresponding antiparticles (isolated neutrons n are unstable with a lifetime of 880 seconds [106], but they can become stable when bound in a nucleus).

Particles that are created in particle colliders and that decay via the weak interaction with lifetimes of 10^{-8} seconds or longer can travel many meters before they decay, due to the relativistic effect of time dilation, and can thus be found in particle collectors. Particles that decay via the strong or electromagnetic interaction are usually too short-lived to be detected themselves [42].

8.11 Exercises for Chap. 8

8.11.1 Consider

$$W = \mathbb{C}^2$$

with the representation

$$\begin{aligned} (\mathrm{SU}(2) \times \mathrm{U}(1)) \times W &\longrightarrow W \\ (A, e^{i\alpha}) \cdot \begin{pmatrix} w_1 \\ w_2 \end{pmatrix} &= A \begin{pmatrix} e^{in_Y\alpha} & 0 \\ 0 & e^{in_Y\alpha} \end{pmatrix} \begin{pmatrix} w_1 \\ w_2 \end{pmatrix}. \end{aligned}$$

Let $V \colon W \to \mathbb{R}$ be a function which is invariant under the action of $\mathrm{SU}(2) \times \mathrm{U}(1)$.

1. Prove that there exists a function $f \colon [0, \infty) \to \mathbb{R}$ such that $V(w) = f(||w||)$, where $||w||^2 = w^\dagger w$.

2. Suppose that V is a polynomial of order at most 4 in the components of w and that V has a minimum, but not in $w = 0$. Prove that $V(w)$ is of the form

$$V(w) = -\mu w^{\dagger} w + \lambda \left(w^{\dagger} w\right)^{2}$$

with certain constants $\mu, \lambda > 0$.

8.11.2 Work out the details in the proof of Theorem 8.1.20 on the existence of unitary gauges in the electroweak theory.

8.11.3 Prove Lemma 8.3.2 on the scalar product on $\mathfrak{su}(2)_L$.

8.11.4 Prove Lemma 8.4.2 on the scalar product on $\mathfrak{su}(3)_C$.

8.11.5 (From [141]) Suppose that the value of the Fermi constant G_F^0 and the electric charge e are known, but the value of the coupling constants g_w, g' and the Weinberg angle θ_W are unknown. Determine lower bounds on the masses of the W- and Z-bosons assuming the electroweak gauge theory.

8.11.6 Derive the weak isospin, weak hypercharge and electric charge of the gauge bosons $W^{\pm}$, Z^0 and γ in Table 8.10.

8.11.7 Prove Theorem 8.5.7 on the subgroup $\mathbb{Z}_6$ in the Standard Model group $\mathrm{SU}(3)_C \times \mathrm{SU}(2)_L \times \mathrm{U}(1)_Y$.

8.11.8 Derive the formulas in Lemma 8.6.1 concerning the action of the electroweak gauge field $W + B$ in the representations $\mathbb{C}^2 \otimes \mathbb{C}_y$ and $\mathbb{C} \otimes \mathbb{C}_y$ of $\mathrm{SU}(2)_L \times \mathrm{U}(1)_Y$.

8.11.9 Prove the formulas for the electroweak interaction vertices of leptons and quarks in Theorem 8.6.2.

8.11.10 Can an electron-positron collider produce hadrons? Which particles can a proton-proton collider produce?

8.11.11 Prove the formulas for the covariant derivative of the Higgs field and for the Higgs potential in Lemma 8.7.1.

8.11.12 Prove the formula for the electroweak Higgs Lagrangian in Theorem 8.7.2.

8.11.13 Prove Lemma 8.7.3 and Proposition 8.7.4 on the curvature of the electroweak gauge field.

8.11.14 Prove Lemma 8.8.1 and Theorem 8.8.2 on the Yukawa Lagrangian for leptons.

8.11.15 Prove Lemma 8.8.3 on the action of $\mathfrak{su}(2)$ on ϕ_c.

8.11.16 Prove Lemma 8.8.4 and Theorem 8.8.6 on the Yukawa Lagrangian for quarks.

8.11.17 The **Georgi–Glashow** SO(3) **model** is described by a Yang–Mills–Higgs Lagrangian over 4-dimensional Minkowski spacetime M with gauge group $G =$

SO(3), Higgs vector space $W = \mathbb{R}^3$ with the fundamental representation of SO(3) and standard scalar product, and Higgs potential

$$V(\phi) = -\mu\phi^T\phi + \lambda\left(\phi^T\phi\right)^2$$

for positive constants μ, λ. Discuss the Higgs mechanism for this theory, determine the masses of the gauge bosons and the explicit form of the Yang–Mills–Higgs Lagrangian in unitary gauge.

Chapter 9
Modern Developments and Topics Beyond the Standard Model

This chapter contains some advanced concepts in particle physics as well as modern developments that aim at going beyond the Standard Model. The topics range from well-established phenomena to more hypothetical theories that predict, for example, the existence of new particles and interactions that so far have not been observed. Most sections in this chapter can be read independently of one another and are only loosely interconnected.

Rather than trying to give a detailed account of these subjects, the intention is to enable the reader to study the extensive research literature him- or herself. Each section is accompanied by some suggestions for further reading. Again, these small guides to the literature do not try to be complete.

9.1 Flavour and Chiral Symmetry

In this section we discuss flavour symmetry and chiral symmetry breaking in QCD (see [36] for more details).

We fix the following data:

- an oriented and time-oriented Lorentzian spin manifold M of even dimension n with metric g of signature $(1, n-1)$ or $(n-1, 1)$
- a spin structure $\text{Spin}^+(M)$ together with complex spinor bundle $S = S_L \oplus S_R \rightarrow M$
- a Dirac form $\langle \cdot , \cdot \rangle$ on the spinor space $\Delta = \Delta_L \oplus \Delta_R$ defined by a matrix A as in Proposition 6.7.13 together with associated Dirac bundle metric $\langle \cdot , \cdot \rangle_S$
- a principal G-bundle $P \rightarrow M$ with compact structure group G of dimension r
- an Ad-invariant positive definite scalar product $\langle \cdot , \cdot \rangle_{\mathfrak{g}}$ on the Lie algebra $\mathfrak{g}$, together with the induced bundle metric $\langle \cdot , \cdot \rangle_{\text{Ad}(P)}$ on the associated vector bundle $\text{Ad}(P)$

M.J.D. Hamilton, *Mathematical Gauge Theory*, Universitext,
https://doi.org/10.1007/978-3-319-68439-0_9

- a complex representation $\rho: G \to \mathrm{GL}(V)$ with associated complex vector bundle $E = P \times_\rho V \to M$
- a G-invariant Hermitian scalar product $\langle \cdot\,, \cdot \rangle_V$ on V with associated bundle metric $\langle \cdot\,, \cdot \rangle_E$ on the vector bundle E. Together with the Dirac form on the spinor bundle S we get a Hermitian scalar product $\langle \cdot\,, \cdot \rangle_{S\otimes E}$ on the twisted spinor bundle $S \otimes E$. We abbreviate $\langle \Psi, \Phi \rangle_{S\otimes E}$ by $\overline{\Psi}\Phi$.

We consider N_f copies E_f, with $f = 1, \ldots, N_f$, of the associated vector bundle E and form the direct sum

$$F = \underbrace{E \oplus \ldots \oplus E}_{N_f}.$$

The different copies of E are called **flavours**. The Hermitian scalar product on F is the direct sum of the scalar products on each copy E_f. The associated Yang–Mills–Dirac Lagrangian is given by

$$\mathscr{L}_{YMD} = \sum_{f=1}^{N_f} \left(\mathrm{Re}\left(\overline{\Psi_f} D_A \Psi_f \right) - m_f \overline{\Psi_f} \Psi_f \right) - \frac{1}{2} \left\langle F_M^A, F_M^A \right\rangle_{\mathrm{Ad}(P)},$$

where Ψ_f is a twisted spinor with values in $S \otimes E_f$ and m_f is the mass of the flavour with index f.

Example 9.1.1 For $M = \mathbb{R}^{1,3}$, $G = \mathrm{SU}(3)$ with the fundamental representation on $V = \mathbb{C}^3$ and six flavours $f = u, d, c, s, t, b$ the Lagrangian $\mathscr{L}_{YMD}$ is the Lagrangian of QCD for quarks of masses m_f. We can consider the Lagrangian $\mathscr{L}_{YMD}$ as a generalization of QCD to an arbitrary compact gauge group G and an arbitrary number of quark flavours. For example, one sometimes considers versions of QCD with gauge group $G = \mathrm{SU}(N_c)$, corresponding to N_c colours.
The Lagrangian $\mathscr{L}_{YMD}$ is gauge invariant under gauge transformations with values in G. However, depending on the quark masses m_f there are additional **global symmetries**, i.e. symmetries which are constant over spacetime. We only consider the Dirac part

$$\mathscr{L}_D = \sum_{f=1}^{N_f} \left(\mathrm{Re}\left(\overline{\Psi_f} D_A \Psi_f \right) - m_f \overline{\Psi_f} \Psi_f \right)$$

of the Lagrangian $\mathscr{L}_{YMD}$, because the global symmetries we consider leave the gauge field A and hence the Yang–Mills Lagrangian invariant.

Lemma 9.1.2 *Suppose that all flavours have the same mass $m_f = m$, for $f = 1, \ldots, N_f$. Let*

$$B \in \mathrm{U}(N_f) = \mathrm{U}(N_f)_V$$

be a (constant, independent of the point in spacetime M) unitary matrix. Then the Dirac Lagrangian $\mathscr{L}_D$ is invariant under the **flavour symmetry**

$$\Psi_{f'}(x) = \sum_{f=1}^{N_f} B_{f'f}\Psi_f(x) \quad \forall x \in M.$$

The flavour symmetry acts through bundle automorphisms of $S \otimes F$, mixing the flavour components.

Recall that the Dirac mass term is equal to

$$-m_f\overline{\Psi_f}\Psi_f = -m_f \mathrm{Re}\left(\overline{\Psi_{Lf}}\Psi_{Rf}\right).$$

The first term in the Dirac Lagrangian is the real part of

$$\overline{\Psi_f}D_A\Psi_f = \overline{\Psi_{Lf}}D_A\Psi_{Lf} + \overline{\Psi_{Rf}}D_A\Psi_{Rf}.$$

Lemma 9.1.3 *Suppose that all flavours have vanishing mass $m_f = 0$, for $f = 1, \ldots, N_f$. Let*

$$(B_L, B_R) \in \mathrm{U}(N_f) \times \mathrm{U}(N_f) = \mathrm{U}(N_f)_L \times \mathrm{U}(N_f)_R$$

be a pair of (constant) unitary matrices. Then the Dirac Lagrangian $\mathscr{L}_D$ is invariant under the **chiral symmetry**

$$\Psi_{Lf'}(x) = \sum_{f=1}^{N_f} B_{Lf'f}\Psi_{Lf}(x) \quad \forall x \in M,$$

$$\Psi_{Rf'}(x) = \sum_{f=1}^{N_f} B_{Rf'f}\Psi_{Rf}(x) \quad \forall x \in M.$$

The chiral symmetry acts through bundle automorphisms of $S_L \otimes F$ and $S_R \otimes F$, mixing the flavours of right-handed and left-handed components separately.

Example 9.1.4 The chiral symmetry is already interesting for a single massless fermion, $N_f = 1$. In this case the chiral symmetry group is

$$\mathrm{U}(1)_L \times \mathrm{U}(1)_R.$$

It is sometimes useful to consider the following subgroups of the chiral symmetry group

$$\mathrm{U}(1)_V = \left\{\left(e^{i\alpha}, e^{i\alpha}\right) \mid \alpha \in \mathbb{R}\right\},$$
$$\mathrm{U}(1)_A = \left\{\left(e^{i\alpha}, e^{-i\alpha}\right) \mid \alpha \in \mathbb{R}\right\},$$

which are called the subgroups of **vector** and **axial** symmetries. The action of the axial symmetry on a spinor $\Psi = \Psi_L + \Psi_R$ is often written in the physics literature as

$$\Psi \longmapsto e^{i\alpha\Gamma_{n+1}}\Psi,$$

where Γ_{n+1} is the physical chirality operator which is $+1$ on Ψ_L and -1 on Ψ_R. As we saw in Lemma 9.1.2 the vector symmetry still exists in the case when the mass m is non-zero.

Example 9.1.5 In QCD the quarks with flavours u, d, s are much lighter than the quarks with flavours c, t, b. If we set $m_u = m_d = m_s = 0$ (this is called the **chiral limit** of QCD), then QCD should have the chiral symmetry $\mathrm{U}(3)_L \times \mathrm{U}(3)_R$.
The chiral symmetry group $\mathrm{U}(N_f)_L \times \mathrm{U}(N_f)_R$ has the following subgroups:

$$\begin{aligned}
\Delta\left(\mathrm{U}(N_f)_L \times \mathrm{U}(N_f)_R\right) &= \left\{(B_L, B_R) \mid B_L, B_R \in \mathrm{U}(N_f), \det B_L = \det B_R\right\},\\
\mathrm{U}(N_f)_V &= \left\{(B, B) \mid B \in \mathrm{U}(N_f)\right\},\\
\mathrm{SU}(N_f)_V &= \left\{(B, B) \mid B \in \mathrm{U}(N_f), \det B = 1\right\},\\
\mathrm{U}(1)_V &= \left\{\left(e^{i\alpha}I, e^{i\alpha}I\right) \mid e^{i\alpha} \in \mathrm{U}(1)\right\},\\
\mathrm{U}(1)_A &= \left\{\left(e^{i\alpha}I, e^{-i\alpha}I\right) \mid e^{i\alpha} \in \mathrm{U}(1)\right\}.
\end{aligned}$$

For the **quantum field theory** associated to QCD in the chiral limit it is known that the full chiral symmetry group $\mathrm{U}(3)_L \times \mathrm{U}(3)_R$ of the classical field theory breaks in two steps to the flavour symmetry group $\mathrm{U}(3)_V$:

$$\mathrm{U}(3)_L \times \mathrm{U}(3)_R \longrightarrow \Delta\left(\mathrm{U}(3)_L \times \mathrm{U}(3)_R\right) \longrightarrow U(3)_V.$$

The reason for the first breaking is that the axial symmetry $\mathrm{U}(1)_A$ does not hold in the quantum theory (because of the **axial anomaly**); note that the discrete subgroup $\mathbb{Z}_{2N_f} = \mathbb{Z}_6 \subset \mathrm{U}(1)_A$ is still contained in $\Delta\left(\mathrm{U}(3)_L \times \mathrm{U}(3)_R\right)$. The second breaking is called **chiral symmetry breaking**, which happens because the vacuum state of QCD is not invariant under the full symmetry group $\Delta\left(\mathrm{U}(3)_L \times \mathrm{U}(3)_R\right)$ (this is an example of spontaneous symmetry breaking of a global symmetry).

Chiral symmetry breaking is not fully understood theoretically, because it happens at low energies (similar to **quark confinement**), where the strong coupling constant is large and perturbation theory is not valid; see Sect. 9.4. However, it can be studied in numerical simulations using lattice QCD and analytically in supersymmetric generalizations of QCD, which are better understood non-perturbatively (see [127, 128]).

The unbroken symmetries of the quantum version of QCD, in addition to gauge symmetry, are therefore

- the abelian symmetry $\mathrm{U}(1)_V$ (related to baryon number conservation)
- the (special) flavour symmetry $\mathrm{SU}(3)_V$.

Baryon number conservation is an exact symmetry of QCD, independent of quark masses. Flavour symmetry, however, in real world QCD, where the quark masses m_u, m_d, m_s are not precisely the same, is only an approximate symmetry, which can nevertheless still be observed in the spectrum of hadrons (mesons and baryons) that are composed of quarks with flavours u, d, s. Historically, flavour symmetry $\mathrm{SU}(3)_V$ preceded the development of QCD with gauge symmetry $\mathrm{SU}(3)_C$ (the dimension 8 of the flavour symmetry group is related to the concept of the *eightfold way*, developed by Murray Gell-Mann in the 1960s).

To summarize, in QCD the Lie group SU(3) appears in two completely different places:

- as the colour gauge group $\mathrm{SU}(3)_C$. Gauge symmetry in QCD is a *local* and *exact* symmetry.
- as the flavour symmetry group $\mathrm{SU}(3)_V$. Flavour symmetry in QCD is a *global* and with non-identical quark masses only *approximate* symmetry.

Remark 9.1.6 **Non-perturbative QCD** is an active and difficult area of research with many open questions. Understanding non-perturbative QCD in particular would mean to understand the structure of hadrons theoretically.

It is not known at present, to mention only one example, how to calculate the mass of the proton and neutron (which are well-known from experiments) analytically from first principles using a "formula" (there are numerical calculations from first principles using **lattice QCD** which are accurate to within a few percents). The proton consists of two up and one down valence quark and the neutron of two down and one up valence quarks. The masses of the proton and neutron are [106]

$$\begin{aligned} m_p &= 938.272\,\mathrm{MeV}, \\ m_n &= 939.565\,\mathrm{MeV}. \end{aligned}$$

However, the sum of the quark rest masses for the proton and neutron are approximately 9 MeV and 11.6 MeV, i.e. they only amount to roughly 1% of the proton and neutron mass. This leads to the remarkable conclusion that 99% of the mass of the visible matter in the universe comes from the binding energy of the gluon field inside the proton and neutron (and hence does not have its origin in the Higgs mechanism) [11, 137].

9.1.1 Further Reading

Reference [21] is an extensive review of strongly coupled QCD. The notes [36] contain a discussion of chiral symmetry breaking. References [44] and [18] are examples of precise calculations of hadron masses using lattice gauge theory.

9.2 Massive Neutrinos

In the Standard Model, neutrinos are assumed massless. According to experiments, however, neutrinos show *oscillations* between different flavours, for example, a free electron neutrino can change in-flight into a muon neutrino or tau neutrino and back (first observed at the Super-Kamiokande detector in 1998; Nobel Prize in Physics 2015 for Takaaki Kajita and Arthur B. McDonald). This can only be explained if neutrinos have different and hence (at least two of them) non-zero masses.

In this section we discuss how mass terms for neutrinos can be added to the Standard Model as well as the phenomenon of neutrino oscillations. The neutrino masses are extremely small compared to the other fermions in the Standard Model and an obvious question is why neutrinos are so light. The famous *seesaw mechanism*, that we discuss in Sect. 9.2.5, is one natural explanation (we follow references [62] and [137] throughout this section).

9.2.1 *Dirac Mass Terms*

To define Dirac mass terms for neutrinos we have to postulate the existence of new particles, **right-handed neutrinos** ν_{iR} for each generation $i = e, \mu, \tau$. Like right-handed electrons, muons and taus, right-handed neutrinos are singlets in the trivial representation of $\mathrm{SU}(3)_C \times \mathrm{SU}(2)_L$. However, because they have zero electric charge, this implies that their weak hypercharge Y is also zero. Hence right-handed neutrinos, if they exist, are **sterile**, i.e. live in the trivial singlet representation of the full Standard Model gauge group

$$G = \mathrm{SU}(3)_C \times \mathrm{SU}(2)_L \times \mathrm{U}(1)_Y$$

and do not interact with other particles via gauge bosons. They can only interact with other particles via the Higgs boson (if their mass is generated by the Higgs mechanism) and gravity. For this reason, sterile neutrinos are one of the candidates to explain *dark matter* in the universe.

If we add right-handed neutrinos to the Standard Model, the particle content of the lepton sector becomes formally very similar to the quark sector, see Table 9.1: there are one left-handed isodoublet and two right-handed isosinglets for each generation. The Dirac mass generation for neutrinos using Yukawa couplings is a straightforward generalization of the mass generation for quarks in Sect. 8.8.2: we would like to determine the most general gauge invariant Yukawa form that generates masses for both electron-type particles and neutrinos. We consider the

Table 9.1 Lepton representations including right-handed neutrinos

Sector	$SU(2)_L \times U(1)_Y$ representation	Basis vectors	Particle	Charges		
				$\mathbb{T}_3$	$\mathbb{Y}$	$\mathbb{Q}$
L_L^1	$\mathbb{C}^2 \otimes \mathbb{C}_{-1}$	$\begin{pmatrix} 1 \\ 0 \end{pmatrix}$	ν_{eL}	$\frac{1}{2}$	-1	0
		$\begin{pmatrix} 0 \\ 1 \end{pmatrix}$	e_L	$-\frac{1}{2}$	-1	-1
L_R^1	$\mathbb{C} \otimes \mathbb{C}_0$	1	ν_{eR}	0	0	0
	$\mathbb{C} \otimes \mathbb{C}_{-2}$	1	e_R	0	-2	-1

following $SU(2)_L \times U(1)_Y$ representation spaces:

$$\bigoplus_{i=1}^{3} \left(\mathbb{C}^2 \otimes \mathbb{C}_{-1}\right) \quad \text{(left-handed)} \tag{9.1}$$

and

$$\bigoplus_{i=1}^{3} ((\mathbb{C} \otimes \mathbb{C}_0) \oplus (\mathbb{C} \otimes \mathbb{C}_{-2})) \quad \text{(right-handed).} \tag{9.2}$$

We write

$$\nu_1^I = \nu_e^I, \nu_2^I = \nu_\mu^I, \nu_3^I = \nu_\tau^I,$$
$$e_1^I = e^I, e_2^I = \mu^I, e_3^I = \tau^I$$

for the irreducible summands in Eqs. (9.1) and (9.2) and indicate the left-handed basis by an index L and the right-handed basis by an index R. We also write for the left-handed isodoublets

$$L_{iL}^I = \begin{pmatrix} \nu_{iL}^I \\ e_{iL}^I \end{pmatrix},$$
$$\overline{L}_{iL}^I = (\overline{\nu}_{iL}^I, \overline{e}_{iL}^I).$$

In complete analogy to Lemma 8.8.4 we have:

Lemma 9.2.1 *For arbitrary complex matrices Y^e and Y^ν the expression*

$$\begin{aligned}\tau_L &= Y^e_{ij}\overline{L}^I_{iL}\phi e^I_{jR} + Y^\nu_{ij}\overline{L}^I_{iL}\phi_c \nu^I_{jR}\\ &= \overline{L}^I_L Y^e \phi e^I_R + \overline{L}^I_L Y^\nu \phi_c \nu^I_R\end{aligned}$$

is an $\mathrm{SU}(2)_L \times \mathrm{U}(1)_Y$*-invariant Yukawa form (the second line is an abbreviation).*
We can find pairs of unitary matrices U^e_L, U^e_R and U^ν_L, U^ν_R that diagonalize the matrices Y^e and Y^ν:

$$U^e_L Y^e U^{e\dagger}_R = D^e = \begin{pmatrix} g_e & & \\ & g_\mu & \\ & & g_\tau \end{pmatrix},$$

$$U^\nu_L Y^\nu U^{\nu\dagger}_R = D^\nu = \begin{pmatrix} g_{\nu_e} & & \\ & g_{\nu_\mu} & \\ & & g_{\nu_\tau} \end{pmatrix},$$

where all entries of the diagonal matrices are real and positive. We define new leptons

$$\begin{pmatrix} \nu_{eL,R} \\ \nu_{\mu L,R} \\ \nu_{\tau L,R} \end{pmatrix} = U^e_{L,R} \begin{pmatrix} \nu^I_{eL,R} \\ \nu^I_{\mu L,R} \\ \nu^I_{\tau L,R} \end{pmatrix},$$

$$\begin{pmatrix} e_{L,R} \\ \mu_{L,R} \\ \tau_{L,R} \end{pmatrix} = U^e_{L,R} \begin{pmatrix} e^I_{L,R} \\ \mu^I_{L,R} \\ \tau^I_{L,R} \end{pmatrix}.$$

These leptons define a new splitting of the representation spaces into direct summands as in Eqs. (9.1) and (9.2).

We also define new neutrinos

$$\begin{aligned}\begin{pmatrix} \nu_{1L,R} \\ \nu_{2L,R} \\ \nu_{3L,R} \end{pmatrix} &= U^\nu_{L,R} \begin{pmatrix} \nu^I_{eL,R} \\ \nu^I_{\mu L,R} \\ \nu^I_{\tau L,R} \end{pmatrix}\\ &= U^\nu_{L,R} U^{e\dagger}_{L,R} \begin{pmatrix} \nu_{eL,R} \\ \nu_{\mu L,R} \\ \nu_{\tau L,R} \end{pmatrix}.\end{aligned}$$

Definition 9.2.2 We call the basis vectors corresponding to the leptons $e, \nu_1, \mu, \nu_2, \tau, \nu_3$ **mass eigenstates** and the basis corresponding to $e, \nu_e, \mu, \nu_\mu, \tau, \nu_\tau$ **weak eigenstates**.

Suppose that the Higgs field ϕ is in unitary gauge,

$$\phi = \frac{1}{\sqrt{2}} \begin{pmatrix} 0 \\ v + H \end{pmatrix}$$

and define the lepton masses

$$m_i = \frac{1}{\sqrt{2}} g_i v$$

for each flavour $i = e, \nu_1, \mu, \nu_2, \tau, \nu_3$.
We then get:

Theorem 9.2.3 (Yukawa Coupling Including Dirac Mass Terms for Neutrinos) *After symmetry breaking the Yukawa Lagrangian for the three lepton generations associated to the Yukawa form in Lemma 9.2.1, including right-handed neutrinos, is given in unitary gauge and in the mass eigenstate basis* $(e, \nu_1, \mu, \nu_2, \tau, \nu_3)$ *by*

$$\begin{aligned} \mathscr{L}_L^D = &-2m_e \mathrm{Re}(\overline{e}_L e_R) - 2m_\mu \mathrm{Re}(\overline{\mu}_L \mu_R) - 2m_\tau \mathrm{Re}(\overline{\tau}_L \tau_R) \\ &- 2m_{\nu_1} \mathrm{Re}(\overline{\nu}_{1L} \nu_{1R}) - 2m_{\nu_2} \mathrm{Re}(\overline{\nu}_{2L} \nu_{2R}) - 2m_{\nu_3} \mathrm{Re}(\overline{\nu}_{3L} \nu_{3R}) \\ &- \frac{2}{v} m_e \mathrm{Re}(\overline{e}_L e_R) H - \frac{2}{v} m_\mu \mathrm{Re}(\overline{\mu}_L \mu_R) H - \frac{2}{v} m_\tau \mathrm{Re}(\overline{\tau}_L \tau_R) H \\ &- \frac{2}{v} m_{\nu_1} \mathrm{Re}(\overline{\nu}_{1L} \nu_{1R}) H - \frac{2}{v} m_{\nu_2} \mathrm{Re}(\overline{\nu}_{2L} \nu_{2R}) H - \frac{2}{v} m_{\nu_3} \mathrm{Re}(\overline{\nu}_{3L} \nu_{3R}) H. \end{aligned} \tag{9.3}$$

The only difference to the Yukawa Lagrangian in Theorem 8.8.2 is that the Yukawa Lagrangian in Theorem 9.2.3 contains (potentially non-zero) masses for the neutrinos and interactions between the neutrinos and the Higgs boson, depicted in the Feynman diagram in Fig. 9.1. We also see that there is in general a non-trivial **neutrino mixing** from the mass eigenstate basis to the weak eigenstate basis.

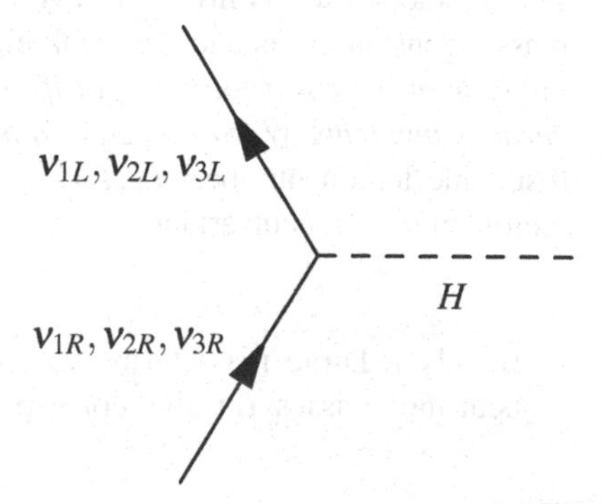

Fig. 9.1 Neutrino-Higgs interaction vertex

Definition 9.2.4 The unitary matrix

$$U_{\text{PMNS}} = U_L^e U_L^{\nu\dagger} \in \text{U}(3)$$

is called the **Pontecorvo–Maki–Nakagawa–Sakata (PMNS) matrix**. The PMNS matrix plays the same role for the **mixing of left-handed neutrinos across generations** as the CKM matrix for the left-handed down quarks.

The PMNS matrix is often written in the following form:

$$U_{\text{PMNS}} = \begin{pmatrix} U_{e1} & U_{e2} & U_{e3} \\ U_{\mu 1} & U_{\mu 2} & U_{\mu 3} \\ U_{\tau 1} & U_{\tau 2} & U_{\tau 3} \end{pmatrix}.$$

This matrix appears in the flavour changing electroweak interaction vertices involving $W^{\pm}$-bosons if we want to write them in the mass eigenstate basis instead of the weak eigenstate basis: consider the following charged current term in Eq. (8.17):

$$-\frac{g_w}{\sqrt{2}}\left(j_{W,l}^{\mu} W_{\mu}^{+} + j_{W,l}^{\mu\dagger} W_{\mu}^{-}\right)$$

with the Hermitian conjugate of the lepton current

$$j_{W,l}^{\mu\dagger} = \overline{e}_L \Gamma^{\mu} \nu_{eL} + \overline{\mu}_L \Gamma^{\mu} \nu_{\mu L} + \overline{\tau}_L \Gamma^{\mu} \nu_{\tau L}.$$

Using the mass eigenstate basis we can write the current equivalently as

$$j_{W,l}^{\mu\dagger} = \sum_{j=1}^{3} \left(\overline{e}_L \Gamma^{\mu} \left(U_{\text{PMNS}}\right)_{ej} \nu_{jL} + \overline{\mu}_L \Gamma^{\mu} \left(U_{\text{PMNS}}\right)_{\mu j} \nu_{jL} + \overline{\tau}_L \Gamma^{\mu} \left(U_{\text{PMNS}}\right)_{\tau j} \nu_{jL}\right). \tag{9.4}$$

The interaction terms involving Z-bosons and photons γ are the same in both the mass eigenstate basis and the weak eigenstate basis. *Similarly to our discussion in the case of quarks it follows that if neutrinos have a Dirac mass, then interaction vertices involving W-bosons can connect leptons from different generations.* As a result the lepton numbers L_e, L_{μ}, L_{τ} are not conserved separately, only the total lepton number L is invariant.

If only a Dirac mass term is present for neutrinos, it is unclear why the neutrino masses (or the corresponding Yukawa couplings) are so much

(continued)

smaller than for the other fermions. If we add to the Lagrangian a Majorana mass term for the right-handed neutrinos, the seesaw mechanism described in Sect. 9.2.5 allows both: a Dirac mass of similar size to the masses of the other fermions *and* a neutrino of very small mass. In addition, the theory then predicts another neutrino of very large mass (that so far has not been observed).

9.2.2 Neutrino Oscillations

In this section we discuss how different masses of neutrinos lead to the phenomenon of *neutrino oscillations*. We just describe the basic idea. Consider the following elementary lemma.

Lemma 9.2.5 *Let E_j and $\mathbf{p}_j$ denote the energies and 3-momenta of relativistic particles with rest mass m_j, where $j = 1, 2$. Suppose that*

$$E_1 = E_2 \quad \textit{and} \quad \mathbf{p}_1 = \mathbf{p}_2.$$

Then $m_1 = m_2$.

Proof Energy and 3-momentum determine the rest mass according to the formula $m^2 = E^2 - \mathbf{p}^2$ (where $c = 1$). This implies the claim. □

To simplify the discussion we only consider two neutrino generations, say the electron neutrino and the muon neutrino (we could similarly analyse the case of three generations). Let $|\nu_j\rangle$ for $j = 1, 2$ denote the mass eigenstates of neutrinos with mass m_j. For a freely moving neutrino in one of the mass eigenstates, the spacetime dependence of the state is given by

$$|\nu_j(\mathbf{x}, t)\rangle = e^{i(\mathbf{p}_j \cdot \mathbf{x} - E_j t)} |\nu_j\rangle.$$

Clearly, the probability of finding the j-th neutrino at the spacetime point $(\mathbf{x}, t)$ in the state $|\nu_j\rangle$ is 1:

$$|\langle \nu_j | \nu_j(\mathbf{x}, t)\rangle|^2 = |e^{i(\mathbf{p}_j \cdot \mathbf{x} - E_j t)}|^2 = 1.$$

On the other hand consider, for example, the electron neutrino ν_e. As a state it decomposes as

$$|\nu_e\rangle = \sum_{j=1}^{2} U_{ej} |\nu_j\rangle,$$

where U is the neutrino mixing matrix corresponding to the PMNS matrix for three generations. We have

$$\langle \nu_e| = \sum_{j=1}^{2} U_{ej}^{*} \langle \nu_j|$$

and

$$|\langle \nu_e|\nu_e\rangle|^2 = \left| \sum_{j=1}^{2} |U_{ej}|^2 \right|^2 = 1$$

since the matrix U is unitary.

The spacetime dependence of the electron neutrino state is given by

$$\begin{aligned} |\nu_e(\mathbf{x}, t)\rangle &= \sum_{j=1}^{2} U_{ej} |\nu_j(\mathbf{x}, t)\rangle \\ &= \sum_{j=1}^{2} U_{ej} e^{i(\mathbf{p}_j \cdot \mathbf{x} - E_j t)} |\nu_j\rangle. \end{aligned}$$

The probability of finding the state at the spacetime point $(\mathbf{x}, t)$ in the state $|\nu_e\rangle$ is

$$|\langle \nu_e|\nu_e(\mathbf{x}, t)\rangle|^2 = \left| \sum_{j=1}^{2} |U_{ej}|^2 e^{i(\mathbf{p}_j \cdot \mathbf{x} - E_j t)} \right|^2 .$$

Proposition 9.2.6 *Suppose that $U_{ej} \neq 0$ for $j = 1, 2$. If*

$$|\langle \nu_e|\nu_e(\mathbf{x}, t)\rangle|^2 = 1 \quad \forall (\mathbf{x}, t) \in \mathbb{R}^4,$$

then $m_1 = m_2$.

Proof Since the matrix U is unitary we have

$$\begin{aligned} |\langle \nu_e|\nu_e(\mathbf{x}, t)\rangle|^2 &= |U_{e1}|^4 + |U_{e2}|^4 + 2|U_{e1}|^2|U_{e2}|^2 \mathrm{Re}\left(e^{i((\mathbf{p}_1 - \mathbf{p}_2) \cdot \mathbf{x} - (E_1 - E_2)t))} \right) \\ &= 1 - 2|U_{e1}|^2|U_{e2}|^2 \left(1 - \mathrm{Re}\left(e^{i((\mathbf{p}_1 - \mathbf{p}_2) \cdot \mathbf{x} - (E_1 - E_2)t))} \right) \right), \end{aligned}$$

which is equal to 1 for all $(\mathbf{x}, t) \in \mathbb{R}^4$ only if $E_1 = E_2$ and $\mathbf{p}_1 = \mathbf{p}_2$. The claim then follows by Lemma 9.2.5. □

This means that if the masses m_1 and m_2 are not the same and $U_{ej} \neq 0$ for $j = 1, 2$, then there will be spacetime points $(\mathbf{x}, t)$ where

$$|\langle \nu_e | \nu_e(\mathbf{x}, t) \rangle|^2 < 1.$$

But this implies that there is a non-zero probability of finding $|\nu_e(\mathbf{x}, t)\rangle$ in the state $|\nu_\mu\rangle$:

$$|\langle \nu_\mu | \nu_e(\mathbf{x}, t) \rangle|^2 \neq 0.$$

These spacetime points are related to the values of the phases $e^{i(\mathbf{p}_j \cdot \mathbf{x} - E_j t)}$ and occur periodically. This is the mechanism of **neutrino oscillations**: there is a non-zero probability of finding a neutrino, which is in a weak eigenstate at one spacetime point $(0, 0)$, in another weak eigenstate at another spacetime point $(\mathbf{x}, t)$.

For more details, see the excellent discussion in Thomson's book [137]. It is possible to show that under certain assumptions the converse to Proposition 9.2.6 holds: if neutrino oscillations occur, then the masses of the neutrinos cannot be the same. These neutrino oscillations are indeed observed in experiments. A detailed calculation shows that using neutrino oscillations only the differences

$$\Delta m_{ij}^2 = m_i^2 - m_j^2$$

can be determined experimentally, but not the absolute values of m_i for $i = 1, 2, 3$.

9.2.3 Experimental Values for the PMNS Matrix and Neutrino Masses

We briefly discuss experimental values for neutrino masses and the PMNS matrix (following [107] and [137, Sect. 13.9]). Since there exists a large number of neutrinos in the universe, neutrino masses potentially have cosmological implications. Recent cosmological measurements set a rough upper limit of

$$\sum_{i=1}^{3} m_i \lesssim 1\text{eV}.$$

Direct measurements of the averaged electron (anti)neutrino mass m_{ν_e}, defined by the square root of

$$m_{\nu_e}^2 = \sum_{j=1}^{3} |U_{ej}|^2 m_j^2,$$

(in particular, for neutrinos from the supernova SN 1987A and β-decay of Tritium ^{3}H) give similar upper bounds [43]. From neutrino oscillation experiments the following is known:

$$\Delta m_{21}^2 = m_2^2 - m_1^2 = 7.53 \pm 0.18 \cdot 10^{-5}\mathrm{eV}^2,$$

$$|\Delta m_{32}^2| = \begin{cases} m_3^2 - m_2^2 = 2.44 \pm 0.06 \cdot 10^{-3}\mathrm{eV}^2 & \text{(assuming normal mass hierarchy)}, \\ m_2^2 - m_3^2 = 2.51 \pm 0.06 \cdot 10^{-3}\mathrm{eV}^2 & \text{(assuming inverted mass hierarchy)}. \end{cases}$$

The sign of Δm_{32}^2 is not known, hence it is not clear whether $m_3 > m_2 > m_1$ **(normal mass hierarchy)** or $m_3 < m_1 < m_2$ **(inverted mass hierarchy)**.

We can write the PMNS matrix in a similar way to the CKM matrix using neutrino mixing angles:

$$U_{\mathrm{PMNS}} = \begin{pmatrix} 1 & 0 & 0 \\ 0 & c_{23} & s_{23} \\ 0 & -s_{23} & c_{23} \end{pmatrix} \begin{pmatrix} c_{13} & 0 & s_{13}e^{-i\delta} \\ 0 & 1 & 0 \\ -s_{13}e^{i\delta} & 0 & c_{13} \end{pmatrix} \begin{pmatrix} c_{12} & s_{12} & 0 \\ -s_{12} & c_{12} & 0 \\ 0 & 0 & 1 \end{pmatrix},$$

where

$$s_{ij} = \sin\theta_{ij}, \quad c_{ij} = \cos\theta_{ij}.$$

The following values are known:

$$\sin^2(\theta_{12}) = 0.304^{+0.014}_{-0.013},$$

$$\sin^2(\theta_{23}) = \begin{cases} 0.51 \pm 0.05 & \text{(assuming normal mass hierarchy)}, \\ 0.50 \pm 0.05 & \text{(assuming inverted mass hierarchy)}, \end{cases}$$

$$\sin^2(\theta_{13}) = 0.0219 \pm 0.0012.$$

The phase δ is unknown. This leads to the following approximate absolute values for the entries of the PMNS matrix:

$$\begin{pmatrix} |U_{e1}| & |U_{e2}| & |U_{e3}| \\ |U_{\mu 1}| & |U_{\mu 2}| & |U_{\mu 3}| \\ |U_{\tau 1}| & |U_{\tau 2}| & |U_{\tau 3}| \end{pmatrix} \approx \begin{pmatrix} 0.83 & 0.54 & 0.15 \\ 0.40 & 0.60 & 0.70 \\ 0.40 & 0.60 & 0.70 \end{pmatrix}.$$

We conclude that the PMNS matrix is clearly much less diagonal than the CKM matrix in Eq. (8.32).

9.2.4 Majorana Mass Terms

Majorana mass terms allow us to introduce a non-zero mass for neutrinos without postulating the existence of right-handed neutrinos. We begin by considering only the first generation. We want to define a Majorana mass term

$$-m_{\nu_e}\mathrm{Re}\,(\tilde{\nu}_{eL}\nu_{eL}),$$

where

$$\tilde{\nu}_{eL} = \nu_{eL}^T C$$

denotes the Majorana conjugate. This expression is not gauge invariant, because the left-handed electron neutrino comes in an isodoublet with the left-handed electron:

$$L_{eL} = \begin{pmatrix} \nu_{eL} \\ e_L \end{pmatrix}$$

with Majorana conjugate

$$\tilde{L}_{eL} = (\tilde{\nu}_{eL}, \tilde{e}_L).$$

We need a generalization of the Higgs mechanism: consider the Higgs doublet

$$\phi = \frac{1}{\sqrt{2}}\begin{pmatrix} \phi^+ \\ \phi^0 \end{pmatrix}.$$

Lemma 9.2.7 *For all matrices $A \in \mathrm{SU}(2)$ the following identity holds:*

$$A^T\sigma_2 A = \sigma_2.$$

Hence the expression

$$\phi^T\sigma_2 L_{eL}$$

is a left-handed Weyl spinor, invariant under $\mathrm{SU}(2)_L \times \mathrm{U}(1)_Y$. Its Majorana conjugate is $-\tilde{L}_{eL}\sigma_2\phi$.

Proof This follows from Lemma 8.8.3, because $A^T = (A^*)^{-1}$ for $A \in \mathrm{SU}(2)$. □

Using the pure uncharged section $\phi^T\sigma_2 L_{eL}$ of the left-handed spinor bundle S_L, we can now form a Majorana mass term:

Proposition 9.2.8 *For constants g_{ν_e} and $\mathcal{M}$ the expression*

$$\mathcal{L}_{L_e}^M = -\frac{2g_{\nu_e}}{\mathcal{M}}\mathrm{Re}\left[\left(\tilde{L}_{eL}\sigma_2\phi\right)\left(\phi^T\sigma_2 L_{eL}\right)\right]$$

is Lorentz and $\mathrm{SU}(2)_L \times \mathrm{U}(1)_Y$ invariant.

Suppose that the Higgs field ϕ is in unitary gauge,

$$\phi = \frac{1}{\sqrt{2}}\begin{pmatrix} 0 \\ v+H \end{pmatrix}$$

and define the neutrino mass

$$m_{\nu_e} = \frac{g_{\nu_e} v^2}{\mathcal{M}}.$$

Corollary 9.2.9 (Majorana Mass Terms for Neutrinos) *After symmetry breaking the Lagrangian $\mathcal{L}^M_{L_e}$ is given in unitary gauge by*

$$\begin{aligned}\mathcal{L}^M_{L_e} &= -m_{\nu_e}\mathrm{Re}\,(\tilde{\nu}_{eL}\nu_{eL})\\ &\quad - \frac{2m_{\nu_e}}{v}\mathrm{Re}\,(\tilde{\nu}_{eL}\nu_{eL})\,H - \frac{m_{\nu_e}}{v^2}\mathrm{Re}\,(\tilde{\nu}_{eL}\nu_{eL})\,H^2.\end{aligned}$$

The term in the first line is the Majorana mass term for the electron neutrino, the terms in the second line are hypothetical interactions between the electron neutrino and the Higgs boson.

The coupling between two scalars and two spinors in the Lagrangian $\mathcal{L}^M_{L_e}$ (or between two spinors and the square of the Higgs boson after symmetry breaking) does not appear in Chap. 7 and is *non-renormalizable*. We conclude that a gauge invariant Majorana mass term for left-handed neutrinos can only be added to the Lagrangian using non-renormalizable interactions. The Lagrangian $\mathcal{L}^M_{L_e}$ is understood as an *effective Lagrangian* that hides new interactions beyond the Standard Model. This can be compared to the non-renormalizable 4-Fermi Lagrangian for the weak force that was hiding gauge interactions (see the beginning of Sect. 8.3).

If g_{ν_e} is dimensionless, then $\mathcal{M}$ must have dimension of mass so that $\mathcal{L}^M_{L_e}$ has the dimension of a Lagrangian. The constant $\mathcal{M}$ is interpreted as a new (large) energy scale. If we assume that $g_{\nu_e} \sim 0.01$ and $v \sim 10^2$ GeV, as in the Standard Model, as well as $\mathcal{M} \sim 10^{11}$ GeV (below a typical Grand Unification scale), then $m_{\nu_e} = 0.01$ eV. This is in agreement with experiments.

Remark 9.2.10 Note that for right-handed sterile neutrinos ν_{eR}, the Majorana mass term

$$-m_{\nu_e}\mathrm{Re}\,(\tilde{\nu}_{eR}\nu_{eR})$$

is already gauge invariant and thus well-defined without the need to introduce the Higgs field. We will combine this term with the Dirac mass term in Sect. 9.2.5.

9.2.5 *Dirac–Majorana Mass Terms and the Seesaw Mechanism*

We again make the assumption from Sect. 9.2.1 that there exist right-handed sterile neutrinos in each of the three lepton generations. We saw that a Dirac mass term can then be generated in a gauge invariant way via the Higgs mechanism. Furthermore, in Remark 9.2.10 we saw that a gauge invariant Majorana mass term for the right-handed neutrinos can be added to the Lagrangian (we ignore the Majorana mass term for the left-handed neutrinos that can only be added in a gauge invariant and renormalizable way by introducing new interactions).

Let m_D be the Dirac mass and M the Majorana mass of the right-handed neutrino (we only consider one generation). Then the Dirac mass term is

$$\mathscr{L}^D = -2m_D \mathrm{Re}(\overline{\nu}_L \nu_R)$$

and the Majorana mass term is

$$\mathscr{L}^M = -M\mathrm{Re}(\tilde{\nu}_R \nu_R).$$

Recall from Sect. 6.7 and Sect. 6.8 the definitions

$$\begin{aligned} \tilde{\psi} &= \psi^T C, \\ \overline{\psi} &= \psi^\dagger A, \\ \psi^C &= B^{-1}\psi^*, \end{aligned}$$

where for 4-dimensional Minkowski spacetime with signature $(+,-,-,-)$

$$\begin{aligned} C &= -C^T = -C^{-1} = i\Gamma^2\Gamma^0, \\ A &= \Gamma^0, \\ B &= B^{-1} = CA = i\Gamma^2. \end{aligned}$$

It follows that

$$\overline{\psi^C} = \tilde{\psi}.$$

Remark 9.2.11 We assume from now for the rest of this section that **components of spinors anticommute**.

Lemma 9.2.12 *The Majorana mass term can be written as*

$$\mathscr{L}^M = -M\mathrm{Re}\left(\nu_R^{CT} C \nu_R^C\right).$$

Proof We calculate

$$\begin{aligned}
\mathrm{Re}\left(\nu_R^{CT} C \nu_R^C\right) &= \mathrm{Re}\left(\nu_R^\dagger (B^{-1})^T C \nu_R^C\right) \\
&= \mathrm{Re}\left(\nu_R^\dagger A \nu_R^C\right) \\
&= \mathrm{Re}\left(\nu_R^\dagger A B^{-1} \nu_R^*\right) \\
&= -\mathrm{Re}\left(\nu_R^\dagger C \nu_R^*\right) \\
&= \mathrm{Re}(\tilde{\nu}_R \nu_R).
\end{aligned}$$

In the last step we used that $(\alpha\beta)^* = -\alpha^*\beta^*$ for Grassmann numbers α, β. □

Lemma 9.2.13 *The Dirac mass term can be written as*

$$\begin{aligned}
\mathscr{L}^D &= -2m_D\mathrm{Re}\left(\nu_R^{CT} C \nu_L\right) \\
&= -m_D\mathrm{Re}\left(\nu_L^T C \nu_R^C + \nu_R^{CT} C \nu_L\right).
\end{aligned}$$

The second line follows from the first, because spinor components anticommute.

Proof We calculate as in Lemma 9.2.12

$$2m_D\mathrm{Re}\left(\nu_R^{CT} C \nu_L\right) = 2m_D\mathrm{Re}\left(\overline{\nu}_R \nu_L\right) = 2m_D\mathrm{Re}\left(\overline{\nu}_L \nu_R\right).$$

□

Proposition 9.2.14 *The sum of the Dirac and Majorana mass term can be written as*

$$\begin{aligned}
\mathscr{L}^{D+M} &= \mathscr{L}^D + \mathscr{L}^M \\
&= -\mathrm{Re}\left((\nu_L^T, \nu_R^{CT}) C \begin{pmatrix} 0 & m_D \\ m_D & M \end{pmatrix} \begin{pmatrix} \nu_L \\ \nu_R^C \end{pmatrix}\right).
\end{aligned}$$

The basis ν_L, ν_R^C is not a mass eigenbasis for $m_D \neq 0$, because the matrix is not diagonal. The following is easy to verify:

Lemma 9.2.15 *The matrix*

$$\begin{pmatrix} 0 & m_D \\ m_D & M \end{pmatrix}$$

has real eigenvalues

$$\frac{M \pm \sqrt{M^2 + 4m_D^2}}{2}.$$

If $M \gg m_D > 0$, then one eigenvalue m_+ is positive and the other eigenvalue $-m_-$ is negative, where

$$m_+ \approx M,$$

$$m_- \approx \frac{m_D^2}{M}.$$

Proposition 9.2.16 (Diagonalization of Dirac–Majorana Mass Terms) *Let R be an orthogonal matrix such that*

$$\begin{pmatrix} 0 & m_D \\ m_D & M \end{pmatrix} = R \begin{pmatrix} m_+ & 0 \\ 0 & -m_- \end{pmatrix} R^T.$$

Define the unitary matrix

$$U = R \begin{pmatrix} 1 & 0 \\ 0 & i \end{pmatrix}$$

and set

$$\begin{pmatrix} \nu_{+L} \\ \nu_{-L} \end{pmatrix} = U^T \begin{pmatrix} \nu_L \\ \nu_R^C \end{pmatrix}.$$

Then the Dirac–Majorana mass term can be written as

$$\mathscr{L}^{D+M} = -\mathrm{Re}\left(m_+ \tilde{\nu}_{+L} \nu_{+L} + m_- \tilde{\nu}_{-L} \nu_{-L}\right).$$

We introduce the spinors

$$\nu_\pm = \frac{1}{\sqrt{2}}\left(\nu_{\pm L} + \nu_{\pm L}^C\right).$$

Then

$$\nu_\pm^C = \nu_\pm,$$

hence $\nu_\pm$ are **Majorana spinors**. Moreover,

$$\overline{\nu}_\pm \nu_\pm = \frac{1}{2}\left(\overline{\nu}_{\pm L}\nu^C_{\pm L} + \overline{\nu^C_{\pm L}}\nu_{\pm L}\right)$$
$$= \mathrm{Re}\,(\tilde{\nu}_{\pm L}\nu_{\pm L})\,.$$

We get:

Corollary 9.2.17 *With the Majorana spinors $\nu_\pm$ the Dirac–Majorana mass term can be written as the Dirac mass term*

$$\mathscr{L}^{D+M} = -m_+\overline{\nu}_+\nu_+ - m_-\overline{\nu}_-\nu_-.$$

The relation between m_+ and m_- for fixed m_D is called a **seesaw** mechanism: if m_+ is large, then m_- is small and if m_- is small, then m_+ is large. The seesaw mechanism allows a Dirac mass m_D (generated by the Higgs mechanism) to be of the same size as for the other fermions of the Standard Model (around 1 GeV) and to have at the same time a small neutrino mass m_- (around 0.01 eV). The theory then predicts another neutrino which has a very large mass m_+ (around 10^{11} GeV).

Considering Dirac–Majorana mass terms for all three generations there is again a phenomenon of neutrino mixing; see [62] for details.

9.2.6 Further Reading

The book [137] by M. Thomson contains many details on neutrino experiments as well as theoretical discussions of neutrino mixing and neutrino oscillations. The book [62] by C. Giunti and C.W. Kim is a comprehensive exposition of neutrino physics, including astrophysical and cosmological implications.

9.3 C, P and CP Violation

In addition to Lorentz symmetries, gauge symmetries and global continuous symmetries (like flavour symmetry), Lagrangians (or actions) can also have **discrete symmetries**. Charge conjugation C and parity inversion P, together with time reversal T, are examples of such symmetries. If the action is not invariant under the composition CP of P followed by C, the theory is said to **violate CP**. It is an important observation that the action of the Standard Model, because of quark mixing and the weak interaction of quarks, indeed violates CP.

Violation of CP in general is linked to the asymmetry between the amount of matter and antimatter in the universe. However, the CP violation coming from the weak interaction of quarks alone does not suffice to explain the observed degree of this asymmetry. There are two other instances in the Standard Model where CP violation can possibly occur: in the interaction of neutrinos, related to neutrino mixing, and perhaps in the strong interaction of quarks if an additional CP violating term is added to the Lagrangian (although this is almost certainly ruled out by experiments, leading to the so-called *strong CP problem*, i.e. to explain why the strong interaction preserves CP).

We only discuss CP violation coming from quark mixing and the weak interaction. Throughout this section we follow the excellent exposition in the book [22], where more details can be found.

9.3.1 The CKM Matrix and the Jarlskog Invariant

We consider the following form of the CKM matrix from Sect. 8.8.3:

$$V_{\text{CKM}} = \begin{pmatrix} V_{ud} & V_{us} & V_{ub} \\ V_{cd} & V_{cs} & V_{cb} \\ V_{td} & V_{ts} & V_{tb} \end{pmatrix}.$$

To understand CP violation in the Standard Model, it will be important to know in which sense the CKM matrix is real or complex. We can change the entries of the CKM matrix without changing the physical content by multiplying any row or column by a phase in U(1). This corresponds to changing the entries by

$$V'_{ij} = e^{i(\alpha_j - \alpha_i)} V_{ij}, \tag{9.5}$$

where $\alpha_i, \alpha_j \in [0, 2\pi)$ are arbitrary angles. Only expressions which are invariant under these rephasings are physically relevant. For example, the term

$$|V_{ij}|^2 = V_{ij} V_{ij}^*$$

is invariant (the experimental values of these norms can be found in Eq. (8.32)). Another invariant term is

$$Q_{ijkl} = V_{ij} V_{kl} V_{il}^* V_{kj}^*,$$

which changes under the rephasings in Eq. (9.5) by the factor

$$e^{i(\alpha_j - \alpha_i)} e^{i(\alpha_l - \alpha_k)} e^{i(\alpha_i - \alpha_l)} e^{i(\alpha_k - \alpha_j)} = 1.$$

Definition 9.3.1 The **Jarlskog invariant** is

$$J = \text{Im}\, Q_{uscb} = \text{Im}(V_{us}V_{cb}V_{ub}^*V_{cs}^*).$$

Proposition 9.3.2 *Suppose that all entries of the CKM matrix V_{CKM} are non-zero. Then the Jarlskog invariant vanishes, $J = 0$, if and only if V_{CKM} can be brought by the rephasings in Eq.* (9.5) *into real form (i.e. all entries of the matrix become real).*

Proof It is possible to show that we can always bring the CKM matrix with rephasings into the form

$$\begin{aligned} V_{\text{CKM}} &= \begin{pmatrix} 1 & 0 & 0 \\ 0 & c_{23} & s_{23} \\ 0 & -s_{23} & c_{23} \end{pmatrix} \begin{pmatrix} c_{13} & 0 & s_{13}e^{-i\delta} \\ 0 & 1 & 0 \\ -s_{13}e^{i\delta} & 0 & c_{13} \end{pmatrix} \begin{pmatrix} c_{12} & s_{12} & 0 \\ -s_{12} & c_{12} & 0 \\ 0 & 0 & 1 \end{pmatrix} \\ &= \begin{pmatrix} c_{12}c_{13} & s_{12}c_{13} & s_{13}e^{-i\delta} \\ -s_{12}c_{23} - c_{12}s_{23}s_{13}e^{i\delta} & c_{12}c_{23} - s_{12}s_{23}s_{13}e^{i\delta} & s_{23}c_{13} \\ s_{12}s_{23} - c_{12}c_{23}s_{13}e^{i\delta} & -c_{12}s_{23} - s_{12}c_{23}s_{13}e^{i\delta} & c_{23}c_{13} \end{pmatrix}, \end{aligned}$$

where

$$s_{ij} = \sin\theta_{ij}, \quad c_{ij} = \cos\theta_{ij}.$$

We get

$$\begin{aligned} J &= \text{Im}Q_{uscb} \\ &= \text{Im}\left(s_{12}s_{23}c_{13}^2s_{13}e^{i\delta}\left(c_{12}c_{23} - s_{12}s_{23}s_{13}e^{-i\delta}\right)\right) \\ &= s_{12}s_{23}c_{13}^2s_{13}c_{12}c_{23}\sin\delta. \end{aligned}$$

Since all entries of the CKM matrix are non-zero, the factor in front of $\sin\delta$ is non-zero. Hence $J = 0$ if and only if $e^{i\delta} = \pm 1$, which happens if and only if the CKM matrix can be brought into real form. □

According to [110] the experimental value for the Jarlskog invariant is

$$J = 3.04^{+0.21}_{-0.20} \cdot 10^{-5}.$$

This is a small but non-zero number, and implies:

Corollary 9.3.3 *The CKM matrix for quark mixing realized in nature cannot be brought into real form by the rephasings in Eq.* (9.5).

Remark 9.3.4 A similar discussion applies to the PMNS matrix describing neutrino mixing. However, as mentioned in Sect. 9.2.3, the analogue of the Jarlskog invariant (or the phase δ) is currently not known for the PMNS matrix. It is likely that $\delta \neq$

$0, \pi$ so that the PMNS matrix cannot be brought into real form by rephasings, but so far this cannot be excluded completely [1].

Remark 9.3.5 Exercise 9.7.1 shows that any unitary 2×2-matrix can be brought into real form by rephasings. This is the reason that led Makoto Kobayashi and Toshihide Maskawa in 1972 (Nobel Prize in Physics 2008) to postulate the existence of a third quark generation.

9.3.2 C and P Transformations

There is some freedom how to define charge conjugation C and parity inversion P in field theories. For the following notation it is useful to think in this section of fields as *quantum fields*, i.e. fields on spacetime with values in the operators on a Hilbert space. Then C and P are unitary operators on this Hilbert space and a field ϕ transforms as

$$\phi \longrightarrow C\phi C^\dagger$$
$$\phi \longrightarrow P\phi P^\dagger.$$

We consider 4-dimensional Minkowski spacetime of signature $(+,-,-,-)$ with the conventions for spinors from Sect. 6.8. Note that according to these conventions

$$\Gamma^0\Gamma^\mu\Gamma^0 = \Gamma^{\mu\dagger} = \Gamma_\mu.$$

We first define **parity inversion** P. On spacetime, P acts by

$$x = x^\mu = (t, \mathbf{x}) \overset{P}{\longmapsto} x_p = (t, -\mathbf{x}) = x_\mu.$$

This implies for partial derivatives

$$\partial_\mu \overset{P}{\longmapsto} \partial^\mu.$$

On a complex scalar field $\phi(x)$ we define

$$P\phi(x)P^\dagger = e^{i\alpha_p}\phi(x_p),$$

which implies

$$P\phi^\dagger(x)P^\dagger = e^{-i\alpha_p}\phi^\dagger(x_p).$$

Here α_p is an arbitrary real constant. On a spinor field $\psi(x)$ we define

$$P\psi(x)P^\dagger = e^{i\beta_p}\Gamma^0\psi(x_P),$$

which implies for $\overline{\psi} = \psi^\dagger \Gamma^0$

$$P\overline{\psi}(x)P^\dagger = e^{-i\beta_p}\overline{\psi}(x_p)\Gamma^0.$$

Here β_p is an arbitrary real constant.

Proposition 9.3.6 (Parity Inversion and Invariance of Actions) *Define parity inversion P on a gauge field A_μ by*

$$PA_\mu(x)P^\dagger = A^\mu(x_p).$$

Then the Klein–Gordon Lagrangian

$$\mathcal{L}_{KG} = \left(\partial^\mu\phi^\dagger + \phi^\dagger A^{\mu\dagger}\right)\left(\partial_\mu\phi + A_\mu\phi\right) - m^2\phi^\dagger\phi,$$

the Higgs Lagrangian

$$\mathcal{L}_H = \left(\partial^\mu\phi^\dagger + \phi^\dagger A^{\mu\dagger}\right)\left(\partial_\mu\phi + A_\mu\phi\right) - V\left(\phi^\dagger\phi\right),$$

the Dirac Lagrangian

$$\mathcal{L}_D = \mathrm{Re}\left(\overline{\psi}i\Gamma^\mu(\partial_\mu + A_\mu)\psi\right) - m\overline{\psi}\psi$$

and the Yang–Mills Lagrangian

$$\begin{aligned}
\mathcal{L}_{YM} &= -\frac{1}{4}F^{Aa}_{\mu\nu}F^{A\mu\nu}_a \\
&= -\frac{1}{4}(\partial_\mu A^a_\nu - \partial_\nu A^a_\mu)(\partial^\mu A^\nu_a - \partial^\nu A^\mu_a) \\
&\quad - \frac{1}{2}f_{abc}(\partial_\mu A^a_\nu - \partial_\nu A^a_\mu)A^{b\mu}A^{c\nu} \\
&\quad - \frac{1}{4}f_{abc}f_{ade}A^b_\mu A^c_\nu A^{d\mu}A^{e\nu}
\end{aligned}$$

transform as

$$\begin{aligned}
P\mathcal{L}_{KG}(x)P^\dagger &= \mathcal{L}_{KG}(x_p), \\
P\mathcal{L}_H(x)P^\dagger &= \mathcal{L}_H(x_p), \\
P\mathcal{L}_D(x)P^\dagger &= \mathcal{L}_D(x_p), \\
P\mathcal{L}_{YM}(x)P^\dagger &= \mathcal{L}_{YM}(x_p).
\end{aligned}$$

In particular, the Klein–Gordon, Higgs, Dirac and Yang–Mills actions (the space-time integrals over the Lagrangians) are invariant under parity inversion.

Proof This is Exercise 9.7.2. □

We now define **charge conjugation** C. On a complex scalar field $\phi(x)$ we define

$$C\phi(x)C^\dagger = e^{i\alpha_c}\phi^*(x),$$

which implies

$$C\phi^\dagger(x)C^\dagger = e^{-i\alpha_c}\phi^T(x).$$

Here α_c is an arbitrary real constant. Let ψ be a spinor field. Recall from Sect. 6.8 that

$$\psi^C = B^{-1}\psi^* = i\Gamma^2\psi^*.$$

We set

$$\begin{aligned} C\psi(x)C^\dagger &= e^{i\beta_c}\psi^C(x) \\ &= e^{i\beta_c}i\Gamma^2\psi^*, \end{aligned}$$

which implies

$$\begin{aligned} C\overline{\psi}(x)C^\dagger &= e^{-i\beta_c}\overline{\psi^C}(x) \\ &= e^{-i\beta_c}i\psi^T(x)\Gamma^2\Gamma^0. \end{aligned}$$

Here β_c is an arbitrary real constant.

Proposition 9.3.7 (Charge Conjugation and Invariance of Actions) *Define charge conjugation C on a gauge field A_μ by*

$$CA_\mu(x)C^\dagger = A_\mu^*(x),$$

where $A_\mu = -A_\mu^\dagger$ is the matrix-valued gauge field in a given unitary representation of the compact gauge group G. Then the Klein–Gordon Lagrangian $\mathscr{L}_{KG}$, the Higgs Lagrangian $\mathscr{L}_H$, the Dirac Lagrangian $\mathscr{L}_D$ and the Yang–Mills Lagrangian $\mathscr{L}_{YM}$ transform as

$$\begin{aligned} C\mathscr{L}_{KG}(x)C^\dagger &= \mathscr{L}_{KG}(x), \\ C\mathscr{L}_H(x)C^\dagger &= \mathscr{L}_H(x), \\ C\mathscr{L}_D(x)C^\dagger &= \mathscr{L}_D(x) + \textit{total derivative}, \\ C\mathscr{L}_{YM}(x)C^\dagger &= \mathscr{L}_{YM}(x). \end{aligned}$$

This implies that the Klein–Gordon, Higgs, Dirac and Yang–Mills actions are invariant under charge conjugation.

Remark 9.3.8 In the case of the Dirac Lagrangian we assume in this proposition that the field operators ψ and $\psi^\dagger$ **anticommute**.

Proof We do the calculation for the Dirac Lagrangian and leave the remaining cases to Exercise 9.7.3. The Dirac mass term transforms as

$$\begin{aligned} Cm\overline{\psi}\psi C &= -m\psi^T\Gamma^2\Gamma^0\Gamma^2\psi^* \\ &= m\psi^T\Gamma^2\Gamma^2\Gamma^0\psi^* \\ &= -m\psi^T\Gamma^0\psi^* \\ &= m\left(\psi^\dagger\Gamma^{0T}\psi\right)^T \\ &= m\overline{\psi}\psi. \end{aligned}$$

In the step from the third to the fourth line we used that ψ^T and ψ^* anticommute. The term involving the Dirac operator transforms as

$$\begin{aligned} C\overline{\psi}i\Gamma^\mu(\partial_\mu + A_\mu)\psi C &= -\psi^T\Gamma^2\Gamma^0 i\Gamma^\mu(\partial_\mu + A_\mu^*)\Gamma^2\psi^* \\ &= \psi^T\Gamma^0\Gamma^2 i\Gamma^\mu\Gamma^2(\partial_\mu + A_\mu^*)\psi^* \\ &= \psi^T\Gamma^0 i\Gamma^{\mu *}(\partial_\mu + A_\mu^*)\psi^* \\ &= -\left((\partial_\mu\psi^\dagger)i\Gamma^{\mu\dagger}\Gamma^{0T}\psi\right)^T - \left(\psi^\dagger A_\mu^\dagger i\Gamma^{\mu\dagger}\Gamma^{0T}\psi\right)^T \\ &= -i\partial_\mu\left(\overline{\psi}\Gamma^\mu\psi\right) + \overline{\psi}i\Gamma^\mu(\partial_\mu + A_\mu)\psi. \end{aligned}$$

From the second to the third line we used that $\Gamma^2\Gamma^\mu\Gamma^2 = \Gamma^{\mu *}$, from the third to the fourth line that ψ^T and ψ^* anticommute and from the fourth to the fifth line that $\Gamma^{\mu\dagger}\Gamma^0 = \Gamma^0\Gamma^\mu$ and $A_\mu^\dagger = -A_\mu$. □

Definition 9.3.9 The transformation *CP* is defined by first applying *P* and then applying *C*.
From the formulas above we see that the transformation *CP* is given on complex scalar fields ϕ by

$$\begin{aligned} (CP)\phi(x)(CP)^\dagger &= e^{i\alpha_{cp}}\phi^*(x_p), \\ (CP)\phi^\dagger(x)(CP)^\dagger &= e^{-i\alpha_{cp}}\phi^T(x_p) \end{aligned}$$

(continued)

Definition 9.3.9 (continued)
and on spinors ψ by

$$(CP)\psi(x)(CP)^{\dagger} = e^{i\beta_{cp}} i\Gamma^2\Gamma^0\psi^*(x_p),$$
$$(CP)\overline{\psi}(x)(CP)^{\dagger} = -e^{-i\beta_{cp}} i\psi^T(x_p)\Gamma^2,$$

where α_{cp}, β_{cp} are arbitrary real numbers.

Corollary 9.3.10 *Define the CP transformation on a gauge field A_μ by*

$$(CP)A_\mu(x)(CP)^{\dagger} = A^{\mu *}(x_p).$$

Then the Klein–Gordon, Higgs, Dirac and Yang–Mills actions are CP invariant.
We shall see in the following section that if we introduce in addition to these Lagrangians a Yukawa Lagrangian, then the complete action may no longer be CP invariant.

9.3.3 CP Violation in the Standard Model

We want to prove the following theorem.

Theorem 9.3.11 (CP Invariance and the Jarlskog Invariant) *The action of the Standard Model (with vanishing neutrino masses) is CP invariant if and only if the Jarlskog invariant J of the CKM quark mixing matrix vanishes.*

Proof We follow the argument in [22] and use anticommuting spinors throughout the proof. We first assume that the action of the Standard Model is CP invariant and consider the Lagrangian of the Standard Model without fixing a particular gauge like the unitary gauge. We write the Higgs field as

$$\phi = \frac{1}{\sqrt{2}}\begin{pmatrix}\phi^+\\ \phi^0\end{pmatrix}$$

and set $\phi^- = \phi^{+*}$. We only need to consider the following parts of the complete Lagrangian:

- From the term $(\nabla^{A\mu}\phi)^{\dagger}\left(\nabla^A_\mu\phi\right)$ in the Higgs Lagrangian the part

$$\mathscr{L}_1 = \frac{1}{2}\left(\partial^\mu\phi^{0*} - \frac{ig_w}{\sqrt{2}}W^{+\mu}\phi^- + \frac{ig_w}{2\cos\theta_W}Z^{0\mu}\phi^{0*}\right)$$
$$\cdot\left(\partial_\mu\phi^0 + \frac{ig_w}{\sqrt{2}}W^-_\mu\phi^+ - \frac{ig_w}{2\cos\theta_W}Z^0_\mu\phi^0\right)$$

- From the Dirac Lagrangian the following charged current part of the electroweak interaction vertex with quarks in the mass eigenstate basis:

$$\mathscr{L}_2 = -\frac{g_w}{\sqrt{2}} \sum_{\alpha=1}^{3} \left(j^{\mu}_{W,q_\alpha} W^+_\mu + j^{\mu\dagger}_{W,q_\alpha} W^-_\mu \right)$$

with

$$j^{\mu}_{W,q_\alpha} = \sum_{k=1}^{3} \overline{u}_{\alpha L} \Gamma^\mu V_{\alpha k} d_{kL}$$

and

$$j^{\mu\dagger}_{W,q_\alpha} = \sum_{k=1}^{3} \overline{d}_{kL} \Gamma^\mu V^*_{\alpha k} u_{\alpha L}$$

as in Eqs. (8.31) and (8.30). Here we use the index α to denote up-type quarks and k to denote down-type quarks. The matrix V is the CKM matrix.

- The Yukawa form for the quarks in Lemma 8.8.4 before symmetry breaking is in the mass eigenstate basis

$$\tau_Q = \left(\overline{u}_L V, \overline{d}_L\right) D^d \phi d_R + \left(\overline{u}_L, \overline{d}_L V^\dagger\right) D^u \phi_c u_R,$$

where V is the CKM matrix. The Yukawa Lagrangian for the quarks before symmetry breaking can then be written as

$$\mathscr{L}_3 = -\frac{1}{\sqrt{2}} \mathrm{Re} \left(\sum_k g_k \overline{d}_{kL} \phi^0 d_{kR} + \sum_\alpha g_\alpha \overline{u}_{\alpha L} \phi^{0*} u_{\alpha R} \right.$$
$$\left. + \sum_{\alpha,k} g_k \overline{u}_{\alpha L} V_{\alpha k} \phi^+ d_{kR} - \sum_{\alpha,k} g_\alpha \overline{d}_{kL} V^*_{\alpha k} \phi^- u_{\alpha R} \right).$$

This is equal to

$$\begin{aligned}
\mathscr{L}_3 = &\frac{1}{2\sqrt{2}} \phi^0 \left(-\sum_k g_k \overline{d}_{kL} d_{kR} - \sum_\alpha g_\alpha \overline{u}_{\alpha R} u_{\alpha L} \right) \\
&+ \frac{1}{2\sqrt{2}} \phi^{0*} \left(-\sum_k g_k \overline{d}_{kR} d_{kL} - \sum_\alpha g_\alpha \overline{u}_{\alpha L} u_{\alpha R} \right) \\
&+ \frac{1}{2\sqrt{2}} \phi^+ \left(-\sum_k g_k \overline{u}_{\alpha L} V_{\alpha k} d_{kR} + \sum_\alpha g_\alpha \overline{u}_{\alpha R} V_{\alpha k} d_{kL} \right) \\
&+ \frac{1}{2\sqrt{2}} \phi^- \left(-\sum_k g_k \overline{d}_{kR} V^*_{\alpha k} u_{\alpha L} + \sum_\alpha g_\alpha \overline{d}_{kL} V^*_{\alpha k} u_{\alpha R} \right).
\end{aligned} \tag{9.6}$$

We now argue as follows (without using $\mathscr{L}_2$): under the CP transformation the field ϕ^+ transforms as

$$(CP)\phi^+(CP)^\dagger = e^{i\xi_W}\phi^-$$

for some real number ξ_W (we do not write the change of argument $x \mapsto x_p$). This implies

$$(CP)\phi^-(CP)^\dagger = e^{-i\xi_W}\phi^+.$$

The Lagrangian $\mathscr{L}_1$ is then CP invariant only if we transform

$$\begin{aligned}
(CP)W_\mu^+(CP)^\dagger &= -e^{i\xi_W}W^{-\mu},\\
(CP)W_\mu^-(CP)^\dagger &= -e^{-i\xi_W}W^{+\mu},\\
(CP)\phi^0(CP)^\dagger &= \phi^{0*},\\
(CP)Z_\mu^0(CP)^\dagger &= -Z^{0\mu}.
\end{aligned}$$

The first line of the Lagrangian $\mathscr{L}_3$ in Eq. (9.6) is invariant under the transformations

$$\begin{aligned}
(CP)u_\alpha(CP)^\dagger &= e^{i\xi_\alpha}i\Gamma^2\Gamma^0u_\alpha^*,\\
(CP)\overline{u}_\alpha(CP)^\dagger &= -e^{-i\xi_\alpha}iu_\alpha^T\Gamma^2,\\
(CP)d_k(CP)^\dagger &= e^{i\xi_k}i\Gamma^2\Gamma^0d_k^*,\\
(CP)\overline{d}_k(CP)^\dagger &= -e^{-i\xi_k}id_k^T\Gamma^2,
\end{aligned}$$

where ξ_α, ξ_k are arbitrary real numbers (and spinor components are anticommuting). Consider the term

$$\phi^+\left(\sum_\alpha g_\alpha\overline{u}_{\alpha R}V_{\alpha k}d_{kL}\right)$$

in $\mathscr{L}_3$. Its CP transform is

$$\begin{aligned}
(CP)\phi^+\left(\sum_\alpha g_\alpha\overline{u}_{\alpha R}V_{\alpha k}d_{kL}\right)(CP)^\dagger &= -e^{i(\xi_W-\xi_\alpha+\xi_k)}\phi^-\left(\sum_\alpha g_\alpha u_{\alpha R}^T\Gamma^0V_{\alpha k}d_{kL}^*\right)\\
&= e^{i(\xi_W-\xi_\alpha+\xi_k)}\phi^-\left(\sum_\alpha g_\alpha\overline{d}_{kL}V_{\alpha k}u_{\alpha R}\right).
\end{aligned}$$

Comparing with the last term in $\mathscr{L}_3$ we see that CP invariance of the Lagrangian implies that

$$V^*_{\alpha k} = e^{i(\xi_W - \xi_\alpha + \xi_k)} V_{\alpha k}.$$

This implies

$$\begin{aligned} Q^*_{uscb} &= V^*_{us} V^*_{cb} V_{ub} V_{cs} \\ &= V_{us} V_{cb} V^*_{ub} V^*_{cs} \\ &= Q_{uscb}. \end{aligned}$$

Hence Q_{uscb} is real and the Jarlskog invariant $J = 0$.

We can argue similarly with the Lagrangian $\mathscr{L}_2$: consider the term

$$\sum_{\alpha,k} \overline{u}_{\alpha L} \Gamma^\mu V_{\alpha k} d_{kL} W^+_\mu .$$

Its CP transform is

$$\begin{aligned} (CP)\left(\sum_{\alpha,k} \overline{u}_{\alpha L} \Gamma^\mu V_{\alpha k} d_{kL} W^+_\mu\right)(CP)^\dagger &= -e^{i(\xi_W - \xi_\alpha + \xi_k)} \sum_{\alpha,k} u^T_{\alpha L} \Gamma^2 \Gamma^\mu \Gamma^2 \Gamma^0 V_{\alpha k} d^*_{kL} W^{-\mu} \\ &= -e^{i(\xi_W - \xi_\alpha + \xi_k)} \sum_{\alpha,k} u^T_{\alpha L} \Gamma^{\mu *} \Gamma^0 V_{\alpha k} d^*_{kL} W^{-\mu} \\ &= e^{i(\xi_W - \xi_\alpha + \xi_k)} \sum_{\alpha,k} \overline{d}_{kL} \Gamma_\mu V_{\alpha k} u_{\alpha L} W^{-\mu}. \end{aligned}$$

Comparing this with the second term in $\mathscr{L}_2$ we conclude again that

$$V^*_{\alpha k} = e^{i(\xi_W - \xi_\alpha + \xi_k)} V_{\alpha k},$$

hence $J = 0$.

Conversely, it can be shown that if $J = 0$, then all terms in the Standard Model Lagrangian are CP invariant. □

Corollary 9.3.12 *Since in nature* $J \approx 3.04 \cdot 10^{-5} \neq 0$, *the action of the Standard Model is not CP invariant.*

A similar discussion can be done with the PMNS matrix describing neutrino mixing. If the corresponding Jarlskog invariant is non-zero, then neutrino mixing also leads to CP violation.

9.3.4 Further Reading

The book [22] by G.C. Branco, L. Lavoura and J.P. Silva is a very good source for numerous details on CP violation. The paper [66] by W. Grimus and M.N. Rebelo is a mathematical reference for CP violation in general gauge theories.

9.4 Vacuum Polarization and Running Coupling Constants

So far we have treated the parameters of the Standard Model, in particular, the coupling constants of the gauge interactions and the masses of the particles, as constants. This is only true at the *classical* or *tree level*, i.e. to zeroth order in the Planck constant $\hbar$. If **quantum corrections** in higher order of $\hbar$ are taken into account, which can be calculated using *loop diagrams* and *renormalization*, it turns out that the masses and the coupling constants depend on an **energy scale** μ with respect to a fixed renormalization scheme. In particular, the coupling constants g become functions $g(\mu)$. These functions are known as **running coupling constants**. In this section we depart from the usual course and discuss (without proofs) this quantum effect, as a preparation for Sect. 9.5 on Grand Unification.

The quantum corrections to the coupling constants are interpreted as **vacuum polarization**. Corresponding loop diagrams are shown in Fig. 9.2, involving a fermion loop, a gauge loop with two 3-boson vertices and a gauge loop with a 4-boson vertex (these diagrams and more details can be found in [125, 137]). For an abelian gauge theory with no direct interaction between gauge bosons, there is only the diagram on the left with a fermion loop.

The dependence on the energy scale is given by the following differential equation

$$\mu\frac{\partial g}{\partial\mu} = \beta(g),$$

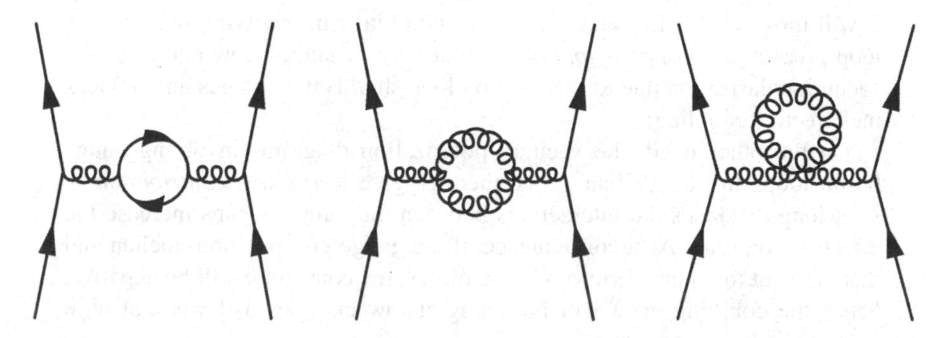

Fig. 9.2 Vacuum polarization diagrams

where β is the so-called **beta function.** It can be proved that to lowest order (one-loop) approximation the beta function is of the form

$$\beta(g) = \frac{1}{4\pi} b g^3$$

for a certain constant b. We get

$$\frac{dg}{g^3} = \frac{1}{4\pi} b \frac{d\mu}{\mu},$$

which can be integrated with respect to an arbitrary reference energy $\mu_0 = M$. This yields:

Proposition 9.4.1 (Running Coupling Constants) *The energy dependence of the coupling constant g is given in one-loop approximation by*

$$\frac{1}{g(\mu)^2} = \frac{1}{g(M)^2} - \frac{1}{2\pi} b \ln \frac{\mu}{M}$$

where b is a constant and M an arbitrary reference energy. If $\ln \frac{\mu}{M}$ *is small, then*

$$g(\mu) \approx g(M) \left(1 + \frac{1}{4\pi} g(M)^2 b \ln \frac{\mu}{M} \right).$$

We see that the sign of b determines whether $g(\mu)$ increases or decreases as the energy scale μ increases:

- if the constant b is *positive*, then $g(\mu)$ *increases* as the energy scale μ increases
- if b is *negative*, then $g(\mu)$ *decreases* as the energy scale μ increases.

Our aim is to give a formula for b depending on the gauge group G and the fermion representations.

It will turn out that the vacuum polarization diagram involving the fermion loop gives a *positive contribution* to b: at long distances (low energies) the vacuum polarization due to the fermion loop shields the charges and reduces the effective coupling.

On the other hand, the vacuum polarization diagrams involving gauge boson loops in non-abelian gauge theories give a *negative contribution* to b: at long distances the interactions between the gauge bosons increase the effective coupling. As a consequence, if the gauge group is non-abelian and there are not too many fermions in the theory, the constant b will be negative, hence the coupling $g(\mu)$ will be strong at low energies and weak at high energies.

9.4.1 Casimir Operators

In this subsection we follow [125, Sect. 25.1]. Suppose that G is a semisimple Lie group with Lie algebra $\mathfrak{g}$ of dimension n. According to Cartan's Criterion 2.4.9 the Killing form $B_{\mathfrak{g}}$ is non-degenerate. Let $\{T_a\}_{a=1}^n$ be an arbitrary basis of $\mathfrak{g}$ and

$$B_{ab} = B_{\mathfrak{g}}(T_a, T_b).$$

The $n \times n$-matrix B is invertible and we denote the entries of the inverse matrix by B^{ab}.

Definition 9.4.2 Let $R\colon G \to \mathrm{GL}(V)$ be a representation of G. Then we define the **Casimir operator** on V by

$$\hat{C}_2(R) = \sum_{a,b=1}^{n} B^{ab} R_*(T_a) \circ R_*(T_b) \in \mathrm{End}(V).$$

If $R = \mathrm{Ad}_G$ is the adjoint representation, then we write $\hat{C}_2(R) = \hat{C}_2(G)$.
It can be shown that the Casimir operator is independent of the choice of basis $\{T_a\}$.

Lemma 9.4.3 *The Casimir operator $\hat{C}_2(R)$ commutes with all generators $R_*(T_a)$ for $a = 1, \ldots, n$.*

Definition 9.4.4 Suppose that R is an irreducible complex representation. Since the Casimir operator commutes with all generators, it follows by Schur's Lemma that

$$\hat{C}_2(R) = C_2(R)\mathrm{Id}_V$$

for some complex number $C_2(R)$, called the **Casimir invariant** of the representation. If G is simple and R the complexified adjoint representation (which is irreducible), we write $C_2(R) = C_2(G)$.

Lemma 9.4.5 *Let $G = \mathrm{SU}(N)$ and $R\colon \mathrm{SU}(N) \to \mathrm{GL}(N, \mathbb{C})$ be the fundamental (defining) representation. Then*

$$C_2(G) = N,$$

$$C_2(R) = \frac{N^2 - 1}{2N}.$$

For more mathematical details on the Casimir operator see, for example, [83, Sect. V.4].

9.4.2 Running Coupling for Gauge Theories with Fermions

Suppose we have an abelian gauge theory with Lie group $G = \mathrm{U}(1)$. We fix the scalar product on $\mathfrak{u}(1)$ such that the vector $i \in \mathfrak{u}(1)$ has length $1/g^2$, where g is the coupling constant. We consider N_f massless Weyl fermions ψ_f in representations

$$\begin{aligned} \rho_{q_f} \colon \mathrm{U}(1) &\longrightarrow \mathrm{U}(1) \\ z &\longmapsto z^{q_f} \end{aligned}$$

of U(1), where $f = 1, \ldots, N_f$ and q_f are integers (winding numbers). We cite the following theorem from quantum field theory:

Theorem 9.4.6 (Vacuum Polarization in Abelian Gauge Theories) *Consider the gauge group $G = \mathrm{U}(1)$. Taking into account the massless fermion one-loop contribution to the vacuum polarization, the constant b is given by*

$$b = \frac{1}{6\pi} \sum_{f=1}^{N_f} q_f^2.$$

References for this formula are [125, Sect. 16.3.3] (for one Dirac fermion of charge $q = 1$) and [80, equation (4.21)] (for arbitrary Weyl fermions).

Example 9.4.7 For QED with gauge group $\mathrm{U}(1)_Q$ and one Dirac electron we get

$$b_Q = \frac{1}{3\pi}.$$

In general, we see that every fermion in abelian gauge theories gives a positive contribution to b.

> Since b_Q is positive in QED, the electric coupling constant $e(\mu)$ increases with increasing energy scale μ.

We can also calculate the vacuum polarization in the case of non-abelian gauge theories: suppose we have a gauge theory with simple Lie group $G = \mathrm{SU}(N_c)$ of dimension $n = N_c^2 - 1$. We fix the Ad-invariant scalar product on $\mathfrak{su}(N_c)$ by

$$-\frac{1}{N_c g^2} B_{\mathfrak{su}(N_c)}(X, Y) = 2\mathrm{tr}(X \cdot Y) \quad \forall X, Y \in \mathfrak{su}(N_c),$$

where g is the coupling constant. We consider N_f massless Weyl fermions ψ_j in representations R_f of G, where $f = 1, \ldots, N_f$. We again cite without proof a theorem from quantum field theory:

Theorem 9.4.8 (Vacuum Polarization in Non-abelian Gauge Theories) *Consider the gauge group $G = \mathrm{SU}(N_c)$. Taking into account the gauge boson and massless fermion one-loop contributions to the vacuum polarization, the constant b is given by*

$$b = -\frac{1}{4\pi}\left(\frac{11}{3} \cdot C_2(G) - \sum_{f=1}^{N_f} \frac{2}{3} \cdot T(R_f)\right)$$

where

$$T(R) = \frac{1}{n} d(R) \cdot C_2(R)$$

and $d(R)$ is the dimension of the representation R. For $G = \mathrm{SU}(N_c)$ with $n = N_c^2 - 1$ we have

$$C_2(\mathrm{SU}(N_c)) = N_c$$

and

$$T(R) = \frac{1}{2}$$

for the fundamental representation R and

$$T(R) = 0$$

for the trivial representation R.

References for this formula (which holds for an arbitrary simple gauge group G with a suitably normalized Ad-invariant scalar product on the Lie algebra) are [125, equation (26.93)], [144, equation (17.5.41)], [85, Part I, equation (2.2)] (with Dirac fermions) and [80, equation (4.29)] (for Weyl fermions). The formula shows that the non-abelian group G itself (the vacuum polarization due to gauge boson loops) gives a negative contribution to b, while the vacuum polarization due to the fermion loops give a positive contribution.

The formulas in the following examples for the Standard Model can also be found in [85, Part I, equation (2.4)].

Example 9.4.9 In QCD we have $G = \mathrm{SU}(3)_C$ and 6 flavours of quarks, i.e. 12 Weyl fermions in the fundamental representation. We get in the limit of massless quarks

$$b_C = -\frac{1}{4\pi}\left(\frac{11}{3}\cdot 3 - \frac{2}{3}\cdot 12\cdot\frac{1}{2}\right) = -\frac{7}{4\pi}.$$

Since b_C is negative in QCD, the strong coupling constant $g_s(\mu)$ gets smaller as the energy scale μ increases. This is known as **asymptotic freedom** and was discovered by David J. Gross, H. David Politzer and Frank Wilczek (Nobel Prize in Physics 2004).

Asymptotic freedom is not only theoretically interesting, but also practically very useful, because it shows that QCD at high energies is amenable to perturbation theory. For low energies the strong coupling constant $g_s(\mu)$ is large, hence QCD becomes non-perturbative.

Example 9.4.10 In the weak interaction described by the group $G = \mathrm{SU}(2)_L$ we have in each generation three quark doublets of colours red, green and blue and one lepton doublet, giving in total 12 doublets of left-handed Weyl fermions in the fundamental representation. We also have right-handed quarks and fermions in the trivial representation that have $T(R) = 0$. In the limit of massless fermions,

$$b_L = -\frac{1}{4\pi}\left(\frac{11}{3}\cdot 2 - \frac{2}{3}\cdot 12\cdot\frac{1}{2}\right) = -\frac{5}{6\pi}.$$

Example 9.4.11 In the hypercharge interaction described by the group $G = \mathrm{U}(1)_Y$ and coupling constant g' we get with correctly normalized coupling constant

$$g = \frac{g'}{6}$$

and charges

$$q = 3y$$

in the limit of massless fermions in three generations of the representations

$$(\mathbf{3},\mathbf{2})_{1/3} \oplus (\mathbf{3},\mathbf{1})_{4/3} \oplus (\mathbf{3},\mathbf{1})_{-2/3} \oplus (\mathbf{1},\mathbf{2})_{-1} \oplus (\mathbf{1},\mathbf{1})_{-2}$$

the coefficient

$$b_Y = \frac{1}{6\pi}\cdot\frac{1}{36}\cdot 3\cdot(6\cdot 1 + 3\cdot 16 + 3\cdot 4 + 2\cdot 9 + 1\cdot 36) = \frac{5}{3\pi}.$$

9.4.3 Experimental Values for Coupling Constants

We discuss some experimental values of coupling constants (following [108, 109]). At the energy scale

$$\mu = m_Z = 91.1876 \pm 0.0021 \text{ GeV},$$

where m_Z is the mass of the Z^0-boson, the electric fine-structure constant, the strong fine-structure constant and the Weinberg angle are given by (with respect to the minimal subtraction renormalization scheme $\overline{\text{MS}}$)

$$\begin{aligned} \alpha_e(m_Z)^{-1} &= 127.950 \pm 0.017, \\ \alpha_s(m_Z) &= 0.1182 \pm 0.0016, \\ \sin^2 \theta_W(m_Z) &= 0.23129 \pm 0.00005. \end{aligned}$$

The coupling constants g are related to the fine-structure constants by $g = \sqrt{4\pi\alpha}$. Together with

$$\begin{aligned} g_w &= \frac{1}{\sin \theta_W} e, \\ g' &= \frac{1}{\cos \theta_W} e, \end{aligned}$$

the coupling constants corresponding to the factors of the Standard Model gauge group $\text{SU}(3)_C \times \text{SU}(2)_L \times \text{U}(1)_Y$ are seen to be:

$$\begin{aligned} g_s(m_Z) &\approx 1.22, \\ g_w(m_Z) &\approx 0.652, \\ g'(m_Z) &\approx 0.357. \end{aligned}$$

Since

$$b_C < b_L < 0 < b_Y$$

it is conceivable that there is some high energy $\mu \gg m_Z$ where the coupling constants g_s, g_w, g' become equal (to make this argument precise one has to rescale g' suitably; see Sect. 9.5.3). This leads to the idea of Grand Unification that we discuss in Sect. 9.5.

9.4.4 Further Reading

The books [80] by T.J. Hollowood, [125] by M.D. Schwartz and [144] by S. Weinberg discuss vacuum polarization. The section by A. Masiero in the book [85] and Sect. 5.5 in the book [96] by R.N. Mohapatra also have short accounts of running coupling constants in connection with Grand Unification.

9.5 Grand Unified Theories

The gauge group of the Standard Model is the compact Lie group

$$G_{\text{SM}} = \text{SU}(3)_C \times \text{SU}(2)_L \times \text{U}(1)_Y.$$

The idea of **Grand Unification** is to unify all forces described by the Standard Model into a simple Lie group with only a single coupling constant. Even though the coupling constants of the strong, weak and electromagnetic interaction are different at energies around 100 GeV, the existence of such a unified theory is not impossible from the outset, because of the quantum effect of running coupling discussed in Sect. 9.4 (this was first realized by H. Georgi, H.R. Quinn and S. Weinberg in 1974 [61]). Strictly speaking, the electroweak interaction, described by the gauge group $\text{SU}(2)_L \times \text{U}(1)_Y$, is not a unification of the weak and electromagnetic interaction in this sense, because it still involves two coupling constants.

After some general remarks, we want to study in this section the Grand Unified Theories described by the simple Lie groups SU(5) and Spin(10). We follow the mathematical reference [9] and the physics references [85] and [96] throughout this section.

9.5.1 Group Theoretic Preliminaries

Definition 9.5.1 We call a Lie group G a **possible Grand Unification group** if it has the following properties:

- G is simple, so that it has only one coupling constant, or G is a product of several copies of the same simple group, where the coupling constant for each factor is set the same (by a discrete symmetry).
- The Lie group G contains (a finite quotient of) the Standard Model group $\text{SU}(3) \times \text{SU}(2) \times \text{U}(1)$.
- G admits complex representations not isomorphic to their complex conjugates; see Sect. 8.5.3.

We would like to find all possible Grand Unification groups. It is more suitable to classify these groups according to *rank* (the maximal dimension of an embedded torus subgroup) than according to dimension. The Standard Model group $SU(3) \times SU(2) \times U(1)$ has rank 4. A possible Grand Unification group therefore must have rank at least 4. We restrict to the three simplest cases where G has rank 4, 5 or 6.

Using the Killing–Cartan Theorem 2.4.23 on the classification of compact simple Lie algebras we can list all simple (and simply connected) compact Lie groups of rank less than or equal to 6:

- rank 1: $SU(2)$
- rank 2: $SU(3)$, $Spin(5)$, G_2
- rank 3: $SU(4)$, $Spin(7)$, $Sp(3)$
- rank 4: $SU(5)$, $Spin(8)$, $Spin(9)$, $Sp(4)$, F_4
- rank 5: $SU(6)$, $Spin(10)$, $Spin(11)$, $Sp(5)$
- rank 6: $SU(7)$, $Spin(12)$, $Spin(13)$, $Sp(6)$, E_6.

The semisimple Lie groups of rank 4 with a single coupling constant are therefore (see the first part of Exercise 2.7.18)

$$\begin{aligned}&SU(2)^4,\\&SU(3)^2, Spin(5)^2, (G_2)^2,\\&SU(5), Spin(8), Spin(9), Sp(4), F_4\end{aligned}$$

(an exponent k denotes the product of k copies of the group). The semisimple Lie groups of rank 5 with a single coupling constant are

$$\begin{aligned}&SU(2)^5,\\&SU(6), Spin(10), Spin(11), Sp(5).\end{aligned}$$

The semisimple Lie groups of rank 6 with a single coupling constant are

$$\begin{aligned}&SU(2)^6.\\&SU(3)^3, Spin(5)^3, (G_2)^3.\\&SU(4)^2, Spin(7)^2, Sp(3)^2.\\&SU(7), Spin(12), Spin(13), Sp(6), E_6.\end{aligned}$$

According to [104, Sect. 5.4] the only compact, simply connected, simple Lie groups which have representations not isomorphic to their complex conjugates are

$$\begin{aligned}&SU(n) \quad \text{for} \quad n \geq 3,\\&Spin(4n+2) \quad \text{for} \quad n \geq 1,\\&E_6.\end{aligned}$$

This implies:

Proposition 9.5.2 *The only possible Grand Unification groups of rank less than or equal to 6 are:*

- *rank* 4*:* $\mathrm{SU}(3)^2$ *and* $\mathrm{SU}(5)$
- *rank* 5*:* $\mathrm{SU}(6)$ *and* $\mathrm{Spin}(10)$
- *rank* 6*:* $\mathrm{SU}(3)^3$, $\mathrm{SU}(4)^2$, $\mathrm{SU}(7)$ *and* E_6.

Here are some references where actual Grand Unified Theories using these groups (except $\mathrm{SU}(3)^2$ and $\mathrm{SU}(4)^2$) have been constructed:

- rank 4:
  - $\mathrm{SU}(5)$: the Georgi–Glashow theory [59] with gauge group $\mathrm{SU}(5)$ from 1974 was the first Grand Unified Theory based on a simple Lie group. In the same paper the Lie group $\mathrm{SU}(3)^2$ is ruled out for physical reasons, leaving $\mathrm{SU}(5)$ as the only GUT group of rank 4.
- rank 5:
  - $\mathrm{Spin}(10)$: the $\mathrm{SO}(10)$ theory, as it is called in physics, was first developed by H. Georgi [58] and H. Fritzsch and P. Minkowski [56] in 1975.
  - $\mathrm{SU}(6)$: there is a theory of A. Hartanto and L.T. Handoko [72] from 2005.
- rank 6:
  - E_6: a Grand Unified Theory with this gauge group arises naturally in heterotic string theory and was first developed by F. Gürsey, P. Ramond and P. Sikivie [68] in 1976.
  - $\mathrm{SU}(3)^3$: there is a theory called *Trinification* with this gauge group proposed by A. de Rújula, H. Georgi and S.L. Glashow [63] in 1984.
  - $\mathrm{SU}(7)$: a Grand Unified Theory based on this Lie group was studied by K. Yamamoto [151] in 1981.

Our aim is to discuss in some detail the Grand Unified Theories corresponding to the simple Lie groups $\mathrm{SU}(5)$ and $\mathrm{Spin}(10)$ (we only consider Grand Unified Theories defined on 4-dimensional Minkowski spacetime).

9.5.2 Embeddings of the Standard Model Gauge Group $G_{\mathrm{SM}}/\mathbb{Z}_6$ into the Simple Lie Groups SU(5) and Spin(10)

We describe how the gauge group of the Standard Model (actually a $\mathbb{Z}_6$ quotient of it) can be embedded into the simple Lie groups $\mathrm{SU}(5)$ and $\mathrm{Spin}(10)$ (we follow [9]).

Definition 9.5.3 For integers $m, n \geq 1$ we set

$$\mathrm{S}(\mathrm{U}(m) \times \mathrm{U}(n)) = \{(A, B) \in \mathrm{U}(m) \times \mathrm{U}(n) \mid \det A \cdot \det B = 1\}.$$

The Lie group $\mathrm{S}(\mathrm{U}(m) \times \mathrm{U}(n))$ is naturally a subgroup of $\mathrm{SU}(m+n)$ under the embedding

$$\begin{aligned} \mathrm{S}(\mathrm{U}(m) \times \mathrm{U}(n)) &\longrightarrow \mathrm{SU}(m+n) \\ (A, B) &\longmapsto \begin{pmatrix} A & 0 \\ 0 & B \end{pmatrix}. \end{aligned}$$

We set

$$G_{\mathrm{SM}} = \mathrm{SU}(3)_C \times \mathrm{SU}(2)_L \times \mathrm{U}(1)_Y.$$

Proposition 9.5.4 (Embedding of $G_{\mathrm{SM}}/\mathbb{Z}_6$ into $\mathrm{SU}(5)$) *The Lie group homomorphism*

$$\begin{aligned} f\colon G_{\mathrm{SM}} &\longrightarrow \mathrm{SU}(5) \\ (g, h, \alpha) &\longmapsto \begin{pmatrix} g\alpha^{-2} & 0 \\ 0 & h\alpha^3 \end{pmatrix} \end{aligned}$$

induces an injective Lie group embedding

$$\bar{f}\colon G_{\mathrm{SM}}/K \hookrightarrow \mathrm{SU}(5)$$

with image

$$\mathrm{S}(\mathrm{U}(3) \times \mathrm{U}(2)) \subset \mathrm{SU}(5).$$

Here $K \cong \mathbb{Z}_6$ is the subgroup

$$K = \left\{ \left(\alpha^2 I_3, \alpha^{-3} I_2, \alpha \right) \in G_{\mathrm{SM}} \mid \alpha \in \mathrm{U}(1), \alpha^6 = 1 \right\}.$$

Proof It is clear that f is a Lie group homomorphism to $\mathrm{U}(5)$. The image is contained in $\mathrm{SU}(5)$, since

$$\begin{aligned} \det \begin{pmatrix} g\alpha^{-2} & 0 \\ 0 & h\alpha^3 \end{pmatrix} &= (\det g)\alpha^{-6}(\det h)\alpha^6 \\ &= 1. \end{aligned}$$

We show that f is surjective onto $\mathrm{S}(\mathrm{U}(3) \times \mathrm{U}(2))$: let

$$\begin{pmatrix} A & 0 \\ 0 & B \end{pmatrix} \in \mathrm{S}(\mathrm{U}(3) \times \mathrm{U}(2)) \subset \mathrm{SU}(5).$$

There exists a complex number $\alpha \in \mathrm{U}(1)$ such that $A\alpha^2 \in \mathrm{SU}(3)$. Consider the matrix $B\alpha^{-3}$. Then

$$\begin{aligned}1 &= (\det A)(\det B)\\ &= (\det A)\alpha^6(\det B)\alpha^{-6}\\ &= \det\left(A\alpha^2\right)\det\left(B\alpha^{-3}\right)\\ &= \det\left(B\alpha^{-3}\right).\end{aligned}$$

Hence $B\alpha^{-3} \in \mathrm{SU}(2)$ and

$$\begin{pmatrix} A & 0 \\ 0 & B \end{pmatrix} = f\left(A\alpha^2, B\alpha^{-3}, \alpha\right).$$

We show that the kernel of f is K: suppose

$$f(g,h,\alpha) = \begin{pmatrix} I_3 & 0 \\ 0 & I_2 \end{pmatrix}.$$

Then

$$\begin{aligned} g &= \alpha^2 I_3,\\ h &= \alpha^{-3} I_2.\end{aligned}$$

Since $g \in \mathrm{SU}(3)$ and $h \in \mathrm{SU}(2)$ we have

$$\begin{aligned} 1 &= \det g = \alpha^6,\\ 1 &= \det h = \alpha^{-6}.\end{aligned}$$

This proves the claim. □

Proposition 9.5.5 *For all $n \geq 2$ there exists a canonical Lie group embedding*

$$h\colon \mathrm{SU}(n) \hookrightarrow \mathrm{Spin}(2n).$$

Proof According to Exercise 1.9.10 there exists an embedding

$$\mathrm{U}(n) \hookrightarrow \mathrm{SO}(2n)$$

and thus

$$\mathrm{SU}(n) \hookrightarrow \mathrm{SO}(2n).$$

Since $\mathrm{SU}(n)$ is simply connected and $\mathrm{Spin}(2n) \to \mathrm{SO}(2n)$ is a covering map, this embedding can be lifted to an embedding

$$\mathrm{SU}(n) \hookrightarrow \mathrm{Spin}(2n).$$

□

Together with Proposition 9.5.4 we get:

Corollary 9.5.6 (Cascade of Grand Unification Groups) *There exists a sequence of Lie group embeddings*

$$G_{\mathrm{SM}}/\mathbb{Z}_6 \hookrightarrow \mathrm{SU}(5) \hookrightarrow \mathrm{Spin}(10).$$

Under the first embedding $G_{\mathrm{SM}}/\mathbb{Z}_6$ *gets identified with* $\mathrm{S}(\mathrm{U}(3) \times \mathrm{U}(2))$.
Together with an embedding of Spin(10) into the exceptional simple Lie group E_6 (see [8, 46]), these groups form the famous cascade of simple Grand Unification groups down to the $\mathbb{Z}_6$ quotient of the Standard Model group:

$$G_{\mathrm{SM}}/\mathbb{Z}_6 \subset \mathrm{SU}(5) \subset \mathrm{Spin}(10) \subset \mathrm{E}_6.$$

In the physics literature (e.g. [126]) the Lie group Spin(10) is sometimes called E_5 and the Lie group SU(5) is called E_4.

Remark 9.5.7 The Lie group Spin(10) actually contains the larger compact embedded Lie group

$$\mathrm{SU}(5) \subset \mathrm{U}(5)/\mathbb{Z}_3 \subset \mathrm{Spin}(10).$$

In the physics literature on Grand Unification the group $\mathrm{U}(5)/\mathbb{Z}_3$ is often denoted by $\mathrm{SU}(5) \times \mathrm{U}(1)$. See Exercise 9.7.4 for more details.

9.5.3 Normalized Hypercharge and Unification of Coupling Constants

We saw in Sect. 9.4 that, due to the quantum effect of running couplings, there is a possibility that the coupling constants of the strong and electroweak interactions become the same at some high energy. We discuss this in more detail (we follow [85] and [96]).

In this subsection we consider a simple Grand Unification group G. Under the embedding of $G_{\mathrm{SM}}/\mathbb{Z}_6$ into G, all basis vectors of G_{SM} must be normalized, because they belong to an orthonormal basis of the Lie algebra $\mathfrak{g}$. This means that with respect to the Killing form of $\mathfrak{g}$, the basis vectors

$$\beta_3' = \frac{i\sigma_3}{2} \in \mathfrak{su}(2)_L,$$

$$\beta_4' = \lambda\frac{i}{6} \in \mathfrak{u}(1)_Y$$

should have the same length, where $\lambda > 0$ is a normalization constant to be determined. We first consider the case $G = \mathrm{SU}(5)$ with the embedding given by Proposition 9.5.4. Under this embedding

$$\beta_3' \longmapsto \begin{pmatrix} 0 & 0 \\ 0 & \frac{i\sigma_3}{2} \end{pmatrix},$$

$$\beta_4' \longmapsto \lambda \begin{pmatrix} -\frac{1}{3} & & & & \\ & -\frac{1}{3} & & & \\ & & -\frac{1}{3} & & \\ & & & \frac{1}{2} & \\ & & & & \frac{1}{2} \end{pmatrix}.$$

According to Exercise 2.7.16 the Killing form of $\mathfrak{su}(5)$ is given by

$$B_{\mathfrak{su}(5)}(X, Y) = 10\mathrm{tr}(X \cdot Y),$$

hence

$$-B_{\mathfrak{su}(5)}(\beta_3', \beta_3') = 5,$$

$$-B_{\mathfrak{su}(5)}(\beta_4', \beta_4') = \frac{25}{3}\lambda^2.$$

A similar argument shows that the basis of $\mathfrak{su}(3)_C$ already has the correct normalization. This implies:

Proposition 9.5.8 *The correctly normalized coupling constants of G_{SM} under the embedding of $G_{\mathrm{SM}}/\mathbb{Z}_6$ into* SU(5) *are given by*

$$g_s, \quad g_w \quad \text{and} \quad g'' = \sqrt{\frac{5}{3}}g'.$$

For a general simple Grand Unification gauge group G we can argue as follows: let ρ denote the representation of G_{SM} on one generation of left-handed fermions and antifermions (see Eq. (8.13)):

$$V_L^i \oplus V_L^{iC} = (\mathbf{3}, \mathbf{2})_{1/3} \oplus (\bar{\mathbf{3}}, \mathbf{1})_{-4/3} \oplus (\bar{\mathbf{3}}, \mathbf{1})_{2/3} \oplus (\mathbf{1}, \mathbf{2})_{-1} \oplus (\mathbf{1}, \mathbf{1})_2.$$

The vectors β_3', β_4' act under ρ_* as

$$\rho_* \beta_3' = i\mathbb{T}_3,$$
$$\rho_* \beta_4' = i\lambda \frac{\mathbb{Y}}{2}.$$

The symmetric bilinear form

$$\begin{aligned} \mathfrak{g} \times \mathfrak{g} &\longrightarrow \mathbb{R} \\ (X, Y) &\longmapsto \operatorname{tr}(\rho_* X \circ \rho_* Y) \end{aligned}$$

is Ad_G-invariant, hence by Theorem 2.5.1 it must be a multiple of the Killing form. We have

$$\begin{aligned} -\operatorname{tr}\left(\rho_* \beta_3' \circ \rho_* \beta_3'\right) &= \operatorname{tr}\left(\mathbb{T}_3^2\right) \\ &= 3 \cdot 2 \cdot \frac{1}{4} + 2 \cdot \frac{1}{4} \\ &= 2, \\ -\operatorname{tr}\left(\rho_* \beta_4' \circ \rho_* \beta_4'\right) &= \frac{1}{4}\lambda^2 \operatorname{tr}\left(\mathbb{Y}^2\right) \\ &= \frac{1}{4}\lambda^2 \left(3 \cdot 2 \cdot \frac{1}{9} + 3 \cdot \frac{16}{9} + 3 \cdot \frac{4}{9} + 2 \cdot 1 + 1 \cdot 4\right) \\ &= \lambda^2 \frac{10}{3}. \end{aligned}$$

Setting

$$\operatorname{tr}\left(\rho_* \beta_3' \circ \rho_* \beta_3'\right) = \operatorname{tr}\left(\rho_* \beta_4' \circ \rho_* \beta_4'\right)$$

we conclude again that

$$\lambda = \sqrt{\frac{3}{5}}.$$

From Sect. 9.4 the correctly normalized running coupling constants are given by

$$\begin{aligned} \frac{1}{g_s(\mu)^2} &= \frac{1}{g_s(M)^2} + \frac{7}{8\pi^2} \ln \frac{\mu}{M}, \\ \frac{1}{g_w(\mu)^2} &= \frac{1}{g_w(M)^2} + \frac{5}{12\pi^2} \ln \frac{\mu}{M}, \\ \frac{1}{g''(\mu)^2} &= \frac{1}{g''(M)^2} - \frac{1}{2\pi^2} \ln \frac{\mu}{M}. \end{aligned}$$

With the approximate values

$$g_s(m_Z) \approx 1.22,$$
$$g_w(m_Z) \approx 0.652,$$
$$g''(m_Z) \approx 0.461$$

at $m_Z \approx 91$ GeV we get (see also Fig. 9.3)

$$g_s(\mu) = g_w(\mu) \approx 0.508 \quad \text{at} \quad \mu \approx 10^{17.68}\,\text{GeV},$$
$$g_s(\mu) = g''(\mu) \approx 0.556 \quad \text{at} \quad \mu \approx 10^{14.53}\,\text{GeV},$$
$$g_w(\mu) = g''(\mu) \approx 0.541 \quad \text{at} \quad \mu \approx 10^{12.96}\,\text{GeV}.$$

We see that the running coupling constants do not exactly match at the same energy, but as a first approximation this is roughly true. In *supersymmetric Grand Unified Theories* (where more particles take part in loop diagrams) the running coupling constants become equal to a much better approximation.

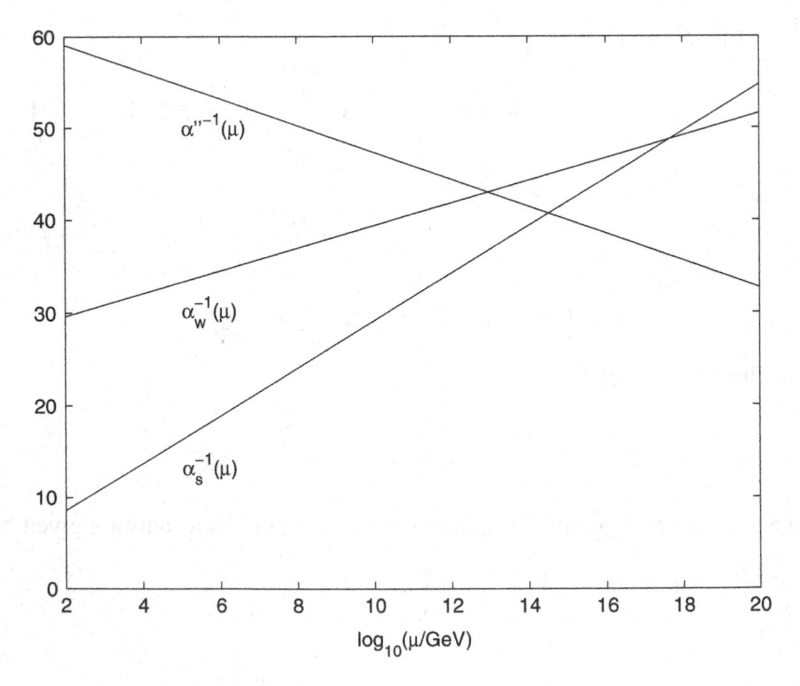

Fig. 9.3 Approximate unification of running fine-structure constants $\alpha = g^2/(4\pi)$

At an energy where $g_w = g''$, we get for the Weinberg angle θ_W the prediction

$$\tan^2 \theta_W = \frac{g'^2}{g_w^2} = \frac{3}{5},$$

hence

$$\sin^2 \theta_W = \frac{3}{8} = 0.375.$$

The experimental value for the Weinberg angle at the much lower energy m_Z is

$$\sin^2 \theta_W = 0.23129 \pm 0.00005.$$

The difference could be explained with a running coupling effect, i.e. quantum corrections.

9.5.4 *The Fermions in the* SU(5) *Grand Unified Theory*

We discuss two specific representations of SU(5) and understand how they decompose under restriction to the subgroup $G_{\text{SM}}/\mathbb{Z}_6$ (we follow [9] and [96]).

Lemma 9.5.9 *Consider the complex conjugate $\bar{\mathbb{C}}^5$ of the fundamental representation of* SU(5)*. Under the restriction to the subgroup $G_{\text{SM}}/\mathbb{Z}_6$ this representation has the branching rule*

$$\bar{\mathbf{5}} = \begin{pmatrix} (\bar{\mathbf{3}}, \mathbf{1})_{2/3} \\ (\mathbf{1}, \bar{\mathbf{2}})_{-1} \end{pmatrix}.$$

Proof Consider the Lie group homomorphism

$$\begin{aligned} f\colon G_{\text{SM}} &\longrightarrow \text{SU}(5) \\ (g, h, \alpha) &\longmapsto \begin{pmatrix} g\alpha^{-2} & 0 \\ 0 & h\alpha^3 \end{pmatrix} \end{aligned}$$

and decompose the column vectors in $\mathbb{C}^5$ into the first three components $\mathbb{C}^3$ and the last two components $\mathbb{C}^2$. On the complex conjugate $\bar{\mathbb{C}}^5$ of the fundamental representation an element $(g, h, \alpha) \in G_{\text{SM}}$ acts on the first three components $x \in \bar{\mathbb{C}}^3$ as

$$(g, h, \alpha) \cdot x = \bar{g}\alpha^2 x$$

and on the last two components $y \in \bar{\mathbb{C}}^2$ as

$$(g, h, \alpha) \cdot y = \bar{h}\alpha^{-3} y.$$

According to Lemma 8.5.1 the representation

$$\alpha \cdot x = \alpha^2 x$$

has $y = 2/3$ and the representation

$$\alpha \cdot y = \alpha^{-3} y$$

has $y = -1$. Therefore the representation $\bar{\mathbb{C}}^5$ splits as

$$\bar{\mathbf{5}} = (\bar{\mathbf{3}}, \mathbf{1})_{\mathbf{2/3}} \oplus (\mathbf{1}, \bar{\mathbf{2}})_{\mathbf{-1}}.$$

□

Note that for the fundamental SU(2) representation $\bar{\mathbf{2}} \cong \mathbf{2}$ by Exercise 2.7.3.

Lemma 9.5.10 *Consider the representation* $\Lambda^2\mathbb{C}^5 \cong \mathbb{C}^{10}$ *of* SU(5). *Under the restriction to the subgroup* $G_{\mathrm{SM}}/\mathbb{Z}_6$ *this representation has the branching rule*

$$\mathbf{10} = \begin{pmatrix} (\bar{\mathbf{3}}, \mathbf{1})_{\mathbf{-4/3}} & (\mathbf{3}, \mathbf{2})_{\mathbf{1/3}} \\ & (\mathbf{1}, \mathbf{1})_{\mathbf{2}} \end{pmatrix} \quad \textit{(antisymmetric).}$$

Proof We identify $\Lambda^2\mathbb{C}^5$ with the complex 5×5 antisymmetric matrices X. Then $M \in \mathrm{SU}(5)$ acts on X by

$$M \cdot X = MXM^T.$$

We write X as

$$X = \begin{pmatrix} 0 & a_3 & -a_2 & b_{11} & b_{12} \\ & 0 & a_1 & b_{21} & b_{22} \\ & & 0 & b_{31} & b_{32} \\ & & & 0 & c \\ & & & & 0 \end{pmatrix} \quad \text{(antisymmetric)}$$

$$= \begin{pmatrix} A & B \\ -B^T & C \end{pmatrix}$$

with

$$A = \begin{pmatrix} 0 & a_3 & -a_2 \\ -a_3 & 0 & a_1 \\ a_2 & -a_1 & 0 \end{pmatrix}, \quad B = \begin{pmatrix} b_{11} & b_{12} \\ b_{21} & b_{22} \\ b_{31} & b_{32} \end{pmatrix}, \quad C = \begin{pmatrix} 0 & c \\ -c & 0 \end{pmatrix}.$$

We consider the Lie group homomorphism

$$\begin{aligned} f\colon G_{\mathrm{SM}} &\longrightarrow \mathrm{SU}(5) \\ (g,h,\alpha) &\longmapsto \begin{pmatrix} g\alpha^{-2} & 0 \\ 0 & h\alpha^3 \end{pmatrix}. \end{aligned}$$

Then (g,h,α) acts on X by

$$\begin{aligned} (g,h,\alpha)\cdot X &= \begin{pmatrix} g\alpha^{-2} & 0 \\ 0 & h\alpha^3 \end{pmatrix} \begin{pmatrix} A & B \\ -B^T & C \end{pmatrix} \begin{pmatrix} g^T\alpha^{-2} & 0 \\ 0 & h^T\alpha^3 \end{pmatrix} \\ &= \begin{pmatrix} gAg^T\alpha^{-4} & gBh^T\alpha \\ -hB^Tg^T\alpha & hCh^T\alpha^6 \end{pmatrix}. \end{aligned}$$

We can directly read off the hypercharges:

$$\begin{aligned} y_A &= -4/3, \\ y_B &= 1/3, \\ y_C &= 2. \end{aligned}$$

It is clear that the $\mathrm{SU}(3)\times\mathrm{SU}(2)$ representation on B is $(\mathbf{3},\mathbf{2})$ and on the 1-dimensional space C it can only be $(\mathbf{1},\mathbf{1})$. Exercise 9.7.5 shows that the representation on A is $(\bar{\mathbf{3}},\mathbf{1})$. □

Comparing with Eq. (8.13) we have:

Theorem 9.5.11 (Fermions in SU(5) Grand Unification) *The representation* $\bar{\mathbf{5}}\oplus\mathbf{10}$ *of* SU(5) *under restriction to the subgroup* $G_{\mathrm{SM}}/\mathbb{Z}_6$ *has the branching rule*

$$\bar{\mathbf{5}}\oplus\mathbf{10} = (\bar{\mathbf{3}},\mathbf{1})_{2/3}\oplus(\mathbf{1},\bar{\mathbf{2}})_{-1}\oplus(\bar{\mathbf{3}},\mathbf{1})_{-4/3}\oplus(\mathbf{3},\mathbf{2})_{1/3}\oplus(\mathbf{1},\mathbf{1})_{2}.$$

The representation on the right is isomorphic to the representation $V_L^i\oplus V_L^{iC}$ *of one full generation of left-handed fermions and antifermions. The explicit*

(continued)

Theorem 9.5.11 (continued)
identification is given by

$$\bar{\mathbf{5}} = \begin{pmatrix} d_r^C \\ d_g^C \\ d_b^C \\ e \\ \nu_e \end{pmatrix}_L$$

$$\mathbf{10} = \begin{pmatrix} 0 & u_b^C & -u_g^C & u_r & d_r \\ & 0 & u_r^C & u_g & d_g \\ & & 0 & u_b & d_b \\ & & & 0 & e^C \\ & & & & 0 \end{pmatrix}_L \quad \textit{(antisymmetric)}.$$

The particle fields which appear in these formulas are the ***current eigenstates*** *that correspond to the weak eigenstates in the Standard Model.*

Remark 9.5.12 We can add a left-handed sterile antineutrino in the trivial singlet representation $\mathbb{C}$ of SU(5). Then the left-handed fermions and antifermions form a 16-dimensional representation

$$\mathbf{1} \oplus \bar{\mathbf{5}} \oplus \mathbf{10} = \mathbb{C} \oplus \bar{\mathbb{C}}^5 \oplus \Lambda^2\mathbb{C}^5.$$

All fermions and antifermions (left-handed and right-handed) together then form the 32-dimensional representation

$$\mathbf{1} \oplus \bar{\mathbf{5}} \oplus \mathbf{10} \oplus \overline{\mathbf{10}} \oplus \mathbf{5} \oplus \mathbf{1} = \Lambda^*\mathbb{C}^5.$$

It is very remarkable that the very simple, decomposable SU(5)-representation $\Lambda^*\mathbb{C}^5$ thus suffices to describe all fermions. This discussion and more details can be found in [9].

9.5.5 *The Fermions in the* Spin(10) *Grand Unified Theory*

The idea of the Grand Unified Theory with Lie group Spin(10) is to look for a representation of Spin(10) that under restriction to the subgroup SU(5) decomposes into the representation $\bar{\mathbf{5}} \oplus \mathbf{10}$. As we saw in Sect. 9.5.4 this representation then decomposes under restriction to $G_{SM}/\mathbb{Z}_6$ into the Standard Model representation.

The most natural representation of the Lie group Spin(10) is the spinor representation. The Weyl spinor representations of Spin(10) have dimension 16. We want to show that these representations can accommodate the 16-dimensional representations of one full generation of fermions and antifermions, including a sterile neutrino, mentioned in Remark 9.5.12 (we follow in this subsection [9]).

We consider more generally for an arbitrary integer $n \geq 2$ the Lie group Spin($2n$). Recall from Sect. 6.4 and Sect. 6.5 that this spin group has a Dirac spinor representation on $\Delta \cong \mathbb{C}^{2^n}$ which is reducible and decomposes into the Weyl spinor representations

$$\Delta = \Delta_+ \oplus \Delta_-,$$

where $\Delta_\pm$ each have half dimension. Note that

$$\Delta \cong \Lambda^*\mathbb{C}^n$$

as complex vector spaces. In fact, we can realize the spinor representation on Δ explicitly on $\Lambda^*\mathbb{C}^n$:

Lemma 9.5.13 (Dirac Spinor Representation of Spin($2n$) **on** $\Lambda^*\mathbb{C}^n$**)** *For $u \in \mathbb{C}^n$ let*

$$\begin{aligned} \alpha(u)\colon \Lambda^*\mathbb{C}^n &\longrightarrow \Lambda^*\mathbb{C}^n \\ \omega &\longmapsto \alpha(u)\omega = u \wedge \omega \end{aligned}$$

and

$$\begin{aligned} \beta(u)\colon \Lambda^*\mathbb{C}^n &\longrightarrow \Lambda^*\mathbb{C}^n \\ \omega &\longmapsto \beta(u)\omega = u \lrcorner \omega. \end{aligned}$$

Then the complex Dirac spinor representation of Cl($2n$) *on $\Delta \cong \Lambda^*\mathbb{C}^n$ is induced by the following Clifford multiplication of vectors*

$$(u, v) \in \mathbb{C}^n \oplus \mathbb{C}^n = \mathbb{C}^{2n}$$

on the exterior algebra $\Lambda^*\mathbb{C}^n$:

$$\begin{aligned} \delta(u, v)\colon \Lambda^*\mathbb{C}^n &\longrightarrow \Lambda^*\mathbb{C}^n \\ \omega &\longmapsto \delta(u, v)\omega = (\alpha(u) - \beta(u))\omega - i(\alpha(v) + \beta(v))\omega. \end{aligned}$$

Proof The original proof is due to [7]. It is easy to check that δ satisfies

$$\{\delta(x), \delta(y)\} = -2q(x, y) \cdot 1 \quad \forall x, y \in \mathbb{C}^{2n}.$$

By the universal property of Clifford algebras there is an induced non-trivial algebra homomorphism

$$\phi\colon \mathbb{C}l(2n) \longrightarrow \mathrm{End}\left(\Lambda^*\mathbb{C}^n\right).$$

Both algebras are isomorphic to $\mathrm{End}\left(\mathbb{C}^{2^n}\right)$. The kernel of ϕ is a two-sided ideal in $\mathbb{C}l(2n) \cong \mathrm{End}\left(\mathbb{C}^{2^n}\right)$ and hence, by a general property of endomorphism algebras, either 0 or $\mathbb{C}l(2n)$. Since ϕ is non-trivial, it follows that ϕ is injective and hence an isomorphism. □

Remark 9.5.14 It can be shown that the subspaces of Weyl spinors correspond to the forms of even and odd degree, where the precise association depends on the integer n; see Exercise 9.7.6.
There is also a natural representation of $\mathrm{SU}(n)$ on $\Lambda^*\mathbb{C}^n$ induced by the fundamental representation on $\mathbb{C}^n$. Consider the Lie group embedding

$$h\colon \mathrm{SU}(n) \hookrightarrow \mathrm{Spin}(2n)$$

from Proposition 9.5.5.

Theorem 9.5.15 *The Dirac spinor representation Δ of* Spin(2*n*) *has under restriction to the subgroup* SU(*n*) *the following branching rule:*

$$\Delta = \Lambda^0\mathbb{C}^n \oplus \Lambda^1\mathbb{C}^n \oplus \ldots \oplus \Lambda^n\mathbb{C}^n.$$

In particular, if the Dirac spinor representation of Spin(2*n*) *on* $\Lambda^*\mathbb{C}^n$ *(which preserves only the parity of the degree) is restricted to* SU(*n*)*, it preserves the integral degree of all forms.*

Proof We need to find an explicit embedding of $\mathrm{SU}(n)$ into $\mathrm{Spin}(2n)$. It suffices to do this on the level of Lie algebras. According to Exercise 1.9.10 the embedding

$$\tau\colon \mathfrak{su}(n) \longrightarrow \mathfrak{so}(2n)$$

can be realized by

$$A_1 + iA_2 \longmapsto \begin{pmatrix} A_1 & A_2 \\ -A_2 & A_1 \end{pmatrix},$$

where A_1, A_2 are real $n \times n$ matrices with

$$A_1^T = -A_1,$$
$$A_2^T = A_2, \quad \mathrm{tr}A_2 = 0.$$

Let $e_1, \ldots, e_{2n}$ be the standard basis of $\mathbb{R}^{2n}$ and let E_{rs} denote the elementary $2n \times 2n$-matrix with a 1 at the intersection of the r-th row and s-th column and zeros elsewhere. We set

$$\epsilon_{rs} = E_{sr} - E_{rs}.$$

Applied to a standard basis vector e_j we get

$$\epsilon_{rs} e_j = \delta_{jr} e_s - \delta_{js} e_r.$$

Then we can expand

$$\begin{aligned} \tau(A_1 + iA_2) = &- \sum_{1 \leq r < s \leq n} \left(A_1^{rs} (\epsilon_{rs} + \epsilon_{r+n,s+n}) + A_2^{rs} (\epsilon_{r,s+n} + \epsilon_{s,r+n}) \right) \\ &- \sum_{1 \leq r \leq n} A_2^{rr} \epsilon_{r,r+n}. \end{aligned}$$

This follows because the sums of the ϵ-matrices in this equation look like (indices on 1 and -1 are matrix indices of the entry)

$$\epsilon_{rs} + \epsilon_{r+n,s+n} = \left(\begin{array}{cc|cc} & -1_{rs} & & \\ 1_{sr} & & & \\ \hline & & & -1_{r+n,s+n} \\ & & 1_{s+n,r+n} & \end{array} \right)$$

$$\epsilon_{r,s+n} + \epsilon_{s,r+n} = \left(\begin{array}{cc|cc} & & & -1_{r,s+n} \\ & & -1_{s,r+n} & \\ \hline & 1_{r+n,s} & & \\ 1_{s+n,r} & & & \end{array} \right)$$

$$\epsilon_{r,r+n} = \left(\begin{array}{c|c} & -1_{r,r+n} \\ \hline 1_{r+n,r} & \end{array}\right).$$

Recall from Sect. 6.5.3 that the Lie algebra of Spin$(2n)$ is given by

$$\mathfrak{spin}(2n) = \text{span}\{e_r e_s \in \text{Cl}(2n) \mid 1 \le r < s \le 2n\}$$

and the isomorphism

$$\lambda_* \colon \mathfrak{spin}(2n) \longrightarrow \mathfrak{so}(2n)$$

maps

$$\lambda_*(e_r e_s) = 2\epsilon_{rs}.$$

It follows that

$$\lambda_*^{-1}\tau(A_1 + iA_2) = -\frac{1}{2}\sum_{1\le r<s\le n}\left(A_1^{rs}(e_r e_s + e_{r+n}e_{s+n}) + A_2^{rs}(e_r e_{s+n} + e_s e_{r+n})\right)$$
$$-\frac{1}{2}\sum_{1\le r\le n} A_2^{rr} e_r e_{r+n}.$$

To apply Lemma 9.5.13 we take for $r \le n$ the basis vector e_r in the first $\mathbb{C}^n$-summand and e_{r+n} in the second summand. Then a calculation shows that for a 1-form $\omega \in \Lambda^1\mathbb{C}^n$ and $1 \le r < s \le n$

$$\frac{1}{2}(e_r e_s + e_{r+n}e_{s+n}) \cdot \omega = -q(e_s, \omega)e_r + q(e_r, \omega)e_s,$$

hence

$$-\frac{1}{2}\sum_{1\le r<s\le n} A_1^{rs}(e_r e_s + e_{r+n}e_{s+n}) \cdot \omega = \sum_{r,s} A_1^{rs} q(e_s, \omega)e_r,$$

where we used that A_1 is skew-symmetric.

Similarly

$$\frac{1}{2}(e_r e_{s+n} + e_s e_{r+n}) \cdot \omega = -i(q(e_s, \omega)e_r + q(e_r, \omega)e_s),$$

$$\frac{1}{2}e_r e_{r+n} \cdot \omega = -i\left(q(e_r, \omega)e_r - \frac{1}{2}\omega\right).$$

This implies

$$-\frac{1}{2}\sum_{1\leq r<s\leq n} A_2^{rs}(e_r e_{s+n} + e_s e_{r+n}) \cdot \omega - \frac{1}{2}\sum_{1\leq r\leq n} A_2^{rr} e_r e_{r+n} \cdot \omega = \sum_{r,s} iA_2^{rs} q(e_s, \omega)e_r$$

since A_2 is symmetric and $\mathrm{tr}A_2 = 0$. We conclude that

$$(\lambda_*^{-1}\tau(A_1 + iA_2)) \cdot \omega = (A_1 + iA_2)\omega \quad \forall \omega \in \Lambda^1\mathbb{C}^n.$$

On the left we have the restriction of the spinor representation to $\mathrm{SU}(n)$ and on the right the fundamental representation of $\mathrm{SU}(n)$. So far we have proved the claim on $\Lambda^1\mathbb{C}^n$, but it is possible to conclude from that the full claim on forms of arbitrary degree. The details are left as Exercise 9.7.7. □

Theorem 9.5.15 and Remark 9.5.14 imply:

Corollary 9.5.16 (Fermions in $\mathrm{Spin}(10)$ **Grand Unification)** *The Dirac spinor representation* **32** *of* $\mathrm{Spin}(10)$ *has under restriction to the subgroup* $\mathrm{SU}(5)$ *the branching rule*

$$\mathbf{32} = \mathbf{1} \oplus \mathbf{5} \oplus \mathbf{10} \oplus \bar{\mathbf{10}} \oplus \bar{\mathbf{5}} \oplus \mathbf{1}.$$

The Weyl spinor representations **16** *and* $\bar{\mathbf{16}}$ *of* $\mathrm{Spin}(10)$ *have under restriction to* $\mathrm{SU}(5)$ *the branching rules*

$$\mathbf{16} = \mathbf{1} \oplus \bar{\mathbf{5}} \oplus \mathbf{10},$$

$$\bar{\mathbf{16}} = \mathbf{1} \oplus \mathbf{5} \oplus \bar{\mathbf{10}}.$$

From Remark 9.5.12 it follows that the Weyl spinor representation **16** of $\mathrm{Spin}(10)$ can accommodate one full left-handed generation of the Standard Model and the Weyl spinor representation $\bar{\mathbf{16}}$ one full right-handed generation, including sterile neutrinos. We get for the first left-handed generation a

(continued)

Corollary 9.5.16 (continued)
decomposition of the form

$$\mathbf{16} = \left(\nu_e^C\right)_L \oplus \begin{pmatrix} d_r^C \\ d_g^C \\ d_b^C \\ e \\ \nu_e \end{pmatrix}_L \oplus \begin{pmatrix} 0 & u_b^C & -u_g^C & u_r & d_r \\ & 0 & u_r^C & u_g & d_g \\ & & 0 & u_b & d_b \\ & & & 0 & e^C \\ & & & & 0 \end{pmatrix}_L .$$

Note that the spinor representations of Spin(10) force us to include a sterile fermion, which can be interpreted as a sterile neutrino. In the case of SU(5) we still had the choice whether we want to add a trivial 1-dimensional representation **1**. In some sense [111] the unification provided by the Spin(10) theory is more complete than for the SU(5) theory, because all (left-handed) fermions of one generation belong to the single, *irreducible* representation **16**.

Remark 9.5.17 See the remark after Exercise 9.7.9 for the U(1) charges under restriction of the spinor representation of Spin(10) to the subgroup $U(5)/\mathbb{Z}_3$.

9.5.6 The Fermions in the E_6 *Grand Unified Theory*

We briefly discuss (without proofs) the Grand Unification group E_6 (see [87] for more details). The Lie group E_6 does not have a non-trivial 15- or 16-dimensional representation, but it has a 27-dimensional representation **27**. Under restriction to the subgroup Spin(10) this representation and its complex conjugate have the branching rules

$$\mathbf{27} = \mathbf{16} \oplus \mathbf{10} \oplus \mathbf{1},$$

$$\bar{\mathbf{27}} = \bar{\mathbf{16}} \oplus \bar{\mathbf{10}} \oplus \mathbf{1}.$$

Here **16** is the spinor representation and **10** the vector representation of Spin(10). As we saw in Sect. 9.5.5, the **16** can accommodate one left-handed generation of the Standard Model, including a sterile neutrino. Similarly the $\bar{\mathbf{16}}$ can accommodate one right-handed generation. The representations **10**, $\bar{\mathbf{10}}$ and **1** correspond to new fermions that so far have not been detected.

9.5.7 Gauge Anomalies

It can be proved that the GUTs constructed above with gauge groups SU(5) and Spin(10) (and E_6) are anomaly free. See [85] for details.

9.5.8 The Gauge Bosons in the SU(5) Grand Unified Theory

We want to study the gauge bosons in the SU(5) Grand Unified Theory (we follow [85]). We first discuss the embedding of the Lie algebra $\mathfrak{g}_{\mathrm{SM}}$ into $\mathfrak{su}(5)$. According to the calculation in Sect. 9.5.3 a correctly normalized basis for $\mathfrak{g}_{\mathrm{SM}}$ is given by

$$\frac{i\lambda_a}{2} \in \mathfrak{su}(3)_C,$$

$$\frac{i\sigma_b}{2} \in \mathfrak{su}(2)_L,$$

$$\sqrt{\frac{3}{5}}\frac{i}{6} = i\sqrt{\frac{1}{60}} \in \mathfrak{u}(1)_Y.$$

Here λ_a are the Gell-Mann matrices and σ_b the Pauli matrices. This implies:

Lemma 9.5.18 *The embedding*

$$f_*\colon \mathfrak{g}_{\mathrm{SM}} \longrightarrow \mathfrak{su}(5)$$

is given by mapping

$$\left(\sum_{a=1}^{8} G_a\frac{i\lambda_a}{2}, \sum_{b=1}^{3} W_b\frac{i\sigma_b}{2}, Bi\sqrt{\frac{1}{60}}\right)$$

to

$$\frac{i}{\sqrt{2}}\begin{pmatrix} \sum_{a=1}^{8} G_a\frac{\lambda_a}{\sqrt{2}} - \sqrt{\frac{2}{15}}BI_3 & 0 \\ 0 & \sum_{b=1}^{3} W_b\frac{\sigma_b}{\sqrt{2}} + \sqrt{\frac{3}{10}}BI_2 \end{pmatrix}$$

We conclude the following:

Theorem 9.5.19 (The SU(5) **Gauge Field)** *The* SU(5) *gauge field* A^μ *with values in* $\mathfrak{su}(5)$ *is given by*

$$A^\mu = \frac{ig_{SU(5)}}{\sqrt{2}} \begin{pmatrix} \multicolumn{3}{c}{\sum_{a=1}^{8} G_a^\mu \frac{\lambda_a}{\sqrt{2}} - \sqrt{\frac{2}{15}} B^\mu I_3} & \begin{matrix} \overline{X}_1^\mu \\ \overline{X}_2^\mu \\ \overline{X}_3^\mu \end{matrix} & \begin{matrix} \overline{Y}_1^\mu \\ \overline{Y}_2^\mu \\ \overline{Y}_3^\mu \end{matrix} \\ \begin{matrix} X_1^\mu \\ Y_1^\mu \end{matrix} & \begin{matrix} X_2^\mu \\ Y_2^\mu \end{matrix} & \begin{matrix} X_3^\mu \\ Y_3^\mu \end{matrix} & \multicolumn{2}{c}{\sum_{b=1}^{3} W_b^\mu \frac{\sigma_b}{\sqrt{2}} + \sqrt{\frac{3}{10}} B^\mu I_2} \end{pmatrix}.$$

Here $g_{SU(5)}$ is the coupling constant of SU(5) and X_j^μ, Y_j^μ are new gauge bosons, corresponding to new forces not present in the Standard Model. We thus get 6 complex (12 real) additional gauge bosons. This is clear, because $\dim SU(5) = 24$ and $\dim G_{SM} = 12$.

From Remark 8.8.9 we expect that the off-diagonal X- and Y-bosons could lead to interesting effects. To understand this in more detail we calculate the relevant part of the Dirac Lagrangian (the same calculation, up to a different choice of signs, can be found in [85]).

Theorem 9.5.20 (*X*- and *Y*-Boson Interaction Vertex) *Consider the left-handed interaction term in the Dirac Lagrangian, given by*

$$i\langle \psi_L, \Gamma^\mu A_\mu \psi_L \rangle.$$

The part of this term involving X- and Y-bosons can be calculated for the first fermion generation as

$$\mathscr{L}_{L,XY} = -\frac{g_{SU(5)}}{\sqrt{2}} j_{LX_i}^\mu X_{i\mu} - \frac{g_{SU(5)}}{\sqrt{2}} j_{LX_i}^{\mu\dagger} \overline{X}_{i\mu} - \frac{g_{SU(5)}}{\sqrt{2}} j_{LY_i}^\mu Y_{i\mu} - \frac{g_{SU(5)}}{\sqrt{2}} j_{LY_i}^{\mu\dagger} \overline{Y}_{i\mu}$$

with the currents

$$\begin{aligned} j_{LX_i}^\mu &= -\epsilon_{ijk} \overline{u}_{jL} \Gamma^\mu u_{kL}^C + \overline{e_L^C} \Gamma^\mu d_{iL} + \overline{d_{iL}^C} \Gamma^\mu e_L, \\ j_{LY_i}^\mu &= -\epsilon_{ijk} \overline{d}_{jL} \Gamma^\mu u_{kL}^C - \overline{e_L^C} \Gamma^\mu u_{iL} + \overline{d_{iL}^C} \Gamma^\mu \nu_{eL}. \end{aligned}$$

In each current the first two terms come from the representation **10** *and the third term from the representation* $\bar{\mathbf{5}}$. *In the first summand there is a sum over indices j, k. Quark indices* 1, 2, 3 *correspond to r, g, b. There are corresponding terms for the second and third generation.*

Feynman diagrams for these interactions are depicted in Figs. 9.4, 9.5, 9.6, 9.7.

In the interaction vertices in Figs. 9.4 and 9.6, an X- or Y-boson pairs a lepton and a quark, hence the X- and Y-bosons are examples of so-called **(vector) leptoquarks**

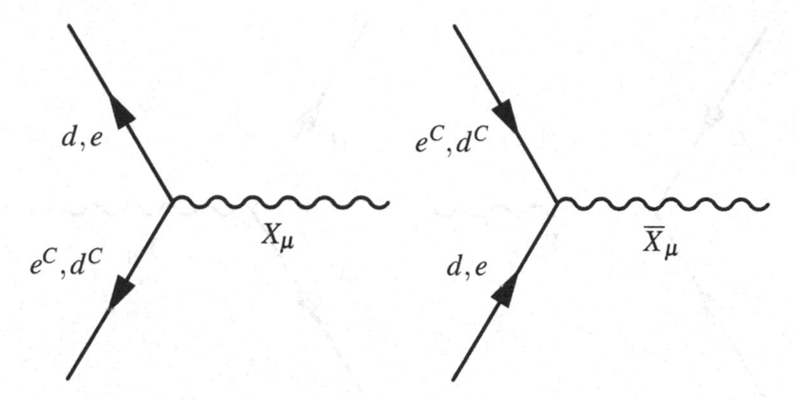

Fig. 9.4 X-boson: quark-lepton vertices

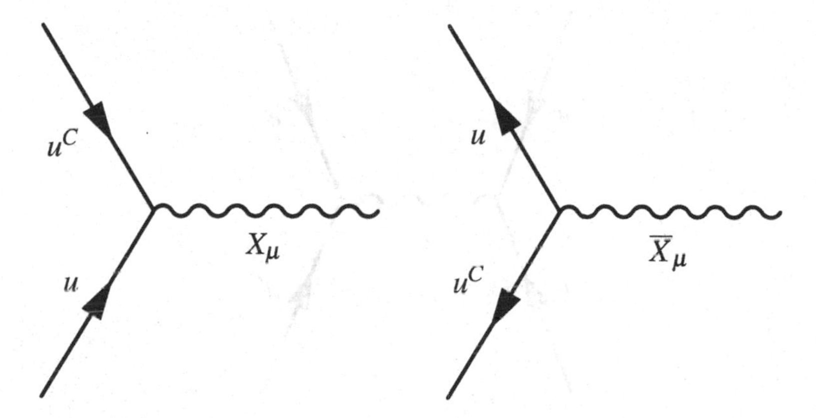

Fig. 9.5 X-boson: quark vertices

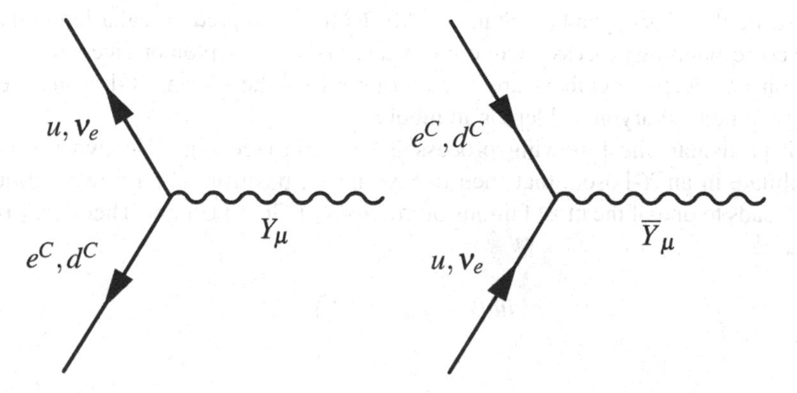

Fig. 9.6 Y-boson: quark-lepton vertices

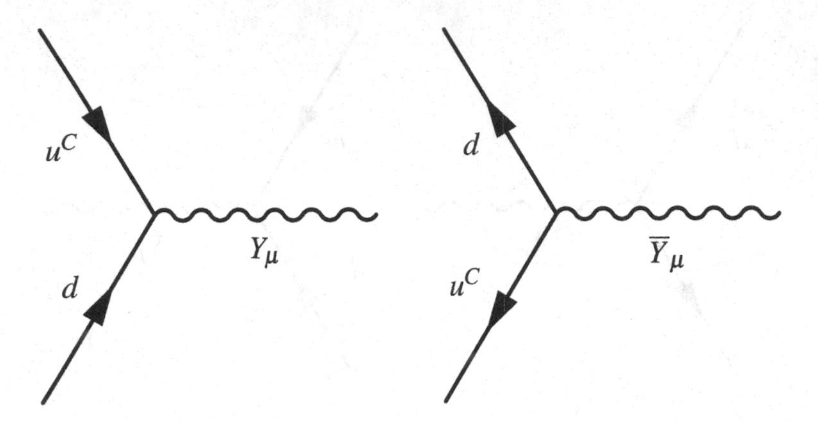

Fig. 9.7 Y-boson: quark vertices

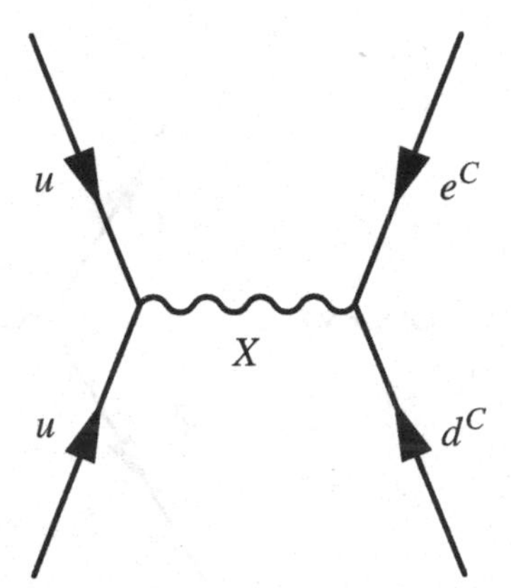

Fig. 9.8 X-boson mediated proton decay

(there are theories beyond the Standard Model that also predict scalar leptoquarks). The corresponding processes can convert a quark into a lepton or vice versa, hence baryon and lepton numbers are not conserved (if the X- and Y-bosons are not assigned both a baryon and lepton number).

In particular, the following process is possible (see Fig. 9.8): two up quarks annihilate in an X-boson, that then decays into a positron and a down antiquark. This leads to one of the most famous predictions of Grand Unified Theories: **proton decay**:[1]

$$p(uud) \longrightarrow \pi^0 \left(dd^C\right) + e^C. \tag{9.7}$$

[1]This is a decay of isolated protons, not to be confused with the weak β^+-decay that can only occur for protons bound in certain nuclei.

The GUT scale of $\sim 10^{15}$ GeV is not directly accessible in experiments and totally out of the reach of present day colliders. However, proton decay as a signature of Grand Unification may in principle be observable, it is just extremely rare. This is related to the extremely heavy mass of the X-boson, making the production of a virtual X-boson via $uu \to X$ very unlikely.

The predicted lifetime of the proton in the SU(5) GUT is $10^{30} - 10^{31}$ years. Experiments show that the lifetime of the proton for the decay into a pion and positron in Eq. (9.7) is longer than $8.2 \cdot 10^{33}$ years [106], hence the minimal version of the SU(5) GUT is already ruled out. In supersymmetric Grand Unified Theories the lifetime of the proton becomes longer and can reach, depending on the model, $10^{34} - 10^{36}$ years or more (see [29] for a nice overview).

A similar discussion applies to the gauge bosons in the Spin(10) theory. Since $\dim \mathrm{Spin}(10) = 45$, we get 33 additional gauge bosons compared to the Standard Model; see [85] for more details.

9.5.9 *Symmetry Breaking and the Higgs Mechanism in the* SU(5) *Grand Unified Theory*

We briefly discuss symmetry breaking in the SU(5) Grand Unified Theory (following [85] and [96]). In the Standard Model, the gauge group

$$G_{\mathrm{SM}} = \mathrm{SU}(3)_C \times \mathrm{SU}(2)_L \times \mathrm{U}(1)_Y$$

is broken to

$$G_{\mathrm{CQ}} = \mathrm{SU}(3)_C \times \mathrm{U}(1)_Q,$$

where according to Eq. (8.4) the unbroken electromagnetic group $\mathrm{U}(1)_Q$ is the subgroup

$$\mathrm{U}(1)_Q = \left\{ \left(\begin{pmatrix} e^{it/2} & 0 \\ 0 & e^{-it/2} \end{pmatrix}, e^{it/6} \right) \middle| \, t \in \mathbb{R} \right\} \subset \mathrm{SU}(2)_L \times \mathrm{U}(1)_Y.$$

The discrete subgroup $K \cong \mathbb{Z}_6 \subset G_{\mathrm{SM}}$ of elements of the form

$$\left(\alpha^2 I_3, \alpha^{-3} I_2, \alpha\right), \quad \alpha \in \mathrm{U}(1), \alpha^6 = 1,$$

is actually a subgroup of G_{CQ} (since $\alpha^{-3} = \alpha^3$) and we get an embedding

$$G_{\mathrm{CQ}}/\mathbb{Z}_6 \subset G_{\mathrm{SM}}/\mathbb{Z}_6.$$

The embedding of $G_{CQ}/\mathbb{Z}_6$ into SU(5) is induced by the homomorphism

$$
\begin{aligned}
\mathrm{SU}(3)_C \times \mathrm{U}(1)_Q &\longrightarrow \mathrm{SU}(5) \\
(g, \beta) &\longmapsto \begin{pmatrix} g\beta^{-2} & 0 \\ 0 & \begin{matrix} \beta^6 & 0 \\ 0 & 1 \end{matrix} \end{pmatrix}
\end{aligned}
\tag{9.8}
$$

which is the restriction of $f\colon G_{SM} \to \mathrm{SU}(5)$ from Proposition 9.5.4 if we identify $\beta \in \mathrm{U}(1)_Q$ with

$$
\left(\begin{pmatrix} \beta^3 & 0 \\ 0 & \beta^{-3} \end{pmatrix}, \beta \right) \in \mathrm{SU}(2)_L \times \mathrm{U}(1)_Y.
$$

Symmetry breaking in the SU(5) GUT occurs in two steps:

$$
\mathrm{SU}(5) \longrightarrow G_{SM}/\mathbb{Z}_6 \longrightarrow G_{CQ}/\mathbb{Z}_6.
$$

The first step

$$
\mathrm{SU}(5) \longrightarrow G_{SM}/\mathbb{Z}_6
$$

can be realized by a Higgs field ϕ with values in

$$
\mathbf{24} = \mathfrak{su}(5)
$$

carrying the adjoint representation of SU(5). To understand this, consider a Higgs condensate $\phi_0 \in \mathfrak{su}(5)$ of the form

$$
\phi_0 = iV' \begin{pmatrix} a & & & & \\ & a & & & \\ & & a & & \\ & & & b & \\ & & & & b \end{pmatrix},
$$

where a, b, V' are real numbers. Since $\phi_0 \in \mathfrak{su}(5)$, we must have

$$
0 = \mathrm{tr}\phi_0 = iV'(3a + 2b).
$$

Setting $V = aV'$ this implies that

$$\phi_0 = iV \begin{pmatrix} 1 & & & & \\ & 1 & & & \\ & & 1 & & \\ & & & -\frac{3}{2} & \\ & & & & -\frac{3}{2} \end{pmatrix} \tag{9.9}$$

with a real constant V.

Proposition 9.5.21 (Symmetry Breaking from $\mathrm{SU}(5)$ **to** $G_{\mathrm{SM}}/\mathbb{Z}_6$**)** *The isotropy group* $\mathrm{SU}(5)_{\phi_0}$ *of the vector* $\phi_0 \in \mathfrak{su}(5)$ *in Eq. (9.9) (with* $V \neq 0$*) under the adjoint representation is the subgroup*

$$\mathrm{S}(\mathrm{U}(3) \times \mathrm{U}(2)) \cong G_{\mathrm{SM}}/\mathbb{Z}_6 \subset \mathrm{SU}(5).$$

Proof This is Exercise 9.7.10. □

The second step

$$G_{\mathrm{SM}}/\mathbb{Z}_6 \longrightarrow G_{\mathrm{CQ}}/\mathbb{Z}_6$$

can be realized by a Higgs field H with values in

$$\mathbf{5} = \mathbb{C}^5$$

carrying the fundamental representation of $\mathrm{SU}(5)$. Consider a Higgs condensate $H_0 \in \mathbb{C}^5$ of the form

$$H_0 = \rho \begin{pmatrix} 0 \\ 0 \\ 0 \\ 0 \\ \frac{1}{\sqrt{2}} \end{pmatrix} \tag{9.10}$$

where ρ is a real number. We then have:

Proposition 9.5.22 (Symmetry Breaking from $G_{\mathrm{SM}}/\mathbb{Z}_6$ **to** $G_{\mathrm{CQ}}/\mathbb{Z}_6$**)** *The isotropy group* $\mathrm{S}(\mathrm{U}(3) \times \mathrm{U}(2))_{H_0}$ *of the vector* $H_0 \in \mathbb{C}^5$ *in Eq. (9.10) (with* $\rho \neq 0$*) under the fundamental representation is the subgroup*

$$G_{\mathrm{CQ}}/\mathbb{Z}_6 \subset G_{\mathrm{SM}}/\mathbb{Z}_6.$$

Proof This is immediate from Eq. (9.8). □

In the model discussed so far, the potential for the Higgs fields is the sum $V(\phi) + V(H)$ of potentials for both fields. For phenomenological reasons it turns out that crossterms of the form $V(\phi, H)$ have to be included. The minimum ϕ_0 then takes the form

$$\phi_0 = iV \begin{pmatrix} 1 & & & & \\ & 1 & & & \\ & & 1 & & \\ & & & -\frac{3}{2} - \frac{\epsilon}{2} & \\ & & & & -\frac{3}{2} + \frac{\epsilon}{2} \end{pmatrix}$$

and thus also breaks $SU(2)_L$.

With the correct Higgs fields determined, it is then possible to calculate the masses of all broken gauge bosons. Finally, suitable Yukawa couplings can be introduced to give masses to the fermions and there is a generalized version of fermion mixing. A similar discussion applies to symmetry breaking in the Spin(10) GUT. Details for these constructions and their physical consequences can be found in [85] and [96].

9.5.10 Further Reading

The article [9] of J. Baez and J. Huerta from 2010 is a mathematical introduction to the representations of the Standard Model and Grand Unified Theories. The review articles [87] by P. Langacker and [131] by R. Slansky summarize the state of the art in Grand Unification in 1981. The book [85] by C. Kounnas et al. from 1984 has a very readable account of the Standard Model and Grand Unification and the book [96] by R.N. Mohapatra from 2003 contains details on Grand Unification and supersymmetry. Shorter expositions of GUTs can be found in the paper [148] by E. Witten from 2002 and the gauge theory books [10] by D. Bailin and A. Love (1993) and [32] by M. Chaichian and N.F. Nelipa (1984).

9.6 A Short Introduction to the Minimal Supersymmetric Standard Model (MSSM)

In this section we give a very brief introduction to supersymmetry. In addition to Lorentz invariance and gauge symmetry, Lagrangians (or actions) of field theories can be *supersymmetric*. The Standard Model itself is not supersymmetric, but can be extended to a (Minimal) Supersymmetric Standard Model by adding *superpartners* to the known particles. Some of these superpartners are potential dark matter candidates. Supersymmetry also offers a solution to the so-called *hierarchy problem* concerning the unnatural smallness of the Higgs boson mass. Furthermore, Grand Unification becomes more realistic (for example, the predicted decay rate of the proton and the unification of coupling constants) under the assumption of supersymmetry.

These are some of the reasons to believe that supersymmetry is realized (as a broken symmetry) in nature. In addition to yielding extensions of the Standard Model, supersymmetric field theories are also theoretically interesting, because they are often better understood non-perturbatively (at strong coupling) than non-supersymmetric field theories.

9.6.1 Graded Lie Algebras and the Supersymmetry Algebra

In this subsection we define graded Lie algebras and the supersymmetry algebra (we follow [53]).

Definition 9.6.1 A **graded Lie algebra** is an algebra, i.e. a vector space $L = L_0 \oplus L_1$ with a bilinear product

$$\cdot : L \times L \longrightarrow L,$$

that has the following properties for all $x_i \in L_i, x_j \in L_j$:

1. **Grading:**

$$x_i \cdot x_j \in L_{i+j \bmod 2}$$

2. **Supersymmetry:**

$$x_i \cdot x_j = -(-1)^{ij} x_j \cdot x_i$$

3. **Super Jacobi identity:**

$$x_k \cdot (x_l \cdot x_m)(-1)^{km} + x_l \cdot (x_m \cdot x_k)(-1)^{lk} + x_m \cdot (x_k \cdot x_l)(-1)^{ml} = 0.$$

The vector subspace L_0 is called the even part and L_1 the odd part.
It follows that the product $\cdot$ is

- antisymmetric on $L_0 \times L_0$ and maps to L_0 (written as $[\cdot, \cdot]$)
- symmetric on $L_1 \times L_1$ and maps to L_0 (written as $\{\cdot, \cdot\}$)
- antisymmetric on $L_1 \times L_0$ and $L_0 \times L_1$ and maps to L_1 (written as $[\cdot, \cdot]$).

The notation $[\cdot, \cdot]$ and $\{\cdot, \cdot\}$ is just a different notation for the product $\cdot$ on the algebra L. On two general elements in L, which have components in both L_0 and L_1, the product is neither symmetric nor antisymmetric and sometimes written as $\{[\cdot, \cdot]\}$. The following is easy to show:

Proposition 9.6.2 (Characterization of Graded Lie Algebras) *Let* $(L, \{[\cdot\,,\cdot]\})$ *be a graded Lie algebra.*

1. *The vector subspace* L_0 *with the product* $[\cdot\,,\cdot]$ *is a Lie algebra.*
2. *The map*

$$\phi\colon L_0 \longrightarrow \mathrm{End}(L_1)$$

   with

$$\phi(x)v = [x, v] \quad \forall x \in L_0, v \in L_1$$

   is a representation of the Lie algebra $(L_0, [\cdot\,,\cdot])$ *on the vector space* L_1.
3. *The map*

$$\{\cdot\,,\cdot\}\colon L_1 \times L_1 \longrightarrow L_0$$

   is a vector space-valued symmetric bilinear form.

Conversely, a Lie algebra $(L_0, [\cdot\,,\cdot])$ *together with a representation* ϕ *of* L_0 *on a vector space* L_1 *and a symmetric bilinear form* $\{\cdot\,,\cdot\}$ *on* L_1 *with values in* L_0 *define a graded Lie algebra if the following identities are satisfied:*

$$\phi(\{u, v\})w + \phi(\{v, w\})u + \phi(\{w, u\})v = 0 \quad \forall u, v, w \in L_1 \tag{9.11}$$

and

$$[x, \{v, w\}] = \{\phi(x)v, w\} + \{v, \phi(x)w\} \quad \forall x \in L_0, v, w \in L_1. \tag{9.12}$$

Proof This is Exercise 9.7.11. □

The supersymmetry algebra that we now discuss is a special graded Lie algebra. We first define the Poincaré algebra: let (V, η) be Minkowski spacetime $V = \mathbb{R}^{1,n}$ of dimension $n + 1$ with Minkowski metric η. Let $\mathfrak{so}(1, n) = \mathfrak{so}(V) = \mathfrak{so}^+(V)$ denote the Lie algebra of the proper orthochronous Lorentz group $\mathrm{SO}^+(V)$. Elements of $\mathfrak{so}(V)$ are denoted in physics by M. They correspond to infinitesimal spacetime rotations.

Definition 9.6.3 The **Poincaré algebra** is a real Lie algebra with underlying vector space

$$\mathfrak{iso}(V) = \mathfrak{so}(V) \oplus V,$$

also denoted by $\mathfrak{iso}(1,n)$. Elements of the subspace V are denoted in physics by P. The commutator on $\mathfrak{iso}(V)$ is a semi-direct product:

1. The commutator on the subspace $\mathfrak{so}(V)$ is the standard one.
2. The commutator on the subspace V vanishes:

$$[P,P'] = 0 \quad \forall P,P' \in V.$$

3. Consider the vector representation

$$v: \mathfrak{so}(V) \longrightarrow \mathrm{End}(V).$$

The commutator between elements of $\mathfrak{so}(V)$ and V maps to V and is defined by

$$[M,P] = v(M)P \quad \forall M \in \mathfrak{so}(V), P \in V.$$

It is easy to check that $\mathfrak{iso}(V)$ is indeed a Lie algebra (in the standard sense). In physics the commutators are often written with respect to a basis for the Lie algebra. The elements of the Poincaré algebra correspond to infinitesimal spacetime rotations and translations.

The Poincaré algebra can be extended to the super-Poincaré algebra by adding a spinor representation.

Definition 9.6.4 The **N = 1 super-Poincaré algebra** or **supersymmetry algebra** is a graded real Lie algebra with underlying vector space

$$\mathfrak{susy}(V) = \mathfrak{iso}(V) \oplus S^* = \mathfrak{so}(V) \oplus V \oplus S^*,$$

also denoted by $\mathfrak{susy}(1,n)$. Here S^* is the dual of a real spinor representation space of minimal dimension of $\mathfrak{so}(V) \cong \mathfrak{spin}(V)$. Elements of S^* are denoted in physics by Q. The multiplication $\{[\cdot,\cdot]\}$ on $\mathfrak{susy}(V)$ is in some sense a graded semi-direct product defined as follows:

1. The multiplication on the even subspace $\mathfrak{iso}(V)$ is the standard commutator in the Poincaré algebra.
2. Consider the dual spinor representation

$$s: \mathfrak{so}(V) \longrightarrow \mathrm{End}(S^*).$$

The multiplication between $\mathfrak{so}(V)$ and the odd subspace S^* is defined by

$$[M,Q] = s(M)Q \quad \forall M \in \mathfrak{so}(V), Q \in S^*.$$

(continued)

Definition 9.6.4 (continued)

3. The multiplication between V and S^* vanishes:

$$[P, Q] = 0 \quad \forall P \in V, Q \in S^*.$$

4. It can be shown that there exists a symmetric bilinear form

$$\Gamma : S^* \times S^* \longrightarrow V$$

that is Lorentz equivariant, i.e. satisfies the identity

$$[M, \Gamma(Q, Q')] = \Gamma([M, Q], Q') + \Gamma(Q, [M, Q']) \tag{9.13}$$

for all $M \in \mathfrak{so}(V), Q, Q' \in S^*$. The multiplication on S^* maps to $V \subset \mathfrak{iso}(V)$ and is defined by

$$\{Q, Q'\} = 2\Gamma(Q, Q') \quad \forall Q, Q' \in S^*.$$

It is easy to check using Proposition 9.6.2 that $\mathfrak{susy}(V)$ is a graded Lie algebra. Equation (9.11) is satisfied, because V acts trivially on S^*, and Eq. (9.12) corresponds to Eq. 9.13. Note that in the super-Poincaré algebra the product

$$\{\cdot, \cdot\} : S^* \times S^* \rightarrow \mathfrak{iso}(V)$$

only maps to the subspace V of $\mathfrak{iso}(V)$. Again, in the physics literature the multiplication on $\mathfrak{susy}(V)$ is usually written with respect to a basis. The elements $Q \in S^* \subset \mathfrak{susy}(V)$ are called **(infinitesimal) supersymmetries**. The supersymmetry algebra $\mathfrak{susy}(V)$ thus consists of infinitesimal spacetime rotations, translations and supersymmetries.

The super-Poincaré algebra for $N \geq 2$ supersymmetries is obtained similarly from the Poincaré algebra by adding N copies of S^*. In this situation there is more flexibility in the definition of the symmetric bilinear form Γ leading to the concept of **central charges**. We will restrict to the case $N = 1$.

9.6.2 Supersymmetric Field Theories

Field theories on Minkowski spacetime $\mathbb{R}^{1,n}$ are usually assumed to be Poincaré invariant. This means that there is a representation of the Poincaré algebra $\mathfrak{iso}(1, n)$ on the fields and the action of the theory is invariant under this algebra.

Supersymmetric field theories on Minkowski spacetime are field theories where the fields have a representation of the super-Poincaré algebra $\mathfrak{susy}(1,n)$ leaving the action invariant. We only consider the case of **rigid supersymmetries** where the spinors $Q \in S^*$ generating the supersymmetries are constant (parallel with respect to the canonical spin covariant derivative) on Minkowski spacetime (the infinitesimal rotations and translations in $\mathfrak{susy}(1,n)$ are constant as well). Just as the infinitesimal isometries in the Poincaré algebra are defined by certain vector fields on spacetime, infinitesimal supersymmetries are defined by (constant) spinors.

Theories with an action invariant under arbitrary supersymmetries (that are **local**, i.e. not necessarily constant) are called **supergravities**. Since the anticommutator of two supersymmetries is basically a vector (translation), theories which are invariant under spacetime-dependent supersymmetries are invariant under the action of all vector fields and hence under all infinitesimal diffeomorphisms. Supergravities are therefore automatically diffeomorphism invariant (at least under diffeomorphisms that can be connected to the identity), i.e. theories of gravity.

This means that spinors can appear in field theories in two different ways:

- (twisted) spinors describe matter particles, i.e. (charged) fermions
- pure (untwisted) spinors can be infinitesimal generators of supersymmetries.

This can be compared to vector fields (or 1-forms) in field theories:

- vector fields with values in the adjoint bundle describe gauge bosons
- pure vector fields (sections of the tangent bundle) describe infinitesimal diffeomorphisms or isometries of the manifold.

Since field theories like the Standard Model involve charged fermions, it seems from this point of view natural to extend the *bosonic* symmetries of spacetime (isometries or diffeomorphisms) by *fermionic* symmetries (rigid or local supersymmetries). According to the Haag–Łopuszański–Sohnius Theorem the only consistent way to do so is by using the super-Poincaré algebra.

One of the interesting things about supersymmetric field theories is that the particle content and the interactions that can occur in the Lagrangian are very restricted. Representations of the Poincaré algebra are given by fields of a certain spin. The elements of the Poincaré algebra map fields of one spin to fields of the same spin. Supersymmetries, on the other hand, map a field of one spin to a field of another spin. This means that fields of different spin are combined in **supersymmetry multiplets** and the elements of the super-Poincaré algebra map a multiplet of one type to a multiplet of the same type. Supersymmetries do not preserve the spin of a field, but they do preserve the type of the supersymmetry multiplet. Only complete supersymmetry multiplets, not just some of their components, can be added to supersymmetric field theories.

For $N = 1$ supersymmetry in $D = 4$ Minkowski spacetime $\mathbb{R}^{1,3}$ there are only two types of supersymmetry multiplets with fields of spin at most 1 (i.e. that do not contain a graviton):

- **vector multiplet** (or **gauge multiplet**) consisting of a vector (spin 1) and a Majorana spinor (spin $\frac{1}{2}$)
- **chiral multiplet** consisting of a Weyl spinor (spin $\frac{1}{2}$) and a complex scalar (spin 0).

Recall that the particles in the Standard Model are of three types:

- vectors (gauge bosons)
- fermions (quarks and leptons)
- scalars (Higgs field and Higgs boson).

It turns out that the particles in the Standard Model (vectors–fermions and fermions–scalars) do not themselves combine to form supersymmetry multiplets. For example, both fields in a vector multiplet have to be associated to the same representation of the gauge group (i.e. to the adjoint representation). Similarly, both fields in a chiral multiplet have to be associated to the same gauge representation.

If we want to extend the Standard Model to a **Minimal** $N = 1$ **Supersymmetric Standard Model (MSSM)**, it follows that we have to add for each particle in the Standard Model a **superpartner** to form either a vector multiplet or a chiral multiplet (see Table 9.2):

- for the gauge bosons (W-bosons, B-boson, Z-boson, photon, gluons) we add Majorana fermions called gauginos (winos, bino, zino, photino, gluinos) to form vector multiplets;
- for the left-handed and right-handed fermions (quarks and leptons) we add complex scalars (squarks, sleptons) to form chiral multiplets;
- for the scalars (Higgs field) we add Weyl fermions (Higgsinos) to form chiral multiplets. In addition to the Higgs isodoublet from the Standard Model with $Y = 1$, now denoted by ϕ_u, we need another Higgs isodoublet with $Y = -1$, denoted by ϕ_d, to ensure cancellation of gauge anomalies in the MSSM.

Using these multiplets it is possible to write down Lagrangians such that the actions are Poincaré invariant and invariant under supersymmetries generated by spinors $Q \in S^*$. The symmetries together satisfy the relations of the super-Poincaré algebra (sometimes only *on-shell*, i.e. if the fields satisfy the equations of motion).

Supersymmetry predicts that all particles in a supersymmetry multiplet have the same mass. Since the superpartners of the particles in the Standard Model so far have not been observed, **supersymmetry (if it exists) must be broken in nature** and the superpartners must be heavier than the known particles.

Table 9.2 Particle content of MSSM (fermions and sfermions repeat in 3 generations)

Standard Model particles			Superpartners		
Name	Field	Spin	Name	Field	Spin
Quarks	$\begin{pmatrix} u_L \\ d_L \end{pmatrix}$	$\frac{1}{2}$	Squarks	$\begin{pmatrix} \tilde{u}_L \\ \tilde{d}_L \end{pmatrix}$	0
	u_R, d_R	$\frac{1}{2}$		$\tilde{u}_R, \tilde{d}_R$	0
Leptons	$\begin{pmatrix} \nu_L \\ e_L \end{pmatrix}$	$\frac{1}{2}$	Sleptons	$\begin{pmatrix} \tilde{\nu}_L \\ \tilde{e}_L \end{pmatrix}$	0
	e_R	$\frac{1}{2}$		$\tilde{e}_R$	0
Higgs	$\begin{pmatrix} \phi_u^+ \\ \phi_u^0 \end{pmatrix}$	0	Higgsinos	$\begin{pmatrix} \tilde{\phi}_u^+ \\ \tilde{\phi}_u^0 \end{pmatrix}$	$\frac{1}{2}$
	$\begin{pmatrix} \phi_d^0 \\ \phi_d^- \end{pmatrix}$	0		$\begin{pmatrix} \tilde{\phi}_d^0 \\ \tilde{\phi}_d^- \end{pmatrix}$	$\frac{1}{2}$
Gluons	G	1	Gluinos	$\tilde{G}$	$\frac{1}{2}$
W-bosons	$W^{\pm}$	1	Winos	$\tilde{W}^{\pm}$	$\frac{1}{2}$
	W^0	1		$\tilde{W}^0$	$\frac{1}{2}$
B-boson	B	1	Bino	$\tilde{B}$	$\frac{1}{2}$
Z-boson	Z^0	1	Zino	$\tilde{Z}^0$	$\frac{1}{2}$
Photon	γ	1	Photino	$\tilde{\gamma}$	$\frac{1}{2}$

The Lagrangian of the MSSM itself is very restricted by demanding supersymmetry. If we want to introduce supersymmetry breaking, however, the Lagrangian becomes much more complicated and involves many additional terms. The precise mechanism of supersymmetry breaking is still under discussion.

Supersymmetry can be combined with Grand Unification and yields, for instance, $N = 1$ supersymmetric extensions of the SU(5) and Spin(10) Grand Unified Theories. Supersymmetric GUTs can also be derived naturally from superstring and M-theory.

9.6.3 Further Reading

The book [53] by D.S. Freed contains a mathematical discussion of supersymmetry. The classical and quantum theory of supersymmetric field theories are covered in the lecture notes [4] by P.C. Argyres. The paper [91] by S.P. Martin is an extensive and readable introduction to supersymmetry and the MSSM. The book [96] by R.N. Mohapatra discusses, among other things, the MSSM and supersymmetric Grand Unification.

9.7 Exercises for Chap. 9

9.7.1 Prove that any unitary 2×2-matrix V can be brought into real form by the rephasings in Eq. (9.5).

9.7.2 Prove Proposition 9.3.6 on the invariance of Lagrangians under parity inversion.

9.7.3 Prove the remaining cases of Proposition 9.3.7 on the invariance of Lagrangians under charge conjugation.

9.7.4 (from [55]) We consider the Lie group $\mathrm{U}(n)$ and the homomorphism

$$\begin{aligned} \sigma \colon \mathrm{U}(n) &\longrightarrow \mathrm{U}(n) \\ A &\longmapsto \det A \cdot A. \end{aligned}$$

On the subgroup $\mathrm{SU}(2n)$ the homomorphism σ is the identity. Let $\iota \colon \mathrm{U}(n) \hookrightarrow \mathrm{SO}(2n)$ be the standard embedding. Suppose that n is odd.

1. Use covering theory to prove that the composition $\iota \circ \sigma$ lifts to a homomorphism

$$\phi \colon \mathrm{U}(n) \longrightarrow \mathrm{Spin}(2n).$$

2. Determine the kernel of σ and the kernel of ϕ. Prove that there is a sequence of Lie group embeddings

$$\mathrm{SU}(n) \hookrightarrow \mathrm{U}(n)/\mathbb{Z}_{(n+1)/2} \overset{\phi}{\hookrightarrow} \mathrm{Spin}(2n).$$

9.7.5 Prove the following statement in Lemma 9.5.10: the representation of $\mathrm{SU}(3)$ on 3×3-antisymmetric matrices A, given by

$$M \cdot A = MAM^T \quad \forall M \in \mathrm{SU}(3),$$

is isomorphic to $\bar{\mathbf{3}}$.

9.7.6 Recall the identification of the spinor representation Δ of $\mathrm{Spin}(2n)$ with a representation on $\Lambda^*\mathbb{C}^n$ from Lemma 9.5.13. Determine in which dimensions n the left-handed (positive) Weyl spinor space Δ_+ corresponds under this identification to the subspace $\Lambda^{\mathrm{even}}\mathbb{C}^n$ or $\Lambda^{\mathrm{odd}}\mathbb{C}^n$ of forms of even and odd degree.

9.7.7 Prove the statement in Theorem 9.5.15 in the remaining case for forms of arbitrary degree in $\Lambda^*\mathbb{C}^n$ (compare with [9]).

9.7.8 Suppose that n is odd. Consider the embedding

$$\phi \colon \mathrm{U}(n)/\mathbb{Z}_{(n+1)/2} \hookrightarrow \mathrm{Spin}(2n)$$

from Exercise 9.7.4 and the Dirac spinor representation $\Delta \cong \Lambda^*\mathbb{C}^n$ of Spin$(2n)$. Let X be the element

$$X = \mathrm{diag}\left(\frac{2i}{n+1}, \ldots, \frac{2i}{n+1}\right) \in \mathfrak{u}(n).$$

Under the homomorphism σ_* this element maps to

$$iA_2 = \mathrm{diag}\,(2i, \ldots, 2i) \in \mathfrak{u}(n)$$

which is then embedded into $\mathfrak{so}(2n)$ as in the proof of Theorem 9.5.15. We want to determine how this element acts on some of the summands of $\Lambda^*\mathbb{C}^n$. Prove that X acts

- on $\Lambda^0\mathbb{C}^n$ by multiplication with $-ni$
- on $\Lambda^1\mathbb{C}^n$ by multiplication with $-(n-2)i$ and
- on $\Lambda^2\mathbb{C}^n$ by multiplication with $-(n-4)i$.

9.7.9 Let $n \geq 2$ be an arbitrary integer and consider the homomorphism

$$\iota \circ \sigma : \mathrm{U}(n)/\mathbb{Z}_{(n+1)/2} \to \mathrm{U}(n) \hookrightarrow \mathrm{SO}(2n)$$

from Exercise 9.7.4. Recall from Exercise 2.7.6 that the complex fundamental representation $V = \mathbb{C}^{2n}$ decomposes under restriction to U(n) into $W \oplus \bar{W}$, where W is the complex fundamental representation of U(n).

Let X be the element in $\mathfrak{u}(n)$ from Exercise 9.7.8 and

$$iA_2 = \mathrm{diag}\,(2i, \ldots, 2i) \in \mathfrak{u}(n).$$

Show that X acts on W by multiplication with $2i$ and on $\bar{W}$ by multiplication with $-2i$.

Remark In the case $n = 5$, taking U(1)-charges with respect to the basis vector X, it follows that the left-handed Weyl spinor and the vector representation of Spin(10) decompose under restriction to U$(5)/\mathbb{Z}_3$ as

$$\mathbf{16} = \mathbf{1}_{-5} \oplus \bar{\mathbf{5}}_3 \oplus \mathbf{10}_{-1},$$

$$\mathbf{10} = \mathbf{5}_2 \oplus \bar{\mathbf{5}}_{-2}.$$

Compare with [131, Table 43].

9.7.10 Prove Proposition 9.5.21 on the isotropy group of $\phi_0 \in \mathfrak{su}(5)$ under the adjoint representation.

9.7.11 Prove Proposition 9.6.2 on graded Lie algebras.

Part III
Appendix

Appendix A
Background on Differentiable Manifolds

From a mathematical point of view, gauge theories are described by a *spacetime* M together with certain *fibre bundles* (principal bundles, associated vector bundles, spinor bundles) over M. Spacetime and fibre bundles are assumed to have the structure of *differentiable manifolds*. Differentiable manifolds in turn are certain *topological spaces* that essentially have the property of being locally Euclidean, i.e. locally look like an open set in some $\mathbb{R}^n$, and that have a differentiable structure, so that we can define differentiable maps (and their derivatives), vector fields, differential forms, etc. on them.

We briefly sketch the definitions of these concepts. More details can be found in any textbook on differentiable manifolds or differential geometry, like [84] and [142].

A.1 Manifolds

A.1.1 Topological Manifolds

Topological manifolds are topological spaces with certain additional structures. They are a first step towards *differentiable manifolds*, which are the main spaces that we will consider in this book.

Definition A.1.1 An **n-dimensional topological manifold**, also called a **topological n-manifold**, is a topological space M such that:

1. M is locally Euclidean, i.e. locally homeomorphic to $\mathbb{R}^n$. This means that around every point $p \in M$ there exists an open neighbourhood $U \subset M$ that is homeomorphic to some open set $V \subset \mathbb{R}^n$ (both open sets with the subspace topology).

M.J.D. Hamilton, *Mathematical Gauge Theory*, Universitext,
https://doi.org/10.1007/978-3-319-68439-0_10

2. M is Hausdorff.
3. M has a countable basis for its topology.

The local homeomorphisms $\phi: M \supset U \to V \subset \mathbb{R}^n$ (and sometimes the subsets U) are called **charts** or **local coordinate systems** for M. Axiom (a) says that we can cover the whole manifold M by charts. Note that the dimension n is assumed to be the same over the whole manifold. Axiom (c) is of a technical nature and usually can be neglected for our purposes. We often denote an n-manifold by M^n.

Example A.1.2 The simplest topological n-manifold is $M = \mathbb{R}^n$ itself. We can cover M by one chart $\phi: \mathbb{R}^n \to \mathbb{R}^n$, given by the identity.

Example A.1.3 Another example of a topological n-manifold is the n**-sphere** $M = S^n$ for $n \geq 0$. We define

$$S^n = \{x \in \mathbb{R}^{n+1} \mid ||x|| = 1\}.$$

Here $||x||$ denotes the Euclidean norm. We endow S^n with the subspace topology of $\mathbb{R}^{n+1}$. It follows that S^n is Hausdorff, compact and has a countable basis for its topology.

We thus only have to cover S^n by charts that define local homeomorphisms to $\mathbb{R}^n$. A very useful choice are two charts given by **stereographic projection**. We think of $\mathbb{R}^n$ as the hyperplane $\{x_{n+1} = 0\}$ in $\mathbb{R}^{n+1}$. We then project a point x in $U_N = S^n \setminus \{N\}$, where N is the north pole

$$N = (0, \ldots, 0, +1) \in S^n \subset \mathbb{R}^{n+1},$$

along the line through N and x onto the hyperplane $\mathbb{R}^n$. It is easy to check that this defines a map

$$\begin{aligned} \phi_N: U_N &\longrightarrow \mathbb{R}^n \\ x &\longmapsto \frac{1}{1 - x_{n+1}}(x_1, \ldots, x_n). \end{aligned}$$

Similarly projection through the south pole

$$S = (0, \ldots, 0, -1) \in S^n \subset \mathbb{R}^{n+1}$$

defines a map on $U_S = S^n \setminus \{S\}$, given by

$$\begin{aligned} \phi_S: U_S &\longrightarrow \mathbb{R}^n \\ x &\longmapsto \frac{1}{1 + x_{n+1}}(x_1, \ldots, x_n). \end{aligned}$$

We can check that ϕ_N and ϕ_S are bijective, continuous and have continuous inverses. Therefore they are homeomorphisms. They define two charts that cover S^n and hence the n-sphere is shown to be a topological manifold.

A.1.2 Differentiable Structures and Atlases

Suppose we have two topological manifolds M and N and a continuous map $f: M \to N$ between them. We want to define what it means that f is *differentiable*. To do so we first have to define a *differentiable (or smooth) structure* on both manifolds.

Definition A.1.4 Let M be a topological n-manifold. Suppose (U, ϕ) and (V, ψ) are two charts of M. We call these charts *compatible* if the **change of coordinates** (or **coordinate transformation**), given by the map

$$\psi \circ \phi^{-1}: \mathbb{R}^n \supset \phi(U \cap V) \longrightarrow \psi(U \cap V) \subset \mathbb{R}^n,$$

is a smooth diffeomorphism between open subsets of $\mathbb{R}^n$, i.e. the homeomorphism $\psi \circ \phi^{-1}$ and its inverse are infinitely differentiable.

Definition A.1.5 Let $\mathscr{A}$ be a set of charts that cover M. We call $\mathscr{A}$ an **atlas** if any two charts in $\mathscr{A}$ are compatible. We call $\mathscr{A}$ a **maximal atlas** (or **differentiable structure**) if the following holds: Any chart of M that is compatible with all charts in $\mathscr{A}$ belongs to $\mathscr{A}$. It can be checked that any given atlas for M is contained in a unique maximal atlas.

Definition A.1.6 A topological manifold M together with a maximal atlas is called a **differentiable** (or **smooth**) **manifold**.

Example A.1.7 The topological manifold $\mathbb{R}^n$ is a differentiable manifold: We have one chart $(\mathbb{R}^n, \text{Id})$, where $\text{Id}: \mathbb{R}^n \to \mathbb{R}^n$ is the identity. Since we only have a single chart, there are no non-trivial changes of coordinates. Therefore $\mathscr{A} = \{(\mathbb{R}^n, \text{Id})\}$ forms an atlas that induces a unique differentiable structure on $\mathbb{R}^n$ (the *standard differentiable structure*).

Example A.1.8 Recall that we defined on the n-sphere S^n two charts (U_N, ϕ_N) and (U_S, ϕ_S). We want to show that these two charts are compatible and hence form an atlas. This atlas is contained in a unique maximal atlas that defines a differentiable structure on the n-sphere (the *standard structure*).

We first have to calculate the inverse of the chart mappings: We have

$$\begin{aligned} \phi_N^{-1}: \mathbb{R}^n &\longrightarrow U_N \\ y &\longmapsto \left(\frac{2y}{1 + ||y||^2}, \frac{||y||^2 - 1}{1 + ||y||^2} \right) \end{aligned}$$

and

$$\begin{aligned}\phi_S^{-1}\colon \mathbb{R}^n &\longrightarrow U_S\\ y &\longmapsto \left(\frac{2y}{1+||y||^2}, \frac{1-||y||^2}{1+||y||^2}\right).\end{aligned}$$

Since

$$\phi_N(U_N \cap U_S) = \phi_S(U_N \cap U_S) = \mathbb{R}^n \setminus \{0\}$$

we get

$$\phi_S \circ \phi_N^{-1}\colon \mathbb{R}^n \setminus \{0\} \to \mathbb{R}^n \setminus \{0\}$$

with

$$\begin{aligned}\phi_S \circ \phi_N^{-1}(y) &= \phi_S\left(\frac{2y}{1+||y||^2}, \frac{||y||^2-1}{1+||y||^2}\right)\\ &= \frac{y}{||y||^2}.\end{aligned}$$

A similar calculation shows that

$$\phi_N \circ \phi_S^{-1}(y) = \frac{y}{||y||^2}.$$

Since these maps are infinitely differentiable, it follows that the charts (U_N, ϕ_N) and (U_S, ϕ_S) are compatible and define a smooth structure on the n-sphere S^n.

Remark A.1.9 In certain dimensions n there exist *exotic spheres*, which are differentiable structures on the topological manifold S^n not diffeomorphic to the standard structure. The first examples have been described by Milnor and Kervaire.

Remark A.1.10 From now we consider **only smooth manifolds**.

Example A.1.11 It is possible to extend the definition of smooth manifolds to include manifolds M with boundary ∂M. We usually consider only manifolds without boundary, even though most concepts in this book also make sense for manifolds with boundary.

Definition A.1.12 A manifold M is called **closed** if it is compact and without boundary.

Definition A.1.13 A manifold M is called **oriented** if it has an atlas $\mathscr{A}$ of charts $\{(U_i, \phi_i)\}$ such that the differential $D_{\phi_i(p)}(\phi_j \circ \phi_i^{-1})$ (represented by the Jacobi matrix) of any change of coordinates has positive determinant at each point.

A.1.3 Differentiable Mappings

We can now define the notion of a differentiable map between differentiable manifolds.

Definition A.1.14 Let M^m and N^n be differentiable manifolds and $f: M \to N$ a continuous map. Let $p \in M$ be a point and (V, ψ) a chart of N around $f(p)$. Since f is continuous, there exists a chart (U, ϕ) around p such that $f(U) \subset V$. We call f **differentiable at** p if the map

$$\psi \circ f \circ \phi^{-1}: \mathbb{R}^m \supset \phi(U) \longrightarrow \psi(V) \subset \mathbb{R}^n$$

is infinitely differentiable (in the usual sense) at $\phi(p)$ as a map between open subsets of $\mathbb{R}^n$.

Remark A.1.15 The property of a map f being differentiable at a point p does not depend on the choice of charts, precisely because all changes of coordinates are diffeomorphisms: if f is differentiable at p for one pair of charts, then it is also differentiable for all other pairs.

Definition A.1.16 We call a continuous map $f: M \to N$ **differentiable** if it is differentiable at every $p \in M$. We call f a **diffeomorphism** if it is a homeomorphism such that f and f^{-1} are differentiable.

Remark A.1.17 All differentiable maps between manifolds in the following will be **infinitely differentiable (smooth)**, also called $\mathscr{C}^\infty$.

Example A.1.18 It is a nice exercise to show that the involution

$$\begin{aligned} i: S^n &\longrightarrow S^n \\ x &\longmapsto -x \end{aligned}$$

is a diffeomorphism.

A.1.4 Products of Manifolds

Let M^m and N^n be differentiable manifolds. Then the Cartesian product $X^{m+n} = M^m \times N^n$ canonically has the structure of a differentiable manifold of dimension $m + n$. We have to define charts for X: Let (U, ϕ) and (V, ψ) be local charts for M and N. Then $(U \times V, \phi \times \psi)$ is a local chart for X, where

$$\begin{aligned} \phi \times \psi: U \times V &\longrightarrow \mathbb{R}^m \times \mathbb{R}^n \\ (x, y) &\longmapsto (\phi(x), \psi(y)). \end{aligned}$$

It can easily be checked that with this definition the changes of coordinates are smooth.

A.1.5 Tangent Space

Suppose M^n is a differentiable manifold and $p \in M$ is a point. An important notion is that of the *tangent space* T_pM of the manifold at the point p. This is something that only exists on *smooth* manifolds and not on *topological* manifolds.

How can we define such a tangent space? To get some intuition, we can first consider the case of a *submanifold* $M \subset \mathbb{R}^d$ of some Euclidean space. The standard definition is that the tangent space in $p \in M$ is the *subspace* of $\mathbb{R}^d$ consisting of all tangent vectors to differentiable curves through p:

$$T_pM = \{\dot{\gamma}(0) \in \mathbb{R}^d \mid \gamma\colon (-\epsilon, \epsilon) \to M \text{ differentiable}, \gamma(0) = p\}.$$

The problem with general manifolds is that they are *a priori* not embedded in any surrounding space, so this notion of tangent vector does not work. However, what we can do, is that instead of taking the tangent vectors in the surrounding space, we take the full set of curves through p in the manifold M and define on this set an equivalence relation that identifies two of them, α and β, if *in a chart* $\phi\colon M \supset U \to \mathbb{R}^n$ they have the same tangent vector in p:

$$\alpha \sim \beta \Leftrightarrow (\phi \,\dot{\circ}\, \alpha)(0) = (\phi \,\dot{\circ}\, \beta)(0).$$

To be equivalent in this sense does not depend on the choice of charts: If we choose another chart $\psi\colon M \supset V \to \mathbb{R}^n$ around p, then the tangent vectors in the charts ϕ and ψ are related by a linear map, the differential $D_{\phi(p)}(\psi \circ \phi^{-1})$ of the change of coordinates. Since the tangent vectors of α and β in chart ϕ are identical, they will thus still be identical in chart ψ. With this equivalence relation we can therefore set:

Definition A.1.19 The **tangent space** of a smooth manifold M^n at a point $p \in M$ is defined by

$$T_pM = \{\gamma \mid \gamma\colon (-\epsilon, \epsilon) \to M \text{ differentiable}, \gamma(0) = p\}/\sim.$$

For the equivalence class of the curve γ in M we write

$$[\gamma] = \dot{\gamma}(0) = \left.\frac{d}{dt}\right|_{t=0} \gamma(t)$$

and call this a **tangent vector**.

Proposition A.1.20 *At any point $p \in M^n$ the tangent space T_pM has the structure of a real n-dimensional vector space.*

Proof Let $\phi\colon U \to \mathbb{R}^n$ be a chart around p. We set

$$\begin{aligned} D_p\phi\colon T_pM &\longrightarrow \mathbb{R}^n \\ [\gamma] &\longmapsto (\phi \,\dot{\circ}\, \gamma)(0). \end{aligned}$$

It can be shown that this is a bijection. We define the vector space structure on T_pM so that this map becomes a vector space isomorphism. This structure does not depend on the choice of chart: If $\psi: V \to \mathbb{R}^n$ is another chart around p, then the following diagram is commutative, where $D_{\phi(p)}(\psi \circ \phi^{-1})$ is a vector space isomorphism:

$$\begin{array}{ccc} T_pM & \xrightarrow{=} & T_pM \\ {\scriptstyle D_p\phi}\downarrow & & \downarrow{\scriptstyle D_p\psi} \\ \mathbb{R}^n & \xrightarrow[D_{\phi(p)}(\psi\circ\phi^{-1})]{} & \mathbb{R}^n \end{array}$$

Hence the identity between T_pM and T_pM defined with the respective vector space structures is a vector space isomorphism. □

Definition A.1.21 The set

$$TM = \bigcup_{p \in M} \{p\} \times T_pM$$

is called the **tangent bundle** of M.
In Sect. 4.5 it is shown that the tangent bundle is an example of a *vector bundle* over M with *fibres* T_pM.

A.1.6 Differential of a Smooth Map

Let $f: M \to N$ be a smooth map between differentiable manifolds. With the tangent space at hand, we can now define the differential of f.

Definition A.1.22 The **differential** D_pf of the map f at a point $p \in M$ is defined by

$$\begin{aligned} D_pf: T_pM &\longrightarrow T_{f(p)}N \\ [\gamma] &\longmapsto [f \circ \gamma]. \end{aligned}$$

Equivalently,

$$\begin{aligned} D_pf: T_pM &\longrightarrow T_{f(p)}N \\ \dot{\gamma}(0) &\longmapsto (f \dot{\circ} \gamma)(0). \end{aligned}$$

The differential is a *well-defined* (independent of choice of representatives for $[\gamma]$) *linear map* between the tangent spaces.

For a vector $X \in T_pM$ we sometimes write

$$f_*X = (D_pf)(X).$$

The differential satisfies the so-called chain rule.

Proposition A.1.23 *The following* **chain rule** *holds for the differential: If $f: X \to Y$ and $g: Y \to Z$ are differentiable maps, then $g \circ f$ is differentiable and at any point $p \in X$*

$$D_p(g \circ f) = D_{f(p)}g \circ D_pf.$$

Corollary A.1.24 *The differential D_pf of a diffeomorphism $f: M \to N$ is at every point $p \in M$ a linear isomorphism of tangent spaces.*

Definition A.1.25 Let $f: M \to N$ be a differentiable map between manifolds.

- A point $p \in M$ is called a **regular point of** f if the differential D_pf is surjective onto $T_{f(p)}N$.
- A point $q \in N$ is called a **regular value of** f if each point p in the preimage $f^{-1}(q) \subset M$ is a regular point.
- The map f is called a **submersion** if every point $p \in M$ is regular.
- The map f is called an **immersion** if the differential D_pf is injective at every point $p \in M$.

Remark A.1.26 Every point of N that is not in the image $f(M)$ is automatically a regular value, because the condition is empty.

Theorem A.1.27 (Sard's Theorem) *For any differentiable map $f: M \to N$ between smooth manifolds M and N the set of regular values is dense in N.*
The following theorem shows that a map f has a certain normal form in a neighbourhood of a regular point.

Theorem A.1.28 (Regular Point Theorem) *Let p be a regular point of the map f. Then there exist charts (U, ϕ) of M around p and (V, ψ) of N around $f(p)$ with*

- $\phi(p) = 0$
- $\psi(f(p)) = 0$
- $f(U) \subset V$

such that the map $\psi \circ f \circ \phi^{-1}$ has the form

$$\psi \circ f \circ \phi^{-1}(x_1, \ldots, x_{n+k}) = (x_1, \ldots, x_n),$$

where $\dim M = n + k$ *and* $\dim N = n$.

Remark A.1.29 The theorem says that in suitable charts the map f is given by the standard projection of $\mathbb{R}^m = \mathbb{R}^n \times \mathbb{R}^k$ onto $\mathbb{R}^n$.

A.1.7 Immersed and Embedded Submanifolds

There are two notions of submanifolds which need to be distinguished.

Definition A.1.30 Let M be a smooth manifold.

1. An **immersed submanifold** of M is the image of an injective immersion $f: N \to M$ from a manifold N to M.
2. An **embedded submanifold** of M is the image of an injective immersion $f: N \to M$ from a manifold N to M which is a homeomorphism onto its image.

In both cases, the set $f(N)$ is endowed with the topology and manifold structure making $f: N \to f(N)$ a diffeomorphism. The difference between embedded and immersed submanifolds $f(N) \subset M$ is whether the topology on $f(N)$ coincides with the subspace topology on $f(N)$ inherited from M or not.
An embedded submanifold can be characterized equivalently as follows:

Proposition A.1.31 *A subset K of an m-dimensional manifold M is an embedded submanifold of dimension k if and only if around each point $p \in K$ there exists a chart (U, ϕ) of M such that*

$$\phi|_{U\cap K}: U \cap K \longrightarrow \phi(U) \cap \left(\mathbb{R}^k \times \{0\}\right) \subset \mathbb{R}^m.$$

Such a chart is also called a ***submanifold chart*** *or* ***flattener*** *for K.*
The regular point theorem implies:

Theorem A.1.32 (Regular Value Theorem) *Let $q \in N$ be a regular value of a smooth map $f: M \to N$ and $L = f^{-1}(q)$ the preimage of q. Then L is an embedded submanifold of M of dimension*

$$\dim L = \dim M - \dim N.$$

A.1.8 Vector Fields

Let M^n be a smooth manifold. A vector field on M is a map X that assigns to each point $p \in M$ a tangent vector $X_p \in T_pM$ in a smooth way. To make this precise let $\phi: M \supset U \to \phi(U) \subset \mathbb{R}^n$ be a chart. We set

$$TU = \bigcup_{p\in U} \{p\} \times T_pM$$

for the tangent bundle of U and define the map

$$\begin{aligned} D\phi: TU &\longrightarrow \phi(U) \times \mathbb{R}^n \\ (p, v) &\longmapsto (\phi(p), D_p\phi(v)). \end{aligned}$$

The map $D\phi$ is on each fibre $\{p\} \times T_pM$ of TU an isomorphism onto $\{\phi(p)\} \times \mathbb{R}^n$.

Definition A.1.33 A **vector field** X on M is a map $X: M \to TM$ such that:

1. $X_p = X(p) \in T_pM$ for all $p \in M$.
2. The map X is differentiable in the following sense: For any chart (U, ϕ) the lower horizontal map in the following diagram

$$\begin{array}{ccc} U & \xrightarrow{X} & TU \\ \phi \downarrow & & \downarrow D\phi \\ \phi(U) & \xrightarrow[D\phi \circ X \circ \phi^{-1}]{} & \phi(U) \times \mathbb{R}^n \end{array}$$

is differentiable (this is just a standard vector field on $\phi(U) \subset \mathbb{R}^n$).

A particularly important set of vector fields is defined by a chart.

Definition A.1.34 Let (U, ϕ) be a chart for M. Then we define at every point $p \in U$ the following vectors:

$$\frac{\partial}{\partial x_\mu}(p) = (D_p\phi)^{-1}(e_\mu), \quad \forall \mu = 1, \ldots, n,$$

where $e_1, \ldots, e_n$ is the standard basis of $\mathbb{R}^n$. We also write

$$\partial_\mu = \frac{\partial}{\partial x_\mu}.$$

For a fixed index μ, as p varies, the vectors $\partial_\mu(p)$ form a smooth vector field ∂_μ on U. We call the vector fields ∂_μ **basis vector fields** or **coordinate vector fields** on U.

Lemma A.1.35 *At each point $p \in U$ the vectors $\partial_1(p), \ldots \partial_n(p)$ form a basis for the tangent space T_pM.*

Proof This is clear, because $D_p\phi: T_pM \to \mathbb{R}^n$ is an isomorphism of vector spaces. □

Proposition A.1.36 *Every smooth vector field X on M can be written on U as*

$$X|_U = \sum_{\mu=1}^{n} X^\mu \partial_\mu \equiv X^\mu \partial_\mu$$

where $X^1, \ldots, X^n: U \to \mathbb{R}$ are smooth real-valued functions on U, called the ***components*** *of X with respect to the basis $\{\partial_\mu\}$.*

Remark A.1.37 The second equality in this proposition is an example of the so-called **Einstein summation convention**.

A.1.9 Integral Curves

Let M be a smooth manifold and X a smooth vector field on M.

Definition A.1.38 A curve $\gamma: I \to M$, where $I \subset \mathbb{R}$ is an open interval around 0, is called an **integral curve** for X through $p \in M$ if

$$\gamma(0) = p \quad \text{and} \quad \dot{\gamma}(t) = X_{\gamma(t)} \quad \forall t \in I.$$

The theory of ordinary differential equations (ODEs) applied in a chart for M shows that:

Theorem A.1.39 *For every point $q \in M$ there exists an interval I_q around 0 and a unique curve $\gamma_q: I_q \to M$ which is an integral curve for X.*
Using a theorem on the behaviour of solutions to ODEs under variation of the initial condition we get:

Theorem A.1.40 *For all $p \in M$ there exists an open neighbourhood U of p in M and an open interval I around 0 such that the integral curves γ_q are defined on I for all $q \in U$. The map*

$$\begin{aligned} \phi_U: U \times I &\longrightarrow M \\ (q, t) &\longmapsto \gamma_q(t) \end{aligned}$$

*is differentiable and is called the **local flow** of X.*

Theorem A.1.41 *Let M be a closed manifold (compact and without boundary). Then there exists a **global flow** of X which is a smooth map*

$$\begin{aligned} \phi: M \times \mathbb{R} &\longrightarrow M \\ (q, t) &\longmapsto \gamma_q(t). \end{aligned}$$

The map

$$\phi_t = \phi(\cdot, t): M \longrightarrow M$$

is a diffeomorphism for all $t \in \mathbb{R}$.

A.1.10 The Commutator of Vector Fields

Let X be a smooth vector field on the manifold M.

Definition A.1.42 The **Lie derivative** L_X is the map

$$L_X: \mathscr{C}^\infty(M) \longrightarrow \mathscr{C}^\infty(M),$$

defined by

$$(L_X f)(p) = (D_p f)(X_p)$$

for all $f \in \mathscr{C}^\infty(M)$ and $p \in M$.
The Lie derivative L_X is the directional derivative of a smooth function along the vector field X: If γ is a curve through p such that $\dot{\gamma}(0) = X_p$, then

$$(L_X f)(p) = (f \dot{\circ} \gamma)(0).$$

Proposition A.1.43 *The Lie derivative is a **derivation**, i.e.*

1. *L_X is $\mathbb{R}$-linear*
2. *L_X satisfies the Leibniz rule:*

$$L_X(f \cdot g) = (L_X f) \cdot g + f \cdot (L_X g) \quad \forall f, g \in \mathscr{C}^\infty(M).$$

Using the Lie derivative we can define the so-called commutator of vector fields.

Theorem A.1.44 *Let X and Y be smooth vector fields on M. Then there exists a unique vector field $[X, Y]$ on M, called the **commutator** of X and Y, such that*

$$L_{[X,Y]} = L_X \circ L_Y - L_Y \circ L_X. \tag{A.1}$$

If in a local chart (U, ϕ) the vector fields are given by

$$X = X^\mu \partial_\mu \quad \text{and} \quad Y = Y^\mu \partial_\mu,$$

then $[X, Y]$ is given by

$$[X, Y] = \left(X^\nu \frac{\partial Y^\mu}{\partial x^\nu} - Y^\nu \frac{\partial X^\mu}{\partial x^\nu} \right) \partial_\mu. \tag{A.2}$$

Theorem A.1.45 *The set of vector field $\mathfrak{X}(M)$ together with the commutator is an (infinite-dimensional) **Lie algebra**, i.e. for all $X, Y, Z \in \mathfrak{X}(M)$ we have:*

- *antisymmetry:*

$$[Y, X] = -[X, Y]$$

- *$\mathbb{R}$-bilinearity:*

$$[aX + bY, Z] = a[X, Z] + b[Y, Z] \quad \forall a, b \in \mathbb{R}$$

- *Jacobi identity:*

$$[X,[Y,Z]]+[Y,[Z,X]]+[Z,[X,Y]]=0.$$

We can calculate the commutator $[X, Y]$ using the flow of X:

Theorem A.1.46 *Let X and Y be smooth vector fields on M, ϕ_t the flow of X and $p \in M$ a point. Then*

$$[X,Y]_p = \left.\frac{d}{dt}\right|_{t=0} (\phi_{-t})_* Y_{\phi_t(p)}.$$

Note that $(\phi_{-t})_* Y_{\phi_t(p)}$ is a smooth curve in T_pM.

A.1.11 Vector Fields Related by a Smooth Map

Definition A.1.47 Let M and N be smooth manifolds and $\phi: M \to N$ a smooth map. Suppose that X is a vector field on M and Y a vector field on N. Then Y is said to be **ϕ-related** to X if

$$Y_{\phi(p)} = (D_p\phi)(X_p) \quad \forall p \in M.$$

Lemma A.1.48 *Let M and N be smooth manifolds, $\phi: M \to N$ a smooth map. Suppose that X and Y are vector fields on M and N and that Y is ϕ-related to X. Then*

$$(L_Y f)\circ\phi = L_X(f\circ\phi) \quad \forall f \in \mathscr{C}^\infty(N).$$

Proposition A.1.49 *Let M and N be smooth manifolds and $\phi: M \to N$ a smooth map. Suppose that X' is ϕ-related to X and Y' is ϕ-related to Y. Then $[X', Y']$ is ϕ-related to $[X, Y]$.*

Definition A.1.50 If $\phi: M \to N$ is a diffeomorphism and X is a smooth vector field on M, then we define a smooth vector field ϕ_*X on N, called the **pushforward** of X under ϕ, by

$$(\phi_*X)_{\phi(p)} = (D_p\phi)(X_p).$$

Note that ϕ_*X is the unique vector field on N that is ϕ-related to X.

Corollary A.1.51 *If $\phi: M \to N$ is a diffeomorphism, then*

$$[\phi_*X, \phi_*Y] = \phi_*[X,Y]$$

for all vector fields X and Y on M.

A.1.12 Distributions and Foliations

We consider some concepts related to distributions and foliations on manifolds (we follow [142] where proofs and more details can be found). Let M be a smooth manifold of dimension n.

Definition A.1.52 A **distribution** D of rank k on M is a collection of vector subspaces $D_p \subset T_pM$ of dimension k for all $p \in M$ which vary smoothly over M, i.e. each $p \in M$ has an open neighbourhood $U \subset M$ so that $D|_U$ is spanned by k smooth vector fields $X_1, \ldots, X_k$ on U.
An equivalent definition is that D is a *subbundle* of rank k of the tangent bundle TM.

Definition A.1.53 A distribution is called **involutive** or **integrable** if for all vector fields X, Y on M with $X_p, Y_p \in D_p$ for all $p \in M$, the vector field $[X, Y]$ on M again satisfies $[X, Y]_p \in D_p$ for all $p \in M$.

Definition A.1.54 A **foliation** $\mathcal{F}$ of rank k on M is a decomposition of M into k-dimensional immersed submanifolds, called **leaves**, which locally have the following structure: around each point $p \in M$ there exists a coordinate neighbourhood diffeomorphic to $\mathbb{R}^n$ such that the leaves of the foliation decompose $\mathbb{R}^n$ into $\mathbb{R}^k \times \mathbb{R}^{n-k}$, with the leaves given by the affine subspaces $\mathbb{R}^k \times \{x\}$ for all $x \in \mathbb{R}^{n-k}$.
It is clear that the tangent spaces to the leaves of a foliation define a distribution. In fact, we have:

Theorem A.1.55 (Frobenius Theorem) *A distribution D defines a foliation $\mathcal{F}$ if and only if D is integrable.*
The following statement is Theorem 1.62 in [142].

Theorem A.1.56 *Let $f: N \to M$ be a smooth map between manifolds, $\mathcal{H}$ a foliation on M and $H \subset M$ a leaf of $\mathcal{H}$. Suppose that f has image in H. Then $f: N \to H$ is smooth.*
This theorem is clear if H is an embedded submanifold of M and only non-trivial if H is an immersed submanifold.

A.2 Tensors and Forms

A.2.1 Tensors and Exterior Algebra of Vector Spaces

We recall some notions from linear algebra. Let V denote an n-dimensional real vector space.

Definition A.2.1 We set

$$V^* = \{\lambda \mid \lambda: V \to \mathbb{R} \text{ is linear}\}$$

for the **dual space** of V. The dual space V^* is itself an n-dimensional real vector space. We call the elements $\lambda \in V^*$ **1-forms** on V.
If $\{e_\mu\}$ is a basis for V we get a **dual basis** $\{\omega^\nu\}$ for V^* defined by

$$\omega^\nu(e_\mu) = \delta^\nu_\mu, \quad \forall \mu, \nu = 1, \ldots, n,$$

where δ^ν_μ is the standard Kronecker delta. Just as we decompose any vector $X \in V$ in the basis $\{e_\mu\}$ as

$$X = X^\mu e_\mu$$

we can decompose any 1-form $\lambda \in V^*$ as

$$\lambda = \lambda_\nu \omega^\nu.$$

(Note the Einstein summation convention in both cases.)

Definition A.2.2 A **tensor** of type (l, k) is a multilinear map

$$T: \underbrace{V^* \times \cdots \times V^*}_{l} \times \underbrace{V \times \cdots \times V}_{k} \longrightarrow \mathbb{R}.$$

In particular, a $(0, 1)$-tensor is a 1-form and a $(1, 0)$-tensor is a vector. The set of all (l, k)-tensors forms a vector space.
We are interested in a particular class of tensors on a vector space V.

Definition A.2.3 We call a $(0, k)$-tensor

$$\lambda: \underbrace{V \times \cdots \times V}_{k} \longrightarrow \mathbb{R}$$

a **k-form** on V if λ is **alternating**, i.e. totally antisymmetric:

$$\lambda(\ldots, v, \ldots, w, \ldots) = -\lambda(\ldots, w, \ldots, v, \ldots)$$

for all insertions of vectors into λ, where only the vectors v and w are interchanged. The set of k-forms on V forms a vector space denoted by $\Lambda^k V^*$.

Remark A.2.4 It follows that for k-forms λ

$$\lambda(\ldots, v, \ldots, v, \ldots) = 0 \quad \forall v \in V$$

and

$$\lambda(v_1, v_2, \ldots, v_k) = 0$$

whenever the vectors $v_1, v_2, \ldots, v_k$ are linearly dependent. In particular, every k-form on V vanishes identically if k is larger than the dimension of V.

Definition A.2.5 Let λ be a k-form and μ an l-form. Then the **wedge product** of $\lambda \wedge \mu$ is the $(k+l)$-form defined by

$$(\lambda \wedge \mu)(X_1, \ldots, X_{k+l})$$
$$= \frac{1}{k!l!} \sum_{\sigma \in S_{k+l}} \operatorname{sgn}(\sigma)\lambda\left(X_{\sigma(1)}, \ldots, X_{\sigma(k)}\right) \cdot \mu\left(X_{\sigma(k+1)}, \ldots, X_{\sigma(k+l)}\right).$$

Here S_{k+l} denotes the set of permutations of $\{1, 2, \ldots, k+l\}$. It can be checked that $\lambda \wedge \mu$ is indeed a $k+l$-form.

Example A.2.6 Let α, β be 1-forms on V. Then

$$(\alpha \wedge \beta)(X, Y) = \alpha(X)\beta(Y) - \alpha(Y)\beta(X)$$

for all vectors $X, Y \in V$.

Lemma A.2.7 *Let V be a vector space of dimension n and $\{\omega^\nu\}$ a basis for V^*. Then the set of k-forms*

$$\omega^{\nu_1} \wedge \cdots \wedge \omega^{\nu_k}, \text{ with } 1 \leq \nu_1 < \nu_2 < \ldots < \nu_k \leq n,$$

is a basis for the vector space of k-forms.

A.2.2 Tensors and Differential Forms on Manifolds

Let M be an n-dimensional smooth manifold. We want to extend the notion of tensors and forms on vector spaces to tensors and forms on M. One possibility is to first define certain *vector bundles* and then tensors and forms as smooth *sections* of these bundles. However, since we define vector bundles in Sect. 4.5, we use here another, equivalent definition for tensors.

Remark A.2.8 In the following all functions and vector fields on M are smooth.

Definition A.2.9 We denote by $\mathscr{C}^\infty(M)$ the ring of all smooth functions $f: M \to \mathbb{R}$. We also denote by $\mathfrak{X}(M)$ the set of all smooth vector fields on M. The set $\mathfrak{X}(M)$ is a real vector space and module over $\mathscr{C}^\infty(M)$ by point-wise multiplication.
We can now define:

Definition A.2.10 A **1-form** λ on the manifold M is a map

$$\lambda: \mathfrak{X}(M) \longrightarrow \mathscr{C}^\infty(M)$$

that is linear over $\mathscr{C}^\infty(M)$, i.e.

$$\lambda(X+Y) = \lambda(X) + \lambda(Y),$$
$$\lambda(f \cdot X) = f \cdot \lambda(X)$$

for all vector fields $X, Y \in \mathfrak{X}(M)$ and functions $f \in \mathscr{C}^\infty(M)$. We denote the set of all 1-forms on M by $\Omega^1(M)$, which is a real vector space and module over $\mathscr{C}^\infty(M)$. The following can be proved:

Proposition A.2.11 *The value of $\lambda(X)(p)$ for a 1-form λ and vector field X at a point $p \in M$ depends only on X_p. Hence if Y is another vector field on M with $Y_p = X_p$, then $\lambda(X)(p) = \lambda(Y)(p)$.*
A proof of this proposition can be found in [142, p. 64]. Similarly we set:

Definition A.2.12 A **tensor** T of type (l, k) on M is a map

$$T: \underbrace{\Omega^1(M) \times \cdots \times \Omega^1(M)}_{l} \times \underbrace{\mathfrak{X}(M) \times \cdots \times \mathfrak{X}(M)}_{k} \longrightarrow \mathscr{C}^\infty(M)$$

that is $\mathscr{C}^\infty(M)$-linear in each entry. A ***k*-form** or **differential form** ω on M is a $(0, k)$-tensor

$$\omega: \underbrace{\mathfrak{X}(M) \times \cdots \times \mathfrak{X}(M)}_{k} \longrightarrow \mathscr{C}^\infty(M)$$

that is in addition alternating (totally antisymmetric). We denote the set of k-forms on M by $\Omega^k(M)$.

Remark A.2.13 An argument similar to the proof of Proposition A.2.11 shows that tensors and k-forms on manifolds have well-defined values at every point $p \in M$. We can therefore insert, for example, in a k-form $\omega \in \Omega^k(M)$ vectors $X_1, \ldots, X_k$ in the tangent space T_pM at any point $p \in M$ and get a real number. We can also speak unambiguously of the value of a tensor or form at a point.

Remark A.2.14 We can define the wedge product $\wedge$ of forms as before by replacing in the definition vectors by vector fields on the manifold. The wedge product is then a map

$$\wedge: \Omega^k(M) \times \Omega^l(M) \longrightarrow \Omega^{k+l}(M).$$

A.2.3 Scalar Products and Metrics on Manifolds

We consider the following definition from linear algebra.

Definition A.2.15 A **scalar product** on the vector space V is a **symmetric non-degenerate** $(0, 2)$-tensor g on V:

$$g(v, w) = g(w, v) \quad \forall v, w \in V \quad \text{(symmetric)}$$
$$g(v, \cdot) \neq 0 \in V^* \quad \forall v \neq 0 \in V \quad \text{(non-degenerate).}$$

The scalar product g is called **Euclidean** if it is positive definite

$$g(v, v) \geq 0 \quad \forall v \in V$$
$$g(v, v) > 0 \quad \forall v \neq 0$$

and **pseudo-Euclidean** otherwise.
We can do the same construction on manifolds.

Definition A.2.16 A **metric** on a smooth manifold M is a $(0, 2)$-tensor g which is a scalar product at each point $p \in M$. The metric is called **Riemannian** if the scalar products g_p are Euclidean and **pseudo-Riemannian** if the scalar products g_p are pseudo-Euclidean, for all $p \in M$.
It can be shown using partitions of unity that every smooth manifold admits a Riemannian metric (but not necessarily a pseudo-Riemannian metric).

A.2.4 The Levi-Civita Connection

Let (M, g) be a pseudo-Riemannian manifold. The **Levi-Civita connection** is a metric and torsion-free, covariant derivative on the tangent bundle of the manifold, i.e. a map

$$\begin{aligned} \nabla \colon \mathfrak{X}(M) \times \mathfrak{X}(M) &\longrightarrow \mathfrak{X}(M) \\ (X, Y) &\longmapsto \nabla_X Y \end{aligned}$$

with the following properties:

1. ∇ is $\mathbb{R}$-linear in both X and Y.
2. ∇ is $\mathscr{C}^\infty(M)$-linear in X and satisfies

$$\nabla_X(fY) = (L_X f)Y + f\nabla_X Y \quad \forall f \in \mathscr{C}^\infty(M), X, Y \in \mathfrak{X}(M).$$

3. ∇ is metric, i.e.

$$L_X g(Y, Z) = g(\nabla_X Y, Z) + g(Y, \nabla_X Z) \quad \forall X, Y, Z \in \mathfrak{X}(M).$$

4. ∇ is torsion-free, i.e.

$$\nabla_X Y - \nabla_Y X = [X, Y] \quad \forall X, Y \in \mathfrak{X}(M).$$

The Levi-Civita connection can be calculated with the following **Koszul formula**:

$$\begin{aligned}2g(\nabla_X Y, Z) &= L_X g(Y,Z) + L_Y g(X,Z) - L_Z g(X,Y)\\ &\quad - g([X,Z],Y) - g([Y,Z],X) + g([X,Y],Z).\end{aligned}$$

A.2.5 Coordinate Representations

We saw above that we can represent every vector field X on a chart neighbourhood U by $X|_U = X^\mu \partial_\mu$, where X^μ are certain functions on U, called components. We want to decompose in a similar way tensors and forms on U. In the physics literature tensors and forms are often given in terms of their components in coordinate systems.

Definition A.2.17 Let U be a chart neighbourhood. We define the set of **dual 1-forms** dx^μ, for $\mu = 1, \ldots, n$, by $dx^\mu(\partial_\nu) = \delta^\mu_\nu$ at each point $p \in U$.

Proposition A.2.18 *Let λ be a 1-form on M. Then we can decompose λ on U as $\lambda|_U = \lambda_\mu dx^\mu$ for certain smooth functions λ_μ on M. Similarly, we can decompose a k-form ω as*

$$\omega|_U = \sum_{1 \leq \nu_1 < \cdots < \nu_k \leq n} \omega_{\nu_1 \ldots \nu_k} dx^{\nu_1} \wedge \cdots \wedge dx^{\nu_k},$$

with smooth functions $\omega_{\nu_1 \ldots \nu_k}$.

Note that these functions, corresponding to the components, depend on the choice of the chart (U, ϕ), while the objects themselves (vectors fields, k-forms) are independent of charts.

A.2.6 The Pullback of Forms on Manifolds

Let $\omega \in \Omega^k(N)$ be a k-form on a manifold N and $f: M \to N$ a smooth map.

Definition A.2.19 The **pullback** of ω under f is the k-form $f^*\omega \in \Omega^k(M)$ on M defined by

$$(f^*\omega)(X_1, \ldots, X_k) = \omega(f_* X_1, \ldots, f_* X_k)$$

for all tangent vectors $X_1, \ldots, X_k \in T_pM$ and all $p \in M$.

Proposition A.2.20 *The pullback defines a map $f^*: \Omega^k(N) \longrightarrow \Omega^k(M)$. We have*

$$f^*(\omega \wedge \eta) = (f^*\omega) \wedge (f^*\eta)$$

for all $\omega \in \Omega^k(N), \eta \in \Omega^l(N)$ and

$$(g \circ f)^* = f^* \circ g^*$$

for all smooth maps $f: M \to N, g: N \to Q$.
The second property follows from the chain rule for the differential of the map $g \circ f$.

A.2.7 The Differential of Forms on Manifolds

The differential is a very important map on forms on a manifold that raises the degree by one.

Theorem A.2.21 *Let M be a smooth manifold. Then there is a unique map*

$$d: \Omega^k(M) \longrightarrow \Omega^{k+1}(M)$$

for every $k \geq 0$, called the ***differential*** *or* ***exterior derivative****, that satisfies the following properties:*

1. *d is $\mathbb{R}$-linear.*
2. *For a function $f \in \Omega^0(M) = \mathscr{C}^\infty(M)$ and a vector field $X \in \mathfrak{X}(M)$ we have $df(X) = L_X f$.*
3. *$d^2 = d \circ d = 0: \Omega^k(M) \to \Omega^{k+2}(M)$.*
4. *d satisfies the following Leibniz rule:*

$$d(\alpha \wedge \beta) = d\alpha \wedge \beta + (-1)^k \alpha \wedge d\beta$$

for all $\alpha \in \Omega^k(M), \beta \in \Omega^l(M)$.

The proof of this fundamental theorem can be found in any book on differential geometry. Let (U, ϕ) be a local chart. If we assume that the differential d has these properties, then it follows that the differential is given on functions f by

$$df = \frac{\partial f}{\partial x^\nu} dx^\nu$$

and on $\Omega^k(M)$ by

$$\begin{aligned} d\omega &= d \sum_{1 \leq \mu_1 < \ldots < \mu_k \leq n} \omega_{\mu_1 \ldots \mu_k} dx^{\mu_1} \wedge \cdots dx^{\mu_k} \\ &= \sum_{1 \leq \mu_1 < \ldots < \mu_k \leq n} \sum_{\nu=1}^{n} \frac{\partial \omega_{\mu_1 \ldots \mu_k}}{\partial x^\nu} dx^\nu \wedge dx^{\mu_1} \wedge \cdots dx^{\mu_k}. \end{aligned}$$

The defining properties of the differential d imply for 1-forms and 2-forms:

Proposition A.2.22 *1. Let $\alpha \in \Omega^1(M)$ be a 1-form. Then*

$$d\alpha(X,Y) = L_X(\alpha(Y)) - L_Y(\alpha(X)) - \alpha([X,Y]) \quad \forall X,Y \in \mathfrak{X}(M).$$

2. Let $\beta \in \Omega^2(M)$ be a 2-form. Then

$$\begin{aligned} d\beta(X,Y,Z) = {} & L_X(\beta(Y,Z)) + L_Y(\beta(Z,X)) + L_Z(\beta(X,Y)) \\ & - \beta([X,Y],Z) - \beta([Y,Z],X) - \beta([Z,X],Y) \quad \forall X,Y,Z \in \mathfrak{X}(M). \end{aligned}$$

The differential is natural under pullback:

Proposition A.2.23 *If $f\colon M \to N$ is a smooth map and $\omega \in \Omega^k(N)$, then $d(f^*\omega) = f^*d\omega$.*

Let M be a compact oriented n-dimensional manifold and $\sigma \in \Omega^n(M)$ a form of top degree. Then there is a well-defined integral

$$\int_M \sigma \in \mathbb{R}.$$

The integral can also be defined if M is non-compact and σ has compact support.

Theorem A.2.24 (Stokes' Theorem)

1. *Let M be a compact n-dimensional oriented manifold with boundary ∂M and $\omega \in \Omega^{n-1}(M)$. Then (with a suitable orientation of the boundary)*

$$\int_M d\omega = \int_{\partial M} \omega.$$

2. *Let M be an n-dimensional oriented manifold (not necessarily compact) without boundary and $\omega \in \Omega^{n-1}(M)$ an $(n-1)$-form with compact support. Then*

$$\int_M d\omega = 0.$$

Appendix B
Background on Special Relativity and Quantum Field Theory

B.1 Basics of Special Relativity

We very briefly recall some basic concepts from special relativity. A very good introduction to the physics and mathematics of special relativity can be found in [95], covering much more than we need.

Special relativity is formulated on **Minkowski spacetime** $M = \mathbb{R}^4$ with a pseudo-Riemannian metric known as **Minkowski metric** η given by (we use units where the speed of light $c = 1$)

$$\eta_{\mu\nu} = \eta(\partial_\mu, \partial_\nu) = \begin{pmatrix} 1 & & & \\ & -1 & & \\ & & -1 & \\ & & & -1 \end{pmatrix}.$$

This choice of signs $(+, -, -, -)$ is called the *West Coast metric*. Sometimes the *East Coast metric* with signature $(-, +, +, +)$ is used instead. The x^μ are the standard coordinates on $\mathbb{R}^4$, also written as

$$x^0 = t, \quad x^1 = x, \quad x^2 = y, \quad x^3 = z$$

or $x^\mu = (t, \mathbf{x})$. Coordinate systems (charts) on Minkowski spacetime correspond to reference frames of moving observers. All inertial systems (unaccelerated orthonormal reference frames, also called Lorentz frames) with the same origin $(0, \mathbf{0})$ are related by **Lorentz transformations**

$$x'^\rho = \Lambda^\rho{}_\mu x^\mu.$$

M.J.D. Hamilton, *Mathematical Gauge Theory*, Universitext,
https://doi.org/10.1007/978-3-319-68439-0_11

Here the Einstein summation convention is understood and Λ is a matrix preserving the metric η

$$\eta(\Lambda v, \Lambda w) = \eta(v, w) \quad \forall v, w \in \mathbb{R}^4,$$

also written as

$$\Lambda^T \eta \Lambda = \eta$$

or

$$\eta_{\rho\sigma} \Lambda^\rho{}_\mu \Lambda^\sigma{}_\nu = \eta_{\mu\nu}.$$

If the origins are different, then the coordinate transformations between inertial systems are given by **Poincaré transformations**

$$x'^\rho = \Lambda^\rho{}_\mu x^\mu + a^\rho,$$

where $a^\rho \in \mathbb{R}^4$ is a constant vector. Poincaré transformations are affine transformations.

Since the Minkowski distance between two points is independent of the chosen inertial frame, two points with distance zero *(lightlike)* in one frame have the same distance zero in any other Lorentz frame, meaning that the speed of light $c = 1$ is the same in any inertial frame.

The basis vectors of Lorentz frames transform as

$$e'_\rho = \left(\Lambda^{-1}\right)^\mu{}_\rho e_\mu,$$

so that the vector $\Delta x^\mu e_\mu$ is invariant:

$$\begin{aligned} \Delta x'^\rho e'_\rho &= \Delta x^\mu \Lambda^\rho_\mu \left(\Lambda^{-1}\right)^\nu{}_\rho e_\nu \\ &= \Delta x^\mu e_\mu. \end{aligned}$$

The same vector a on spacetime can be expressed in the frame e_μ or the frame e'_ρ:

$$a^\mu e_\mu = a = a'^\rho e'_\rho.$$

This implies that

$$a'^\rho = \Lambda^\rho{}_\mu a^\mu.$$

Similarly, a 1-form ω can be expressed in the frame dx^μ or the frame dx'^ρ:

$$\omega_\mu dx^\mu = \omega = \omega'_\rho dx'^\rho.$$

This implies that

$$\omega'_\rho = \left(\Lambda^{-1}\right)^\mu{}_\rho \omega_\mu.$$

If a is a vector with components a^μ in the frame e_μ and we set

$$a_\nu = \eta_{\nu\mu} a^\mu,$$

then a_ν transforms as the components of a 1-form. Similarly, if we define the matrix $\eta^{\mu\nu}$ as the inverse of the matrix $\eta_{\mu\nu}$ (in the Minkowski case this is the same matrix) and if ω_μ are the components of a 1-form in the frame dx^μ, then

$$\omega^\nu = \eta^{\nu\mu} \omega_\mu$$

transforms as the components of a vector. This is the idea behind lowering and raising indices, which can be extended to arbitrary tensors.

Let $u \in \mathbb{R}^4$ be the velocity vector of a particle of mass $m > 0$. Going to the rest frame of the particle, the vector u has components $u = (1, \mathbf{0}) = (1, 0, 0, 0)$, which implies that $\eta(u, u) = 1$ independent of the chosen frame. In any frame we define the 4-momentum

$$mu = (E, \mathbf{p}),$$

where E is the energy and $\mathbf{p}$ the 3-momentum of the particle in that frame. Then $\eta(u, u) = 1$ implies

$$m^2 = E^2 - \mathbf{p}^2.$$

If we introduce again the speed of light c, then $\eta(u, u) = c^2$,

$$mu = \left(\frac{E}{c}, \mathbf{p}\right),$$

and

$$m^2 c^4 = E^2 - \mathbf{p}^2 c^2.$$

Relativistic theories of physics are theories formulated in Minkowski spacetime whose laws are invariant under Poincaré transformations. This means that the laws of physics are independent of where and when experiments are performed (invariance under space and time translations), how the experiments are oriented in space (invariance under rotations) and whether they are performed in different inertial systems moving with constant velocity (invariance under Lorentz boosts). For example, the principle of relativity claims that the laws of physics are the same here and in the Andromeda galaxy, they are the same now and in 1 million years and they are the same on board two spacecrafts flying with arbitrary, constant velocities in different directions.

B.2 A Short Introduction to Quantum Field Theory

From classical gravity and electromagnetism we are used to thinking of matter as particles and interactions as carried by fields. However, according to quantum field theory, matter and interactions can both be described by particles *and* fields. Quantum field theory can be thought of as a unification of the concepts of classical fields and point particles and thus as a unification (in some sense) of interactions and matter (supersymmetric quantum field theories are a unification of both concepts in an even stronger sense). The remarkable consequences of this approach are that forces between matter particles can be reduced to couplings between different types of fields and that symmetry groups, such as gauge symmetries, can act through representations on both interaction and matter fields.

In the following sections we briefly want to discuss the basics of quantum field theory and the relation between particles and fields. Our intention is to give a short overview and interpretation, without any calculations or trying to be mathematically rigorous. We also assume a basic familiarity with quantum mechanics.

B.2.1 Quantum Field Theory and Quantum Mechanics

Quantum field theory (QFT) is a quantum theory, in some sense similar to quantum mechanics (QM):

- A quantum system has a **Hilbert space** V with a Hermitian scalar product $\langle\cdot|\cdot\rangle$. Elements of the vector space V are **state vectors (states)** $|v\rangle$ (we normalize these vectors to unit norm). We think of the state of the system as being time-dependent $|v,t\rangle$ (Schrödinger picture). However, we can equivalently think of the states as being time-independent and instead the operators as being time-dependent (Heisenberg picture, usually preferred in QFT).
- We cannot measure the state of a system directly, we can only measure the value of **observables**, described by Hermitian operators A on V. If $|v\rangle$ is an eigenvector of A with eigenvalue a,

$$A|v\rangle = a|v\rangle,$$

  and we measure the observable A if the system is in the state $|v\rangle$, then the value is the eigenvalue a. For an arbitrary state $|w\rangle$, the expectation value of the observable A is related to $\langle w|A|w\rangle$.

- We are also interested in **transition amplitudes** between states, given by scalar products $\langle w|v\rangle$. The amplitudes determine transition probabilities (the probability that the system in the state $|v\rangle$ is found after a measurement in the state $|w\rangle$) by taking the absolute value squared of this complex number.
- There is a Hermitian **Hamiltonian operator** H which determines the evolution of states between times t_0 and t (by convention $t_0 = 0$): we define the unitary operator

$$U(t,0) = e^{-\frac{i}{\hbar}Ht},$$

where $\hbar$ is the Planck constant (note that the exponential of a skew-Hermitian operator is unitary). Then time evolution of states is given by

$$|v,t\rangle = U(t,0)|v,0\rangle.$$

One of our aims is to determine the time evolution operator $U(t,0)$. Ideally we would like to diagonalize H, i.e. find an eigenbasis for H of states $|n\rangle$ of energy E_n,

$$H|n\rangle = E_n|n\rangle,$$

because such states have a very simple time evolution:

$$|n,t\rangle = e^{-iE_nt}|n,0\rangle,$$

where $e^{-iE_nt} \in \mathrm{U}(1)$ is just a complex number of absolute value 1. In general, in an interacting theory, this will be practically impossible.
- We can change from the Heisenberg picture to the Schrödinger picture and vice versa as follows: the Schrödinger-type operator is the Heisenberg-type operator taken at $t_0 = 0$:

$$A_S = A_H(0).$$

The time evolution of the Heisenberg-type operator is then given by the Hamiltonian:

$$A_H(t) = e^{\frac{i}{\hbar}Ht}A_S e^{-\frac{i}{\hbar}Ht}.$$

So far everything should be familiar from QM. We now discuss what is peculiar about QFT.

B.2.2 Free Quantum Field Theory on 0-Dimensional Space

Suppose that space is 0-dimensional and consists only of a single point. A real-valued field at this point is just a time dependent real number $\phi(t)$. The simplest type of quadratic Lagrangian for this field is

$$\mathscr{L} = \frac{1}{2}(\dot{\phi})^2 - \frac{1}{2}m^2\phi^2.$$

This Lagrangian is known as the **harmonic oscillator**. The Euler–Lagrange equation for this Lagrangian is the ordinary differential equation

$$\ddot{\phi} + m^2\phi = 0.$$

The Hilbert space of the associated quantum theory can be described as follows: let

$$\mathscr{H} = \bigoplus_{n=0}^{\infty} \mathbb{C}.$$

The basis states corresponding to this direct sum decomposition are denoted by

$$|n\rangle, \quad n = 0, 1, 2, 3, \ldots$$

The vector space $\mathscr{H}$ is the Hilbert space of the harmonic oscillator.

- The states $|n\rangle$ are eigenvectors for the Hamiltonian H with energy E_n growing linearly with n. These states are interpreted as the *discrete set* of different vibrational modes of the field at the point.
- There is a Hermitian **number operator** N (an observable) whose eigenvectors are $|n\rangle$ with eigenvalue n:

$$N|n\rangle = n|n\rangle.$$

B.2.3 Free Quantum Field Theory on d-Dimensional Space

Free quantum field theories (and to a certain degree, weakly interacting, perturbative quantum field theories) have an interpretation in terms of **particles**.

Canonical Quantization

We consider the case of field theories on d-dimensional Euclidean space (for simplicity we assume $d = 3$). A real-valued field is now a real function $\phi(t, \mathbf{x})$

depending on time t and the space coordinate $\mathbf{x}$. The simplest type of quadratic Lagrangian for this field is the **Klein–Gordon Lagrangian**.

$$\mathscr{L} = \frac{1}{2}(\partial^\mu \phi)(\partial_\mu \phi) - \frac{1}{2}m^2\phi^2. \tag{B.1}$$

The Euler–Lagrange equation for this Lagrangian is the linear wave equation

$$\partial^\mu \partial_\mu \phi + m^2 \phi = 0,$$

called the Klein–Gordon equation.

The Hilbert space of the associated quantum field theory can be described as follows: let V_1 be the Hilbert space of a single free **bosonic** particle. It is spanned by the basis states $|\mathbf{p}\rangle = |1_\mathbf{p}\rangle$, where $\mathbf{p} \in \mathbb{R}^3$ is the momentum of the particle, related to its energy by $m^2 = E^2 - \mathbf{p}^2$. In these states the particle is totally delocalized in position space. States where the particle is localized both in momentum and position space with a certain minimal width (given by Heisenberg's uncertainty principle) can be obtained as linear combinations of the states $|\mathbf{p}\rangle$, called **wave packets**.

A general construction in quantum theory implies that the Hilbert space of n indistinguishable particles of the same type is given by

$$V_n = \mathrm{Sym}^n V_1.$$

n-particle states are thus (linear combinations of) symmetrized tensor products of 1-particle states. We then form the (bosonic) **Fock space**

$$\mathscr{F} = \mathrm{Sym}^* V_1 = \bigoplus_{n=0}^{\infty} \mathrm{Sym}^n V_1$$

that contains states with an arbitrary number of particles. It turns out that the Fock space $\mathscr{F}$ is a suitable Hilbert space for the quantum field theory described by the Klein–Gordon Lagrangian (B.1).

- A basis for the Fock space is given by states

  $$|n_{\mathbf{p}_1}, \ldots, n_{\mathbf{p}_r}\rangle, \quad n_{\mathbf{p}_i} = 0, 1, 2, 3, \ldots,$$

  where $\mathbf{p}_i \in \mathbb{R}^3$ are vectors and $n_{\mathbf{p}_i}$ is the number of particles of momentum $\mathbf{p}_i$. The total number of particles in this state is

  $$n = \sum_{i=0}^{r} n_{\mathbf{p}_i}.$$

- These states are eigenvectors of the Hamiltonian H with energy growing linearly with the numbers $n_{\mathbf{p}_i}$ and they again correspond to different vibrational modes of the field.
- The basis state $|0\rangle$ of the 1-dimensional space $V_0 \cong \mathbb{C}$, where all $n_{\mathbf{p}_i}$ are zero, is called the **vacuum**. The vacuum is the unique eigenstate of the Hamiltonian of eigenvalue 0.
- For every vector $\mathbf{p}$ there is a Hermitian **number operator** $N_\mathbf{p}$ with eigenvectors

$$N_\mathbf{p}|n_\mathbf{p}, n_{\mathbf{p}_1}, \ldots, n_{\mathbf{p}_r}\rangle = n_\mathbf{p}|n_\mathbf{p}, n_{\mathbf{p}_1}, \ldots, n_{\mathbf{p}_r}\rangle, \quad \text{where} \quad \mathbf{p}_i \neq \mathbf{p} \quad \forall i = 1, \ldots, r.$$

  The number operator is an observable which returns the number of particles of a given momentum in a given quantum state.
- The classical field $\phi(\mathbf{x})$ becomes in the QFT a field of Schrödinger-type operators $\hat{\phi}(\mathbf{x})$, depending on the space point $\mathbf{x}$ (more precisely, an operator-valued distribution), that all act on the same Hilbert space V. This field $\hat{\phi}$ of operators is called the **quantum field**. Together with the adjoint quantum field $\hat{\phi}^\dagger$ it creates and annihilates particles in the point $\mathbf{x}$, i.e. adds or removes these particles from the state in the Hilbert space. In the Heisenberg picture the field depends on the point $(t, \mathbf{x})$ in spacetime.

Similarly, the Fock space V for a free **fermionic** field can be generated using antisymmetrized tensor products:

$$\mathscr{F} = \Lambda^* V_1 = \bigoplus_{n=0}^{\infty} \Lambda^n V_1.$$

In this case the numbers $n_\mathbf{p}$ can only take the values 0 or 1.

These descriptions of the Fock spaces make it clear that the Hilbert space in quantum field theory is infinite-dimensional in two ways: the one-particle space V_1 is infinite-dimensional because the vector space $\mathbb{R}^3$ of momenta $\mathbf{p}$ has infinitely many elements (this is related to the fact that space is continuous and infinitely extended). In addition there is the infinite direct sum over the number of particles $N \in \mathbb{N}_0$ that we already encountered in the case of quantum mechanics of the harmonic oscillator (for the harmonic oscillator the vector space corresponding to V_1 is 1-dimensional).

B.2.4 Unitary Representation of the Poincaré Group

As one of the general axioms of QFT on 4-dimensional Minkowski spacetime we assume that the Hilbert space of the quantum theory carries a **unitary representation** of the universal covering group $\mathrm{SL}(2,\mathbb{C}) \ltimes \mathbb{R}^{1,3}$ of the Poincaré group. Note that the non-compact simple Lie group $SL(2,\mathbb{C})$ does not admit non-trivial finite-dimensional unitary representations according to Theorem 2.1.44.

B.2.5 Interacting Quantum Field Theories

A typical question in QFT is to calculate *scattering amplitudes*: Suppose we send in an (idealized) collider a total number of n particles with certain momenta $\mathbf{p}_i$ and want to determine the probability that we find after collision n' particles with certain momenta $\mathbf{p}'_j$. This process is governed by the laws of quantum theory: in general we can only calculate a *probability* for the process or transition to happen, we cannot predict the outcome completely, even if we know the initial state exactly. Since the numbers n and n' as well as the types of particles and their momenta can be different, certain particles get created and others annihilated in the scattering process.[1]

To describe interacting QFTs, like the ϕ^4-theory with Lagrangian

$$\mathscr{L} = \frac{1}{2}(\partial^\mu\phi)(\partial_\mu\phi) - \frac{1}{2}m^2\phi^2 - \frac{1}{4!}\lambda\phi^4,$$

we make the following assumptions, known as the **interaction picture**:

- The Hilbert space V of the interacting theory is the same Hilbert space as in the free theory (the Fock space). The Schrödinger-type field operators (operator-valued distributions) are also the same as in the free theory.
- The Hamiltonian H can be calculated from the Lagrangian $\mathscr{L}$ of the field theory, expressed through the fields, that we collectively denote by ϕ. The Hamiltonian of the interacting theory is of the form

$$H = H_0 + H_I,$$

  where H_0 is the Hamiltonian of the free theory and H_I is the interaction part. Since the Hamiltonian H of the interacting theory is different from the Hamiltonian H_0 of the free theory, we expect the vacuum state $|\Omega\rangle$ of the interacting theory to be different from the vacuum state $|0\rangle$ of the free theory, even though both states (under our assumption) are elements of the same Fock space V.
- The time-dependence of the Heisenberg-type field operators will be different in the free and interacting theories, because the Hamiltonians are different. One considers two types of time-dependent quantum fields: the *Heisenberg picture field* is given the time evolution according to the full Hamiltonian H and the *interaction picture field* is given the time evolution according to the free Hamiltonian H_0.

[1] Bound states, like atoms and hadrons, are described in QFTs using other methods.

Scattering of particles in a collider can now be described as follows. We assume that for time $t \to -\infty$ in the distant past and for time $t' \to +\infty$ in the distant future the particles are far apart and can be considered as free. We can then think of the collections of particles that we send into and get out of the collider as states in the Hilbert space V:

$$|v,t\rangle = |n_{\mathbf{p}_1},\ldots,n_{\mathbf{p}_r},t\rangle,$$
$$|v',t'\rangle = |n_{\mathbf{p}_1'},\ldots,n_{\mathbf{p}_s'},t'\rangle.$$

We want to calculate the scalar product

$$\langle v',t'|U(t',t)|v,t\rangle.$$

If we know these scalar products, we can calculate the transition amplitudes between any two states, because the momentum states form a basis for the Hilbert space. Since the time evolution operator U is defined via the Hamiltonian H, it follows that the scalar product for different ingoing and outgoing states will be non-zero only if the action of H on particle states creates and annihilates certain particles.

More precisely, in a free field theory the Hamiltonian is quadratic in the fields ϕ and can be diagonalized. The eigenbasis is just given by the particle momentum states

$$|n_{\mathbf{p}_1},\ldots,n_{\mathbf{p}_r}\rangle$$

which are fixed by the action of H up to multiplication by the eigenvalue (the total energy E of the collection of particles). It follows that in a free field theory particles do not get created or annihilated. The transition amplitude between different particle momentum states is zero and all scattering processes are trivial: we get the same particles with the same momenta out that we sent into the collider. The vibrational modes of the field described by these states are constant, independent of time.

In an interacting theory the Hamiltonian contains **anharmonic terms**, i.e. terms of order three or higher in the fields ϕ. Such Hamiltonians lead to non-trivial creation and annihilation of particles and thus to non-trivial scattering processes. Heuristically, the vibrational modes of the fields change with time and, since the fields are coupled and the corresponding equations of motion are **non-linear wave equations**, the vibrations of one field can start vibrations of another field.

The description so far assumed that the Hilbert space and the action of the Schrödinger-type field operators are the same for the interacting theory as for the free theory, only the vacuum state and the Hamiltonian have changed. This assumption is merely a first approximation and actually not consistent, according to *Haag's Theorem*: Schrödinger-type quantum fields for a free and an interacting theory cannot be the same; see, for example, [116, p. 391]. To define quantum field theories in a mathematically rigorous way is the aim of *constructive quantum field theory (CQFT)* and *algebraic quantum field theory (AQFT)*.

B.2.6 Path Integrals

Transition amplitudes can also be calculated using **path integrals**. Path integrals for interacting theories, in particular, gauge theories, are more convenient than canonical quantization. The path integral approach to quantum theory was originally developed by P.A.M. Dirac and R.P. Feynman.

To understand the idea of path integrals, recall that standard integrals are integrals of functions over finite-dimensional vector spaces (or finite-dimensional manifolds). Path integrals are integrals of functions (also called functionals) over *infinite-dimensional* vector spaces (or infinite-dimensional manifolds). These vector spaces arise naturally as the spaces of all fields of a certain type on spacetime, i.e. the space $\mathscr{C}^\infty(M, W)$, where M is spacetime and W is the vector space in which the field ϕ takes values (more generally, we could consider the space of all sections of a vector bundle over M). Path integrals over the infinite-dimensional vector space $\mathscr{C}^\infty(M, W)$ can be approximated by standard integrals over a finite-dimensional vector space if spacetime is replaced by a lattice with finite lattice spacing $a > 0$ (and finite extension).

The path integrals that appear in QFT are of the form

$$G(x_1, \dots, x_n) = \frac{1}{\mathscr{N}} \int \mathscr{D}\phi\, \phi(x_1) \dots \phi(x_n) \exp\left(\frac{i}{\hbar} \int_M \mathscr{L}(\phi)\, \mathrm{dvol}\right),$$

called **Green's functions** or **correlators**. Here $x_1, \dots, x_n \in M$ are points in spacetime, ϕ is the field, $\mathscr{D}\phi$ is the **path integral measure** on the space $\mathscr{C}^\infty(M, W)$ and $\mathscr{L}$ is the Lagrangian of the field theory. The number $\mathscr{N}$ is a normalization constant. It is clear that (say for a complex scalar field ϕ, taking value in the complex numbers $W = \mathbb{C}$) for fixed points $x_1, \dots, x_n$ the map

$$\begin{aligned} F\colon \mathscr{C}^\infty(M, \mathbb{C}) &\longrightarrow \mathbb{C} \\ \phi &\longmapsto \phi(x_1) \dots \phi(x_n) \exp\left(\frac{i}{\hbar} \int_M \mathscr{L}(\phi)\, \mathrm{dvol}\right) \end{aligned}$$

is a function (functional) on the vector space $\mathscr{C}^\infty(M, \mathbb{C})$. The path integral

$$G(x_1, \dots, x_n) = \frac{1}{\mathscr{N}} \int \mathscr{D}\phi\, F(\phi)$$

is the integral of this function over the infinite-dimensional space $\mathscr{C}^\infty(M, \mathbb{C})$.

Notice that the field ϕ here is the classical field, not the quantum field of operators. The approach to QFT using path integrals is independent of the approach using Hilbert spaces and quantum fields. However, it can be shown that if one knows all the Green's functions, then the Hilbert space together with the quantum fields can be reconstructed (this is known as the *Wightman Reconstruction Theorem*).

In general, it is very difficult to calculate or even define these path integrals precisely. Usually, this can only be done in the case of the free field with a quadratic Lagrangian $\mathscr{L}$. Scattering amplitudes can be calculated from Green's functions using the *LSZ reduction formula*, named after H. Lehmann, K. Symanzik and W. Zimmermann.

B.2.7 Series Expansions

The actual calculation of scattering amplitudes is a formidable task and often can only be done approximately, using power series expansions. In general, the Green's functions are functions of the coupling constant(s) g and the Planck constant $\hbar$:

$$G = G(x_1, \ldots, x_n, g, \hbar).$$

There are mainly two types of series expansions:

Perturbation Theory

Perturbation theory works if the coupling constant g is small so that the full Lagrangian $\mathscr{L}$ is a small perturbation

$$\mathscr{L} = \mathscr{L}_0 + g\mathscr{L}_{\text{int}}$$

of the free Lagrangian $\mathscr{L}_0$. The Green's functions for $\mathscr{L}_0$ are known and the Green's functions for $\mathscr{L}$ can be calculated in a series expansion in orders of g, by expanding the exponential

$$\exp\left(\frac{ig}{\hbar}\int_M \mathscr{L}_{\text{int}}(\phi)\,\text{dvol}\right)$$

in a power series in orders of g and then interchanging the path integral and the infinite sum (this step is mathematically *not* justified [50]). The terms in the power series expansion are described by **Feynman diagrams**. With the order of g increasing in each step by 1, the terms in the series expansion for a process with fixed external lines (in-coming and out-going particles) are called **leading order (LO), next-to-leading order (NLO), next-to-next-to-leading order (NNLO)**, and so on.

In perturbation theory, the full interacting Lagrangian is treated as a small perturbation of the free Lagrangian. Since the states of a free quantum field have an interpretation in terms of particles, it makes sense to think of perturbation theory as describing **(weakly) interacting particles**. Feynman diagrams, that depict these interactions, are the hallmark of perturbative quantum field theory.

Internal lines in Feynman diagrams represent intermediate **virtual particles** which are off-shell, i.e. do not satisfy the mass energy relation $m^2 = E^2 - \mathbf{p}^2$ (even though momentum and energy are conserved at each vertex). In contrast to the in- and out-state particles, virtual particles are therefore not "real" particles and cannot be detected (the photons that mediate interactions between electrons are different from the photons that we can see or detect with cameras).

A problem of perturbation theory is that the perturbation expansion in orders of the coupling constant g actually converges only for $g = 0$, i.e. the radius of convergence of the power series is zero (there is an argument due to Freeman Dyson that a QFT cannot be well-defined for negative values of the coupling constant g, hence the expansion around $g = 0$ must have radius of convergence equal to zero). This implies that the perturbation expansion only makes sense as an **asymptotic expansion**: up to a certain optimal order of g the series expansion approximates the Green's function better and better, but then, adding terms of higher order, the series expansion starts getting worse and eventually diverges.

Semi-Classical Approximation

Semi-classical approximation can be used if the Planck constant $\hbar$ is a (relatively) small number. The power series for the Green's functions is then an expansion in orders of $\hbar$. The lowest term of order zero is the **classical contribution** and terms in higher order of $\hbar$ are **quantum corrections**. In Feynman diagrams the classical contribution corresponds to **tree diagrams**, whereas quantum corrections correspond to **loop diagrams**. With respect to path integrals the semi-classical approximation is an expansion around the critical points of the Lagrangian, i.e. the classical solutions of the field equations (for $\hbar \to 0$ the path integral localizes at these classical solutions).

Non-Perturbative Quantum Field Theories

Note that we do not claim that the Green's functions are analytic in g or $\hbar$. Most smooth functions are, of course, not analytic, because analytic functions are determined everywhere in their domain of definition by their values in an arbitrarily small neighbourhood of the center of expansion. The series expansions in QFT will therefore only be approximately accurate for small values of g and $\hbar$ and unusable if these parameters are large. In particular, if g is large, the QFT is called **non-perturbative**.

The term *non-perturbative* is essentially a synonym for *non-analytic*. Non-perturbative effects in QFT will typically become dominant if the coupling constant g is large. The particle interpretation breaks down (at least for the fundamental fields) for strongly interacting, non-perturbative QFTs.

B.2.8 Renormalization

The calculation of the contribution of Feynman diagrams with loops[2] involves certain integrals that can diverge and lead to infinite Green's functions. The idea of **renormalization** is to *absorb* the infinities that occur in the Green's functions into the parameters (in particular, the masses and coupling constants), which then become infinite themselves, while the Green's functions become finite. For this to work the parameters have to go in "the right way" to infinity, so that the Green's functions stay finite. More precisely, the parameters are no longer constants, but certain functions of a *cutoff*, and go to infinity when the cutoff is removed, whereas the Green's functions remain finite.

Alternatively, renormalization can be understood as adding to the original Lagrangian of the field theory a counterterm Lagrangian that cancels the divergences of the Green's functions. If we include terms (interactions) of the same form as the counterterms in the original Lagrangian, then adding counterterms is equivalent to a renormalization of parameters. A QFT is called **renormalizable** if finitely many counterterms are needed to cancel the divergences and **non-renormalizable** if infinitely many counterterms are needed. Non-renormalizable theories contain infinitely many different types of interactions and infinitely many parameters, but can still be useful (cf. [125, Sect. 21.2.2]).

The process of renormalization can be explained with a classical analogy, first observed by M. Abraham and H. Lorentz: the electric field of a charged point particle is of the form

$$\mathbf{E}(r) = \frac{\alpha \mathbf{r}}{r^3}$$

[2]Only loop diagrams require renormalization.

where $\alpha \neq 0$ is some constant and $\mathbf{r}$ is the radial vector. The energy density u of the electric field is proportional to $|\mathbf{E}|^2$, hence of the form

$$u(r) = \frac{\beta}{r^4}$$

with a constant $\beta \neq 0$. It follows that the total energy of the field is

$$\int_{\mathbb{R}^3} \frac{\beta}{r^4} \mathrm{dvol} = 4\pi\beta \int_0^\infty \frac{1}{r^2} dr.$$

This integral, when extended all the way to 0, is infinite. It follows that a charged point particle, like an electron, has an infinite energy in its electric field. If this energy is added to the bare rest mass of the electron via $E = mc^2$, corresponding to an *electromagnetic mass*, the total mass becomes infinite, which seems like a contradiction.

The idea is to set the bare (unobservable) rest mass m_B of the electron equal to $-\infty$, so that when we add the infinite energy due to the electric field the total (observable) mass m becomes finite. We define a cutoff $\epsilon > 0$ and set

$$I(\epsilon) = 4\pi\beta \int_\epsilon^\infty \frac{1}{r^2} dr,$$

which is finite for all $\epsilon > 0$. This is called **regularization** of the divergent integral. We also define

$$m_B(\epsilon) = m - \frac{I(\epsilon)}{c^2},$$

where m is the observed mass of the electron, known from experiments. This is called **renormalization** of the mass. The bare mass is thus a function of the cutoff ϵ and goes to $-\infty$ if we let $\epsilon \to 0$. However, the total mass is now

$$m_B(\epsilon) + \frac{I(\epsilon)}{c^2}$$

which is constant and equal to the finite mass m for all $\epsilon > 0$. We see that we have hidden the infinity from the divergent integral $I(0)$ in the renormalization of the mass m_B.

In general, the divergences encountered in QFTs can be traced to two aspects of space: space is continuous (leading to UV divergences) and space is infinitely extended (leading to IR divergences). Both aspects imply that QFTs, which describe time-dependent fields defined on space, have to deal with systems with infinitely many degrees of freedom, the crucial difference to QM. A QFT can be regularized by introducing cutoffs: a UV cutoff essentially means to reduce space to a lattice with finite lattice spacing $a > 0$ (corresponding to an upper cutoff on the norm $|\mathbf{p}|$

of the momentum) and an IR cutoff means to consider the theory in a finite volume $V < \infty$ of space (corresponding to a discrete set of momenta). Both regularizations together reduce the QFT to a system with finitely many degrees of freedom (in continuous time), essentially a version of QM.

B.2.9 Further Reading

Introductory accounts of quantum field theory are [51] and [86]. Extensive discussions from a physics point of view can be found, for example, in [124, 125, 132] and [143–145]. Mathematically rigorous discussions can be found in the classic references [17, 69, 134] and in [38, 45, 82]. Perturbation theory, semi-classical approximation and renormalization are very well and comprehensibly explained from a mathematical point of view in the lecture notes [50].

References

1. Abe, K. et al. (The T2K Collaboration): First combined analysis of neutrino and antineutrino oscillations at T2K. arXiv:1701.00432
2. Adams, J.F.: On the non-existence of elements of Hopf invariant one. Ann. Math. **72**, 20–104 (1960)
3. Adams, J.F.: Vector fields on spheres. Ann. Math. **75**, 603–632 (1962)
4. Argyres, P.C.: An introduction to global supersymmetry. Lecture notes, Cornell University 2001. Available at http://homepages.uc.edu/~argyrepc/cu661-gr-SUSY/index.html
5. Atiyah, M.F.: K-Theory. Notes by D.W. Anderson. W.A. Benjamin, New York/Amsterdam (1967)
6. Atiyah, M.F., Bott, R.: The Yang–Mills equations over Riemann surfaces. Phil. Trans. R. Soc. Lond. A **308**, 523–615 (1983)
7. Atiyah, M.F., Bott, R., Shapiro, A.: Clifford modules. Topology **3**, 3–38 (1964)
8. Baez, J.C.: The octonions. Bull. Am. Math. Soc. (N.S.) **39**, 145–205 (2002); Erratum: Bull. Am. Math. Soc. (N.S.) **42**, 213 (2005)
9. Baez, J., Huerta, J.: The algebra of grand unified theories. Bull. Am. Math. Soc. (N.S.) **47**, 483–552 (2010)
10. Bailin, D., Love, A.: Introduction to Gauge Field Theory. Institute of Physics Publishing, Bristol/Philadelphia (1993)
11. Ball. P.: Nuclear masses calculated from scratch. Nature, published online 20 November 2008. doi:10.1038/news.2008.1246
12. Barut, A.O., Raczka, R.: Theory of Group Representations and Applications. Polish Scientific Publishers, Warszawa (1980)
13. Baum, H.: Spin-Strukturen und Dirac-Operatoren über pseudoriemannschen Mannigfaltigkeiten. Teubner Verlagsgesellschaft, Leipzig (1981)
14. Baum, H.: Eichfeldtheorie. Springer, Berlin/Heidelberg (2014)
15. Berline, N., Getzler, E., Vergne, M.: Heat Kernels and Dirac Operators. Springer, Berlin/Heidelberg (2004)
16. Bleecker, D.: Gauge Theory and Variational Principles. Addison-Wesley Publishing Company, Reading, MA (1981)
17. Bogolubov, N.N., Logunov, A.A., Todorov, I.T.: Introduction to Axiomatic Quantum Field Theory. W. A. Benjamin, Reading, MA (1975)
18. Borsanyi, Sz. et al: Ab initio calculation of the neutron-proton mass difference. Science **347**(6229), 1452–1455 (2015)
19. Bott, M.R.: An application of the Morse theory to the topology of Lie-groups, Bull. Soc. Math. France **84**, 251–281 (1956)

M.J.D. Hamilton, *Mathematical Gauge Theory*, Universitext,
https://doi.org/10.1007/978-3-319-68439-0

20. Bourguignon, J.-P., Hijazi, O., Milhorat, J.-L., Moroianu, A., Moroianu, S.: A Spinorial Approach to Riemannian and Conformal Geometry. European Mathematical Society, Zürich (2015)
21. Brambilla, N. et al.: QCD and strongly coupled gauge theories: challenges and perspectives. Eur. Phys. J. C **74**, 2981 (2014)
22. Branco, G.C., Lavoura, L., Silva, J.P.: CP Violation. Oxford University Press, Oxford (1999)
23. Bredon, G.E.: Introduction to Compact Transformation Groups. Academic Press, New York/London (1972)
24. Bröcker, T., tom Dieck, T.: Representations of Compact Lie Groups. Springer, Berlin/Heidelberg/New York (2010)
25. Bröcker, T., Jänich, K.: Einführung in die Differentialtopologie. Springer, Berlin/Heidelberg/New York (1990)
26. Bryant, R.L.: Metrics with exceptional holonomy. Ann. Math. **126**, 525–576 (1987)
27. Bryant, R.L.: Submanifolds and special structures on the octonians. J. Differ. Geom. **17**, 185–232 (1982)
28. Budinich, R., Trautman, A.: The Spinorial Chessboard. Springer, Berlin/Heidelberg (1988)
29. Bueno, A. et al.: Nucleon decay searches with large liquid Argon TPC detectors at shallow depths: atmospheric neutrinos and cosmogenic backgrounds. JHEP **0704**, 041 (2007)
30. Čap, A., Slovák, J.: Parabolic Geometries I: Background and General Theory. American Mathematical Society, Providence, RI (2009)
31. CERN Press Release: CERN experiments observe particle consistent with long-sought Higgs boson. Available at http://press.cern/press-releases/2012/07/cern-experiments-observe-particle-consistent-long-sought-higgs-boson
32. Chaichian, M., Nelipa, N.F.: Introduction to Gauge Field Theories. Springer, Berlin/Heidelberg/New York/Tokyo (1984)
33. Cheng, T.-P., Li, L.-F.: Gauge Theory of Elementary Particle Physics. Oxford University Press, Oxford (1988)
34. Chevalley, C.: Theory of Lie Groups I. Princeton University Press, Princeton (1946)
35. Chevalley, C.: The Algebraic Theory of Spinors and Clifford Algebras. Collected Works, Vol. 2. Springer, Berlin/Heidelberg (1997)
36. Chivukula, R.S.: The origin of mass in QCD. arXiv:hep-ph/0411198
37. Clay Mathematics Institute: Millenium problems. Yang–Mills and mass gap. Available at http://www.claymath.org/millennium-problems/yang--mills-and-mass-gap
38. Costello, K.: Renormalization and Effective Field Theory. Mathematical Surveys and Monographs, Vol. 170. American Mathematical Society, Providence, RI (2011)
39. Darling, R.W.R.: Differential Forms and Connections. Cambridge University Press, Cambridge (1994)
40. D'Auria, R., Ferrara, S., Lledó, M.A., Varadarajan, V.S.: Spinor algebras. J. Geom. Phys. **40**, 101–129 (2001)
41. Derdzinski, A.: Geometry of the Standard Model of Elementary Particles. Springer, Berlin/Heidelberg (1992)
42. Dissertori, G., Knowles, I., Schmelling, M.: Quantum Chromodynamics. High Energy Experiments and Theory. Oxford University Press, Oxford (2003)
43. Drexlin, G., Hannen, V., Mertens, S., Weinheimer, C.: Current direct neutrino mass experiments. Adv. High Energy Phys. **2013**, Article ID 293986 (2013)
44. Dürr, S. et al.: Ab initio determination of light hadron masses. Science **322**, 1224–1227 (2008)
45. Duncan, A.: The Conceptual Framework of Quantum Field Theory. Oxford University Press, Oxford (2013)
46. Dynkin, E.B.: Semisimple subalgebras of the semisimple Lie algebras. (Russian) Mat. Sbornik **30**, 349–462 (1952); English translation: Am. Math. Soc. Transl. Ser. 2 **6**, 111–244 (1957)
47. Elliott, C.: Gauge Theoretic Aspects of the Geometric Langlands Correspondence. Ph.D. Thesis, Northwestern University (2016)

48. Englert, F., Brout R.: Broken symmetry and the mass of gauge vector mesons. Phys. Rev. Lett. **13**, 321–323 (1964)
49. Figueroa-O'Farrill, J.: Majorana Spinors. Lecture Notes. University of Edinburgh (2015)
50. Flory, M., Helling, R.C., Sluka, C.: How I learned to stop worrying and love QFT. arXiv:1201.2714 [math-ph]
51. Folland, G.B.: Quantum Field Theory. A Tourist Guide for Mathematicians. American Mathematical Society, Providence, Rhodes Island (2008)
52. Freed, D.S.: Classical Chern–Simons theory, 1. Adv. Math. **113**, 237–303 (1995)
53. Freed, D.S.: Five Lectures on Supersymmetry. American Mathematical Society, Providence, RI (1999)
54. Freedman, D.Z., Van Proeyen, A.: Supergravity. Cambridge University Press, Cambridge (2012)
55. Friedrich, T.: Dirac Operators in Riemannian Geometry. American Mathematical Society, Providence, RI (2000)
56. Fritzsch, H., Minkowski, P.: Unified interactions of leptons and hadrons. Ann. Phys. **93**, 193–266 (1975)
57. Geiges, H.: An Introduction to Contact Topology. Cambridge University Press, Cambridge (2008)
58. Georgi, H.: The state of the art – gauge theories. In: Carlson, C.E. (ed.) Particles and Fields – 1974: Proceedings of the Williamsburg Meeting of APS/DPF, pp. 575–582. AIP, New York (1975)
59. Georgi, H., Glashow, S.L.: Unity of all elementary-particle forces. Phys. Rev. Lett. **32**, 438–441 (1974)
60. Georgi, H.M., Glashow, S.L., Machacek, M.E., Nanopoulos, D.V.: Higgs Bosons from two-gluon annihilation in proton-proton collisions. Phys. Rev. Lett. **40** 692 (1978)
61. Georgi, H., Quinn, H.R., Weinberg, S.: Hierarchy of interactions in unified gauge theories. Phys. Rev. Lett. **33**, 451–454 (1974)
62. Giunti, C., Kim, C.W.: Fundamentals of Neutrino Physics and Astrophysics. Oxford University Press, Oxford (2007)
63. Glashow, S.L.: Trinification of all elementary particle forces. In: 5th Workshop on Grand Unification, Providence, RI, April 12–14, 1984
64. Glashow, S.L., Iliopoulos, J., Maiani, L.: Weak interactions with lepton-hadron symmetry. Phys. Rev. D **2**, 1285–1292 (1970)
65. Gleason, A.M.: Groups without small subgroups. Ann. Math. **56**, 193–212 (1952)
66. Grimus, W., Rebelo, M.N.: Automorphisms in gauge theories and the definition of CP and P. Phys. Rep. **281**, 239–308 (1997)
67. Guralnik, G.S., Hagen, C.R., Kibble, T.W.B.: Global conservation laws and massless particles. Phys. Rev. Lett. **13**, 585–587 (1964)
68. Gürsey, F., Ramond, P., Sikivie, P.: A universal gauge theory model based on E_6. Phys. Lett. B **60**, 177–180 (1976)
69. Haag, R.: Local Quantum Physics. Fields, Particles, Algebras. Springer, Berlin/Heidelberg/New York (1996)
70. Hall, B.C.: Lie Groups, Lie Algebras and Representations. An Elementary Introduction. Springer, Cham Heidelberg/New York/Dordrecht/London (2016)
71. Halzen, F., Martin, A.D.: Quarks and Leptons. An Introductory Course in Modern Particle Physics. Wiley, New York/Chichester/Brisbane/Toronto/Singapore (1984)
72. Hartanto, A., Handoko L.T.: Grand unified theory based on the SU(6) symmetry. Phys. Rev. D **71**, 095013 (2005)
73. Harvey, R., Lawson, H.B.: Calibrated geometries. Acta Math. **148**, 47–157 (1982)
74. Hatcher, A.: Vector bundles and K-theory. Version 2.1, May 2009
75. Heeck, J.: Interpretation of lepton flavor violation. Phys. Rev. D **95**, 015022 (2017)
76. Higgs, P.W.: Broken symmetries and the masses of gauge bosons. Phys. Rev. Lett. **13**, 508–509 (1964)

77. Hilgert, J., Neeb, K.-H.: Structure and Geometry of Lie Groups. Springer, New York/Dordrecht/Heidelberg/London (2012)
78. Hirsch, M.W.: Differential Topology. Springer, New York/Berlin/Heidelberg (1997)
79. Hoddeson, L., Brown, L., Riordan, M., Dresden, M. (ed.): The Rise of the Standard Model: Particle Physics in the 1960s and 1970s. Cambridge University Press, Cambridge (1997)
80. Hollowood, T.J.: Renormalization Group and Fixed Points in Quantum Field Theory. Springer, Heidelberg/New York/Dordrecht/London (2013)
81. Husemoller, D.: Fibre Bundles. Springer, New York (1994)
82. Klaczynski, L.: Haag's Theorem in renormalisable quantum field theory. Ph.D. Thesis, Humboldt Universität zu Berlin (2015)
83. Knapp, A.W.: Lie Groups Beyond an Introduction. Birkhäuser, Boston/Basel/Berlin (2002)
84. Kobayashi, S., Nomizu, K.: Foundations of Differential Geometry, Vol. I. Interscience Publishers, New York/London (1963)
85. Kounnas, C., Masiero, A., Nanopoulos, D.V., Olive, K.A.: Grand Unification with and Without Supersymmetry and Cosmological Implications. World Scientific, Singapore (1984)
86. Lancaster, T., Blundell, S. J.: Quantum Field Theory for the Gifted Amateur. Oxford University Press, Oxford (2014)
87. Langacker, P.: Grand unified theories and proton decay. Phys. Rep. **72**, 185–385 (1981)
88. Lawson, H.B. Jr., Michelsohn, M.-L.: Spin Geometry. Princeton University Press, Princeton, NJ (1989)
89. Lee, J.M.: Introduction to Smooth Manifolds. Springer, New York/Heidelberg/Dordrecht/London (2013)
90. Leigh, R.G., Strassler, M.J.: Duality of $Sp(2N_c)$ and $SO(N_c)$ supersymmetric gauge theories with adjoint matter. Phys. Lett. B **356**, 492–499 (1995)
91. Martin, S.P.: A supersymmetry primer. arXiv:hep-ph/9709356
92. Mayer, M.E.: Review: David D. Bleecker, Gauge theory and variational principles. Bull. Am. Math. Soc. (N.S.) **9**, 83–92 (1983)
93. Meinrenken, E.: Clifford Algebras and Lie Theory. Springer, Berlin/Heidelberg (2013)
94. Milnor, J.: On manifolds homeomorphic to the 7-sphere. Ann. Math. **64**, 399–405 (1956)
95. Misner, C.W., Thorne, K.S., Wheeler, J.A.: Gravitation. W. H. Freeman and Company, New York (1973)
96. Mohapatra, R.N.: Unification and Supersymmetry. The Frontiers of Quark-Lepton Physics. Springer, New York/Berlin/Heidelberg (2003)
97. Montgomery, D., Zippin, L.: Small subgroups of finite-dimensional groups. Ann. Math. **56**, 213–241 (1952)
98. Moore, J.D.: Lectures on Seiberg–Witten Invariants. Springer, Berlin/Heidelberg/New York (2001)
99. Morgan, J.W.: The Seiberg–Witten Equations and Applications to the Topology of Smooth Four-Manifolds. Princeton University Press, Princeton, NJ (1996)
100. Mosel, U.: Fields, Symmetries, and Quarks. Springer, Berlin/Heidelberg (1999)
101. Naber, G.L.: Topology, Geometry and Gauge Fields. Foundations. Springer, New York (2011)
102. Naber, G.L.: Topology, Geometry and Gauge Fields. Interactions. Springer, New York (2011)
103. Nakahara, M.: Geometry, Topology and Physics, 2nd edn. IOP Publishing Ltd, Bristol/Philadelphia (2003)
104. O'Raifeartaigh, L.: Group Structure of Gauge Theories. Cambridge University Press, Cambridge (1986)
105. Patrignani, C. et al. (Particle Data Group): 2016 Review of particle physics. Chin. Phys. C **40**, 100001 (2016). http://www-pdg.lbl.gov/
106. Patrignani, C. et al. (Particle Data Group): 2016 Review of particle physics. Particle listings. Chin. Phys. C **40**, 100001 (2016). http://www-pdg.lbl.gov/
107. Patrignani, C. et al. (Particle Data Group): 2016 Review of particle physics. Particle listings. Neutrino mixing. Chin. Phys. C **40**, 100001 (2016). http://www-pdg.lbl.gov/
108. Patrignani, C. et al. (Particle Data Group): 2016 Review of particle physics. Reviews, tables, and plots. 1. Physical constants. Chin. Phys. C **40**, 100001 (2016). http://www-pdg.lbl.gov/

109. Patrignani, C. et al. (Particle Data Group): 2016 Review of particle physics. Reviews, tables, and plots. 10. Electroweak model and constraints on new physics. Chin. Phys. C **40**, 100001 (2016). http://www-pdg.lbl.gov/
110. Patrignani, C. et al. (Particle Data Group): 2016 Review of particle physics. Reviews, tables, and plots. 12. The CKM quark-mixing matrix. Chin. Phys. C **40**, 100001 (2016). http://www-pdg.lbl.gov/
111. Patrignani, C. et al. (Particle Data Group): 2016 Review of particle physics. Reviews, tables, and plots. 16. Grand unified theories. Chin. Phys. C **40**, 100001 (2016). http://www-pdg.lbl.gov/
112. Pich, A.: The Standard Model of electroweak interactions. arXiv:1201.0537 [hep-ph]
113. Quigg, C.: Gauge Theories of the Strong, Weak, and Electromagnetic Interactions. Westview Press, Boulder, Colorado (1997)
114. Robinson, M.: Symmetry and the Standard Model. Mathematics and Particle Physics. Springer, New York/Dordrecht/Heidelberg/London (2011)
115. Roe, J.: Elliptic Operators, Topology and Asymptotic Methods. Longman Scientific & Technical, Harlow (1988)
116. Roman, P.: Introduction to Quantum Field Theory. Wiley, New York/London/Sydney/Toronto (1969)
117. Royal Swedish Academy of Sciences: The official web site of the nobel prize. https://www.nobelprize.org/nobel_prizes/physics/
118. Royal Swedish Academy of Sciences: Asymptotic freedom and quantum chromodynamics: the key to the understanding of the strong nuclear forces. Advanced information on the Nobel Prize in Physics, 5 October 2004. https://www.nobelprize.org/nobel_prizes/physics/laureates/2004/advanced.html
119. Royal Swedish Academy of Sciences: Class of Physics. Broken symmetries. Scientific Background on the Nobel Prize in Physics 2008. https://www.nobelprize.org/nobel_prizes/physics/laureates/2008/advanced.html
120. Royal Swedish Academy of Sciences: Class of Physics. The BEH-mechanism, interactions with short range forces and scalar particles. Scientific Background on the Nobel Prize in Physics 2013. https://www.nobelprize.org/nobel_prizes/physics/laureates/2013/advanced.html
121. Royal Swedish Academy of Sciences: Class of Physics. Neutrino oscillations. Scientific Background on the Nobel Prize in Physics 2015. https://www.nobelprize.org/nobel_prizes/physics/laureates/2015/advanced.html
122. Rudolph, G., Schmidt, M.: Differential Geometry and Mathematical Physics. Part I. Manifolds, Lie Groups and Hamiltonian Systems. Springer Netherlands, Dordrecht (2013)
123. Rudolph, G., Schmidt, M.: Differential Geometry and Mathematical Physics. Part II. Fibre Bundles, Topology and Gauge Fields. Springer Netherlands, Dordrecht (2017)
124. Ryder, L.H.: Quantum Field Theory. Cambridge University Press, Cambridge (1996)
125. Schwartz, M.D.: Quantum Field Theory and the Standard Model. Cambridge University Press, Cambridge (2014)
126. Seiberg, N.: Five dimensional SUSY field theories, non-trivial fixed points and string dynamics. Phys. Lett. B **388**, 753–760 (1996)
127. Seiberg, N., Witten, E.: Electric-magnetic duality, monopole condensation, and confinement in $N = 2$ supersymmetric Yang–Mills theory. Nucl. Phys. B **426**, 19–52 (1994). Erratum: Nucl. Phys. B **430**, 485–486 (1994)
128. Seiberg, N., Witten, E.: Monopoles, duality and chiral symmetry breaking in $N = 2$ supersymmetric QCD. Nucl. Phys. B **431**, 484–550 (1994)
129. Sepanski, M.R.: Compact Lie Groups. Springer Science+Business Media LLC, New York (2007)
130. Serre, J.-P.: Lie Algebras and Lie Groups. 1964 Lectures given at Harvard University. Springer, Berlin/Heidelberg (1992)
131. Slansky, R.: Group theory for unified model building. Phys. Rep. **79**, 1–128 (1981)
132. Srednicki, M.: Quantum Field Theory. Cambridge University Press, Cambridge (2007)

133. Steenrod, N.: The Topology of Fibre Bundles. Princeton University Press, Princeton (1951)
134. Streater, R.F., Wightman, A.S.: PCT, Spin and Statistics, and All That. Princeton University Press, Princeton, NJ (2000)
135. Tao, T.: Hilbert's Fifth Problem and Related Topics. Graduate Studies in Mathematics, Vol. 153. American Mathematical Society, Providence, RI (2014)
136. Taubes, C.H.: Differential Geometry. Bundles, Connections, Metrics and Curvature. Oxford University Press, Oxford (2011)
137. Thomson, M.: Modern Particle Physics. Cambridge University Press, Cambridge (2013)
138. Vafa, C., Zwiebach, B.: N = 1 dualities of *SO* and *USp* gauge theories and T-duality of string theory. Nucl. Phys. B **506**, 143–156 (1997)
139. van den Ban, E.P.: Notes on quotients and group actions. Fall 2006. Universiteit Utrecht
140. Van Proeyen, A.: Tools for supersymmetry. arXiv:hep-th/9910030
141. van Vulpen, I.: The Standard Model Higgs boson. Part of the Lecture Particle Physics II, University of Amsterdam Particle Physics Master 2013–2014
142. Warner, F.W.: Foundations of Differentiable Manifolds and Lie Groups. Springer, New York (2010)
143. Weinberg, S.: The Quantum Theory of Fields, Vol. I. Foundations. Cambridge University Press, Cambridge (1995)
144. Weinberg, S.: The Quantum Theory of Fields, Vol. II. Modern Applications. Cambridge University Press, Cambridge (1996)
145. Weinberg, S.: The Quantum Theory of Fields, Vol. III. Supersymmetry. Cambridge University Press, Cambridge (2005)
146. Wess, J., Bagger, J.: Supersymmetry and Supergravity, 2nd edn. Princeton University Press, Princeton, NJ (1992)
147. Wilczek, F.: Decays of heavy vector mesons into Higgs particles. Phys. Rev. Lett. **39**, 1304 (1977)
148. Witten, E.: Quest for unification. arXiv:hep-ph/0207124
149. Witten, E.: Chiral ring of $Sp(N)$ and $SO(N)$ supersymmetric gauge theory in four dimensions. Chin. Ann. Math. **24**, 403 (2003)
150. Witten, E.: Newton lecture 2010: String theory and the universe. Available at http://www.iop.org/resources/videos/lectures/page_44292.html Cited 20 Nov 2016
151. Yamamoto, K.: SU(7) Grand Unified Theory. Ph.D. thesis, Kyoto University (1981)
152. Zhang, F.: Quaternions and matrices of quaternions. Linear Algebra Appl. **251**, 21–57 (1997)
153. Ziller, W.: Lie Groups. Representation theory and symmetric spaces. Lecture Notes, University of Pennsylvania, Fall 2010. Available at https://www.math.upenn.edu/~wziller/

Index

M.J.D. Hamilton, *Mathematical Gauge Theory*, Universitext,
https://doi.org/10.1007/978-3-319-68439-0

Printed in the United States
By Bookmasters